Ideas and Methods in Mathematical Analysis,
Stochastics, and Applications

Ideas and Methods in Mathematical Analysis, Stochastics, and Applications

In Memory of Raphael Høegh-Krohn (1938-1988)
Volume 1

Edited by

Sergio Albeverio
Ruhr-Universität Bochum, Germany

Jens Erik Fenstad
University of Oslo, Norway

Helge Holden
University of Trondheim, Norway

Tom Lindstrøm
University of Oslo, Norway

CAMBRIDGE
UNIVERSITY PRESS

Published by the Press Syndicate of the University of Cambridge
The Pitt Building, Trumpington Street, Cambridge CB2 1RP
40 West 20th Street, New York, NY 10011-4211, USA
10, Stamford Road, Oakleigh, Victoria 3166, Australia

First published 1992

Printed in MEXICO

Library of Congress cataloguing in publication data applied for

British Library cataloguing in publication data applied for

ISBN 0-521-41929-8 hardback

La filosofia è scritta in questo grandissimo libro che continuamente ci sta aperto innanzi a gli occhi (io dico l'universo), ma non si può intendere se prima non s'impara a intender la lingua, e conoscer i caràtteri, ne' quali è scritto. Egli è scritto in lingua matematica, e i caràtteri son triangoli, cerchi, ed altre figure geometriche, senza i quali mezi è impossibile a intenderne umanamente parola; senza questi è un aggirarsi vanamente per un oscuro laberinto.

GALILEO GALILEI

Døyr fe;
døyr frendar;
døyr sjølv det same.
Men ordet om deg
aldri døyr
vinn du eit gjetord gjevt.

HÅVAMÅL

CONTENTS

Volume II

Ideas and Methods in Quantum and Statistical Physics

The contribution by F. Guerra will be published elsewhere.

Raphael Jan Høegh–Krohn
1938 – 1988

PREFACE

Raphael Høegh-Krohn, the mathematician and physicist, died suddenly on January 24, 1988, a few weeks before his fiftieth birthday. At the time of his death, Høegh-Krohn and his many collaborators were involved in a large number of projects; some near completion, but others either in rapid development or still on the starting blocks. In September of 1988, a symposium on "Ideas and Methods in Mathematics and Physics" was held at the University of Oslo with a twofold purpose; partly to honor the memory of Raphael Høegh-Krohn and to celebrate his scientific achievements, and partly to stimulate his friends and collaborators to continue the work he had left unfinished. The present volumes grew out of that symposium; many of the papers in it are based on talks given in Oslo at the time, while others were sent to us by scientists who could not attend the conference for practical reasons, but who still wanted to contribute in some way. A few were commissioned by the editors to give a more complete picture of the breadth and depth of Høegh-Krohn's contribution to science.

The most prominent characteristic of Raphael Høegh-Krohn as a scientist was probably the unique mixture of knowledge and intuition which made it possible for him to grasp the significance and true meaning of new developments almost immediately. Although his technical skills were formidable, mere technical improvements never interested him much; he was always looking for the underlying ideas and the correct physical interpretation of the results. Many of us have heard him at conferences where his comments after a lecture could in just a few seconds explain the true relevance of the speaker's result to central problems in physics and mathematics, and outline the most promising directions for future research. At the Oslo Symposium there were moments when everybody seemed to be waiting for his voice from the back of the room; waiting for him to trace the hidden connections, to tie up the loose ends, to put the emphasis where it belonged, or simply for him once more to suffuse the subject with his enthusiasm. We shall have to do without his help in the future, but we hope that these volumes dedicated to his memory will make the task a little easier – it is written by many of his closest collaborators and some of his most esteemed colleagues in a spirit close to his own. As with his own papers, the range of subjects may seem daunting, but there is a unifying theme; the quest for the right language for the formulation of physical laws.

To Raphael Høegh-Krohn physics and mathematics were intimately connected; he knew the physical significance of every little part of his formulas – the sign of a constant or the size of an exponent always had an immediate physical meaning to him – and, on the other hand, every physical phenomenon seemed to him to be waiting for a mathematical description. Particularly close to his heart were solvable models; models which are simple enough to admit an analytic solution, yet sufficiently general to describe the typical behaviour of a system. He never tired of emphasizing the wealth of invaluable information which can first be extracted from solvable models and later extended to more general situations. In solvable models he saw the interplay between the arts of physics and mathematics at its finest; the deepest insight of the physicist is needed to create a simplified model retaining all important features of the original phenomenon, and the greatest skill of the mathematician is required in solving the ensuing equations explicitly. But this love for the specific and the explicit never stopped him from generalizing and unifying; he was a Grand Master of formal calculations, and on entering his office one would usually find oneself in the midst of dazzling formulas whose physical content was bursting with significance, but whose mathematical status was far from secure. Much of his work went into making rigorous sense out of these formal calculations, and in this he was helped as much by his ebullient optimism as by his enormous technical skills; he liked to tease more timid souls by telling them that "if you can do it for analytic step functions, you can do it for anything". We hope the present volume achieves the kind of balance Raphael Høegh-Krohn was always looking for; the balance between the general and the specific, between the heuristic and the rigorous, between physical content and mathematical form.

As we dedicate this book to the memory of Raphael Høegh-Krohn, we would like to thank everybody who has helped in its production; the contributors, the participants in the Oslo Symposium, the staff at the mathematics departments in Bochum, Oslo and Trondheim, and the staff at Cambridge University Press, in particular Dr. A. Harvey. But first and foremost our thoughts go to those left behind, those who have had to come to terms with a loss even greater than ours – to all the members of his family.

Bochum, Oslo and Trondheim
May 1991

Sergio Albeverio   Jens Erik Fenstad   Helge Holden   Tom Lindstrøm

Bibliography of Raphael Høegh–Krohn

1963

1. *Some mathematical problems in the theory of the pi-meson-nuclear interactions*, Mag. scient. Thesis, Aarhus University, Denmark

1966

2. *Partly gentle perturbations with applications to perturbations by creation & annihilation operators*, Ph. D. Thesis, Courant Institute of Mathematical Sciences, New York University

1967

3. *Partly gentle perturbation with application to perturbation by annihilation-creation operators*, Proceedings of the National Academy of Sciences, 58 , 2187–2192

1968

4. *Partly gentle perturbations with application to perturbations by annihilation-creation operators*, Communications on Pure and Applied Mathematics, 11 , 313–342

5. *Gentle perturbations by annihilation-creation operators*, Communications on Pure and Applied Mathematics, 11 , 343–357

6. *Asymptotic fields in some models of quantum field theory I*, Journal of Mathematical Physics, 9 , 2075–2080

7. *The group of local automorphisms of the Minkofsky space*, Matematisk seminar no 1, University of Oslo (unpublished)

1969

8. *Asymptotic fields in some models of quantum field theory II*, Journal of Mathematical Physics, 10 , 639–643

9. *Boson fields under a general class of cut-off interactions*, Communications in Mathematical Physics, 12 , 216–225

10. *Boson fields under a general class of local relativistic invariant interactions*, Communications in Mathematical Physics, 14 , 171–184

11. *A local relativistic boson field with the $\lambda|\varphi|$ interaction*, Communications in Mathematical Physics, 14 , 185–194

1970

12. *Asymptotic fields in some models of quantum field theory III* , Journal of Mathematical Physics, 11 , 185–188

13. *Boson fields with bounded interaction densities,* Communications in Mathematical Physics, 17 , 179–193

14. *On the scattering operator for quantum fields,* Communications in Mathematical Physics, 18 , 109–126

1971

15. *A general class of quantum fields without cut-offs in two space-time dimensions,* Communications in Mathematical Physics, 21 , 244–255

16. *On the spectrum of the space cut-off :$P(\varphi)$: Hamiltonian in two space-time dimensions,* Communications in Mathematical Physics, 21 , 256–260

1972

17. *Hypercontractive semigroups and two dimensional self-coupled Bose fields,* Journal of Functional Analysis, 9 , 121–180 (with B. Simon)

18. *Infinite dimensional analysis with applications to self-interacting boson fields in two space-time dimensions,* in: "Open House for Functional Analysts", Various Publications Series, no 23, Institute of Mathematics, Aarhus University, paper no 8, 16 pp.

1973

19. *Uniqueness of the physical vacuum and the Wightman functions in the infinite volume limit for some nonpolynomial interactions,* Communications in Mathematical Physics, 30 , 171–200 (with S. Albeverio)

20. *The scattering matrix for some non-polynomial interactions I,* Helvetica Physica Acta, 46 , 504–534 (with S. Albeverio)

21. *The scattering matrix for some non-polynomial interactions II,* Helvetica Physica Acta, 46 , 535–545 (with S. Albeverio)

22. *Asymptotic series for the scattering operator and asymptotic unitarity of the space cut-off interactions,* Il Nuovo Cimento, 18A , 285–307 (with S. Albeverio)

1974

23. *The Wightman axioms and the mass gap for strong interactions of exponential type in two-dimensional space-time*, Journal of Functional Analysis, 16 , 39–82 (with S. Albeverio)

24. *Relativistic quantum statistical mechanics in two-dimensional space-time*, Communications in Mathematical Physics, 38 , 195–224

25. *A remark on the connection between stochastic mechanics and the heat equation*, Journal of Mathematical Physics, 15 , 1745–1747 (with S. Albeverio)

1975

26. *Homogeneous random fields and statistical mechanics*, Journal of Functional Analysis, 19 , 242–272 (with S. Albeverio)

27. *Interacting relativistic Boson fields in the De Sitter universe with two space-time dimensions*, Communications in Mathematical Physics, 44 , 265–278 (with R. Figari, Ch. R. Nappi)

1976

28. **Mathematical Theory of Feynman Path Integrals**, Lecture Notes in Mathematics, Volume 523, Springer-Verlag, Berlin, 139 pp. (with S. Albeverio)

29. *Relativistic quantum statistical mechanics*, in: "Mathematical Physics and Physical Mathematics", (Eds. K.Maurin, R. Raczka) Reidel Publ. Co, Dordrecht, and PWN - Polish Scientific Publishers, Warszawa, pp. 483-497

30. *Quasi invariant measures, symmetric diffusion processes and quantum fields*, in: "Proceedings of the International Colloquium on Mathematical Methods of Quantum Field Theory", Editions du CNRS Colloques Internationaux du Centre National de la Recherche Scientifique, No. 248, pp. 11–59 (with S. Albeverio)

31. *Canonical quantum fields in two space-time dimensions*, Preprint no 4, University of Oslo (with S. Albeverio) (unpublished)

1977

32. *Oscillatory integrals and the method of stationary phase in infinitely many dimensions, with applications to the classical limit of quantum mechanics I*, Inventiones Mathematicae, 40 , 59–106 (with S. Albeverio)

33. *Ergodic actions by compact groups on von Neumann algebras*, Preprint 77/P. 961, CNRS, Luminy, France (with S. Albeverio) (unpublished)

34. *Dirichlet forms and diffusion processes on rigged Hilbert spaces*, Zeitschrift für Wahrscheinlichkeitstheorie und verwandte Gebiete, 40 , 1–57 (with S. Albeverio)

35. *Energy forms, Hamiltonians and distorted Brownian paths*, Journal of Mathematical Physics, 18 , 907–917 (with S. Albeverio, L. Streit)

36. *Hunt processes and analytic potential theory on rigged Hilbert spaces*, Annales de l'Institut Henri Poincaré, B13 , 269–291 (with S. Albeverio)

37. *Dirichlet forms and Markov semigroups on C^*–algebras*, Communications in Mathematical Physics, 56 , 173–187 (with S. Albeverio)

38. *The structure of diffusion processes*, Preprint 77/P. 958, CNRS-Luminy, France (with S. Albeverio) (unpublished)

1978

39. *Remarks on the energy representation of Sobolev-Lie groups*, in: "Group Theoretical Methods in Physics, Proceedings 1977", (Eds. P. Kramer, A. Rieckers), Lecture Notes in Physics, Volume 79 , Springer-Verlag, Berlin, pp. 359–360 (with S. Albeverio)

40. *Spectral properties of positive maps on C^* -algebras*, Journal of the London Mathematical Society, 17 , 345–355 (with D. E. Evans)

41. *Topics in infinite dimensional analysis*, in: "Mathematical Problems in Theoretical Physics, Proceedings, Rome, 1977", (Eds. G. Dell'Antonio, S. Doplicher, G. Jona-Lasinio), Lecture Notes in Physics, Volume 80 , Springer-Verlag, Berlin, pp. 279–302 (with S. Albeverio)

42. *The energy representation of Sobolev-Lie groups*, Compositio Mathematica, 36 , 37–52 (with S. Albeverio)

43. *Frobenius theory for positive maps of von Neumann algebras*, Communications in Mathematical Physics, 64 , 83–94 (with S. Albeverio)

1979

44. *The method of Dirichlet forms*, in: "Stochastic Behaviour in Classical and Quantum Systems, Volta Memorial Conference, Como, 1977", (Eds. G. Casati, J. Ford), Lecture Notes in Physics, Volume 93, Springer-Verlag, Berlin, pp. 250–258 (with S. Albeverio)

45. **Feynman Path Integrals. Proceedings, Marseille 1978**, Lecture Notes in Physics, Volume 106, Springer–Verlag, Berlin, 452 pp. (joint editor with S. Albeverio. Ph. Combe, G. Rideau, M. Sirugue–Collin, M. Sirugue, R. Stora)

46. *Feynman path integrals and the corresponding method of stationary phase*, in: "Feynman Path Integrals Proceedings, Marseille, 1978", (Eds. S. Albeverio, Ph. Combe, R. Høegh-Krohn, G. Rideau, M. Sirugue-Collin, M. Sirugue, R. Stora), Lecture Notes in Physics , Volume 106, Springer–Verlag, Berlin, pp. 3–57 (with S. Albeverio)

47. *Singular perturbations and nonstandard analysis*, Transactions of the American Mathematical Society, 252 , 275–295 (with S. Albeverio, J.E. Fenstad)

48. *Stationary measures for the periodic Euler flow in two dimensions*, Journal of Statistical Physics, 20 , 585–595 (with S. Albeverio, M. de Faria)

49. *Uniqueness and the global Markov property for Euclidean fields. The case of trigonometric interactions*, Communications in Mathematical Physics, 68 , 95–128 (with S. Albeverio)

50. *Some results on the exponential interaction in 2 or more dimensions*, Communications in Mathematical Physics, 70 , 187–192 (with S. Albeverio, G. Gallavotti)

51. *The exponential interaction in* $\mathbb{R}^n$, Physics Letters, B83 , 177–178 (with S. Albeverio, G. Gallavotti)

52. *The global Markov property for Euclidean and lattice fields*, Physics Letters, B84 , 89–90 (with S. Albeverio)

1980

53. *A class of explicitly soluble, local, many-center Hamiltonians for one-particle quantum mechanics in two and three dimensions, I*, Journal of Mathematical Physics, 21 , 2376–2385 (with A. Grossmann, M. Mebkhout)

54. *The one particle theory of periodic point interacions. Polymers, monomolecular layers, and crystals*, Communications in Mathematical Physics, 77 , 87–110 (with A. Grossmann, M. Mebkhout)

55. *Regularization of Hamiltonians and processes*, Journal of Mathematical Physics, 21, 1636–1642 (with S. Albeverio, L. Streit)

56. *Poisson processes on groups and Feynman path integrals*, Communications in Mathematical Physics, 77 , 269–288 (with Ph. Combe, R. Rodriguez, M. Sirugue, M. Sirugue-Collin)

57. *Uniqueness and global Markov property for Euclidean fields and lattice systems*, in: "Quantum Fields – Algebras, Processes", (Ed. L. Streit), Springer-Verlag, Wien, pp. 303–329 (with S. Albeverio)

58. *Martingale convergence and the exponential interaction in $\mathbb{R}^n$*, in: "Quantum Fields – Algebras, Processes", (Ed. L. Streit), Springer-Verlag, Wien, pp. 331–335 (with S. Albeverio)

59. *Summary*, in: "Quantum Fields – Algebras, Processes", (Ed. L. Streit), Springer-Verlag, Wien, pp. 441–444

60. *Dynamical semigroups and Markov processes on $C^\star$- algebras*, Journal für die reine und angewandte Mathematik, 319, 25–37 (with S. Albeverio, G. Olsen)

61. *Ergodic actions by compact groups on $C^\star$– algebras*, Mathematische Zeitschrift, 174, 1–17 (with S. Albeverio)

62. *Stationary phase for the Feynman integral and the trace formula*, in: "Functional Integration", (Eds. J.P. Antoine, E. Tirapegui), Plenum, New York, pp. 23–41 (with S. Albeverio, Ph. Blanchard)

63. *Feynman formula and Poisson processes for gentle perturbations*, in: "Functional Integration", (Eds. J.P. Antoine, E. Tirapegui), Plenum, New York, pp. 53–64 (with Ph. Combe, R. Rodriguez, M Sirugue, M. Sirugue-Collin)

64. *Let us talk about the weather–A model for large scale atmospheric phenomena*, Proc. Les Embiez 1980 Conference "Disordered Systems", (with S. Albeverio) (unpublished)

1981

65. *Some Markov processes and Markov fields in quantum theory, group theory, hydrodynamics and $C^\star$-algebras*, in: "Stochastic Integrals, Durham, July 1980", (Ed. D. Williams), Lecture Notes in Mathematics, Volume 851, Springer-Verlag, Berlin, pp. 497–540 (with S. Albeverio)

66. *Uniqueness of Gibbs states and global Markov property for Euclidean fields*, in: "Random Fields — Rigorous Results in Statistical Mechanics and Quantum Fields Theory, Proceedings Colloquium, Esztertgom 1979", (Eds. J. Fritz, J.L. Lebowitz, D. Szász), Colloquia Mathematica Societatis Janos Bolyai, 27, Vol.I., North-Holland, Amsterdam, pp. 37–64 (with S. Albeverio)

67. *Stochastic methods in quantum theory and hydrodynamics*, in: "New Stochastic Methods in Physics", (Eds. C. De Witt-Morette, K.D. Elworthy), Physics Reports 77, pp. 193–214 (with S. Albeverio)

68. *Generalized Poisson processes in quantum mechanics and field theory*, in: "New Stochastic Methods in Physics", (Eds. C. De Witt-Morette, K.D. Elworthy), Physics Reports, 77, pp. 221–233, (with Ph. Combe, R. Rodriguez, M. Sirugue-Collin)

69. *Feynman path integrals, the Poisson formula and the theta function for the Schrödinger operators,* in: "Trends in Applications of Pure Mathematics to Mechanics, Vol.III", (Ed. R.J. Knops), Pitman, Boston, pp. 1–21 (with S. Albeverio, Ph. Blanchard)

70. *The global Markov property for lattice fields,* Journal of Multivariate Analysis, 11 , 599–607 (with S. Albeverio, G. Olsen)

71. *Classification of C^*–algebras admitting ergodic actions of the two-dimensional torus,* Journal für die reine und angewandte Mathematik, 328 , 1–8 (with T. Skjelbred)

72. *Compact ergodic groups of automorphisms,* Annals of Mathematics, 114 , 75–86 (with M. B. Landstad, E. Størmer)

73. *Point interactions as limits of short range interactions,* Journal of Operator Theory, 6 , 313–339 (with S. Albeverio)

74. *Irreducibility and reducibility for the energy representation of the group of mappings of a Riemannian manifold into a compact semisimple Lie group,* Journal of Functional Analysis, 41 , 378–396 (with S. Albeverio, D. Testard)

75. *A class of exactly solvable three-body quantum mechanical problems and the universal low energy behavior,* Physics Letters, 83A , 105–109 (with S. Albeverio, T.T. Wu)

76. *On the universal low energy limit in non relativistic scattering theory,* Acta Physica Austriaca Supplementum, 23 , 577–585 (with S. Albeverio, F. Gesztesy)

1982

77. *Some applications of functional integration,* in: "Mathematical Problems in Theoretical Physics, Proceedings International Association of Mathematical Physics, Conference, Berlin 1981", (Ed. R. Schrader, R. Seiler, D.A. Uhlenbrock), Lecture Notes in Physics, Volume 15, Springer-Verlag, Berlin, pp. 265–275 (with S. Albeverio, Ph. Blanchard)

78. *Perturbations of free evolutions and Poisson measures,* in: "Operator Algebras and Applications", Proceedings Symposia Pure Mathematics, Volume 38 , Part 2 , American Mathematical Society, Providence, pp. 509–512 (with Ph. Combe, R. Rodriguez, M. Sirugue, M. Sirugue-Collin)

79. *Feynman path integral and Poisson processes with piecewise classical paths,* Journal of Mathematical Physics, 23 , 405–411 (with Ph. Combe, R. Rodriguez, M. Sirugue-Collin)

80. *Zero mass, 2 dimensional, real time sine Gordon model without u.v. cut-offs,* Annales de l'Institut Henri Poincaré, 37A , 115–127 (with Ph. Combe, R. Rodriguez, M. Sirugue, M. Sirugue-Collin)

81. *Compactness and the maximal Gibbs state for random Gibbs fields on a lattice,* Communications in Mathematical Physics, 84 , 297–327 (with J. Bellissard)

82. *Feynman path integrals and the trace formula for Schrödinger operators,* Communications in Mathematical Physics, 83 , 49–76 (with S. Albeverio, Ph. Blanchard)

83. *The low energy expansion in nonrelativistic scattering theory,* Annales de l'Institut Henri Poincaré, A37, 1–28 (with S. Albeverio, F. Gesztesy)

84. *Some remarks on Dirichlet forms and their applications to quantum mechanics,* in: "Functional Analysis in Markov Processes, Proceedings Katata Workshop and Kyoto Conference", (Ed. M. Fukushima), Lecture Notes in Mathematics, Volume 923, Springer-Verlag, Berlin, pp. 120–132 (with S. Albeverio)

85. *Diffusions, quantum fields and groups of mappings,* in: "Functional Analysis in Markov Processes, Proceedings Katata Workshop and Kyoto Conference", (Ed. M. Fukushima), Lecture Notes in Mathematics, Volume 923, Springer-Verlag, Berlin, pp. 133–145 (with S. Albeverio)

86. *The spectrum of the three-dimensional Kronig-Penney model with random point defects,* Advances in Applied Mathematics, 3 , 435–440 (with S. Albeverio, W. Kirsch, F. Martinelli)

1983

87. *Processus de diffusion, confinement et formation de "jet-streams" dans la nébuleuse protosolaire,* Proceedings III. Encontro de Fisica e Matématica, Coimbra 1982, Preprint no. TH. 3536 - CERN (with S. Albeverio, Ph. Blanchard) (unpublished)

88. *Factorial representations of path groups,* Journal of Functional Analysis, 51 , 115–131 *Note,* ibid., 52 , 148 (with S. Albeverio, D. Testard, A. Vershik),

89. *The $\frac{1}{r}$ expansion for the critical multiple well problem,* Communications in Mathematical Physics, 91 , 65–73 (with M. Mebkhout)

90. *The spectrum of defect periodic point interactions,* Letters in Mathematical Physics, 7 , 221–228 (with H. Holden, F. Martinelli)

91. *The short range expansion,* Advances in Applied Mathematics, 4 , 402–421 (with H. Holden, S. Johannesen)

92. *Charged particles with short range interactions,* Annales de l'Institut Henri Poincaré, A38 , 263–293 (with S. Albeverio, F. Gesztesy, L. Streit)

93. *Local relativistic invariant flow for quantum fields,* Communications in Mathematical Physics, 90 , 329–351 (with S. Albeverio, Ph. Blanchard, Ph. Combe, M. Sirugue)

94. *A stochastic model for the orbits of planets and satellites — an interpretation of Titius-Bode law,* Expositiones Mathematicae, 4 , 365–373 (with S. Albeverio, Ph. Blanchard)

95. *Low-energy parameters in nonrelativistic scattering theory,* Annals of Physics, 148 , 308–326 (with S. Albeverio, D. Bollé, F. Gesztesy, L. Streit)

96. *The reflection from a semi-infinite crystal of point scatterers,* Physics Letters, 94A , 42–44 (with J.E. Avron, A. Grossmann)

1984

97. *Newtonian diffusions and planets, with a remark on non-standard Dirichlet forms and polymers,* in: "Stochastic Analysis and Applications, Swansea, 1983" (Eds. A. Truman, D. Williams), Lecture Notes in Mathematics, Volume 1095 , Springer-Verlag, Berlin, pp. 1–24 (with S. Albeverio, Ph. Blanchard)

98. *Local and global Markoff fields,* Reports in Mathematical Physics, 19 , 225–248 (with S. Albeverio)

99. *The short-range expansion in solid state physics,* Annales de l'Institut Henri Poincaré, A41 , 335–362 (with H. Holden, S. Johannesen)

100. *Schrödinger operators with point interactions and short range expansions,* Physica, 124A , 11–27 (with S. Albeverio)

101. *Some exactly solvable models in quantum mechanics and the low energy expansions,* in: "Procedings of the Second International Conference on Operator Algebras, Ideals and their Applications in Theoretical Physics, Leipzig 1983", (Eds. H. Baumgärtel, G. Lassner, A. Pietsch, A. Uhlmann) Teubner, Leibzig, pp. 12–28 (with S. Albeverio, F. Gesztesy, H. Holden)

102. *Perturbation of resonances in quantum mechanics,* Journal of Mathematical Analysis and Application, 101 , 491–513 (with S. Albeverio)

103. *Diffusion fields, quantum fields and fields with values in Lie groups,* in: "Advances in Probability, Stochastic Analysis and Applications", (Eds. M. Pinsky), M. Dekker, New York, pp. 1–98 (with S. Albeverio)

104. *Model dependence of Coulomb corrected scattering lengths,* Physical Review, C29 , 680–683 (with S. Albeverio, L.S. Ferreira, F. Gesztesy, L. Streit)

105. *On point interactions in one dimension,* Journal of Operator Theory, 12 , 101–126, Errata; ibid., 13 , (1985) 419 (with S. Albeverio, F. Gesztesy, W. Kirsch)

106. *Scattering by impurities in a solvable model of a 3–dimensional crystal,* Journal of Mathematical Physics, 25 , 1327–1334 (with S. Albeverio, M. Mebkhout)

107. *Perturbations of the Laplacian supported by null sets, with applications to polymer measures and quantum fields,* Physics Letters, A104 , 396–400 (with S. Albeverio, J.E. Fenstad, W. Karwowski, T. Lindstrøm)

108. *Markov cosurfaces and gauge fields,* Acta Physica Austriaca Supplementum, 26 , 211–231 (with S. Albeverio, H. Holden)

109. *Factoriality of representations on the group of paths of $SU(n)$,* Journal of Functional Analysis, 57 , 49–55 (with S. Albeverio, D. Testard)

110. *The resonance expansion for the Green's function of the Schrödinger and wave equations,* in: "Resonances – Models and Phenomena", (Eds. S. Albeverio, L. Ferreira, L. Streit), Lecture Notes in Physics, Volume 211 , Springer-Verlag, Berlin, pp. 105–127, (with S. Albeverio)

111. *Zonal wind structure of the planetary atmosphere. A unified stochastic approach,* Preprint no 40, ZiF Bielefeld, project no 2 (with S. Albeverio, Ph. Blanchard, Ph. Combe, R. Rodriguez, M. Sirugue, M. Sirugue-Collin) (unpublished)

112. *Magnetic bottles in a dirty environment: A stochastic model for radiation belts,* Preprint no 54, ZiF Bielefeld, project no 2 (with S. Albeverio, Ph. Blanchard, Ph. Combe, R. Rodriguez, M. Sirugue, M. Sirugue-Collin) (unpublished)

1985

113. *The multiple well problem-asymptotic behavior of the eigenvalues and resonances,* in: "Trends and Developments in the Eighties, (Bielefeld, 1982/1983)", (Eds. S. Albeverio, Ph. Blanchard), World Scientific, Singapore, pp. 244–272 (with M. Mebkhout)

114. *Reduction of nonlinear problems to Schrödinger or heat equations: formation of Kepler orbits, singular solutions for hydrodynamical equations,* in "Stochastic Aspects of Classical and Quantum Systems, Proceedings Marseille 1983", (Eds. S. Albeverio, Ph. Combe, M. Sirugue-Collin), Lecture Notes in Mathematics , Volume 1109 , Springer-Verlag, Berlin, pp. 189–206 (with S. Albeverio, Ph. Blanchard)

115. *A remark on dynamical semigroups in terms of diffusion processes,* in: "Quantum Probability and Applications, Heidelberg 1984" (Eds. L. Accardi, W. von Waldenfels), Lecture Notes in Mathematics, Volume 1136 , Springer-Verlag, Berlin, pp. 40–45 (with S. Albeverio)

116. *Euler flows, associated generalized random fields and Coulomb systems,* in: "Infinite Dimensional Analysis and Stochastic Processes", (Ed. S. Albeverio), Research Notes in Mathematics, Volume 124 , Pitman, Boston-London, pp. 216–244 (with S. Albeverio, D. Merlini)

117. *Markov processes on infinite dimensional spaces, Markov fields and Markov cosurfaces*, in: "Stochastic Space-Time Models and Limit Theorems", (Eds. L. Arnold, P. Kotelenez), D. Reidel, Dordrecht, pp. 11–40 (with S. Albeverio, H. Holden)

118. *Diffusions sur un variété Riemannienne: barrières infranchissables et applications*, in: "Colloque en l'Honneur de Laurent Schwartz", Vol.2, Astérisque 132 , 181–201 (with S. Albeverio, Ph. Blanchard)

119. *The short-range expansion for multiple well scattering theory*, Journal of Mathematical Physics, 26 , 145–151 (with H. Holden, M. Mebkhout)

1986

120. *The Fermi surface for point interactions*, Journal of Mathematical Physics, 27 , 385–405 (with H. Holden, S. Johannesen, T. Wentzel-Larsen)

121. *Brownian motion, Markov cosurfaces, Higgs fields*, in: "Fundamental Aspects of Quantum Theory, Proc. Como. Conf. 1985", (Eds. V. Gorini, A. Frigerio), Plenum Publ., New York, pp. 95–104 (with S. Albeverio)

122. *Random fields with values in Lie groups and Higgs fields*, in: "Stochastic Processes in Classical and Quantum Systems, Ascona 1985", (Eds. S. Albeverio, G. Casati, D. Merlini), Lecture Notes in Physics, Volume 262 , Springer-Verlag, Berlin, pp. 1–13 (with S. Albeverio, H. Holden)

123. **Nonstandard Methods in Stochastic Analysis and Mathematical Physics**, Academic Press, New York, 514 pp. (with S. Albeverio, J.E. Fenstad, T. Lindstrøm). Russian translation by A.K. Zvonkin, M.A. Shubin, Mir, Moscow, 1990

124. *Stochastic Lie group-valued measures and their relation to stochastic curve integrals, gauge fields and Markov cosurfaces*, in: "Stochastic Processes in Mathematics and Physics", (Eds. S. Albeverio, Ph. Blanchard, L. Streit), Lecture Notes in Mathematics, Volume 1158 , Springer-Verlag, Berlin, pp.1–24 (with S. Albeverio, H. Holden)

125. *Path space measure for the Liouville quantum field theory and the construction of relativistic strings*, Physics Letters, B174 , 81–86 (with S. Albeverio, S. Paycha, S. Scarlatti)

126. *Euclidean Markov fields and relativistic quantum fields from stochastic partial differential equations in four dimensions*, Physics Letters, B177 , 175–179 (with S. Albeverio)

127. *Strata and voids in galactic structures—A probabilistic approach*, BiBoS-Preprint no 154 (with S. Albeverio, Ph. Blanchard, M. Mebkhout) (unpublished)

128. *A stochastic model for plasma dynamics*, BiBoS Preprint no 196, (with D. Gandolfo, R. Rodriguez) (unpublished)

129. *The Riemann problem for a scalar hyperbolic conservation law in several space dimensions*, in: "Proceedings from the Seminar on Reservoir Description and Simulation with Emphasis on EOR Holmenkollen, Oslo", Institute for Energy Technology, Scientific Software-Intercomp Ltd, England (with N.H. Risebro) (unpublished)

130. *The Schrödinger operator for a particle in a solid with deterministic and stochastic point interactions*, in: "Schrödinger Operators, Aarhus 1985", (Ed. E. Balslev), Lecture Notes in Mathematics, Volume 1218, Springer-Verlag, Berlin, pp. 1–38 (with S. Albeverio, F. Gesztesy, H. Holden, W. Kirsch)

1987

131. *Quaternionic non abelian relativistic quantum fields in four space-time dimensions*, Physics Letters, B189 , 329–336 (with S. Albeverio)

132. *Point interactions in two dimensions: basic properties, approximations and applications to solid state physics*, Journal für die reine und angewandte Mathematik, 380 , 87–107 (with S. Albeverio, F. Gesztesy, H. Holden)

1988

133. *Some recent interactions between mathematics and physics in connection with generalized random fields*, in: "Proceedings 1st World Congress of the Bernoulli Society, Tashkent 1986", (Eds. Yu. Prokhorov, V.V. Sazonov), Volume 1, Probability Theory and Applications, VNU-Press, pp.455–472 (with S. Albeverio)

134. *A numerical method for first order nonlinear scalar conservation laws in one-dimension*, Computers and Mathematics with Applications, 15 , 595–602 (with H. Holden, L. Holden)

135. *Stochastic multiplicative measures, generalized Markov semigroups and group valued stochastic processes and fields*, Journal of Functional Analysis, 78 , 154–184 (with S. Albeverio, H. Holden)

136. **Solvable Models in Quantum Mechanics**, Springer-Verlag, Berlin, 452 pp., (with S. Albeverio, F. Gesztesy, H. Holden),
Russian translation by Yu.A. Kuperin, K.A. Makarov, V.A. Geilerk, Mir, Moscow 1991

137. *Construction of interacting local relativistic quantum fields in four space-time dimensions*, Physics Letters, B200 , 108–114 , Errata, ibid., B202 , 621 (with S. Albeverio)

138. *The Riemann problem for a single conservation law in two space dimensions*, Preprint no 14, University of Oslo, (with N.H. Risebro) (unpublished)

1989

139. *Stochastic flows with stationary distribution for two-dimensional inviscid fluids*, Stochastic Processes and their Applications, 31 , 1–31 (with S. Albeverio)

140. *Representation and construction of multiplicative noise*, Journal of Functional Analysis, 87 , 250–272 (with S. Albeverio, H. Holden, T. Kolsrud)

141. *A remark on the formation of crystals at zero temperature*, in: "Stochastic Methods in Mathematics and Physics", (Eds. R. Gielerak, W. Karwowsky), World Scientific, Singapore, pp.211–220 (with S. Albeverio, H. Holden, T. Kolsrud, M. Mebkhout)

142. *A covariant Feynman–Kac formula for unitary bundles over Euclidean space*, in: "Stochastic Partial Differential Equations and Applications II", (Eds. G. Da Prato, L. Tubaro), Lecture Notes Mathematics, Volume 1390 , Springer-Verlag, Berlin, pp. 1–12 (with S. Albeverio, H. Holden, T. Kolsrud)

143. *Uniqueness and global Markov property for Euclidean fields. The case of general polynomial interactions*, Communications in Mathematical Physics, 123 , 377–424 (with S. Albeverio, B. Zegarlinski)

144. *Uniqueness of Gibbs states for general $P(\varphi)_2$–weak coupling models by cluster expansion*, Communications in Mathematical Physics, 121 , 683–697 (with S. Albeverio, B. Zegarlinski)

145. *Construction of Higgs-like fields in two dimensions*, Physics Letters, B222 , 263–268 (with S. Albeverio, H. Holden, T. Kolsrud)

146. *Covariant Markovian random fields in four space-time dimensions with nonlinear electromagnetic interaction*, in: "Applications of Self-Adjoint Extensions in Quantum Physics, Dubna 1987", (Eds. P. Exner, P. Šeba), Lecture Notes in Physics, Volume 324 , Springer-Verlag, Berlin, pp.69–83 (with S. Albeverio, K. Iwata)

147. *Point interaction Hamiltonians for crystals with random defects*, in: "Applications of Self-Adjoint Extensions in Quantum Physics, Dubna 1987" (Eds. P. Exner, P. Šeba), Lecture Notes in Physics , Volume 324 , Springer-Verlag, Berlin, pp.87–99 (with S. Albeverio, R. Figari, F. Gesztezy, H. Holden, W. Kirsch)

148. *A probability measure for random surfaces of arbitrary genus and bosonic strings in 4 dimensions*, Nuclear Physics B (Proc. Suppl.), 6 , 180–182 (with S. Albeverio, S. Paycha, S. Scarlatti)

149. *A global and stochastic analysis approach to bosonic strings and associated quantum fields*, Preprint no 47, SFB-237, Bochum, (with S. Albeverio, S. Paycha, S. Scarlatti) (unpublished)

1990

150. *A class of N hyperbolic conservation laws*, Journal of Differential Equations, 84 , 73–99 (with L. Holden)

151. *Classification and construction of quasisimple Lie algebras*, Journal of Functional Analysis, 89 , 106–136 (with B. Torresani)

1992

152. *Some Euclidean integer-valued random fields with Markov properties*, in: "Ideas and Methods in Mathematical Analysis, Stochastics, and Applications", (Eds. S. Albeverio, J. E. Fenstad, H. Holden, T. Lindstrøm), Cambridge University Press, Cambridge, pp. 93–114 (with S. Albeverio, D. Surgailis)

153. *Schrödinger operators with potentials supported by null sets*, in: "Ideas and Methods in Quantum and Statistical Physics", (Eds. S. Albeverio, J. E. Fenstad, H. Holden, T. Lindstrøm), Cambridge University Press, Cambridge, pp. 63–95 (with S. Albeverio, J. E. Fenstad, W. Karwowski, T. Lindstrøm)

154. *Morphology and classification of galaxies. A stochastic model*, in: "Ideas and Methods in Quantum and Statistical Physics", (Eds. S. Albeverio, J. E. Fenstad, H. Holden, T. Lindstrøm), Cambridge University Press, Cambridge, pp. 447–460 (with S. Albeverio, Ph. Blanchard, D. Gandolfo, M. Mebkhout)

155. **Noncommutative Distributions – Unitary Representations of Gauge Groups**, M. Dekker, New York, (with S. Albeverio, J. Marion, D. Testard, B. Torresani) (in preparation)

On the Scientific Work of Raphael Høegh–Krohn

Sergio Albeverio
Fakultät für Mathematik
Ruhr–Universität Bochum
D–4630 Bochum 1
Germany

"...
allá detrás de una
frágil pared de vientos,
de cielos y de años."
(P. Salinas)

0. A short glimpse at Raphael's life

Raphael was born in Ålesund on the West Coast of Norway on the 10th of February 1938.
He grew up in this small town during war time. He often talked of his childhood there: playing with his sister and two brothers in dangerous situations or watching from the windows as the sky was illuminated just before the planes would come and bomb. He was struck by the contrast betweeen the hardship of war and the innocence of playing children. He spoke of his parents staying up late at night discussing what to do to survive while he was in his bed trying to figure out ways to help. He used to spend the summertime in Norddal, a small village near Ålesund. He was very attached to these places and at times would have liked to write about them, but refrained from it, fearing he might become "too romantic".
He moved to Oslo in 1956 to study mathematics and physics. He was an enthusiastic student, motivated by the idea of mathematics as the best way to grasp what one can of the world.
In 1959 he took his Cand. mag. degree in Oslo, and from 1959 to 1960 he did his military service at Forsvarets forskningsinstitutt, Kjeller. Raphael's work was led by Ragnar Frisch who later became a Nobel Laureate in Economics. Raphael was greatly impressed by him and remembered his enthusiasm, as well as his scepticism, for the numerical implementation on computers of optimization for large scale economic systems (in fact, for the Egyptian economy, following a contract Frisch had at that time with the Egyptian government).
The period from 1960 to 1963 Raphael spent in Aarhus, Denmark, studying theoretical physics and there took his Mag. science degree [1][0], under the direction of P. Kristensen, in 1963 (his written Thesis was finished in August '62). This work discussed the order of singularity of certain singular functions in the pseudo scalar π–meson theory, presented

[0] The numbered references refer to Raphael Høegh–Krohn's bibliography

in the LSZ–quantum field theoretical formalism. Within the limits of such a subject one already recognizes in this work the strength of Raphael's mathematical analysis.

1963–66 Raphael had a Research Fellowship at the Courant Institute of New York University. There he married Miftah, with whom he would later have four children (Ludvig, Helga, Solveig, Renée). All his life he loved to be with his family; he had an extremely broad and deep knowledge of natural sciences (especially paleontology, geology, but also biology, as well as physics and astrophysics) and loved to share it with his family, observing nature with them, and interpreting it.

1966–1969 Raphael was back in Oslo, as a lecturer, and from 1969–1970 he spent two years in Princeton (with a fellowship from the Norwegian Government, to do research in the group of A. S. Wightman). After this period at Princeton he settled in Oslo as a lecturer again, then as a Professor and finally as a full Professor in 1982. In 1977 he received the Fridtjof Nansen Prize of the Norwegian Academy of Science and Letters "for applications of mathematics to physics", and in 1979 he became a Member of that Academy. In 1976 he was a member of Ludwig Streit's Research Project "Mathematical Problems of Quantum Dynamics" at the Center for Interdisciplinary Research (ZiF) of the University of Bielefeld. He spent the academic year 1977–1978 at the Centre de Physique Théorique, CNRS, Marseille – and from that time on he started regularly spending long periods at the Université de Provence (where he was nominated as a Professor), at the Université d'Aix–Marseille at Luminy and at the Centre de Physique Théorique of the CNRS, Marseille. He spent the winter term 1978–79 as a visiting Professor at the Mathematics Department of the University of Bielefeld and made many other shorter visits later on at this University and at ZiF (Project no. 2: "Mathematics and Physics", 1983–84), at the Ruhr–University at Bochum and at the Bielefeld–Bochum Stochastic Research Centre. He also made many shorter visits to other universities in countries all over the world. He was extremely active in the faculty and, by the breadth and depth of his knowledge and his clear judgement, his advice was asked steadily from all sides. In the last few years of his life he invested a lot of his energies in building up a program of studies and a research group in industrial mathematics at the University of Oslo.

He was always full of ideas, projects and activities, in both theoretical and applied science. Nobody would have believed it possible that he would depart so early, as he looked so strong and energetic. He was planning to celebrate his 50th birthday "among friends, with a little good mathematics ...". Tragically, on January 24, 1988, a few weeks before his birthday, he suffered heart failure, alone in his car, during a trip to the West Coast near his home town on the mountains he liked so much. The news of his death was a terrible blow from which it is very hard to recover. He was a great man and an extremely original, intuitive and gifted mathematician. He had a unique combination of high technical skills, intelligence and imagination and he generously shared his wealth of ideas with his many co-workers. It was a greatly enriching human experience to come into close contact with him. He was always concerned with making others feel at ease and his optimism was really contagious. Everything in life, culture and nature interested him deeply and he had original ideas about everything he considered. He always put friendship first, considering it a much superior and more precious value than any particular achievement in life or science. More should really be written on the life and personal sides of this truly exceptional man, however, here I shall try to concentrate on his admirable scientific work. This work is of great inspiration and ranges through practically the entire spectrum of mathematics and physics. He always stressed the unity of our knowledge, based on the analysis of few natural models and basic simple facts. He had a great faith in reason as a unifying force; one of his dreams was one day to write a compact encyclopedia, concentrating in a few volumes all the essential knowledge of the world surrounding us. Longing for great synthesis (quantum theory/relativity, Markov fields/geometry) he followed the inspiration of Galilei and Klein in looking for basic forms of our world and intuition, stressing basic geometrical objects (like Lie algebras and groups) and basic "solvable models" (like the ones he studied in the theory of Markov fields and in quantum theory).

I shall now try to lead you through the main themes of Raphael's work, hoping to manage at least to give a glimpse of its richness and multiformity which is permeated by a splendid unity.

1. Early work on spectral theory and scattering theory for models of quantum fields and Schrödinger operators

As already mentioned, during the period 1963–1966 Raphael was in New York to work on his Ph.D. studies at the Courant Institute of Mathematical Sciences. He spoke enthusiastically about this time: the big town, the atmosphere at the Courant, the sense of being in the middle of true mathematics – he greatly admired his teacher, K.O. Friedrichs; he used to talk about Friedrichs' lectures, apparently unprepared on purpose and thus real creative events. In any case this was the style which suited Raphael best. He was a great teacher, capable of inventing a whole lecture on the spot (before a lecture Raphael used to retire in his office or sit alone somewhere and think or look through the window or go for a walk. Only rarely did he sit down and write up notes for the lectures...).
While in New York, he had to pass an English examination to be admitted to graduate school. He liked to tell the story about his interview with Lipman Bers (if I remember correctly), who asked whether he could speak broken English, "the only truly international language" – after verification that this was the case, he was admitted as a Ph.D. student at the Courant. Raphael liked his studies there – attending courses by K.O. Friedrichs, P. D. Lax and K. Symanzik (and passing an examination in a very friendly atmosphere with the latter, the examination consisting in a chat together – Raphael having been the only listener to the lecture. He learned a lot, in fact he probably learned and developed here on his own the tricks which made him a master of formal computations). He used to speak of these people and the atmosphere at Courant as something unique for analysis (and the best of mathematics for Raphael was always analysis, understood in a very broad sense. I remember a lecture he gave in Oslo on "history of analysis", in which Darwin and Mendeleyev were represented ...)

The title of the Ph.D. Thesis, written under the direction of K.O. Friedrichs and presented in October 1966, is "On partly gentle perturbation with application to perturbation by creation & annihilation operators" [2]. Three publications came out of it: one in the

Proceedings of the National Academy of Sciences in 1967 [3] and two in Communications of Pure and Applied Mathematics in 1968, [4], [5].

The broad framework of the Thesis is Friedrichs' theory of perturbations, as part of the perturbation theory of the continuum spectrum.

Friedrichs (1938, 1948) introduced stationary methods for constructing intertwining operators for a pair of operators in Hilbert space or Banach space (wave operators). These were the origins of stationary scattering theory (pursued later by many authors, including L. Faddeev, T. Kato, see e.g. [A-J-S], [D-S], [R-Si], [B-W], [Ka1] and references therein). He also introduced gentle perturbations, studied later by O. Ladyženskaja and L. Faddeev (1958, 1964), J. Schwartz (1960), P. Rejto (1963, 1964), T. Ushijima (1966), K. Mochizuki (1967), S.C. Lin (1969). An extension to "partly gentle operators" was given by P. Rejto (1967), J.S. Howland (1967) and especially Raphael. The theory was later extended particulary by T. Kato and S.T. Kuroda (1971) to an "abstract theory of scattering", see e.g. [R-Si] and [B-W].

Raphael gives precise mathematical form to the concept of partly gentle perturbation of operators in a Hilbert space (this also extends related work by Rejto in 1963–64).

Let H_0 be a self–adjoint operator in a Hilbert space $\mathcal{H}$ and let V be a H_0–bounded self–adjoint operator on $\mathcal{H}$ (cf. e.g. [Ka1] for definitions). Raphael shows that if $H_\lambda = H_0 + \lambda V$ and $\sigma(H_\lambda)$ denotes the spectrum of H_λ then if H_λ is gentle on some subset I of $\sigma(H_\lambda)$ then there is a $\tilde{\lambda}_0 > 0$ s.t. for all $|\lambda| \leq \tilde{\lambda}_0$ there exists a partial isometry U_λ on $\mathcal{H}$ s.t. $E^I_\lambda H_\lambda = U^*_\lambda E^I_0 H_0 U_\lambda$, with E^I_0 resp. E^I_λ the spectral projections on I associated with H_0 resp. H_λ. One has $U^*_\lambda U_\lambda = E^I_\lambda$, $U_\lambda U^*_\lambda = E^I_0$. An application of this result is that for $H_0 = -\Delta$ in $L^2(\mathbb{R}^d)$, $d \geq 3$, and $V \in L^{\frac{3}{2}}(\mathbb{R}^3)$ for $d = 3$ resp. $V \in (L^\infty \cap L^1)(\mathbb{R}^d)$ for $d > 3$, one has that H_λ is gentle for all $\lambda \in \mathbb{R}$ and there exists $\lambda_0 > 0$ s.t. for all $|\lambda| \leq \lambda_0$ there exists a unitary mapping U_λ analytic in λ such that $U_\lambda H_\lambda U^*_\lambda = H_0$ (more general results in all dimensions $d > 3$ had been proved independently by T. Kato the same year [Ka2]). Another application of Raphael's result is the fact that for $V \in (L^\infty \cap L^1)(\mathbb{R}^d)$, $d \geq 3$ the wave operators $W^\pm_\lambda = \lim_{t \to \pm\infty} e^{it\, H_\lambda} e^{-it\, H_0}$ exist and the S–matrix $S_\lambda \equiv W^-_\lambda (W^+_\lambda)^*$ is analytic in λ, unitary for all $|\lambda| \leq \lambda_0$.

As we shall see Raphael maintained all his life a basic interest in Schrödinger operators. Let me mention now some other results of his Thesis which concern scattering theory in quantum field models (this work was going to be continued by Raphael roughly until '73). In the fifties Friedrichs had been studying (at the same time as I. Segal) mathematical problems of the foundation of quantum field theory and functional integration [F-S]. A book based on lectures given at the University of Boulder, Colorado, in 1960, had just appeared in 1965 [Fr2] when Raphael started working on his Thesis. Friedrichs, as well as Segal, stressed quantum field theory as mechanics for (relativistic) systems of infinitely many degrees of freedom. This exploits the fact that a classical field equation like the Klein–Gordon one

$$(\,\Box + m^2)u = 0 \tag{1.1}$$

with $\Box = \frac{\partial^2}{\partial t^2} - \Delta_x$ denoting the D'Alembertian ($t \in \mathbb{R}$ being time, $x \in \mathbb{R}^s$ the space, m being the (positive) mass), can formally be looked upon as a Newton equation for a degree of freedom $t \to u(t, \cdot)$ with values in functions (or distributions) in x, with "force" at time t in x, $K(u(t,x)) \equiv (\Delta - m^2)u(t,x)$ (with $\Delta \equiv \Delta_x$). Now recall how the quantization of a classical particle moving on $\mathbb{R}^d$ under a quadratic potential $V(x) = \frac{1}{8}xA^2x - tr\frac{A}{4}$ can be achieved. One can take as Hilbert space $L^2(\mathbb{R}^d, dx) \cong L^2(\mathbb{R}^d, d\mu)$, with $d\mu(x) = Ce^{-\frac{1}{2}xAx}dx$, $C \equiv \left(\int e^{-\frac{1}{2}xAx}dx\right)^{-1}$, the normal distribution of mean zero and covariance A^{-1}, and as operator representing the position $u(t) = e^{it\tilde{H}}\, u(0)e^{-it\tilde{H}}$, with $u(0)$ multiplication by x and $H = -\frac{1}{2}\Delta + V(x)$ as a self-adjoint operator in $L^2(\mathbb{R}^d, dx)$, $\tilde{H} = -\frac{1}{2}\Delta + \frac{1}{2}Ax \cdot \nabla$ in $L^2(\mathbb{R}^d, d\mu)$. Similarly, for quantum fields over the space $\mathbb{R}^s$ one can take as Hilbert space $L^2(\mathcal{S}'(\mathbb{R}^s), d\mu)$, with μ the Gaussian measure on $\mathcal{S}'(\mathbb{R}^s)$ with mean 0 and covariance A^{-1}, with $A = \sqrt{-\Delta_x + m^2}$. The position $u(0)$ is then realized as the coordinate function $< g, X >$, $g \in \mathcal{S}(\mathbb{R}^s), X \in \mathcal{S}'(\mathbb{R}^s)$. The Fock space structure (described e.g. in [Fr1], [Kas1], [Gl-J], [Si1], [Al1], [Mey]) yields the isomorphism

$$L^2(\mu_0^0) = \bigoplus_{n=0}^{\infty} L^{2(n)}(\mu) \cong \exp(L^2(\tilde{\mathbb{R}}^s)) \equiv \bigoplus_{n=0}^{\infty} L^2(\tilde{\mathbb{R}}^s)^{\otimes^s n}, \tag{1.2}$$

with $\exp K$ denoting the symmetric tensor algebra (Fock space) over the Hilbert space K and $L^{2(n)}(\mu)$ the subspace of $L^2(\mu)$ generated by the n-th generalized Hermitian polynomials (n-th Wiener–Ito chaos with respect to μ). The function 1 goes into the vector $\Omega_0 = (1, 0, ..., 0, ...)$ in $\exp(L^2(\tilde{\mathbb{R}}^s))$ and the projection of $\prod_{i=1}^{n} < g_i, X >$ into $L^{2(n)}(\mu)$, suitably normed by multiplications by $\frac{1}{\sqrt{n!}}$, is the vector $S_n(\frac{\tilde{g}_1}{\sqrt{\nu}} \otimes \ldots \otimes \frac{\tilde{g}_n}{\sqrt{\nu}})$, where S_n means symmetrization, with $\nu(k)^2 \equiv k^2 + m^2$, $\tilde{g}(k)$ being the Fourier transform of $g \in \mathcal{S}(\mathbb{R}^s)$, evaluated at $k \in \tilde{\mathbb{R}}^s$. The field operator $< g, X >$ is then represented in $\exp(L^2(\tilde{\mathbb{R}}^s))$ by $\frac{1}{\sqrt{2}}a^*(\frac{\tilde{g}}{\sqrt{\nu}}) + a(\frac{\bar{\tilde{g}}}{\sqrt{\nu}})$, with a, a^* annihilation resp. creation operators defined by $(a(k)\psi)^{(n)}(k_1, \ldots, k_n) = \sqrt{n+1}\psi^{(n+1)}(k_1, \ldots, k_n, k)$, $a(k)\Omega_0 = 0$, $a^*(k)$ its adjoint, $^-$ meaning conjugation.

The Hamiltonian is represented in $L^2(\mu)$ by the Ornstein–Uhlenbeck operator H_μ (the self-adjoint operator in $L^2(\mu)$ associated with the classical Dirichlet form $\frac{1}{2}\int |\nabla F|^2 d\mu$ given by μ, cf. [A-R4]), and the associated semigroup in the Fock space $\exp L^2(\tilde{\mathbb{R}}^s)$ by $\exp(\frac{1}{2}A)$.

One then has that the self-adjoint operator $< g, \Phi(t) > \equiv e^{itH_\mu} < g, X > e^{-itH_\mu}$ in the Hilbert space $L^2(\mu)$ solves the relativistic equation $(\Box + m^2)\Phi(t,x) = 0$, with $< g, \Phi(t) > = \int \Phi(t,x)g(x)dx$, in distributional sense.

In this framework the interaction is defined heuristically by

$$\tilde{V}(X(0)) \equiv \int_{\mathbb{R}^s} v(X(0,x))dx \tag{1.3}$$

with $v(\alpha) = \sum_{M=0}^{\infty} a_M \alpha^M$, $\alpha \in \mathbb{R}$, a_M being given suitable coefficients. v is called interaction density.

Considering only the term of order M we have $\tilde{V}(X(0)) = a_M \int X(0,x)^M dx$, and this has the heuristic form

$$\tilde{V}(X(0)) = a_M \sum_{j=0}^{M} W_{j,M}(X(0)) \ , \tag{1.4}$$

with

$$W_{j,M}(x) \equiv \int \frac{e^{-i\sum_{l=1}^{M} k_l x}}{\prod_{i=1}^{M}\sqrt{\nu(k_i)}} a^*(k_1)...a^*(k_j)a(-k_{j+1})\dots a(-k_M)dk_1\dots dk_M \ . \tag{1.5}$$

Let us also introduce $\tilde{W}_{j,M}$ defined similarly as $W_{j,M}$ but with the numerical factor $\frac{e^{-i\sum_{l=1}^{M} k_l x}}{\prod_{i=1}^{M}\sqrt{\nu(k_i)}}$ multiplied by $\chi(k_1,\dots,k_M)$, with χ a continuous function of compact support (called "ultraviolet cut-off").

Let $\tilde{V}_\chi(X(0)) = a_M \sum_{j=0}^{M} \tilde{W}_{j,M}(X(0)) \equiv \int \tilde{v}(X(0,x))\,dx$. Raphael studies in the above publications [2]-[5] "ultraviolet space cut-off" Hamiltonians for quantum fields of the form $H_{g,\chi} \equiv H_\mu + \tilde{V}_{g,\chi}$, in $L^2(\mu)$, with

$$\tilde{V}_{g,\chi}(X(0)) \equiv \int \tilde{v}(X(0,x))g(x)dx \ , \tag{1.6}$$

with g a "space cut-off" i.e. bounded measurable function of compact support. He shows that $H_{g,\chi}$ is a well-defined self-adjoint lower bounded operator in the Fock-space $L^2(\mu)$, with the spectrum consisting of eigenvalues and a continuum of the form $[\tilde{m},\infty)$ (for some finite $\tilde{m}$), and defines the scattering quantities for the pair $(H_{g,\chi}, H_\mu)$.

He also shows that if the model is so modified that there are no vacuum polarization terms i.e. $\tilde{W}_{0,M} = \tilde{W}_{M,M} = 0$ then, even in the case $\chi = 1$, $H_{g,\chi}$ is gentle on any bounded interval of the real axis. The wave operators $W_\pm \equiv s-\lim_{t\to\pm\infty} e^{itH}e^{-itH_0}$ exist. Moreover $E^I H_{g,\chi} E^I$ and $E_0^I H_\mu E_0^I$ are unitary equivalent for all $|\lambda| \le \lambda_0(I)$, for some $\lambda_0(I) > 0$. Furthermore, at fixed energy ω, there exists $\lambda_0(\omega)$ s.t., for all $|\lambda| \le \lambda_0(\omega)$, the S-matrix is unitary and strongly analytic on λ.

As we mentioned in Sect. 0, after obtaining his Ph.D. Raphael returned to Oslo in '66, where he took up a position as lecturer. In the years '66–'70 he continued his work on scattering theory in models of quantum fields (although he was well aware of axiomatic developments, he had a natural scepticism about them, he always stressed the importance of models as the best means to get true insights).

A paper submitted in '67 and which appeared in '69, [8] treats the case of the above Hamiltonian with $g = 1$ under the assumption that it contains no terms which are creating and annihilating one particle. In this paper asymptotic creation and annihilating operators are constructed.

In [6] asymptotic annihilation-creation operators $a_\pm^\sharp = s-\lim_{t\to\pm\infty} e^{-itH}e^{itH_0}a^\sharp e^{-itH_0}e^{itH}$ for models describing fermions, with H_0 a free Hamiltonian and H a space and ultraviolet cut-off Hamiltonian, are constructed, asymptotic states are introduced and a decomposition of the Hilbert space as tensor product of an asymptotic Fock space and a zero mass particle space exhibited (the symbol $a^\sharp$ stands for creation or annihilation operator).

Edward Nelson introduced in 1964 a model ("Nelson's model") treating, in the physical 4–dimensional space–time, non–relativistic nucleons Yukawa coupled with relativistic mesons. This model was studied by J. Cannon in 1968 in his Princeton Ph.D. Thesis. In [12] Raphael constructs asymptotic creation–annihilation operators for mesons. Models of this type were further studied by J.P. Eckmann in 1970.

At the end of '69 I came to Princeton for a two year stay on a Research Fellowship from the Swiss National Science Foundation. It took me some time to adjust to the new place: I had been teaching in a high–school for a year, and had gone through a very difficult period of my life. The encounter with Raphael, who was spending 1969–1971 at the same University, gave me an entirely new confidence in mathematics. I had seen Raphael in some seminars; he had a beautiful open smile and talked about mathematics as something entirely natural, as somebody does who really grew up with it.

Once I decided to knock at the door of his office to ask him about scattering theory for nucleons in Nelson's model (a problem I had just started to work on). He was kind, open and extremely bright unlike anyone else I had ever encountered – he said we would discuss it further but not today – at the same time he was putting on his shoes – he must rush home, since it was his wedding anniversary ... (he had married Miftah one year before). From that moment a wave of sympathy grew between us which later led to our brotherly friendship and scientific cooperation for over 18 years, until Raphael died suddenly.

In '71 I worked on Eckmann's version of Nelson's model in particular constructing the asymptotic nuclear field, and the S–matrix, in the spirit of Raphael's work [Al2] (some related results were obtained independently in Zürich by J. Fröhlich, at that time a student of K. Hepp, who then continued the investigations also studying the case of mass zero particles, see also [Sl]).

The interest in the construction of models of relativistic quantum fields which was alive in work by Friedrichs [Fr1], [Fr2] and Segal [Se] (see also [G-S], [M-Se]), and was decisively enhanced by the fundamental estimate of Nelson [Nel2], had spread, in the late sixties, mainly under the influence of A.S. Wightman, to Princeton and to the Courant Institute in New York. Intensive activity on the construction of models of relativistic quantum fields (what came to be called constructive field theory) had been started (work by J.T. Cannon, J. Glimm, A. Jaffe, O. Lanford).

Raphael immediately showed the ease he had as a mathematician looking in an unconventional way at such models. He constructed models with various interactions, cf. Sect. 2. In two papers which received much attention from the main body of constructivists in quantum field theory [14], [16] Raphael shows that in $P(\varphi)_2$ models with space cut–off the spectrum of the space cut–off Hamiltonian H_g (defined by $H_{g,\chi}$ with $\tilde{V}_g \equiv V_{g,1}$ given by (1.6) with $\chi = 1$) consists of a finite number of eigenvalues between the infimum E_g of the spectrum and the threshold $E_g + m$ of an absolutely continuum spectrum $[E_g + m, \infty)$ (which is also the essential spectrum).

He did this by constructing asymptotic creation–annihilation operators and proving their commutation relations with H_g. These results were also extended to the model with exponential interaction described in Sect. 2 (cf. [Al1, Sect. II.4]).

Asymptotic expansions in power series of the coupling constant λ for the asymptotic creation–annihilation operators and the S–matrix were given for several models in the work of Raphael and myself [22], [19], [20], [21]. Related developments were obtained in a series of papers by Kato and Mugibayashi [Ka-M1], [Ka-M2], see also [Pru].
Later on some of the techniques were taken up again in work by Baumgärtel and Wollenberg [B-W].

2. Early work, by Hamiltonian methods, on constructive quantum field theory, and work on hypercontractive semigroups

2.1. Time evolution in several models of quantum fields

During his stay at Princeton University 1969–1971 Raphael continued his work on the construction and study of Hamiltonians for quantum fields, coming closer to the activities going on in that domain at Princeton (E. Nelson, B. Simon), New York (J. Glimm) and Harvard (A. Jaffe). He studied two types of models distinguished by the type of interaction term given by a real–valued function v on $\mathbb{R}$ (cf. (1.3)):

a) v bounded and continuous or, more specially, v Fourier transform of a bounded measure [9], [10], [13], [14].
b) v of modulus type i.e. $v(\alpha) = \lambda|\alpha|$, [11].
c) v of exponential type, [15].

As to a) he demonstrated that the models with space cut–off g and suitable ultraviolet cut–offs χ exist. By looking at Heisenberg fields he was able to show that the cut–offs can be removed, but that removal of the ultraviolet–cut–off yields free fields. This was a neat mathematical result showing the necessity of renormalization for obtaining a non trivial limit when the regularization is removed. Later on this type of models was studied further by Euclidean methods (see [19] and Sect. 3 below) and by oscillatory integral methods (see [46], [93] and Sect. 7):

b) The model $v(\alpha) = \lambda|\alpha|$ in $\mathbb{R}^4$ is handled by Raphael by methods similar to those of a), and convergence of Heisenberg fields is proven.

c) Of fundamental importance is Raphael's paper on the exponential interaction [15], a model coined the "Høegh–Krohn model" in the literature, cf. [Si1]. For this case v is exponential, more precisely $v(\alpha) = \int e^{\alpha\beta} d\nu(\beta)$, with ν a positive measure with support in $(-\sqrt{2\pi}, \sqrt{2\pi})$. The Hamiltonian is defined in Fock space as $H_g = H_0 + V_g$, with V_g the space cut–off interaction, with space cut–off g and $H_0 = H_\mu$ (cf. Sect. 1) the free Hamiltonian. Raphael shows that the Hamiltonian is well–defined and essentially self–adjoint on a dense domain $D \subset D(H_0) \cap D(V_g)$ in the Fock space. In addition he shows that the generated semigroup e^{-tH_g} is the strong limit of the semigroup $e^{-tH_{g,\chi}}$ generated by the corresponding space and ultraviolet cut–off Hamiltonian $H_{g,\chi}$ when the ultraviolet cut–off χ is removed. Moreover he shows that the kernel of e^{-tH_g} is positive ergodic, and e^{-tH_g} is a contraction semigroup on all L^p–spaces (for all $1 \le p \le \infty$). By these methods Raphael shows the uniqueness of the ground state (vacuum) Ω_g with energy E_g and the existence of at most finitely many eigenvalues in $[E_g, E_g + m - \varepsilon]$ for any $\varepsilon > 0$. The proof exploits the hypercontractivity of e^{-tH_o} (this is the origin of a well-known work of Raphael with Barry Simon [17] on general hypercontractive semigroups, cf. Sect. 2.2). He also shows that the limit as the space cut-off is removed, $g \to 1$, of Heisenberg fields exists, by adapting C^*-algebra methods used by Segal and Glimm-Jaffe for the φ_2^4 model. In the limit he gets a locally Fock representation. A more direct and specific control of the limit was achieved later [23], by Euclidean methods, in '73, cf. Sect.4.

This model turned out to be important in several other contexts (like gauge group representations (cf. Sect. 15), string theory, where the mass zero version of the model appears as quantization of the "Liouville model", and gravity theory (cf. Sect. 17)). While working on such models Raphael was at the time fully aware that in domains like general relativity non polynomial interactions are more important than polynomial ones and this strongly motivated his research.

2.2. Hypercontractive semigroups

In the second part of his stay in Princeton Raphael began to work with Barry Simon on hypercontactive semigroups [17]. They introduced the definition of a contraction self-adjoint semigroup P_t on a $L^2(\nu)$-space, with ν a probability measure, as *hypercontractive* if it is a contraction on $L^p(\nu)$ for all $1 \leq p \leq \infty$ and there exists a $t > 0$ such that P_t is bounded for $L^2(\nu)$ into $L^4(\nu)$. This covers the examples of Glimm, such that $P_t\, 1 = 1$, $P_t \restriction \{1\}^\perp \leq e^{-mt}$ (with $\perp$ meaning orthogonal complement), or Segal, $P_t = e^{-tH_\mu}$.

This led to important work providing abstract, unified and simplified versions of techniques used on space cut-off models of quantum fields (cf. Sect. 2.1). The methods provided among other things proofs of essential self-adjointness of Hamiltonians (e.g. for $V \in L^p(\nu)$ for some $p > 2$, $e^{-tV} \in L^1(\nu)$ for all $t > 0$, $H_0 + V$ is essentially self-adjoint on $D \subset D(H_0) \cap D(V)$) of relevance in the study of the infinite volume limits of quantum field models by operator techniques. The paper was very influential (see e.g. the review article by Davies, Gross, and Simon in these volumes). The words written in that paper "... The result seem to us to be a theory not only of great physical interest, but also of considerable mathematical beauty ..." are most pertinent. Mathematical beauty was intrinsic for Raphael – understood not as superficial beauty, but as an inner beauty. He often talked about his dream of writing mathematics as a symphony or a good piece of chamber music. He came to admire the essentiality, pureness of Mozart's musical language more and more. He often compared mathematics and music – and at times would say that mathematics at its best can even be superior to music, however he couldn't be sure with Mozart.

To give an idea of the results of [17] let us mention that they prove that in $P(\varphi)_2$-models with space cut-off $g \in L^1 \cap L^2(I\!R)$, $g \geq 0$ the space-cut-off Hamiltonian H_g is essentially self-adjoint on a domain $D \subset D(H_0) \cap D(V_g)$, the interaction V_g is in L^p for all $1 \leq p < \infty$ and is such that $e^{-tV_g} \in L^1(\mu)$ for all $t > 0$. Raphael had, as already mentioned, used these tools to prove that $H_0 + V_g$ has discrete spectrum in $[E_g, E_g + m)$ and to show that the essential spectrum coincides with the absolutely continuous spectrum $[E_g + m, \infty)$.

From the above results on V_g and e^{-tV_g}, it follows that Rayleigh–Schrödinger expansion in λ as $\lambda \downarrow 0$ holds for the ground state energy E_λ of $H_0 + \lambda V_g$ and for the corresponding eigenfunction Ω_λ. Examples of applications in the study of non-relativistic Schrödinger operators were also given, e.g. a proof that for potentials V satisfying $V \geq a|x|^2 - b$, $a > 0$, $\int |V|^2 e^{-\sqrt{a}|x|^2} dx < \infty$, the Hamiltonian $-\Delta + V$ is essentially self-adjoint on $D(-\Delta) \cap D(V)$. Another result concerns N-body Hamiltonians with 2-body potentials $V_{ij} \in R + L^\infty$ (with R denoting the Rollnik class), for which Raphael and Simon showed that if a ground state exists it is positive almost everywhere and unique. For further developments of this important paper see e.g. [18], [Gro2], [Gro3], [L-C], [R-Si] (Vol IV), [Si2], and as already mentioned, the review paper by Davies, Gross, Simon in these volumes. Hypercontractivity has been also discussed in the context of the theory of Dirichlet forms and symmetric Markov processes, see e.g. [Gro4], [Car-S], [Eck], [S-Z], a topic which would become of major interest for Raphael after '74.

3. The first control of infinite volume limits in quantum field theory by probabilistic (Euclidean) statistical mechanical methods: bounded interaction models

In the summer '71 Raphael returned to Oslo, and invited me to take up a visiting lectureship at Oslo, starting in the beginning of '72. So began my long stay in Norway – Raphael did everything possible to make it a wonderful experience. And indeed it was. He had to give up one thing early: teaching me and my wife to behave properly on skis as he did (he was behaving on skis as one expected him to do, being a strong Norwegian, who grew up on the west coast).

We began our joint work on trigonometric interactions by Euclidean methods in the spring of '72, under the influence of new developments by Nelson and Guerra. Shortly before I left Princeton Nelson's work about the Markov property of free fields started to become known. I attended Nelson's class where this work was being prepared. As a consequence also of discussions I had with Francesco Guerra who had gotten the basic convergence result for the free energy [Gu1] using Nelson's methods, just before I left I became enthusiastic about Nelson's methods [Nel3]. These are Euclidean methods, which had been used before in a largely heuristic way by K. Symanzik and were immediately familar to Raphael, who had attended Symanzik's lectures and had a strong background in probability theory which he had acquired in his years of studies in Aarhus and New York.
It should also be noted that Euclidean methods are intrinsically probabilistic and the foundations of integration theory in infinite dimensional spaces had been established after the pioneering studies of Friedrichs and Segal, through fundamental work of L. Gross [Gro1] and of Sazanov and Minlos (see [G-V]).
We considered [19] the model of boson fields with interaction density $\lambda v(\alpha)$, with λ a real constant and v the Fourier transform of a bounded measure ν (this is a model Raphael had already discussed in [9], [10]) (the same class occurs in later work by Raphael and co-workers on homogeneous fields and statistical mechanics [26], the global Markov property [49], Feynman path integrals [28], [32], [46], [93] and low temperature expansions for

classical statistical mechanical systems [157]; moreover it has been used recently in other contexts, like in filter theory [A-A-F] and semiclassical expansions [O-P-C]). The heuristic measure, in the case of a space–time cut-off $g : \mathbb{R}^d \to \mathbb{R}$ and an ultraviolet cut-off χ is given by

$$d\mu_{g,\chi}(X) \equiv \frac{e^{-\lambda \int_{\mathbb{R}^d} v(X*\chi)g(x)dx} d\mu_0(X)}{\int e^{-\lambda \int_{\mathbb{R}^d} v(X*\chi)g(x)dx} d\mu_0(X)} ,$$

with μ_0 the Euclidean Nelson's free field measure [Nel4], the normal distribution of mean zero and covariance $(-\Delta + m^2)^{-1}$, with Δ the Laplacian on space–time $\mathbb{R}^d$, X the coordinate function in $\mathcal{S}'(\mathbb{R}^d)$. The sine–Gordon model corresponds to ν being a sum of delta functions in $\pm\alpha_0$ (for some real α_0). Similar models had been studied before, by other methods, by Streater and Wilde [S-Wil]. In our joint paper, [19] Raphael and I managed to give the first result of constructive quantum field theory about uniqueness of the infinite volume limit and the vacuum in a model (for $|\lambda|$ small). The space time dimension was arbitrary, but we kept an ultraviolet cut-off. We managed to express the Schrödinger and Wightman functions of the model (and later also scattering functions [20]-[22]) by quantities of classical statistical mechanics. In this paper we exploited what became later known as the sine–Gordon transformation (which, as we realized later, extended an analogous transformation used in [E-L] for a one space–time dimensional model), which gives an isomorphism between classical statistical mechanics of a gas with Yukawa (or Coulomb) interactions and Euclidean quantum field theory. Via Nelson's analytic continuation procedure one recovers, after removal of the ultraviolet cut-off [Frö2], the Lorentz–invariant functions. It should be remarked that at the same time Guerra–Rosen–Simon were working on the lattice approach to quantum field theory [G-R-S1], [G-R-S2], and ideas of Wilson were becoming influential in connecting quantum field theory with statistical mechanics (this gave constructive content to previous suggestions by Nakano, Schwinger and Symanzik on Euclidean methods).

4. Euclidean methods for Høegh-Krohn's model and refined results on singularity of Gaussian measures, with applications to the representation theory of infinite dimensional groups

At the end of '72 I left Oslo for a one year stay in Naples, but with a firm determination to continue my collaboration with Raphael. In the spring of '73 we managed [23] to present the complete construction of a non trivial model of relativistic local quantum fields satisfying all postulates of the general theory [Jo], [S-W], [B-L-T]. This is the model whose Hamiltonian with space cut-off had been constructed by Raphael [15]. We managed to remove the space cut-off (thermodynamic limit) by developing for this model the lattice approximation first introduced as a tool in constructive quantum field theory by Nelson in lectures given in Princeton (and partly published in [Nel5]) and Guerra–Rosen–Simon, from whom we received preprints in the spring of '73 (this work was then published in [G-R-S1]). Nelson–Guerra–Rosen–Simon's fundamental ideas offered another relation (besides the tranformation [19] between sine-Gordon models and Coulomb gas models mentioned in Sect. 3) between classical statistical mechanics and quantum field theory, one between lattice systems and Euclidean quantum fields on a lattice. The main tools provided by this relation are inequalities for expectations with respect to certain Gaussian measures, called "ferromagnetic correlation inequalities". (The relation between Euclidean fields and lattice models of classical statistical mechanics corresponds to a finite difference method; it combines two important tools, the Euclidean method and a method of numerical analysis. The utility of this framework for heuristic or numerical computation and simulations was stressed afterwards by many physicists, in particular Wilson; it has become a standard tool for e.g. "renormalization group" calculations and Monte Carlo simulations, especially in "quantum chromodynamics". It is also related to the heuristic "Regge calculus", which recently has found some mathematical implementation, by a method of simplicial approximation of quantum fields on manifolds [A-Z1], [A-Z2]).

Høegh-Krohn's model is defined by a Euclidean measure, cf. Sect. 3, with interaction given first in a bounded open subset Λ of $\mathbb{R}^2$ with ultraviolet cutoff χ of the type

$$d\mu_{\Lambda,\chi} = \frac{\exp(-\lambda V_{\Lambda,\chi})\, d\mu_0}{\int \exp(-\lambda V_{\Lambda,\chi})\, d\mu_0}, \ \lambda \geq 0 \tag{4.1}$$

$$V_{\Lambda,\chi} = \int_\Lambda : v(X_\chi(x)) : dx, \tag{4.2}$$

with $v(s)$ of the form $v(s) = \int e^{\alpha s} d\nu(\alpha)$, with ν a positive σ-finite measure on $\mathbb{R}$ with support contained in $(-\sqrt{4\pi}, \sqrt{4\pi})$, and satisfying $\nu(d\alpha) = \nu(-d\alpha)$. : : is the Wiener–Ito–Wick ordering (with respect to the Gaussian measure μ_0 described in Sect.3) defined by

$$: v(X_\chi(x)) :\equiv e^{-\frac{\alpha^2}{2} G_\chi(0)} e^{\alpha X_\chi(x)} , \tag{4.3}$$

with the ultraviolet cut-off functions X_χ, G_χ defined by

$$X_\chi \equiv X * \chi \ , \ G_\chi = \chi * G * \chi$$

It is shown in [23] that $V_{\Lambda,\chi} \geq 0$, $V_{\Lambda,\chi} \in L^2(\mu_0)$, and for $\alpha^2 < 4\pi$ the strong limit of $V_{\Lambda,\chi}$ as $\chi \to 1$ exists in $L^2(\mu_0)$ and yields a positive function $V_\Lambda \in L^2(\mu_0)$.

The restriction $\alpha^2 < 4\pi$ is used in proving $V_\Lambda \in L^2(\mu_0)$ and is based on the fact that, due to the behavior $G(x-y) \sim -\frac{1}{2\pi} \ln|x-y|$ of the kernel $G(x-y)$ on the diagonal (with $G = (-\Delta)^{-1}$) one has

$$\int_{\mathbb{R}^2} \int_{\mathbb{R}^2} g(x) g(y) e^{\alpha^2 G(x-y)} dx dy < \infty \tag{4.4}$$

for all $g \in L^1(dx) \cap L^2(dx)$.

We also showed later on [58] that

$$e^{-V_{\Lambda,\chi}(X)} = \int_\Lambda \mu_0(X_\chi - \alpha G_\chi(x)) dx \tag{4.5}$$

is a positive submartingale which converges μ_0-almost surely and in $L^1(\mu_0)$ as $\chi \to 1$ to

$$e^{-V_\Lambda(X)} = \int_\Lambda \mu_0(X - \alpha G(x))dx \ . \tag{4.6}$$

To control the thermodynamic limit $\Lambda \uparrow \mathbb{R}^2$, as mentioned above, we used a lattice approximation, together with ferromagnetic inequalities. In particular, we showed in [23] that with

$$\mu_\Lambda \equiv \left(\int e^{-V_\Lambda} d\mu_0 \right)^{-1} e^{-V_\Lambda} \mu_0$$

and E_{μ_Λ} the expectation with respect to μ_Λ, one has that

$$E_{\mu_\Lambda} \left(\prod_{i=1}^n < h_i, X > \right)$$

is monotone in Λ for all $h_i \geq 0$ such that the expectation exists, and converges to a non-trivial limit. The latter is non-trivial in as much as it is bounded below by

$$0 \neq E_{\mu_\Lambda}^{\partial\Lambda} \left(\prod_{i=1}^n < h_i, X > \right) ,$$

where $E_{\mu_\Lambda}^{\partial\Lambda}$ is the expectation with respect to the probability measure $\mu_\Lambda^{\partial\Lambda}$ defined as μ_Λ but with μ_0 replaced by $\mu_0^{\partial\Lambda}$, the Gaussian measure of mean zero and covariance $(-\Delta_{\partial\Lambda} + m^2)^{-1}$, with $\Delta_{\partial\Lambda}$ the Laplacian with Dirichlet boundary conditions on the boundary $\partial\Lambda$ of Λ. All Wightman axioms for a relativistic theory with mass gap ($\geq m^2$) have been verified in [23] to hold.

Remarks 1. Høegh-Krohn's model seems to be, historically, the first non-trivial model in which all axioms have been verified (for weak coupling $P(\varphi)_2$ models the fundamental results of Glimm–Jaffe–Spencer using cluster expansions, however, were obtained at about the same time). It is also the simplest model in quantum field theory for which the whole construction can be given.

2. The construction of the Høegh-Krohn model is entirely non perturbative. In fact, higher order terms in perturbation theory diverge even after renormalization (cf. [15]).

3. The paper contains many further results: the infinite volume limit is restricted however, to $\alpha^2 < 16/\pi$ (instead of the natural limitation $\alpha^2 < 4\pi$) because the lattice approximation had been shown to converge only under such a restriction. Also the measure ν was taken to be even. These assumptions were later removed by Fröhlich and Park [F-P1], [F-P2] see also [123] for a construction of the model by methods of nonstandard analysis.

4. For the discussion of the global Markov property for this model, see Sect. 14. Nelson's axioms have been verified, cf. [103].

5. The existence of the model also for $\alpha^2 < 8\pi$ (and Λ bounded) has been achieved by Kusuoka (in these volumes). (In this case $V_\Lambda \notin L^2(\mu_0)$, but still $V_\Lambda \in L^{1+\varepsilon}(\mu_0)$ for some $\varepsilon > 0$).

6. See Sect. 17 for the connection with string theory.

7. A discussion of the connection with representation theory of current algebras can be found in Sect. 14.

The first type of triviality result for quantum fields (which attracted a lot of interest later on in connection with polynomial models, cf. e.g. [Gl-J] and references therein) was obtained by Raphael and myself in this model for dimension $d \geq 4$ [58] and later on (1977/78) extended in our work with G. Gallavotti [50], [51].

More precisely, exploiting (4.6), one shows that for $d \geq 3$ or $d = 2$, $|\alpha| > \alpha_0$, $\alpha_0 \geq 2\sqrt{\pi}$, all moments of $\mu_{\Lambda,\chi}$ converge to those of μ_0. Moreover the measures $\int_\Lambda \mu_0(\cdot - \alpha G(x))dx$, and $\mu_0(\cdot)$ have disjoint supports [58]. Osipov showed that one can take $\alpha_0 = \sqrt{8\pi}$ [Os1].

The result on absolute continuity of $\int_\Lambda \mu_0(\cdot - \alpha G(x))dx$ with respect to μ_0 for $d = 2$, $|\alpha|^2 < 4\pi$ (resp. 8π) can actually be looked upon as a result for Gaussian measures, not deducible from previous general results on Gaussian measures. It has been formulated as such in greater generality in connection with studies of the energy representation, see [74].

For a recent reference concerning these topics, with applications to representation theory, see also [Wal] and below, under Sect. 14. A particular result in this line is the following [58]: Let $\mathcal{H}$ be a real separable Hilbert space and $\underline{\mathcal{H}} \subset \mathcal{H} \subset \bar{\mathcal{H}}$ an $\mathcal{H}$-rigging. Let $A \geq 0$ be a Hilbert-Schmidt operator on $\mathcal{H}$, with eigenvalues λ_i and normalized eigenfunctions e_i. Let ν be a probability measure on $\bar{\mathcal{H}}$. If $(x, Ax) = \sum \lambda_i (e_i, x)^2 = \infty$ for ν a.e. $x \in \bar{\mathcal{H}}$, then $\nu \times \mu_{\mathcal{H}}$ is orthogonal to $\mu_{\mathcal{H}}$, where $\mu_{\mathcal{H}}$ is the standard normal probability distribution on $\mathcal{H}$. This result was later discussed further and improved by Wick [Wic].

5. Quantum statistical mechanics and quantum fields on De Sitter universe. Models of classical and quantum statistical mechanics

In '73 Raphael also started working in quantum statistical mechanics. In the beginning of '74 I was back in Oslo, with the help of a fellowship from the Norwegian Research Council for Science and the Humanities. Raphael was at that time very busy with two papers which came out in '74 [24], [29] in which he considered Gibbs states for a quantum mechanical harmonic oscillator perturbed by a potential V, i.e. for a "V-anharmonic oscillator". The Hamiltonian is then given as a self-adjoint realization of

$$H = H_0 + V, \text{ with } H_0 = -\frac{1}{2}\Delta + \frac{1}{2}(x, A^2 x) - \frac{1}{2} \text{ tr } A$$

in $L^2(I\!R^n,\ dx) \equiv \mathcal{H}$, with A an $N \times N$ symmetric matrix such that $A \geq c1\!\!1$ for some $C > 0$, tr is its trace, (,) the scalar product in $I\!R^N$, and V is a lower bounded function (the potential) and such that $H_0 + V$ is essentially self-adjoint on the domain of $D(H_0)$.

For $\beta > 0$ Raphael defined $\omega_\beta(B) \equiv (\text{ tr } e^{-\beta H})^{-1} \text{ tr } (Be^{-\beta H})$ for any $B \in \mathcal{B}(\mathcal{H})$, the algebra of bounded operators on $\mathcal{H}$. ω_β is a normal state on $\mathcal{B}(\mathcal{H})$ called the "Gibbs state" (for the temperature $\frac{1}{\beta}$) for the V-anharmonic oscillator.

Raphael then gave a probabilistic representation of ω_β in terms of an expectation with respect to a Brownian loop, using

$$\text{tr } e^{-\beta H} = \int E^{\beta}_{(x,x)} e^{-\int_0^\beta U(X(s))ds} dx \ ,$$

with

$$U(x) \equiv \frac{1}{2}(x, A^2 x) - \frac{1}{2} \text{ tr } A + V(x) \ .$$

$E^{\beta}_{(x,x)}$ is the expectation with respect to a (multi-dimensional) Brownian loop going from $x \in I\!R^N$ at "time 0" to x at "time β". By a Feynman–Kac formula he then proves that

$$\text{tr } e^{-\beta H} = \left|1 - e^{-\beta A}\right|^{-1} E^{\beta} \left[e^{-\int_0^\beta V(X(x))ds}\right] ,$$

with $|\quad|$ the determinant and E^β the expectation with respect to the standard normal distribution $\mu_{0,\beta}$ associated with the Hilbert space of paths γ: $[0, \beta] \to I\!R^N$ with $\gamma(0) = \gamma(\beta)$, with norm (squared) $\int_0^\beta \left[|\dot\gamma|^2 ds + (\gamma(s), A^2\gamma(s))\right] ds$ (and corresponding scalar product). A corresponding formula is then written for ω_β, i.e. the Gibbs state is expressed by expectations with respect to the Gaussian measure $\mu_{0,\beta}$. Moreover Raphael stresses the translation invariance and KMS properties

$$\omega_\beta(B\alpha_t(C)) = \omega_\beta(\alpha_{-t}(B)C) = \omega_\beta(B\alpha_{t-i\beta}(C))$$

for any $B, C \in \mathcal{B}(\mathcal{H})$, with

$$\alpha_t(B) \equiv e^{-itH} B e^{itH} , \ t \in I\!R \ .$$

Later on this property received a lot of attention in the study of algebraic foundations of quantum statistical mechanics, see e.g. [Kas2]. In fact, in this approach intrinsic characterizations of KMS states are given.

We shall return below to this topic, but let us first indicate how brilliantly Raphael managed to extend his considerations to the case of relativistic quantum fields. In the case of free quantum fields in s space dimensions the space $I\!R^N$ is replaced by $L^2(I\!R^s)$ and the matrix A by the pseudodifferential operator $\sqrt{-\Delta + m^2}$, Δ being the Laplacian in $L^2(I\!R^s)$. The paths $X(t)$ are replaced by fields $X(t)(x)$, $x \in I\!R^s$, with values in $I\!R$, and the covariance operator $\left(-\frac{d^2}{ds^2}\right)^{-1}$ on $L^2(S^\beta; I\!R^N)$ is replaced by the covariance operator $\left(-\frac{d^2}{dt^2} - \Delta + m^2\right)^{-1}$ on $L^2(S^\beta \times I\!R^s)$ (where S^β stands for the torus $[0, \beta]$ obtained by taking the interval $[0, \beta]$ and identifying the boundary points 0 and β).

The corresponding Gibbs state ω_β again exists, and can be expressed by expectations with respect to the Gaussian measure $\mu_{0,\beta}$ of mean zero and above covariance. In the case

$s = 1$, interactions can again be introduced by replacing the probability measure $\mu_{0,\beta}$ by, for any $l > 0$:

$$\mu_{l,\beta} \equiv Z_l^{-1} e^{-\int_0^\beta \int_{-l}^{l} :v(X(s)(y)):dsdy} d\mu_{0,\beta}(X) ,$$

with

$$Z_l \equiv \int e^{-\int_0^\beta \int_{-l}^{l} : v(X(s)(y)):dsdy} d\mu_{0,\beta}(X)$$

with v a polynomial or exponential type function as in Sect. 4 (in fact this can be extended also to trigonometric functions, of the type of those of Sect. 3, see e.g. [102]).

The limit $l \to \infty$ exists and gives a probability measure $\mu_{\infty,\beta}$ on $\mathcal{S}'(S^\beta \times \mathbb{R})$, translation invariant and strongly mixing with respect to space translations, i.e. translation induced by

$$X(s)(x) \longrightarrow X(s)(x+a), \; a \in \mathbb{R} .$$

The corresponding Gibbs state ω_β is translation invariant, KMS, strongly mixing, locally Fock. These KMS states have been used later on, see e.g. [Fröl], [K-L], [Sew]. Moreover Raphael proves a duality theorem of the following type: the Schwinger functions S_β^∞ corresponding to the probability measure $\mu_{\infty,\beta}$ and those, S_∞^β corresponding to a zero temperature quantum field in a periodic box of length β in the spatial direction, are dual in the sense that, for all $s_i, x_i \in S^\beta$

$$S_\infty^\beta(s_1, x_1, ..., s_n, x_n) = S_\beta^\infty(x_1, s_1, ..., x_n, s_n) .$$

Also the essential observation is made that whereas ω_β is obviously not relativistic invariant, the "time variable" in S_∞^β playing a different role from the "space variable" in $\mathbb{R}$ (or $\mathbb{R}^s$ in the $v = 0$ case), but that the relativistic invariant zero temperature state in De Sitter space has the same form. This corresponds to the idea that the curvature of "time" given by the torus of length $\frac{\beta}{2\pi}$ is balanced by a curvature of space, in a De Sitter universe of radius $R = \frac{\beta}{2\pi}$. This led to fundamental work in 1974 by Raphael and two young co-workers from Naples, Rodolfo Figari and Chiara Nappi [27].

First let me briefly describe this work. Let us consider a two dimensional De Sitter space which can be realized as the hyperboloid $\mathcal{H}_R$ embedded in $\mathbb{R}^3$, where $\mathcal{H}_R$ is described by

$$\mathcal{H}_R = \left\{(\xi_1, \xi_2, \xi_3) \in \mathbb{R}^3 | \xi_1^2 + \xi_2^2 - \xi_3^2 = R^2\right\}$$

where R is a positive number, and metric

$$ds^2 = d\xi_3^2 - d\xi_1^2 - d\xi_2^2 .$$

One can introduce "polar coordinates" φ, t such that

$$\xi_1 = R\, Sh \frac{\varphi}{R}, \; \xi_2 = R\, Ch\, \frac{t}{R} \cos \frac{\varphi}{R} ,$$

$$\xi_3 = R \sin \frac{t}{R} \cos \frac{\varphi}{R}, \; t \in \mathbb{R}, \; -\pi \frac{R}{2} \le \varphi \le \pi \frac{R}{2} ,$$

so that for an observer situated at $\varphi = 0$, $t = 0$ the "visible part" is covered. ds^2 is then expressed by $ds^2 = \cos^2 \frac{\varphi}{R} dt^2 - d\varphi^2$.($Sh, Ch$ denotes respectively the hyperbolic sinus and cosinus).

Let us consider the temperature zero state given by a heuristic probability measure obtained by perturbing a "flat measure" by a Gibbs factor $e^{-W(X)}$, the action $W(X)$ being given by

$$W(X) = \int_{-\infty}^{\infty} \int_{-\frac{\pi}{2}R}^{\frac{\pi}{2}R} L\left(X, \frac{\partial X}{\partial \varphi}, \frac{\partial X}{\partial t}\right) \sqrt{|g|} d\varphi dt \ ,$$

with

$$L \equiv \frac{1}{2}\left[g^{ik} \frac{\partial}{\partial x_i} X \frac{\partial}{\partial x_k} X - m^2 X^2\right] \ ,$$

with $ds^2 = g^{ik} dx_i dx_k$ (with the usual summation convention) and x_0, x_1 such that

$$ds^2 = \left[1 + \frac{(x^1)^2 - (x^0)^2}{4R^2}\right]^{-1} \left((dx^0)^2 - (dx^1)^2\right) \ .$$

Let us call μ_0 this probability measure and μ the probability measure constructed, for $d = 2$, from μ_0 by a perturbation given by an interaction term

$$\int : v\left(X(\varphi, t)\right) : d\varphi dt$$

with as usual v a polynomial or exponential type function as in Sect. 4 or a trigonometric one, as in Sect. 3.

Raphael, Figari and Nappi constructed the corresponding Schwinger- and Wightman-functions and showed their invariance properties under the De Sitter group for $\mathcal{H}_R$.

I said this was a fundamental observation. In fact, essentially the same discovery was made at about the same time by Hawking [Haw], who made it the basis of many physical (and at times very popular) considerations. Many subsequent authors have unfortunately been unaware of [27], see [Fu-R], [Ful] for recent surveys of the subject. It should be noted that in those years some papers discussing the mathematical construction of quantum fields or curved manifolds started to appear [Bou], [Bi-W], [Ish], [Di], see also [Ful], [Sew], [DeA-D-G], [Str2]. The main point in the physical considerations mentioned above is that the curvature effect can be looked upon as a "thermalization" effect which leads to creation of particles in the vicinity of black holes or very massive bodies.

Raphael was fully aware of these consequences but dismissed them, in his usual characteristic attitude, as speculations (although he terribly liked speculations, and was brilliant at making them himself, he was careful not to give them too much weight before he could provide full mathematical proofs).

When I arrived again in Oslo at the beginning of '74, I started working with Raphael on models of statistical mechanics, partly based on the ideas I have just described. Our results got written up in a paper which appeared in [26].

Let me describe the main results, connecting them to some later and current investigations. We start with the description of lattice systems in quantum statistical mechanics. Let $\Lambda \subset \mathbb{Z}^d$ be bounded and let for $n \in \Lambda$, $\mathcal{H}_n \equiv L^2(\mathbb{R}^m)$, for some fixed $m \in \mathbb{N}$. Let $\mathcal{H}_\Lambda \equiv \bigotimes_{n\in\Lambda} \mathcal{H}_n$. Define the free quantum Hamiltonian $H_0(\Lambda)$ for lattice variables attached to Λ by

$$H_0(\Lambda) = -\frac{1}{2}\sum_{n\in\Lambda} \Delta_n + \frac{1}{2}\sum_{n,n'\in\Lambda} x_n A(n-n')x_{n'} ,$$

where Δ_n is the Laplacian acting in $L^2(\mathbb{R}^m)$, $x_n \in \mathbb{R}^m$ and $A(n-n')$ is a $|\Lambda|\times|\Lambda|$-strictly positive definite symmetric matrix depending only on $n-n'$ and decreasing as $|n-n'| \to \infty$. Let f be an element of $\mathcal{F}(\mathbb{R}^{mk})$, (the Banach algebra of Fourier transforms of bounded complex measures on $\mathbb{R}^{mk}$), $k \in \mathbb{N}$, and consider for $\lambda \in \mathbb{R}$, $a_i \in \mathbb{Z}^d$, $i = 1, ..., k$:

$$H(\Lambda) \equiv H_0(\Lambda) + V(x),\ V(x) \equiv \lambda \sum_{n\in\Lambda} f(x_{n+a_1}, ..., x_{n+a_k}),\ x \equiv (x_n,\ n \in \Lambda)$$

We showed in [26] that the $\Lambda \uparrow \mathbb{Z}^d$ limit ω_β, the corresponding Gibbs state, exists and calling α_t, $t \in \mathbb{R}$, the limit automorphism group, ω_β is a KMS state for this group, at least for $|\lambda|$ small enough, in which case we also showed convergence of the cluster expansion. These results were continued in [Gl-K], [Ko2].

A second type of model we considered was the case of classical mechanical Gibbs states for crystals, defined by the probability measure

$$d\mu_\Lambda(x) = Z_\Lambda^{-1} e^{-\beta V_\Lambda(x)} \prod_{n\in\Lambda} dx_n ,$$

with

$$Z_\Lambda \equiv \int e^{-\beta V_\Lambda(x)} \prod_{n\in\Lambda} dx_n .$$

Again we showed the existence of the limit Gibbs state $\mu_{\mathbb{Z}^d}$, obtained by letting $\Lambda \uparrow \mathbb{Z}^d$, at least for $|\lambda|$ sufficiently small.

We also discussed the approach of the quantum mechanical model to the classical model, when Planck's constant (correctly introduced in above formulae) is made to approach zero, from positive values. This type of models has been studied quite extensively, see e.g. [F-P2].

Raphael never lost interest in crystals, even though it was only in the last few years of his life that he started working intensively again on this topic. He was indeed fascinated by the very existence of crystals and wanted to prove their existence from the fundamental classical statistical mechanical Gibbs Ansatz. I remember a whole night we spent walking through the streets of an almost deserted Moscow in the fall of '87, while Raphael conducted an intense discussion about his attempts to prove their existence. He had a tremendous capacity for making long explicit calculations in his head and also had a very vivid geometrical imagination. The walk went eratically, with Raphael stopping from time to time to look for some stones in a field near the street to use them to illustrate molecule configuration to me, or using a stick to draw some figure on the sidewalk. Sometimes he realized how crazy we would look to other people watching us and made jokes how crazy scientists can be in general.

What was so fascinating about him was his capacity for looking at a problem from all sides. Most of the time he foresaw the solution long before he could explain it to anybody. He had a realistic view of mathematics; he liked to talk of discoveries, "seeing the truth", rather than in terms of creations. He was a very religious man, in a rather unorthodox way, however. One evening during that same period he once described to me a fascinating view of the development of the world, a mixture – as he pointed out himself – of science (he was a man of considerable culture and he had an extremely vivid imagination in many other fields outside mathematics and physics) and poetry (he liked poetry very much, and he knew much of the old Norse literature by heart).

But back to the crystals. In the work he began to do in the last couple of years he was examining classical statistical mechanics described by a classical Hamilton function of non relativistic particles interacting by 2-body forces

$$H_N = \sum_{i=1}^{N} \frac{p_i^2}{2m_i} + \sum_{i<j} V(x_i - x_j) \ ,$$

attempting to discuss the formation of crystals at low temperatures. Part of this work is written up in [141], see also [A-G-H-K-M]. First putting the particles in a bounded box $\Lambda \subset \mathbb{R}^3$, their partition function is defined by

$$Z_{\Lambda,\beta,N} \equiv c_{\beta,\Lambda} \int_{\Lambda^N} e^{-\beta H_N} \prod_{i=1}^{N} dx_i \ .$$

By assuming $V(x) = V(-x)$, V of compact support and suitable behavior at the origin, one can profitably introduce new real variables y_λ associated with lattice points λ in a Bravais lattice Y in $\mathbb{R}^3$,

$$Y = \left\{ \sum_{i=1}^{3} n_i a_i \in \mathbb{R}^3 \mid n = (n_1, n_2, n_3) \in \mathbb{Z}^3 \right\} \ ,$$

with a_i a given basis in $\mathbb{R}^3$. Writing the interaction as a function of $\{y_\lambda | \lambda \in \Lambda \cap Y\}$, $y_\lambda = x + \lambda$, and decomposing it accordingly, one can discuss the limit of $\frac{1}{|\Lambda|} \ln Z_{\Lambda,\beta,N}$ as $N \to \infty$, $\Lambda \uparrow \mathbb{R}^3$, for large enough β (low temperature). For results in this direction, see [141] and [A-G-H-K-M].

Raphael had many other projects in this direction concerning the quantum mechanical case, in particular attempting to give a new simple explanation of the phonon spectrum and of superconductivity phenomena.

Unfortunately, due to his untimely death these projects were only partly realized.

6. Stochastic mechanics and applications

In the spring of '73 I stayed at Raphael's house during a short visit to Oslo from Naples. At night, as usual, after the family had gone to bed, we would sit in the kitchen and begin to work. Raphael liked those hours very much; we would work almost all night through, stopping only from time to time to heat up water for tea or coffee. One of those evenings we were discussing Nelson's stochastic mechanics. We were both fond of it, perhaps in a somewhat naive way, at least we did not like the epistemological "Verbote" of the standard Copenhagen interpretation of quantum mechanics and much of the ideology around it – so Nelson's stochastic mechanics [Nel1], [Nel7], [Nel8] (see also e.g. [Gu2], [Gu3]) appeared to us as a philosophically much more pleasing description of a certain class of phenomena. We had come to think that the nodes of the wave function would yield a confinement mechanism, the singularity of the corresponding drift keeping the process away from the node. We were very excited about this phenomenon and started speculating about applications even outside quantum mechanics. Watching a drop of water which had formed in the middle of the hot plate, where we had heated our water, oscillate with nodes and maxima at regular places, Raphael exclaimed: "look, there they are, our nodal surfaces!" As usual, with his great imagination, he started describing all kinds of situations in nature where similar phenomena would arise.

The results we obtained were written up first in [25] and reformulated and expanded subsequently with a variety of applications (Nagasawa also discovered the phenomenon independently later on and gave applications particularly in biology and sociology, starting in the early eighties [Nag1], [Nag2]). Let me explain a little the idea following the general setting of [118].
Let M be a Riemannian manifold of dimension d and $X(t), t \in I\!R_+$ a diffusion process on M, with generator of the form $G_t = \frac{1}{2}\Delta + \beta^i D_i$, with $\beta^i(t,x), i = 1, ..., d,\ t \in I\!R_+, x \in M$ a given vector field (assumed to be sufficiently smooth) and D_i the covariant derivative in the direction i, Δ the Laplace–Beltrami operator on M.
Let $\rho(x,t),\ t \in I\!R_+,\ x \in M$ be the density of the probability law of $X(t)$ with respect to the Riemannian volume measure $\sigma(dx)$. Assume β^i is such that ρ is differentiable. Then

$$\frac{\partial \rho}{\partial t} = \frac{1}{2}\Delta\rho - \text{div } (\rho\beta).$$

Let $\eta^*(t), t \in I\!R_+$ be the time reversed diffusion with generator $\frac{1}{2}\Delta - \beta^{*i}D_i$,with $\beta^* = \beta - \frac{1}{2}\nabla \ln \rho$. Define $u \equiv \frac{1}{2}(\beta - \beta^*) = \frac{1}{2}\nabla \ln \rho$, then $\frac{1}{2}\Delta\rho =$ div (ρu). Moreover

$$\frac{\partial \rho}{\partial t} = -\text{ div } (\rho v),$$

with $v \equiv \frac{1}{2}(\beta + \beta^*)$. Furthermore we have $\beta(t, X(t)) = D_+X(t)$ and $\beta^*(t, X(t)) = D_-X(t)$, with $D_\pm$ being the forward resp. backward (stochastic) derivatives (i.e. $D_+X(t) = \lim_{h\downarrow 0} E\left(\frac{X(t+h)-X(t)}{h} | X(t)\right)$, with $E(\cdot|\cdot)$ denoting conditional expectation). Nelson showed, see also e.g.[A-F-K-S], [Car1], [Car2], [B-G], [Z-M] that in the case $\rho > 0$, if we set

$$a(X(t)) \equiv \frac{1}{2}(D_+D_- + D_-D_+)X(t)$$

we get

$$a(X(t)) = \frac{\partial v}{\partial t} + \nabla v \cdot v - \nabla u \cdot u - \frac{1}{2}\Delta_{DR} u,$$

with Δ_{DR} the De Rham-Laplace operator on M and ∇ the gradient operator. If we assume furthermore that v is a gradient field then we discover that the equation

$$a = -\nabla V$$

for some given real valued (smooth bounded) potential V on M is equivalent to the time-dependent Schrödinger equation

$$i\frac{\partial}{\partial t}\psi = H\psi,$$

with $H = -\frac{1}{2}\Delta + V$, ψ being related to ρ and hence to β, u, v, and thus to $a(X(t))$, by $\psi(t,x) = \sqrt{\rho(t,x)}e^{iS(t,x)}$ (in which case $v = \nabla S$). In general however ρ and thus ψ might have zeros. Let us consider in this case the stationary situation where ψ, ρ, S are independent of t. The nodal manifold is

$$U_\psi \equiv \{x \in M | \psi(x) = 0\}$$

In this case $v = 0$ and $\beta = u$. The drift β has then a repulsive singularity on U_ψ and the process $X(t)$ gets confined to regions separated by U_ψ (see [25], [34], [35], [A-F-K-S]), i.e. it is no longer ergodic but decomposes into ergodic components running in regions on M separated by U_ψ. For further studies in this direction see [B-C-Z]. This phenomenon has become important also in the theory of Dirichlet form, see e.g. [118], [A-F-K-S] and references therein.

The mechanism by which such a confinement arises is easy to see once one accepts the starting point, namely that there are phenomena in nature which are, at least approximatively, described by stochastic equations of the above form. One can motivate this a bit further by e.g. following arguments given by us or Nagasawa [Nag2] (in the context of the Berstein–Schrödinger processes discussed in [Zam]), and e.g. [A-Za]. It should be noted however, that the phenomenon is more general than formulated above, in fact a diffusion coefficient can be assumed to be present, and the drift need not be of the above special form. However in this case the equation satisfied by ψ is different (more complicated), so that for this reason the simpler model was taken in the applications. It is indeed puzzling that such a model appears to describe well phenomena in different areas, like meteorology [111], [A-B-G-S], [A-B-C-R-S-S] astrophysics [87], [94], [97], [114], [127], plasma physics [112], [128]. Raphael was very fond of these applications, and we feel that still more work should be done to clarify their meaning and extend further the range of applications.

7. Oscillatory integrals in infinite dimensions (Feynman path integrals)

In the summer of '74 Raphael and I started to work hard on putting the theory of Feynman path integrals on a solid mathematical basis. We hesitated first between two possibilities: a configuration space approach and a momentum space approach. We decided to develop the former one first. I will discuss this one first. I shall write later in this section on the other approach.

When we started we soon had the basic idea of using a Parseval kind of formula and Fourier transforms of measures. We actually discovered later in the course of our work that this idea had already been used before by K. Ito (1961/67) [Ito], J. Tarski (1971/74) and C. De Witt-Morette (1972/74). In any case, we decided to write a monograph on the topic, systematically developing the theory on the basis of a mathematical definition of the integral, and providing applications. This work appeared in '76 [28]; although at that time we had already finished the development of a mathematical method of stationary phase for our integrals, we did not include that in the book, but rather published it separately, in Inventiones Mathematicae (1977) [32] (probably our decision was not wise, that paper apparently escaped the attention of several authors who only came across the book). Later on our approach was complemented and continued in several directions. As far as the basic properties of the integral are concerned we refer to Cameron-Storvick [Ca-S], Johnson and Lapidus [J-L], Bromley, Kannan, Karandikar, Kallianpur [K-B], [K-K-K], Mandrekar [Man], Watanabe [Wa], see also [Ex], [Mal], [Hu-M], [A14], [F-P-S], [Fuj], [Mo], [Pap], [Sm-K], [123] for recent discussions. An important application of our approach was obtained in [62], [69], [82], where Raphael along with Philippe Blanchard and myself proved a trace formula for Schrödinger operators for the first time.

Finally let me mention that one of the reasons Raphael and I had originally studied such integrals were applications to the construction of quantum fields directly in Minkowski space: for a partial success this in direction see [45], [93]. Let me now briefly give an idea of our approach.

The basic idea is to carry over to the infinite dimensional case the Parseval relation valid on $\mathbb{R}^m$, for arbitrary $m \in \mathbb{N}$:

$$(2\pi i)^{-m/2} \int_{\mathbb{R}^m} e^{\frac{i}{2}|x|^2} f(x)dx = \int_{\mathbb{R}^m} e^{-\frac{i}{2}|\alpha|^2} \hat{\mu}_f(d\alpha) \tag{7.1}$$

for any function f on $\mathbb{R}^m$ which is the Fourier transform of a complex measure $\hat{\mu}_f$ on $\mathbb{R}^m$. Calling $\mathcal{F}(\mathbb{R}^n)$ the space of such functions, one realizes that in the case where $\mathbb{R}^m$ is replaced e.g. by a separable Hilbert space $\mathcal{H}$ the corresponding space $\mathcal{F}(\mathcal{H})$ is also well-defined, as well as the r.h.s. of (7.1). Taking this as a definition of the l.h.s. in the infinite dimensional case, one arrives at an integral of the form

$$I(f) = \int_{\mathcal{H}} e^{-\frac{i}{2}|\alpha|^2} \hat{\mu}_f(d\alpha),$$

also written as $\tilde{\int}_{\mathcal{H}} e^{\frac{i}{2}|x|^2} f(x)dx$.

We then showed that $f \to I(f)$ is a linear, normalized, continuous functional, when $\mathcal{F}(\mathcal{H})$

is given its natural topology, using the total variation norm of the complex measure $\hat{\mu}_f$. We proved a Fubini theorem for iterated integrals as well as various approximation formulae (the latter were also extended in subsequent work, e.g. [E-T1], [E-T2], [C-R-R-S], [Za]). These results can easily be extended to the case where the phase function $|x|^2$ is replaced by a quadratic form (x, Bx) with B symmetric with an at least partial inverse, see [28]. The Hilbert space can be replaced by a Banach space. We already gave applications to quantum mechanics in [28], by showing that the heuristic (non–mathematically defined) integrals considered by Feynman, and called Feynman path integrals, can be rigorously defined in terms of the above oscillatory integrals, it suffices in these applications to find a suitable Hilbert space $\mathcal{H}$ (or a Banach space). E.g. the solution of the Schrödinger equation on $\mathbb{R}^d$

$$ih\frac{\partial}{\partial t}\psi(t,x) = \left(-\frac{h^2}{2m}\Delta + \frac{\lambda}{2}(x, A^2x) + V(x)\right)\psi(t,x) \tag{7.2}$$

with $\lambda \geq 0, h > 0$ (Planck's constant divided by 2π), A^2 a positive and symmetric $d \times d$–matrix, $(,)$ the scalar product in $\mathbb{R}^d$, $V \in \mathcal{F}(\mathbb{R}^d)$, can be given by the above mathematically defined Feynman path integrals. We also gave the representation of scattering quantities in term of such integrals and extended the theory to the infinite dimensional setting, with applications in quantum field theory. The detailed discussion of thermal states is also noteworthy in this setting, a topic to which a certain amount of attention has been given in recent years in connection with the quantization of general relativity (see Sect. 5.). See [A-Sch] for an application to the construction of the abelian Chern–Simons model.

In September 1975 we finished a long paper containing the mathematical realization of certain heuristic ideas of Dirac and Feynman which were at the basis of the very introduction and heuristic use of Feynman path integrals. This realization of stationary phase in infinite dimensions, permitted a detailed study of the way the smallness of Planck's constant permits relation of quantum mechanical quantities to corresponding classical ones (see [32]).

Mathematically, the basic work which has to be done is to find a suitable extension of the classical method of stationary phase for oscillatory integrals to the infinite dimensional setting. However, it is to be noticed that the way the integrals have to be defined in infinite dimensions, for integrable functions in the class $\mathcal{F}(\mathcal{H})$, also requires an adequate reformulation of the finite dimensional theory, see [Rez1], [Rez2], [A-B1], [A-B2] (in which also a few minor slips in the original work [32] are corrected). Raphael and I took a lot of pleasure in this, also getting involved in discussions about catastrophy theory. I remember Raphael taking every opportunity in the cafeteria of the Mathematics building in Oslo to show us, with an almost childish joy, various singularities appearing on the surface of the liquid in a cup of coffee. As always, he liked the beauty of mathematical concepts and how they fit nature. He would get extremely nervous and disappointed when people would try forcefully to impose inadequate mathematical structures to subjects just to show off: he resented this almost as a personal offence.

To return to the method of stationary phase, the results were of the following type. Consider oscillatory integrals of the form

$$\widetilde{\int_{\mathcal{H}}} e^{\frac{i}{2h}|x|^2} e^{-\frac{i}{h}W(x)} g(x)dx \tag{7.3}$$

with $h > 0$, $W, g \in \mathcal{F}(\mathcal{H})$, with $\mathcal{H}$ a real separable Hilbert space.
The analogy with the finite dimensional situation ($\mathcal{H} = I\!R^m$) suggests obtaining asymptotic expansions in powers of h around stationary (i.e. critical) points of the phase function $x \to \frac{1}{2}|x|^2 - W(x)$, if there are any, and obtaining an expression vanishing faster than any power of h if the are no critical points. The character of the expansion is different according to the type of stationary point (whether isolated, occurring in families, degenerate or not). All of this can indeed be implemented by a suitable Morse theory in infinite dimensions, as done in [32] (with subsequent improvements and extensions in [Rez1], [Rez2], [A-B1], [A-B2]; this also gives the connection with catastrophy theory alluded to above).
Applications to the discussion of the behavior of solutions of the initial problem for Schrödinger equation, considering Planck's constant h to be small were already given in the original paper.
In later years, the interest in detailed discussions of precisely this behavior increased strongly in physics in connection with an extensive study on the heuristic level of "quantum chaos", (it is appropriate to point out here that the above paper provides a mathematical framework for work on this topic, at least concerning the initial value problem).

By the same methods one also obtains a representation formula for the kernel of the fundamental solution of Schrödinger equation, and the problem of relating the spectrum of the Hamiltonian to the periodic orbits of the underlying classical Hamiltonian system asymptotically for h in a neighborhood of zero arises. Heuristically this was pointed out forcefully by Gutzwiller. These heuristic methods have received a lot of attention in recent years, especially in the physics literature, (see e.g. [Gu] and references therein), and again in connection with discussions of quantum chaos versus classical chaos. Some of these mathematical developments can be found in a paper Raphael wrote with Philippe Blanchard and myself in '81 [82], where we gave the first mathematical proof of a trace formula for Schrödinger operators. The idea is very simple: by our definition of oscillatory integrals in infinite dimensions the kernel $e^{-\frac{i}{h}tH}(x,y)$, $x, y \in I\!R^d$, with H of the above form, can be written as

$$e^{-\frac{i}{h}tH}(x,y) = \widetilde{\int_{\mathcal{H}_{x,y}}} e^{\frac{i}{h}|\gamma|^2} e^{-\frac{i}{h}W^t_{x,y}(\gamma)} d\gamma, \tag{7.4}$$

for a certain Hilbert space $\mathcal{H}_{x,y}$ of paths $\gamma : [0,t] \to I\!R^d$ such that $\gamma(0) = x$, $\gamma(t) = y$ and a certain phase function $W^t_{x,y}(\cdot)$, and for all t outside a certain discrete set. Thus one obtains a formula for the trace

$$\theta_h(t) \equiv \sum_n e^{-\frac{i}{h}t\lambda_n} = \operatorname{tr} e^{-\frac{i}{h}tH} = \int_{I\!R^d} e^{-\frac{i}{h}tH}(x,x)dx =$$

$$= \int_{I\!R^d} dx \widetilde{\int_{\mathcal{H}_{x,x}}} e^{\frac{i}{h}|\gamma|^2} e^{-\frac{i}{h}W^t_{x,x}(\gamma)} d\gamma$$

where λ_n are the eigenvalues of H (we assume here $\lambda > 0$ in (7.2)).
Applications of the method of stationary phase in infinite dimensions developed by Raphael and myself [32] yield then the trace formula in the form of a rigorous asymptotic expansion of the theta function for Schrödinger operators, $\theta_h(t)$, in terms of contributions given by periodic orbits of the underlying classical Hamiltonian system.
The relation with Maslov quantization formulae was also discussed in [82], see also [A-B2], [A-BB-L] for further developments along these lines.
Raphael was very proud of these results and fascinated by their relations with the Selberg trace formula, the theory of automorphic functions and number theory (see also [A-A], [Gu] in this relation). He had many original ideas on this topic, and I am sure he would have returned to work in this direction if he had been given a chance.

In the fall of '77 Raphael and I moved with our families to spend a year at the Centre de Physique Théorique, CNRS, Marseille following an invitation by Stora and Kastler. We returned there often, even after the Centre was moved to Luminy. In fact, Raphael developed a strong friendship with Mohammed Mebkhout, Doyen of the Faculté des Sciences of the Université d'Aix-Marseille II, Luminy, and regularly spent long periods there, working with him and several other theoreticians, in particular Michel Sirugue (who succumbed to cancer shortly before Raphael died). Raphael took up a Professorship at the Université de Provence, from 1982 to 1985. In '77 the presence of a group of mathematical physicists working at the CNRS and interested in paths integrals (in particular Philippe Combe, Roger Rodriguez, Madeleine Sirugue-Collin and Michel Sirugue) motivated the organization of a seminar on Feynman path integrals. G. Rideau reported on some work by Chebotarev and V. Maslov (cf. [C-M]) who had developed essentially the same idea as Raphael and I had contemplated doing in Oslo, concerning a momentum space approach to Feynman path integrals. Raphael told me that perhaps this was the approach Varadhan had been sketching before however we could not locate the publication, so we had to put it aside. When Chebotarev–Maslov's papers appeared, Raphael felt free to start thinking about the topic again and started a large project in this direction mainly with the above mentioned mathematical physicists in Marseille (whom Blanchard and myself occasionally joined).
Let me briefly report on this work, referring to existing survey papers like, for example, [B-C-S-SC] for more details. In [56] Raphael, in collaboration with Combe, Rodriguez, Sirugue and Sirugue-Collin, gave a representation of the solution of Schrödinger's equation (7.2) (with $\lambda = 0$) in the form (momentum space variable $p \in I\!R^d$) :

$$\left(e^{-itH}\hat{\psi}\right)(p) = E\left\{e^{-i\int_0^t h_0(p-p(s))ds}\,\hat{\psi}(p-p(0))\right\},$$

where $\hat{}$ means Fourier transform and the expectation is with respect to a Poisson process $p : [0,t] \to I\!R^d$, with $p(t) = 0$, associated with $V \in \mathcal{F}(I\!R^d)$; h_0 is the function $h_0(p) = \frac{|p|^2}{2}$. This is actually a special case of a more general result in which $I\!R^d$ is replaced by a group G; H_0 (represented above by the function h_0) is a self-adjoint operator affiliated with the Banach algebra $L^\infty(\mathcal{X}, \mu)$, where μ is a probability measure on a topological space $\mathcal{X}$, quasi-invariant with respect to the action of the group G,

$$V = \int_G U(g) d\nu(g),$$

with ν a complex measure on G and $U(\cdot)$ a continuous unitary representation of G in $L^2(\mathcal{X}, \mu)$. Important special cases are:

1) $G = I\!R^{2N}$
2) $G = G_0 \times G_1, G_1$ a subgroup of the dual of G_0, where G_0 is the group of all finite subsets of a countable set with symmetric difference as group law. In this case one obtains the Schrödinger evolution of spin systems (see [68], [78]).

In further collaborations, extensions [79] and applications to quantum fields are given. In [78], [80] the case of a zero mass two dimensional sine-Gordon field, without ultraviolet cut-off, in a space-time bounded domain $\Lambda \subset I\!R^2$ (2-dimensional space–time), given by a probability measure of the form (3.1) with $m = 0$, $v(x) = \int \cos(\alpha x + \theta) d\nu(\alpha) 0 \leq \theta$, $\operatorname{supp} \nu \subset (-\sqrt{2\pi}, \sqrt{2\pi})$ is studied. The matrix elements of the time evolution operators are expressed in terms of the above type of Poisson integrals and related to the grand canonical partition function of a neutral Coulomb gas in two dimensions, thus to the type of models studied by Raphael and myself in '72 [19]. Moreover it is shown that for suitable coherent states ψ in Fock space the limit $\lim_{\chi \to 1}(\psi, e^{itH_0} e^{-it(H_0 + \lambda V_{g,\chi})} \psi)$ exists and is given by an expansion with respect to a Poisson measure.
In connection with a conference on Feynman path integrals taking place in Marseille in the summer of '78, Raphael and I worked on providing a representation of the Lorentz group by automorphisms on a Banach space and gave a solution using our mathematically defined configuration space oscillatory integrals [45]. This work was continued and expanded in '81, when Raphael worked with Blanchard, Combe, Sirugue and myself on a paper [93] presenting the construction of a local relativistic invariant flow associated with quantum fields with sine-Gordon interaction. In fact, using the momentum space oscillatory integrals discussed above, we obtained a representation of the Poincaré group by a group of non trivial automorphisms on a Banach space generated by the Weyl algebra in all space–time dimensions, so that the fields, represented through a representation of the Weyl algebra of commutation relations, transform covariantly under the Poincaré group. The method used is a combination of the above representation of Schrödinger unitary groups in terms of Poisson type integrals with the same method that Raphael and I had used in our study of the Coulomb gas [19].
Raphael and co-worker's analysis of Poisson processes in quantum theory stimulated further research, see e.g. [B-C-S-SC], [B-R] (and references therein).

8. The theory of Dirichlet forms and its applications in quantum theory

During '74-'75 Raphael and I tried hard, partly in connection with our work on Feynman path integrals, to construct models of non–trivial interacting quantum fields.

Throughout that work we were often reflecting on foundations of quantum field theory, in particular on the representation of Weyl commutation relations and canonical quantum fields. We presented a lecture at a conference in Marseille '75 on the representation of the Hamiltonian in quantum theory in terms of quadratic forms of the type $\frac{1}{2}\int \nabla f \cdot \nabla g d\mu$, for f, g smooth bounded cylinder functions and μ some (probability) measure on some infinite dimensional space.

In writing up our paper [30], we discovered some work by M. Fukushima [Fu4] which made us begin a whole new program of investigation. I still remember vividly the moment in which Raphael entered my office, and after showing him my discovery of Fukushima's paper, he said with a smile full of joy, as he had been relieved from all worries: "Come, let us go to my office and work it out". Very soon we saw a whole world opening up. In fact, we realized that we could develop a theory of Dirichlet forms over infinite dimensional spaces, by extending Fukushima's finite dimensional theory in an appropriate way. The latter theory has its roots in Hilbert space methods in classical potential theory, in particular the work of Beurling and Deny in the late fifties. Fukushima developed the theory systematically and made probabilistic tools available for it by his construction of associated Hunt processes, see [Fu1], [Sil].

In the second part of '75 we started working in this direction systematically. We presented versions of the infinite dimensional theory, with applications to quantum field theory, in [30], [31], [34], [36] and, in collaboration with Streit, we found a new way to look at the finite dimensional case with applications to non relativistic quantum mechanics [35]. The continuation of our work on Dirichlet forms and their applications was also stimulated by Raphael's and my stay at ZiF, as members of the research project "Mathematical Problems of Quantum Dynamics" led by Streit.

Let me go over very briefly some basic elements of the theory of Dirichlet forms.

Let us consider a Markov (strongly continuous contraction) semigroup $P_t, t \geq 0$ on some L^2-space, say $L^2(\mu)$, with μ some positive measure. Markov means $0 \leq u \leq 1 \rightarrow 0 \leq P_t u \leq 1$ for all $t \geq 0$, $u \in L^2(\mu)$. Assume that P_t is symmetric. There is a bijective correspondence between symmetric Markov semigroups and certain closed, bilinear, positive forms on $L^2(\mu) \times L^2(\mu)$, called Dirichlet forms E, given by $D(E) = D(\sqrt{-A})$, and, for all $u, v \in D(E)$

$$E(u,v) = (\sqrt{-A}u, \sqrt{-A}v),$$

with A the generator of P_t and $(,)$ the scalar product in $L^2(\mu)$. The Markov property of P_t corresponds to the Dirichlet property of E, i.e.

$$E(T(f), T(f)) \leq E(f,f)$$

for all real functions T on $\mathbb{R}$ such that

$$|T(\alpha) - T(\beta)| \leq |\alpha - \beta|,\ T(0) = 0.$$

It was shown by Fukushima that if μ is a Radon measure on a locally compact separable space E, there is a Hunt process $X(t)$ such that $P_t u(x) = E^x\left(u(X(t))\right)$ for μ-a.e. $x \in E$, with E^x the expectation with respect to the process started at time 0 from x. Moreover there is a natural capacity associated with a given Dirichlet form and $x \to E^x(u(X(t))$ is a quasi continuous (i.e. modulo sets with capacity zero) modification of $P_t u(x)$. Finally a general decomposition theorem of Dirichlet forms, associated semigroups and processes, in a diffusion part, a jump part and a killing part, is available.

A particular case is given by so-called classical Dirichlet forms ([A-R1]) which are essentially the Dirichlet forms associated with diffusions over $\mathbb{R}^d$. For $u, v \in C_0^\infty(\mathbb{R}^d)$ they are given by

$$E(u,v) = \frac{1}{2}\int \nabla u \cdot \nabla v d\mu \tag{8.1}$$

in $L^2(\mu)$. Raphael and I managed to extend a good part of the theory of forms of this type for μ a Q-quasi-invariant probability Radon measure on a rigged Hilbert space $Q \subset \mathcal{H} \subset Q'$ (with $\mathcal{H}$ real separable, Q countably normed nuclear and Q' its topological dual), μ being assumed in addition to satisfy a weak regularity condition (namely be such that 1 is in the domain of the adjoint of the partial derivative operator in any direction of a base in $\mathcal{H}$ made of elements from Q). We were the first ones, to our knowledge, to verify that the associated process satisfies in the weak sense a stochastic equation with initial measure μ (i.e. solves the corresponding martingale problem).

We were very pleased to be able to prove, after hard work, that interacting quantum fields in two space time dimensions (with interactions of the polynomial or exponential type) have time-zero measures which satisfy the above assumptions, thereby providing very interesting examples for our general theory. We then went on to prove further properties of the associated processes and fields, like e.g. the smoothness of μ (as a consequence of infinite dimensional ellipticity) and provide ergodic decompositions. This was done basically in two papers [34], [36] which appeared in '77.

We managed, and it was our main aim at that time, to prove that the Hamiltonians of interacting quantum fields in two space time dimensions can be described in terms of these forms (as well as the free Hamiltonians, in arbitrary dimensions). By so doing we incidentally had to develop some tools which later on, in the setting of (abstract) Wiener spaces, were also to appear as aspects of the "Malliavin calculus" (we reported on this work at the famous stochastic analysis calculus conference in Durham, '80 [65]).

Our work was continued in '82 and directed essentially towards getting better path properties for the process by Kusuoka [Kus2], in which the state space is a Banach space (Fukushima had suggested the topics to Kusuoka, on the basis of our work). Also Krée, who was a referee of our paper [34], advised Paclet to work on the subject [Pa]. Furthermore a treatment of infinite dimensional Dirichlet forms is also contained in the book [123] on nonstandard analysis which Raphael coauthored, see Sect. 10. Later on the work was continued by myself in collaboration with Röckner [A-R1], [A-R2], [A-R3], [A-R4], Kusuoka and Ma Zhiming [A-K-R], [A-M-R], [A-M] as well as by Schmuland [Sch], Lyons, Song Shiqui, Fan Rhuzong and in the framework of Hida's white noise calculus by Hida, Kuo, Potthoff, Röckner, Streit, and myself, see e.g. [A-H-P-R-S1], [A-H-P-R-S2], [H-K-P-S], [Ra]. See also e.g. [Dy1], [Dy2], [Dy3], [B-H], [A-F-H-M-R].

We would also like to mention that an early reference on Dirichlet forms over infinite dimensional spaces is [Gro4], who, assuming closability, was interested in the analytic part: associating logarithmic Sobolev inequalities to them (see also the survey paper by Davies,

Gross and Simon, these volumes, and [Gro5], [S-Z] for new developments).
Nowadays the theory of Dirichlet forms plays an increasingly important role, which certainly would have pleased Raphael very much.
The work of Raphael, Streit and myself [35] basically concerned problems in the theory of finite dimensional Dirichlet forms of diffusion type and its relations with (nonrelativistic) quantum theory. We especially investigated classical Dirichlet forms by providing first concrete closability criteria of their restriction on C_0^∞ functions in terms of properties of μ. We gave several examples of Dirichlet forms and discussed ergodic properties of the associated processes and semigroups, making the distinction clear between ergodicity with respect to time translations and space translations (as a special case of the infinite dimensional theory developed in [30], [34], [36]). We also discussed the stochastic equation satisfied in the weak sense by the process (more general than a semimartingale) with given invariant measure μ (this ran independently and parallel to the more general Fukushima's Doob-Meyer type decomposition [Fu1], and answered a question raised by Ezawa, Klauder and Shepp [E-K-S]).
Also, the fundamental relation between Dirichlet forms and Schrödinger theory, already studied in [30], was pursued systematically. To explain it quickly, let $d\mu = \varphi^2 dx$ on $I\!R^d$ with $\varphi > 0$ smooth, then E defined by (8.1) is closable in $L^2(\mu)$. Let H_μ be the negative of the generator of the associated semigroup, i.e. $P_t = e^{-tH_\mu}$ in $L^2(\mu)$, $t \geq 0$. Then $\varphi H_\mu \varphi^{-1}$ is a self-adjoint operator H in $L^2(I\!R^d, dx)$, and $H = -\frac{1}{2}\Delta + V$ with $V = \frac{1}{2}\frac{\Delta\varphi}{\varphi}$. Whereas H_μ also exists when φ is not C^2, it cannot be written in this case as $-\frac{1}{2}\Delta + V$; so that in this case it is called a generalized Schrödinger operator (cf. [A-B-R]).
A typical example discussed in [35], is $\varphi(x) = \frac{e^{4\pi\alpha|x|}}{|x|}$, $x \in I\!R^3 \setminus \{0\}, \alpha \in I\!R$, in which case $\varphi[H_\mu - (4\pi\alpha)^2]\varphi^{-1}$ is the generalized Schrödinger operator rigorously describing a perturbation of $-\frac{1}{2}\Delta$ by a point interaction (renormalized delta function) situated at the origin of $I\!R^3$. This type of models was going to receive a lot of attention later on in connection with Raphael's work on point interactions, see [136] and Sect. 11 below.
In a subsequent paper [55] with Streit and myself, Raphael studied the limit of Dirichlet forms and associated Hamiltonians, i.e. the situation which arises when one takes a sequence of classical Dirichlet forms given by measures $d\mu_n = \varphi_n^2 dx$, such that $\varphi_n \to \varphi$ in a suitable sense.
This work was continued in [A-K-S], for a recent application see also e.g. [I-O]. An important uniqueness question both for finite and infinite dimensional Dirichlet forms raised by Raphael and myself in [84], [85] was treated in [Wie], [Ta1], [Ta2], [R-Zh]. For recent surveys about Dirichlet forms see [Al3], [AR2], [Osh].

Certainly through the work initiated by Raphael the interaction of the theory of Dirichlet forms with quantum theory became quite strong, see e.g. [44], [107], [A-B-G-S], [A-K-R-S], [Ca], [R-Si], [Pan], [Ta1], [Ta2], [J-L], [J-M-S], [Br1], [Br2], [Br3], [Fu2], [Fu3], [Ko1], [Li-S], [A-G-K-S], [Osh], [Ka-M], [Ko-K], [A-F-H-M-R], [A-R-S], [B-K], [R-S].

9. Operator algebras: ergodic theory, automorphism groups, non-commutative Dirichlet forms

9.1. Non-commutative Dirichlet forms

The study of problems in operator algebras is one of the centres of activity in the Institute of Mathematics at the University of Oslo. A group had been active for many years around E. Alfsen and E. Størmer. Raphael always closely followed the activities of this group, in his way: he was extremely quick and could grasp on the spot by an "intuition stroke" what one would normally do by a series of systematic deductions and studies. He was often skeptical about too abstract developments, however he could change his mind and enthusiastically start to work out consequences if he saw the possibility of some striking applications or developments.

In 1976, after having worked on Dirichlet forms, he had the idea to extend the theory to the non-commutative case (and in this way also get closer contact with the above quoted groups). We wrote then a paper [37]. In this work the state space $\mathcal{H}$ of our previous study of classical Dirichlet forms, is replaced by a C^*-algebra $\mathcal{A}$, the measure μ (giving the Hilbert space $L^2(\mu)$ of the theory) is replaced by a faithful (lower semicontinuous) trace τ on $\mathcal{A}$. The Hilbert space $L^2(\mu)$, with its natural order "$\leq$", is replaced by $L^2(\mathcal{A},\tau) \equiv \{x \in \mathcal{A} | \tau(x^* x) < \infty\}$, with the order "$\leq$" of closed operators.

A Dirichlet form E on $L^2(\mu)$ is replaced by a sesquilinear positive form E on a dense domain $D(E)$ of $L^2(\mathcal{A},\tau)$, having the contraction property $E(T(x), T(x)) \leq E(x,x)$ for all hermitian x in $D(E)$ and $T : D(E) \longrightarrow I\!R$ with $T(0) = 0$, $|T(x-y)| \leq |x-y|$. The symmetric Markov semigroup on $L^2(\mu)$ of the commutative theory, uniquely associated with the commutative Dirichlet form E, is replaced by a semigroup ϕ_t on $L^2(\mathcal{A},\tau)$, symmetric and such that $0 \leq x \leq 1 \rightarrow 0 \leq \phi_t(x) \leq 1$, for all $x \in L^2(\mathcal{A},\tau)$.

It turns out that in the case where $\mathcal{A}$ is a subalgebra of the space $\mathcal{B}(\mathcal{K})$ of all bounded operators on a Hilbert space $\mathcal{K}$ to the classical Dirichlet form

$$E(u,v) = \frac{1}{2} \int_{\mathcal{H}} \nabla u \cdot \nabla v d\mu$$

there corresponds the form

$$E(x,x) = \text{ tr } (x^2 M) + \sum_{i=1}^{\infty} \text{ tr } ([x_i, m_i]^* [x_i, m_i]) \ ,$$

with M a self-adjoint positive operator and m_i-bounded operators such that $\text{tr } (m_i^* m_i) < \infty$.

Also it turns out that in a certain sense a special class of Markov semigroups which we called completely Markov semigroups, and which are in 1-1 correspondence with "completely Dirichlet forms" correspond to commutative Markov processes. Our work was related to the theory of quantum dynamical semigroups (developed by Davies, Gorini, Kossakowski, Sudarshan, Lindblad and others, see also Sect. 9.3 below), non equilibrium quantum statistical mechanics and quantum probability (see e.g. [A-W]).

Recently our work was extended in connection with the theory of non-commutative processes in work by Davies, Lindsay [Da-L], and Applebaum, Goldstein, and Sauvageot.

9.2. Non-commutative Perron-Frobenius ergodic theory

During the academic year '76 – '77 David Evans was visiting Oslo, and Raphael started working with him on an extension of Perron-Frobenius theory for positive maps on non-commutative matrix algebras (instead of the classical case of ordered vector spaces). Among the main results let us mention the following [40]:

Let ϕ be a positive irreducible linear map on a finite dimensional C^*-algebra $\mathcal{A}$. Let

$$r_x \equiv \sup\{\rho \in \mathbb{R} | \rho x \leq \phi(x)\} .$$

Then r_x attains its maximum r at a strictly positive element $z \in A$ (so that $r_z = r$). r (the spectral radius of ϕ or characteristic number of ϕ) is a simple eigenvalue of ϕ with eigenvector z. Moreover z gives a faithful state. If ϕ is a "Schwarz" (e.g. 2-positive) map of $\mathcal{A}$ (i.e. $\phi(1) = 1$, $\phi(x^*x) \geq \phi(x)^*\phi(x)$ for all $x \in \mathcal{A}$), then the intersection of the spectrum of ϕ with S^1 is a discrete subgroup Γ of S^1. Each eigenvalue of ϕ in Γ is simple and the eigenvectors are scalar multiples of unitary elements in $\mathcal{A}$. Corresponding detailed results are given for irreducible affine maps on the state space $S(\mathcal{A})$ of $\mathcal{A}$ (whose structure is also studied in details).

In the spring of '77 I joined Raphael in this study, and in May '77 we wrote a paper [43] on a further extension of the Perron-Frobenius theory to non-commutative von Neumann algebras (extending in particular in this direction work by Størmer on automorphism groups on commutative C^*-algebras).

Let ϕ_t be a completely Markov semigroup (2-positive actually suffices) on a von Neumann algebra M. We showed in particular that if ϕ_t is ergodic then the spectrum of the infinitesimal generator is contained in the upper half plane of the complex plane. The possible eigenvalues α form a subgroup Γ of the additive group $\mathbb{R}$, which is either dense in $\mathbb{R}$ or discrete. The spectrum of ϕ_t is invariant under Γ. The eigenvectors ϕ_α correspond to unitary operators u_α in M and $\alpha \to u_\alpha$ is a multiplier representation of Γ, unitary if Γ is cyclic. The restriction of ϕ_t to the von Neumann subalgebra generated by the u_α, $\alpha \in \Gamma$ yields a group of ergodic automorphisms and the corresponding restriction of the vector states invariant for ϕ_t is a trace.

Later related work on these problems is contained in papers by Accardi, Davies and especially Frigerio and Verri, see e.g. [Fr1], [Fr2], [F-V].

9.3 Dynamical semigroups on C^*-algebras

In '77 I was offered a permanent job in Bielefeld. At that time there was no permanent position available in Oslo. The decision was a very difficult one for us both to take – we had been working so closely for many years and a separation seemed almost impossible, both personally and scientifically. But I had no other choice, so I took the job in Bielefeld. Raphael also spent a winter term in Bielefeld, after we returned from a stay in Marseille ('77-'78). Afterwards we learned to develop a new system of communication and interaction, with frequent trips to and from Oslo and many common stays abroad, so that we managed to maintain a vibrant collaboration until Raphael's death in '88. In '79 Raphael, with Gunnar Olsen (a member of the "operator algebra group" in Oslo) and myself, again took up the study of dynamical semigroups on C^*-algebras which we had already initiated in connection with the study of non-commutative Dirichlet forms. In [60] we considered the algebra $\mathcal{A} = M_n$ of $n \times n$ matrices with a completely positive semigroup ϕ_t of maps from $\mathcal{A}$ into $\mathcal{A}$, called a dynamical semigroup. We looked at ϕ_t as a non-commutative analogue of a Markov process in the sense that ϕ_t is a non-commutative analogue of the random endomorphisms associated with the commutative process X_t^z running on a state space Z, starting at $z \in Z$, defined by

$$f \in C(Z) \to X_t^z(f)(\cdot) \equiv f(X_t^z(\cdot)) \ .$$

(the study of these random endomorphisms received a lot of attention later on, in connection with stochastic differential equations, see e.g. [Ku]).

Let L be the infinitesimal generator of ϕ_t. We prove that L is characterized by $\sum_{i=1}^N (ad\beta_i)^2$ for some natural number N and some $\beta_i \in su(N) \simeq$ aut $\mathcal{A}$ (where we use that the group aut $\mathcal{A}$ of * automorphisms of $\mathcal{A} = M_n$ is $SU(n)$). This uses essentially results by Stinespring on the characterization of completely positive maps. Identifying the Lie algebra $su(N)$ of $SU(N)$ with the left invariant vector fields on $\mathcal{A}$, so that there corresponds to β_i such a vector field X_i, we see that to a symmetric left invariant diffusion X_t^e on Aut $\mathcal{A} \simeq SU(N)$ with generator $\sum_{i=1}^N X_i^2$, started at the origin $e \in$ Aut $\mathcal{A}$ there corresponds the symmetric dynamical semigroup ϕ_t on $\mathcal{A}$ and one has

$$\phi_t(a) = E(X_t^e(a)) \ ,$$

for all $a \in \mathcal{A}$ which relates a semigroup on a non-commutative C^*-algebra with an ordinary commutative diffusion X_t^e with values in a matrix group. This is also taken over to the state space $S(\mathcal{A})$ of $\mathcal{A}$: for $\sigma \in S(\mathcal{A})$, $\sigma \circ X_t^e$ is a Markov process on $S(\mathcal{A})$, started in σ, one has $\sigma \circ \phi_t = E(\sigma \circ X_t^e)$ and $\sigma \circ X_t^e$ runs on points inasmuch as the extreme points of $S(\mathcal{A})$ go into themselves under $\sigma \circ X_t^e$; moreover $\sigma \to \sigma \circ X_t^e$ is an affine map of $S(\mathcal{A})$.

In my opinion these are basic results relating commutative diffusions with non-commutative ones. These results deserve further investigation in connection with newer developments in non-commutative probability (see e.g. [115], [Kum-M], [D-W] and the contribution by Accardi and Gang in these volumes).

9.4 Ergodic actions of compact groups on C^*-algebras

In '77 Raphael and I, motivated by an observation of Størmer [Stø], began investigating the action of compact groups by automorphisms on C^*-algebras and W^*-algebras. The minimal (irreducible) objects in this case are the so-called ergodic actions (i.e. such that the only invariant elements in the algebra are the scalar multiples of the identity). [33], which was produced as a Preprint in Marseille in '79, is the first of several papers by Raphael and co-workers ([61], [71], [72]), and others, on the subject. Raphael's motivation for such an investigation was, as in other areas, the strive to find simple basic elements behind the proliferation of phenomena (he admired e.g. Lie groups as well as Klein's points of view on classifying geometries, as models to be imitated in all areas).
The first basic ideas and results on the subject were already derived in [33], which remained unpublished however, due to a gap in the proof of a conjecture of Størmer [Stø].
[33] contains e.g. the principle which illustrates that in order to classify the ergodic actions of G it suffices to classify the ergodic actions of all closed subgroups on factors. Moreover if G is abelian, Raphael and I show in [33] (also partly contained in [61]) that an ergodic action is completely described by a homomorphism $h : \hat{G} \to G$ (with $\hat{G}$ the dual of G) which is antisymmetric ($< h(\alpha), \alpha >= 1$ for all $\alpha \in \hat{G}$). This we did in detail for $G = T^n$.
In [61], we gave a complete classification of actions, by automorphisms, of compact abelian groups on C^*-algebras, a study extended later by Olesen, Pedersen and Takesaki [O-P-T]. We showed that there exists a unique invariant state which is a faithful trace. Each irreducible representation of G occurs with multiplicity at most 1. Particular cases were then studied:

$$G = (S^1)^{|J|} \text{ with } |J| = 1, 2 \; .$$

For $|J| = 1$ there is only one ergodic action, given by left translations. For $|J| = 2$ there is a 1-dimensional family of ergodic actions, characterized by a family of bicharacters indexed by an irrational number λ and a natural connection with the unique hyperfinite factor II_1 was found. This investigation was continued in '80 in work of Raphael with Tor Sjelbred [71]. Here it is shown that if α is any continuous ergodic effective action by automorphisms of the 2-torus on a C^*-algebra $\mathcal{A}$, then $\mathcal{A}$ is isomorphic to a certain sub-C^*-algebra $\mathcal{A}_\lambda$ of $\mathcal{B}(L^2(\mathbb{R}))$ generated by the operators a, b defined, for $f \in L^2(\mathbb{R})$, by

$$(af)(x) \equiv f(x + \lambda), \; (bf)(x) \equiv e^{2\pi i x} f(x) \; ,$$

λ irrational, $x \in \mathbb{R}$, and α is given on $\mathcal{A}_\lambda$ by

$$\alpha_g(a^{n_1} b^{n_2}) = \gamma(g) a^{n_1} b^{n_2} \; ,$$

with

$$\gamma(g) =< (n_1, n_2), g >, (n_1, n_2) \in \hat{G} \simeq \mathbb{Z}^2 \; ,$$

$<,>$ being the dualization between G and $\hat{G}$. Moreover if $\mathcal{A}_\lambda \simeq \mathcal{A}'_\lambda$, then $e^{2\pi i \lambda} = e^{\pm 2\pi i \lambda'}$. Further studies in this direction were done by Rieffel. In fact C^*–algebras of the form $\mathcal{A}_\lambda$ are the prototype of non–commutative tori, which are the basic examples of non-commutative differential geometry, see e.g. [B-Z], [Bra], [D-R], [Wat], [Rie1], [Rie2], [Kas3].

Until his death, Raphael maintained a vivid interest for these structures, which he considered to be the beginning of a truly non-commutative analysis (at some point around '77, he gave seminars on Heisenberg's matrix mechanics reading Heisenberg's original papers and was struck by his mathematical originality in considering canonical relations as a new, non-commutative starting point for both mathematics and physics).

If $\alpha : G \to Aut(M)$ is an ergodic action, of a compact group by automorphisms on a W^*-algebra M then $\tau(a) = \int \alpha_g(a)dg$ gives a faithful normal state on M. Størmer had shown earlier that τ is a trace if G is abelian, and conjectured that this was true in general. The proof of this conjecture attempted in [33] was not correct, but in [72] Raphael, in collaboration with Landstad and Størmer, succeded in giving a proof, which turned out to be much more complicated that one would expect from the short proof given by Størmer in the abelian case. A detailed knowledge of the decomposition of a n-fold tensor product of irreducible representations of G was needed. The authors, studying the ergodic actions by $*$ automorphisms of a compact (not necessarily abelian) group G on a unital C^*-algebra $\mathcal{A}$, managed to show that the unique G-invariant state is a trace. If $\mathcal{A}$ is a von Neumann algebra, then τ is finite and injective.

More details about ergodic actions of compact groups are given in the contributions by Landstad and Størmer in these volumes. The view one also finds in [33] that a C^*-algebra $\mathcal{A}$ on which G acts ergodically, can be considered as a non-commutative ("quantized") version of the commutative algebra $C(G)$, has turned out to be very fruitful and leads, for instance, to connections with solutions of the quantum Yang-Baxter equation (see e.g. [Dri], [Was1], [Was2]).

10. Nonstandard analysis

At an early stage of his scientific life Raphael was already fascinated by nonstandard analysis. He mentioned Skolem-Robinson's work to me when we met in Princeton in '70, as giving a framework in which one could in principle justify all "correct heuristic computations" we liked to do in our conversations and dreams, e.g. in connection with the construction of quantum fields. Certainly the presence in Oslo of a strong tradition in mathematical logic, which has its origin in the work of Skolem, represented by J. E. Fenstad, S. Aanderaa, D. Normann and their students, made a most stimulating background for coming in contact with nonstandard analysis. Raphael would often sit in the cafeteria and present his newest "heuristic computations" and challenge his colleagues from mathematical logic to give a meaning to them using nonstandard tools. We often discussed in more serious moods the possibility of starting some systematic work in this direction. It was eventually Ed Nelson's seminal AMS-address "Internal set theory" from 1976 [Nel6] which made us go over to the active phase. Besides presenting the theory in his very original way, Nelson discussed a couple of examples, including one using infinitesimals to parametrize one point perturbations of the Laplacian in $\mathbb{R}^3$ which were familiar, in a heuristic or indirect way, to us. So we decided to organize a joint seminar on nonstandard analysis with Jens Erik Fenstad at the end of '76. Among the participants were Bent Birkeland, Dag Normann and Tom Lindstrøm. It was soon realized that although

Nelson's original approach did not give a total classification of the point interactions, a different parametrization would give the full result, and this quickly lead to a paper on singular perturbations and nonstandard analysis [47]. In this paper we showed that the Schrödinger operator in $L^2(I\!R^3, dx)$ describing the perturbation of the Laplacian by an interaction concentrated on a single point 0, "point interaction at the origin", can be described as standard part (in the resolvent sense) of the internal operator $-\Delta + \lambda_\varepsilon \delta_\varepsilon(x)$ in ${}^*L^2(I\!R^3, dx)$, with $\lambda_\varepsilon = -4\pi\varepsilon^2[1 - \alpha\varepsilon + \beta\varepsilon^2]$, ε a positive infinitesimal, α, β near standard numbers, δ_ε a smooth internal realization of the δ-function (corresponding to a standard δ-sequence of support in a ball of radius ε). The standard part is independent of β and only depends on α. We also extended the method to the case of finitely many point interactions. The solution of this problem greatly pleased Raphael, he used to say that nonstandard analysis was a natural tool to "sharpen our eyes", through it we see things which we would not otherwise uncover or would not dare to look at.

Another topic in the paper [47] is a treatment of Sturm-Liouville problems with singular coefficients, in particular of the form $-\frac{d^2}{dx^2} + \mu(x)$, where μ is a measure. It is shown that a separation theorem for the eigenfunctions which holds by transfer for a suitable nonstandard realization of μ carries over to the standard part. These results influenced further work, by Birkeland [Bi], Birkeland and Normann [B-N] and Persson [Per].

The late 70's were good years for nonstandard analysis. Just to mention a few of the developments closest to Raphael's interests, in addition to Nelson's influential address, the first version of Keisler's groundbreaking monograph on nonstandard stochastic analysis started circulating, Lindstrøm's Thesis [Li], the intriguing work of Benoit, Callot, Diener, and Diener [B-C-D-D] on limit cycles of ordinary differential equations appeared, Arkeryd started his profound work on the Boltzmann equation [Ark], Lawler [Law] proved rigorous results about self-avoiding random walks, Helms and Loeb [H-L] and Hurd [Hur] obtained the first results in nonstandard statistical mechanics. In this atmosphere of progress and success, the Oslo group decided to write a book on applied nonstandard analysis. The original plan was to write a brief text containing an introduction to nonstandard analysis plus a few substantial applications chosen to whet people's appetite, but the book soon developed in a different direction. When it finally appeared in 1986 under the title "Nonstandard Methods in Stochastic Analysis and Mathematical Physics" [123] about half the contents were original research on a variety of subjects such as measure extensions, stochastic differential equations, Dirichlet forms, singular perturbations of operators, lattice models in statistical mechanics, and quantum fields. Since Raphael was involved mainly with the parts dealing with perturbations of operators, statistical mechanics and quantum fields, we shall concentrate on these topics here.

A natural generalization of point interactions is to ask when a self-adjoint operator A on a space $L^2(X, m)$ has a nontrivial perturbation supported on a given set C of m-measure zero. Chapter VI of [123] gives a sufficient condition for this to be the case; basically there should be a measure ρ supported on C such that

$$G_\alpha(x,y)G_\alpha(x,z) \quad \text{is} \quad dm(x) \otimes d\rho(y) \otimes d\rho(z)\text{-integrable} \tag{10.1}$$

(here $G_\alpha(x, y)$ is the resolvent kernel, and there are some additional, technical conditions that need not concern us here). Under the stronger condition that

$$G_\alpha(x,y) \text{ is } d\rho(x) \otimes d\rho(y)\text{-integrable,} \tag{10.2}$$

the perturbations can be realized as closures of forms

$$(Au,v) + \int_C \lambda(x)u(x)v(x)d\rho(x) \tag{10.3}$$

where λ is a function supported on C, but in the general case the "coupling constant" λ has to be chosen infinitesimal (and negative) in a sense that can be made precise with nonstandard analysis. The existence of non-trivial point interactions with $A = -\Delta$ in dimension $d \leq 3$ (with noninfinitesimal coupling constant only if $d = 1$) is an immediate consequence of these results, and as a more substantial application it was shown that perturbations of the Laplacian supported by Brownian paths exist for $d \leq 5$ (and with noninfinitesimal coupling constants for $d \leq 3$). There are now several other ways to approach the perturbation problem avoiding nonstandard analysis ([Br2], [Br3], [Te], [Ch], [153], contributions by Brasche and Teta and by Raphael with Fenstad, Karwowski, Lindstrøm and myself, in these volumes), and they all seem to indicate that the conditions (10.1) and (10.2) above are quite canonical. For the Laplacian, Brasche [Br3] has linked these conditions to the Hausdorff dimension and capacity (with respect to the kernel $|x-y|^{d-4}$) of the set C, thereby showing that condition (10.1) is (almost) necessary as well as sufficient in this case.

One advantage of the nonstandard approach to the perturbation problem is that in the nonstandard world the perturbed operator is generated by the form

$$(Au,v) + \int_C \lambda(x)u(x)v(x)d\rho(x)$$

even when λ is infinitesimal, and hence we can compute the associated semigroup using a nonstandard Feynman-Kac formula. When C is a Brownian path $B(\omega,\cdot)$, we get

$$T^t_\omega f(x) = E_x[f(B'(t))\exp(-\int_{[0,t]}\int_{[0,t]} \lambda_\omega(B'(r))\delta(B(r),B'(s))drds)] \tag{10.4}$$

where the expectation is with respect to a new Brownian motion B' independent of B. The expression on the right is closely related to the Edwards model for polymers (except that we want the two Brownian motions to be the same); see [123] for discussion of the relationship with Westwater's work on polymers. As will become clear later, there is also a close relationship between this formula and quantum field theory. An alternative nonstandard approach to the Edwards model (based on infinitesimal random walks) has been developed by Stoll [Sto].

The basic idea in the nonstandard theory of lattice statistical mechanics is that states on an unbounded standard lattice can be represented by states in a bounded but infinitely large nonstandard lattice. Hence one can work with boundary conditions at infinity in a natural and intuitive way. In Chapter VII of [123] this idea is used to analyze the global Markov property of Gibbs states of infinite systems (the exposition is partly based on contributions by Kessler [Ke1], [Ke2]). The result is a detailed and rather complete picture of relationship between the global Markov property and various uniqueness, continuity, and measurability conditions.

From a nonstandard point of view, a Euclidean quantum field can be thought of as a continuous spin system on a lattice with infinitesimal spacing. This makes it possible to give a very concrete presentation of the theory based on hyperfinite dimensional Gaussian measures, avoiding much of the technical machinery necessary in the standard approach. Since hyperfinite dimensional Gaussian measures can be manipulated in exactly the same way as finite dimensional Gaussian measures, the nonstandard theory turns out to be a very suitable tool for making rigorous sense out of heuristic calculations. Two examples of this are worked out in [123]; first a hyperfinite representation of gauge fields, and then a detailed discussion of the relationship between polymer models and quantum fields (Symanzik's program [Sym 2]). Let us give a brief summary of the second topic.

The basic observation is that the square of the free field can be represented as a Poisson random field of local times of Brownian loops. In the "discrete" nonstandard setting (where a Brownian loop is just the loop of a random walk with infinitesimal increments), these concepts are unproblematic and the relationship is established through a simple comparison of Laplace transforms. Adding an (even) interaction to the field corresponds to adding terms counting the intersections between the Brownian loops, and then renormalizing in the nonstandard setting, this relationship is established through a somewhat lengthy but totally elementary calculation involving Gaussian densities. The result is a nonstandard version of Dynkin's representation formula.

With this representation, checking the nontriviality of an interaction is in principle just a question of showing that the new contributions are not cancelled by the renormalization, but in practice this is, of course, extremely difficult. What is interesting in the present context is that the new interaction terms in the nonstandard representation formula are partly of the form (10.4) above (intersection of distinct paths) and partly of the same form but with $B' = B$ (self-intersection of paths). Since we have no control over the self-intersection terms, it is difficult to say much about the quartic self interaction model in four dimensions, Φ_4^4 (where these terms make a nontrivial contribution), but the control which (10.4) gives us over the other terms, makes it possible to get partial results about the corresponding model $(\Phi_{(1)}^2\Phi_{(2)}^2)_4$ with two independent fields $\Phi_{(1)}$, $\Phi_{(2)}$ (but even in this case there are difficulties due to the fact that we do not know whether the infinitesimal coupling "constant" λ is actually constant, i.e. independent of the position on the Brownian path, see [123] for details). The nonstandard approach to quantum fields thus gives a certain hope that $(\Phi_{(1)}^2\Phi_{(2)}^2)$ and (perhaps even) Φ_4^4 with negative and infinitesimal coupling constants may be nontrivial, but in that case the hardest and most difficult work is still do be done.

A more comprehensive survey of nonstandard methods in mathematical physics can be found in [Al5]. Raphael never abandoned the idea of the utility of nonstandard methods, in fact we planned work using these methods in string theory, cf. Sect. 17, and in the study of gauge fields, cf. Sect. 16.

11. Point interactions

Raphael was fascinated by solvable models all his life. He regarded the study of models as the real strength of mathematics; he distrusted largely abstract methods deprived of interesting models, although he himself was a great master of abstract thinking. His deep philosophical–religious attitude towards the world, including the one of mathematics, made him believe in *discovering* as uncovering, rather than creation (although he had also a great confidence in the analysis of formalisms as tools, he strongly believed that mathematical truth is a vision, based on a substratum outside the formalism; how often he would say I *see* the proof, it is just a matter of writing it down in words, and this might take a terribly long time ...). He liked the word *research* (search, and search again) rather than the German "Forschung" or Norwegian "forskning". Solvable models greatly fascinated Raphael because they embodied an extraordinary amount of information in compact formulae, which would allow us to see better and further, and to make educated guesses for new lines of developments. He often said that our knowledge of quantum mechanics was based on two solvable models, the harmonic oscillator and the Coulomb potential. The first being a positive potential giving a purely discrete spectrum, the latter a negative, long-range potential with an infinite number of bound states and a continuous spectrum. Point interactions provide a third important solvable model, with zero-range, finitely many bound states and a continuous spectrum, perfectly suited to model phenomena in solid state physics.

The first mathematical discussion of point interactions is due to Berezin and Faddeev [Be-F] in 1961. Raphael and I had already discussed point interactions during our stay in Princeton in '70. In fact, he had given the study of the model with finitely many point interactions to an M. Sc. student, S. Øgrim [Øg]. I was already a little familiar with point interactions through my Ph.D. Thesis at ETH under the direction of R. Jost in which I studied a model of 3 particles with interactions of the boundary conditions type. Despite our common interest in these problems, we did not start really working on this topic at that time, only later did we come back to it, in connection with our work on Dirichlet forms [35] and our interest in nonstandard analysis, as described in Sect. 10. In '78, after accepting the position in Marseille, Raphael started a systematic study of point interactions with several co-workers. Over the years Raphael invested a tremendous amount of energy in this work. He was fascinated by the concrete calculations which went with it, and by the new insight he was winning. Much of the work was then written up in a book, which appeared in '88 [136], only a week after he died. Raphael and I had the great chance to be supported by the great engagement of two young co-workers, Streit's student Fritz Gesztesy and Raphael's student Helge Holden [Ho1], [Ho2]. Besides solid state physics applications concern nuclear physics, the theory of electromagnetism, wave guides, and antennas as well as the theory of polymers and quantum fields. (For a description of some of the early heuristic applications of point interactions to atomic physics we refer to the book by Demkov and Ostrovskii [D-O].) To describe all this work in details, going through all the papers written by Raphael on the topic would take too much space. Let me try to give at least some idea of it, referring to the mentioned book [136] for more details.

Let $Y \subset \mathbb{R}^3$ denote the discrete set where we want to locate the point interactions with strength, i.e. renormalized coupling constant, $\alpha = (\alpha_y,\ y \in Y)$, $\alpha_y \in I\!R$. Then the Hamiltonian with point interactions at Y, denoted by $-\Delta_{Y,\alpha}$, is characterized by its resolvent, viz.

$$\left(-\Delta_{Y,\alpha} - k^2\right)^{-1}(x,x') = G_k(x-x') - \sum_{y,y' \in Y} \Gamma_{Y,\alpha}(k)^{-1}_{y,y'} G_k(x-y) G_k(x-y')$$

with $\Gamma_{Y,\alpha}(k)^{-1}_{y,y'}$ the kernel at y, y' of the inverse of the closed operator $\Gamma_{Y,\alpha}(k)$ in $l^2(Y)$, where

$$\Gamma_{Y,\alpha}(k) = \left(\frac{ik}{4\pi} - \alpha_y\right)\delta_{yy'} + \tilde{G}_k(y-y')\ ,$$

with $\tilde{G}_k(z) \equiv G_k(z) \equiv (-\Delta - k^2)^{-1}(z)$ for $z \neq 0$, $\tilde{G}_k(z) = 0$ for $z = 0$, $\text{Im}k > 0$. In [47], [53] this operator is studied systematically in the case when Y is finite. The ground is also prepared in this paper for the fundamental paper [54] in which cases where $|Y| = \infty$ are considered, with important applications to the study of crystals and other systems of interest in solid state physics. Before we come to this important case, let us shortly mention a few other developments concerning the case $|Y| < \infty$. The realisation of $-\Delta_{Y,\alpha}$ as *norm* resolvent–limit of Schrödinger operators with regularized local potential $-\Delta + \lambda(\varepsilon)\varepsilon^{-2}V(x/\varepsilon)$ as $\varepsilon \downarrow 0$ has been obtained, as well as expansions in powers of ε [73], [76], [83], [91]. It was realized that these expansions depend essentially on the presence or absence of zero energy resonances of $-\Delta + V$. By a scaling argument this "short range expansion" can be related to a "low energy expansion".
Spectral quantities like eigenvalues, resonances, scattering amplitudes associated with $-\Delta_{Y,\alpha}$ have been studied from this point of view (asymptotic expansions in ε) (see [136] and references therein). These results have also been extended to the case with $|Y| = \infty$ as well as to one and two dimensions, see e.g. [99], [100], [101], [105], [132].

An extension of this result to the case where a point interaction is perturbed by the addition of a Coulomb interaction has also been worked out by Raphael in collaboration with Bollé, Ferreira, Gesztesy, Streit and myself, with relevant applications in nuclear physics [92], [95], [104].

In '80 a collaboration was started in Marseille with T.T. Wu, and continued in Oslo and Bielefeld, concerning a solvable model of three particles in interaction by 2–body point interactions (a model for which Raphael, Streit and myself had already defined a positive Hamiltonian [35], however via "effective three–body interactions", by the use of Dirichlet forms). This led to the announcement [75] concerning in particular exact computations of the Efimov effect (lower unboundedness of the spectrum of the Hamiltonian of 3 particles, interacting by 2–body potentials). Unfortunately detailed proofs were not given; our attempts at them by nonstandard analysis were interrupted, although Raphael always wanted to finish this work some day ...
The case $|Y| = \infty$, of infinitely many centers, was first analyzed in the fundamental work of Raphael with A. Grossmann and M. Mebkhout [54]. In particular the case when Y is a lattice, is considered. (The corresponding model in one space dimension is the celebrated Kronig-Penney model [K-P] from 1931.) In this case one has a one–electron model of an ordered solid. Besides defining the Hamiltonian and describing its resolvent by exploiting

the lattice symmetries, statements about the band structure of the spectrum and scattering by impurities [106] are made, see also [136] (and references therein) for improvements. Also short range expansions for such systems have been obtained by Raphael in collaboration with Holden and Johannesen [99], with a detailed computation of the Fermi surfaces [120]. This is in fact the first mathematical computation of a Fermi surface from first principles.

Other cases with infinitely many centers but different symmetries had also been studied by Raphael and co-workers [96], see [136] and references therein. An important application of these methods, combined with methods of the theory of random Schrödinger Hamiltonians, is in work concerning random defects in crystals and alloys. This work was carried out in [86], [90], [130] by Raphael in collaboration with F. Gesztesy, H. Holden, W. Kirsch, F. Martinelli and myself. A surprising result in this direction is that starting from random Hamiltonians and creating or switching off points one after the other does not change the spectrum. I distinctly remember Raphael first carrying out this computation, and a sketch of the corresponding proof in his head, when we were walking back to the Campus of Luminy all the way from the center of Marseille, after a good dinner. In a sense Raphael never stopped thinking mathematics – even in the middle of what looked to be completely relaxed moments, far away of any preoccupation with work, he was able to announce new mathematical facts he had just worked out or "seen".

Investigation of certain models for porous media (diffusion obstructed by a large number of small balls) has been put in relation with models with point interaction, and the latter have also provided new insights in this domain, see e.g. [147], [F-H-T], [F-T]. Let us also recall that, as already mentioned in Sect. 10, interactions supported on curves have been studied, see [B-K], and contributions by Brasche and Teta and by Raphael, with Fenstad, Karwowski, Lindstrøm and myself in these volumes, as well as on certain wave guides and branched structures by P. Exner and P. Šeba in Dubna, see e.g. contribution in these volumes and [E-S1], [E-S2], [E-S3], or polymers [123], [Ch]. Strong activity began around Pavlov in Leningrad: studying point interaction models in solid state physics, electromagnetism and nuclear physics, see e.g. [K-P-K-M-M-Y] and references therein. All this work was influenced by Raphael's pioneering work on models with point interactions, at least indirectly. Finally let me also mention the possibility of using point interactions as limiting cases of approximating smooth potentials, like e.g. in the already mentioned study of the Efimov effect and in the study of "multiple well potentials" (see e.g. Klaus and Simon [K-S], [Kla]), i.e. Hamiltonians of the form

$$-\Delta + \sum_{i=1}^{N} V(x - \frac{1}{\varepsilon} x_i)$$

$\varepsilon > 0$ small (in the sense of the study of eigenvalues, eigenvectors, resonances, for ε small). The latter was done by Raphael with Holden and Mebkhout in [89], [113], [119] providing explicit asymptotic formulae, and in connection with a resonance expansion (as substitute of the usual expansion in generalized eigenfunctions) for Schrödinger operators, by Raphael and myself in '84 [102], [110].

After publication of the book by Raphael with Gesztesy, Holden and myself, many new developments in forms of applications occurred; in addition to the ones already quoted, let me mention e.g. work concerning models of "wave chaos" [A-S].

12. Work on hydrodynamics and plasma physics

Raphael's work in hydrodynamics started in connection with the Ph.D. Thesis of Margarida de Faria, [Far1], [Far2], a student of mine at the Université d'Aix-Marseille II. The main results of the thesis were published in [48]. Two other papers were written by Raphael in close connection with this topic, one with me, practically finished in 1979, but published much later [139] and one with Merlini and myself, with some applications in plasma physics as well as meteorology, especially in connection with the vortex model [116].

Let me shortly describe some of the results. One considers Euler's equation of hydrodynamics in a domain $\Lambda \subset \mathbb{R}^2$ with piecewise C^1 boundary:

$$\frac{\partial}{\partial t}u = -(u \cdot \nabla)u - f$$

with rot $f = 0$, $\nabla \cdot u = 0$, f being the external force, $u = (u_1, u_2) \in \mathbb{R}^2$ the velocity field of the fluid, ∇ the gradient operator. The boundary condition is that the component of u in the normal direction to $\partial\Lambda$ is 0 (if Λ is simply connected, then rot $f = 0 \Leftrightarrow f = \nabla p$ for some p, in which case one recovers, in particular when $\Lambda = \mathbb{R}^2$, the standard version of Euler's equation.) By assumption (here 2-dimensionality is essential!) there exists a real-valued function ϕ on Λ such that $u = \nabla^\perp \phi$, with $\nabla^\perp \equiv (-\partial_2, \partial_1)$. It is easily seen that Euler's equation for u is equivalent with the equation

$$\partial_t \Delta\phi = -(\nabla^\perp \phi) \cdot \nabla\Delta\phi, \qquad \text{i.e. } \partial_t \text{ rot } u = -u \cdot \nabla \text{ rot } u$$

for ϕ, with boundary condition ϕ = constant on each component of $\partial\Lambda$ (rot $u = \partial_1 u_2 - \partial_2 u_1$). If u is a solution of Euler's equation, then the energy $H(u) \equiv \frac{1}{2}\int u^2 dx$ and the enstrophy $S(u) \equiv \frac{1}{2}\int_\Lambda (\text{rot } u)^2 dx$ are independent of time t.

If $\partial\Lambda$ is connected, H and S can be looked upon as functionals of $\nabla^\perp\phi$ resp. $\Delta\phi$. Heuristically then, the measures

$$d\mu_{\beta,\gamma}(u) = Z_{\beta,\gamma}^{-1} e^{-\beta H(u)} e^{-\gamma S(u)} d(\text{rot } u)$$

with

$$Z_{\beta,\gamma} \equiv \int e^{-\beta H(u)} e^{-\gamma S(u)} d(\text{rot } u) \ ,$$

(and $d(\text{rot } u)$ a heuristic "flat measure"), β, γ non negative constants, should be invariant under the flow $u(0) \to u(t)$ given by the solution of Euler's equation, since H, S are invariant and $d(\text{rot } u)$ is also a heuristic invariant measure, as seen from Euler's equation.

In [48] and [139] a meaning is given to the measures $\tilde{\mu}_{\beta,\gamma}$ related to the above candidates $\mu_{\beta,\gamma}$ as well-defined probability measures: for $\beta = 0, \gamma > 0$ $\tilde{\mu}_{0,\gamma} = \otimes N(0; \frac{1}{\gamma})$, where $N(0; \frac{1}{\gamma})$ is the 1-dimensional Gaussian measure with mean 0 and covariance $\frac{1}{\gamma}$, the product being taken for so many copies of $N(0; \frac{1}{\gamma})$ as there are members of the complete orthogonal system in $L^2(\Lambda)$ given by the eigenfunctions f_n to the eigenvalues $-\lambda_n$ of the Dirichlet Laplacian Δ on Λ.

Let

$$: H(u) :\equiv \frac{1}{2} \sum_{n=0}^{\infty} \frac{1}{\lambda_n} \left(< f_n, u >^2 - \frac{1}{\gamma} \right) ,$$

then $: H(u) :$ is a well-defined measurable function with respect to $\tilde{\mu}_{0,\gamma}$ and in fact

$$\exp\left[-\beta : H(u) :\right] \in L^1(\tilde{\mu}_{0,\gamma})$$

for all $\beta > -\lambda_1$. We define

$$d\tilde{\mu}_{\beta,\gamma} \equiv \frac{\exp[-\beta : H(\cdot) :] d\tilde{\mu}_{0,\gamma}}{\int e^{-\beta : H(\cdot) :} d\tilde{\mu}_{0,\gamma}}$$

In [48], [67], [139], [116] it is also shown that the corresponding Liouville equation describing the time evolution of cylinder functions of the form $F(< f_1, \text{rot } u >, ..., < f_n, \text{rot } u >)$ for $F \in C^\infty(\mathbb{R}^n; \mathbb{R})$, as well as an L^2-flow in the sense of Koopman-von Neumann are well-defined. A result of infinitesimal invariance of $\tilde{\mu}_{\beta,\gamma}$ is given in [139], [116], namely that $\int BF d\tilde{\mu}_{\beta,\gamma} = 1$, where B is the generator of the flow. The question of the full invariance was discussed in these references and in [C-G], and a full solution given later on in [A-C]. Questions of support of the measures $\tilde{\mu}_{\beta,\gamma}$ were discussed in [B-F1], [B-F2], [Wel]. Very recently an interpretation of the measures $\tilde{\mu}_{\beta,\gamma}$ in terms of nonstandard analysis has been given [Ca-C]. Extensions to stochastically perturbed Navier–Stokes equations have been obtained in [Cr], [A-C]. See also [Hab2] in which ergodic properties are discussed.

These mathematical results should be compared with many heuristic discussions in the literature e.g. [K-M], [On], [Lee]. In particular the necessity of the above renormalization of the energy ($H(u)$ being $\tilde{\mu}_{\beta,\gamma}$-a.s. infinite, differing almost surely from $: H(u) :$ by an infinite constant) is clearly shown by the mathematical results.

In connection with a talk given by Raphael at a meeting in Les Embiez, Raphael and I also looked into the Lin-Onsager vortex model of hydrodynamics, see [64] and [116]. Here we are given n degrees of freedom $x_i(t)$, $t \geq 0$ in $\mathbb{R}^2$ solving the ordinary differential equation

$$a_i \dot{x}_i = (\nabla^\perp)_i \sum_{\substack{j=1 \\ j \neq i}}^{n} a_j G(x_i(t); x_j(t))$$

with initial condition $x_i(0) = x_i$, $a_i \in \mathbb{R}$, $G = (-\Delta)^{-1}$ being the Newton kernel. Then u defined by rot $u(t, dx) \equiv \sum_{i=1}^n a_i \delta_{x_i(t)}(dx)$ solves in a generalized sense Euler's equation (see e.g. [M-P]). The system of the x_i is a Hamiltonian system with energy $\sum_{i \neq j}^n a_i a_j G(x_i - x_j)$. The Poisson-type measure with characteristic functional of the type

$$E(e^{i<f,\cdot>}) = e^{\int_\Lambda \int_{\mathbb{R}} (e^{i\alpha f(x)} - 1) d\nu(\alpha) dx}$$

$\Lambda \subset \mathbb{R}^2$ bounded, $f \in C_0^\infty(\mathbb{R}^2)$, ν a discrete measure on $\mathbb{R}$ is invariant under the flow given by the vortex equations and under the Euler flow. This had been anticipated in

[B-F1],[B-F2], [64] and proven in [D-P]. The statistical mechanics of the vortex model is related to the one of the Coulomb model, the sine-Gordon model, as well as guiding center plasma, as discussed in [C-M1], [C-M2], [116] (cf. also Sect. 3). See also [B-P-P] for the relation between the vortex model and the above Gaussian model.

As already mentioned in Sect. 6, the problem of plasma dynamics was also discussed by Raphael, with Gandolfo and Rodriguez in [128], see also Gandolfo [Gan]. In this work a stochastic model for plasma dynamics is given. According to the ideas described briefly in Sect. 6, the dynamics of charged particles in magnetic structures is modelled by a diffusion process of the type of stochastic mechanics and an interpretation of the Bohr-like character of the relation between diffusion constant and magnetic field is given.

13. Work on conservation laws

Around '81 Raphael had the idea of creating a research group at the University of Oslo in applied mathematics with the aim of studying the mathematical problems connected with oil reservoir simulation. In addition he engaged himself with the great enthusiasm and energy typical of anything in which he believed, into creating a new direction of study in "industrial mathematics" in Oslo. He directed seminars himself and with his communicative enthusiasm he soon had a lot of students and co-workers around him. He also took a genuine interest in research in this direction, in fact he had been fascinated for a long time by nonlinear problems, and he also had, besides being an eminent theoretician, insight into numerical problems. He used to say that he so strongly engaged in this direction primarily to provide better chances for jobs for gifted young scientists in Norway. As a matter of fact, in the last years of his life these activities occupied an increasing portion of his time. The ease by which he could grasp a problem, even one far from his own previous domain of activity, and quickly suggest solutions, combined with his natural "Ausstrahlung", which induced great confidence around him, allowed him to be very successful in directing the research group and getting increasing financial support, even from the industry. Despite the apparent "nonchalance" and ease by which he could administrate, his real gifts and inclinations (as he himself often described them) were in research. The conflict between research, which he continued to pursue with undiminished intensity, and administration, which intruded more and more into his life, created an increasing tension in him during his last years.

But let us go back to the beginning. In '82, together with Helge Holden, Raphael started a grand project with the ultimate goal of developing a computer simulator for a full scale petroleum reservoir based on deep mathematical insight into the models describing the flow in porous media. For a long time there had been commercial simulators available, but it was thought in the applied mathematics community that most of them were based on old-fashioned numerical and mathematical techniques. James Glimm and Oliver McBryan, then at Courant Institute of Mathematical Sciences, New York University, had already started a project with a similar aim in 1979. The concept of front tracking is fundamental in their approach. In the reservoir there are sharp transitions, called *fronts*, from the region with oil to the region with water. The idea is to treat these entities as independent variables in the simulator. Raphael and Holden followed the same idea, and formed a group consisting of Lars Holden, Kyrre Bratvedt, Frode Bratvedt, Christian F. Buchholz, Nils Henrik Risebro, Leif Alm and Daniel Gandolfo (a former student of Raphael in Marseille). The group was supported by Norsk Data, Norwegian Research Council for Science and the Humanities (NAVF), the Royal Norwegian Council for Technical and Industrial Research (NTNF) and VISTA (a joint project between Den norske stats oljeselskap, Statoil, and the Norwegian Academy of Science and Letters).

The standard mathematical model of a petroleum reservoir consists of a conservation law, which expresses the conservation of mass in the system, and a pressure equation. All the theoretical work by Raphael was done on conservation laws. A conservation law can be written as

$$u_t + \nabla \cdot f(u) = 0 \ . \tag{13.1}$$

In this case $u(x,t) \in I\!R^N$, $x \in I\!R^n$, $t \geq 0$, denotes saturation, i.e. relative volume fraction, of the constituents. The first work concentrated on the Cauchy problem for the scalar one-dimensional case, i.e. the non-viscous Buckley–Leverett equation

$$u_t + f(u)_x = 0, \qquad u(x,0) = u_0(x) \ .$$

This equation had been studied for a long time, see [Sm], and existence, uniqueness and stability of solution under general conditions had been proven. Raphael, in collaboration with Helge and Lars Holden discovered that if one approximated f by a polygon, i.e. a continuous, piecewise linear function, and approximated the initial data by a step function, one could easily construct the approximate solution, and this would converge to the solution of the original equation. This approach can be used to provide an alternate proof of existence, uniqueness and stability for this equation. Furthermore it gives a very efficient numerical method ("hyperfast" was the term Raphael liked to use). Later on they realized that the idea of considering a polygonal approximation had already been suggested by Dafermos [Da], but only used in some simpler cases. The numerical technique was published in [134], and the full proofs appear in the contribution by Holden and Holden in these volumes.

This approach could also be used to study the more difficult problem of a scalar equation in more than one dimension. Consider the scalar equation (13.1) in two dimensions, viz.

$$u_t + f(u)_x + g(u)_y = 0 \ .$$

The proof of existence of solutions to this equation is due to [C-S], [Kr]. However this proof gives little insight into the qualitative behavior of the solution. Therefore one has to study in detail the solution of the Riemann problem, which is the Cauchy problem with a special choice of initial data. In two dimensions Riemann initial data consists of data that are constant on a finite number of wedges focused at a single point in the (x,y)-plane. Guckenheimer [Guc] had initiated the analysis of this problem. Using the above approach based on polygonal approximations to the flux function, Raphael and Risebro were able to construct the solution of the Riemann problem [129], [138]. The extension of the method to N dimensions was given later on by Risebro [Ri].

While the theory in one dimension is fairly well-developed, the theory for systems is much more incomplete. Much effort has been invested in the study of the Riemann problem where the initial data has the simple form in one dimension

$$u(x,0) = \begin{cases} u_L & \text{if } x < 0 \\ u_R & \text{if } x > 0 \end{cases}$$

where u_L, u_R are constants. The classical theory for the Riemann problem is due to Lax [La], proving that the Riemann problem has a unique (weak) solution in the small, i.e. if the initial states are close, provided the system is strictly hyperbolic and genuinely nonlinear. This was later extended by Liu [Liu1], [Liu2]. He replaced the assumption of genuine nonlinearity by monotonicity assumptions on the flux function, and proved existence and uniqueness of the (weak) solution for any Riemann initial data. Together with L. Holden, Raphael found a general $N \times N$ hyperbolic system for which one can solve the Riemann problem without the assumption of genuine nonlinearity or strict hyperbolicity [150]. The system has the following form

$$\partial_i u_t + f_i(u_1, \ldots, u_i)_x = 0, \quad i = 1, \ldots, N.$$

The special feature of this system is that the derivative of the flux function f is a lower triangular matrix. Hence the system is strictly hyperbolic iff the elements on the diagonal of Df are distinct, and genuine nonlinearity is equivalent to $\frac{\partial^2 f_i}{\partial u_i^2} \neq 0$ for $i = 1, \ldots, N$. It is shown that the system has a unique solution generically when one uses the travelling wave entropy condition. By explicit examples they show that neither the Lax nor the Oleinik–Liu entropy condition are sufficient for this system, furthermore that the solution does not necessarily depend continuously on the data.

When Raphael died suddenly, the project was still in the beginning stage, and the actual implementation and scope of the computer code has necessarily changed since then. A presentation of some of the results can be found in the contribution by Bratvedt *et al.* in these volumes.

14. Random fields: the global Markov property and the study of Gibbs states

In previous sections we have already mentioned Raphael's strong interest in the theory of random fields and quantum fields. His interest was already present during his stay in Aarhus where he met several probabilists, including Dinges, Donsker, and Krickeberg. Certainly his thinking on the subject was influenced by Symanzik's lectures in New York and at the Varenna School [Sym] in '68. In any case, as we mentioned in Sect. 3, when Nelson's work on the free field first appeared, Raphael was already familiar with the subject. The basic problem in Nelson's approach to quantum field theory is proving the Markov property with respect to $(d-1)$-dimensional hyperplanes of the invariant random fields describing the interactions (heuristically given in terms of a density function v by

(4.1), (4.2) with $\Lambda = \mathbb{R}^d$, $\chi = 1$). This problem remained open for about 8 years, until it was solved by Raphael and myself for the first time for the sine-Gordon model (v of trigonometric type) in two space time dimensions ($d = 2$) in '79 [49]. At the same time we gave a proof in collaboration with G. Olsen of the global Markov property for the case of discrete random fields (over discrete sets and with finitely many values), a problem which was apparently overlooked in the existing literature on random fields [52], [70]. The proof of the global Markov property was also obtained independently by H. Föllmer [Fö1]. The result holds in a "uniqueness region" (high temperature, Dobrushin condition) for the corresponding Gibbs state.

Exploitation of uniqueness, in fact of a stronger uniqueness for conditional Gibbs states, was the essential tool which was missing in many previous erroneous proofs which had been circulating, over many years, in manuscript or preprint form by various authors (the essential difference between local and global Markov fields was pointed out by C. Newman [New]).

Since ours and Föllmer's '79 work much activity has gone into extending the proof to other situations. As far as quantum fields are concerned, the proof was extended to a special case of the Høegh-Krohn model (v of exponential type, cf. Sect. 4) by Gielerak [Gie] and to the general case of this model by Zegarlinski [Ze1] in '84. Strong trigonometric interactions were considered by Gielerak in '88, see contribution in these volumes. The original problem, which had kept many leading theorists busy proving the Markov property for a model with polynomial interaction in two space time dimensions ($P(\varphi)_2$–model), was eventually solved by Raphael in collaboration with Boguslav Zegarlinski and myself, [143], almost twenty years after it had been first posed. At the same time the methods yield uniqueness of Gibbs states, see also [144].
In the solution of the problem, in addition to new methods like FKG–order in the continuum, practically all available tools of constructive quantum field theory were used, together with original ideas of Raphael and myself [49], as well as those of Dobrushin and Minlos [D-M], of expressing conditional expectations of functionals of Gaussian generalized random fields (of the Nelson free Markov field type) through translations by generalized solutions of Dirichlet boundary problems with distributional boundary data. The study of such boundary value problems was also taken up later on in work by Röckner and Zegarlinski [Rö1], [Rö2], [Rö3], [Rö4], [R-Z], [Ze2], Dynkin [Dy1], [Dy2], Kolsrud [Ko], Rozanov [Ro]. The problems on discrete structures, like lattices, are simpler. Raphael discussed with Jean Bellissard in '81 [81] uniqueness results for Gibbs states of continuous spin systems on lattices, however their proof of the global Markov property was incomplete and was achieved later by Zegarlinski [Ze3], [Ze4] see also [123], where an idea of Goldstein [Gol] was used.
Further, extensive work on problems related with the structure of Gibbs states and the Markov property, especially in lattice models, is contained in a number of references, all of which are influenced in one way or another by Raphael's original ideas, see e.g. [41], [77], [98], [Ke1], [Ke2], [Es], [Geo], [Fö2]. Since Zegarlinski and I have a review article on the subject in these volumes, I would like to limit myself here to a couple of remarks.
Why is the proof of the global Markov property of Euclidean random fields important?

The main point – which Raphael liked to stress – is a philosophical one. Let me describe it for random fields associated with quantum fields. If such a Euclidean random field has the Markov property with respect to the $\{x^0 = 0\}$–hyperplane (denoting by $x = (x^0, \vec{x}) \in \mathbb{R} \times \mathbb{R}^{d-1}$ a point in $\mathbb{R}^d$), then one can look at it as a time reversal symmetric Markov process with state space of (generalized) functions of $\vec{x}$ and time parameter x^0. Its transition symmetric Markov semigroup has generator $-H$ with H the Hamiltonian for the associated quantum field. H is a self-adjoint operator in $L^2(\mu_0)$ with μ_0 ("ground state measure", "vacuum") being the restriction of the probability measure μ describing the distribution of the Euclidean random fields to the σ–algebra generated by the zero fields. This is entirely similar to the situation one has in non relativistic quantum mechanics (of finitely many degrees of freedom). Using results of [34] and [A-R3] in this way, the Hamiltonian appears as the positive self–adjoint operator associated with a Dirichlet form ε (coinciding with the classical Dirichlet forms ε_0 given by μ_0, cf. Sect. 8, on smooth cylinder functions; in the case of Høegh-Krohn's model, ε coincides with the maximal Dirichlet form described in the contribution by Kusuoka and myself in these volumes, hence it is a realization of an "infinite dimensional Schrödinger operator" in the proper setting). Moreover the Hilbert space for the physical model is in this case spanned by the time zero fields, just as in the non relativistic case. Finally, the full picture of canonical quantum field theory holds here, realizing the heuristic program initiated by Heisenberg, Pauli, Jordan, Wentzel, Araki, Coester–Haag together with Nelson's axioms [Nel3], [Si1]. In such models causality has then the strongest possible form. Thus there are self-adjoint position variables $X(\varphi)$ (the time zero field, $\varphi \in \mathcal{S}(\mathbb{R}^{d-1})$) and $\pi(\varphi)$ (the conjugate momentum) satisfying the Weyl commutation relations and the Hamiltonian equation of motions. $\left\{\exp(iX(\varphi))1 | \varphi \in \mathcal{S}(\mathbb{R}^{d-1})\right\}$ is dense in $L^2(\mu_0)$, $X(\varphi)$, $\pi(\varphi)$ are essentially self-adjoint on smooth cylinder function, and one has a strongly continuous unitary representation of the Lorentz group in $L^2(\mu_0)$. Using the $X(\varphi)$ field and $\exp(itH)$ one can get relativistic fields which are quantized versions of classical fields satisfying a field equation of the type $(\,\square + m^2)\Phi + v'(\Phi) = 0$, with v the real-valued function on $\mathbb{R}$ giving the density of the interaction (e.g. $v(\alpha) = \lambda \cos \alpha$, $\lambda, \alpha \in \mathbb{R}$, $|\alpha| < 2\pi$, for the sine–Gordon equation, cf. Sect. 3). Note that for v of the above types also the other generator of the Lorentz group admits a representation in terms of Dirichlet forms, cf. [103]. Moreover, it is observed in [52], [57], [66] that Λ can be indentified with the generator of the Tomita automorphism of the relevant von Neumann algebra (implementing axiomatic results of [Bi-W] in these models). Raphael was particularly proud of having contributed in an essential way to this reconciliation of quantum theory and relativity theory whereby quantum field theory appears as quantum mechanics for infinitely many degrees of freedom, at least in two space time dimensions.

Let us close this section by mentioning another type of homogeneous Markov fields Raphael studied. In fact, this type of Markov fields fascinated him as a sort of continuum version of Ising fields. We discovered them around '75, but the paper was only recently finished when new stimulus came from work by Arak and Surgailis [Ar-S], [Ar-S]. The latter was essentially restricted to two dimensions, whereas the model Raphael and I had considered was valid in all dimensions. This gave rise to the paper [152], which I will only briefly describe. Let ν be the classical Gibbs state of a gas of particles interacting by pair potentials.

For any finite subset γ of $\mathbb{R}^d$ and $x \in \mathbb{R}^d$ we consider

$$\xi(x)(\gamma) = \sum_{y \in \gamma} \chi(| x - y | \leq r),$$

χ being the indicator function and $r \geq 0$ a parameter. In particular, if the pair potential has a hard core of radius r, then $\xi(x)(\gamma)$ is $0-1$ valued over $\mathbb{R}^d$. In all cases ξ has the global Markov property in the following sense: for any disjoint closed sets Λ_1,$\Lambda_2 \subset \mathbb{R}^d$ separated by $\Lambda_0 = \mathbb{R}^d - (\Lambda_1 \cup \Lambda_2)$, the σ-algebras $\Sigma_{\Lambda_i} = \sigma\{\xi(x), x \in \Lambda_i\}$ associated with the Λ_i, $i = 1, 2$ are conditionally independent given the σ-algebra Σ_{Λ_0}. In the proof, the global Markov property of the Gibbs measure ν is used. For $d \geq 3$ ξ has also the sharp global Markov property with respect to a large class of $(d-1)$-dimensional hypersurfaces, in the sense that $\Sigma_{\Lambda_1}, \Sigma_{\Lambda_2}$ are conditionally independent given Σ_{Λ_0} for $\Lambda_0 = \partial\Lambda_i$ such a $(d-1)$-dimensional hypersurface. Here, however, the case where Λ_0 is a hyperplane is excluded, rather, Λ_0 should be "spherically irregular" in the sense of being "non flat, non spherical in any neighborhood of its points". The proof uses essentially geometric facts, see [152] for details. For yet other types of homogeneous Markov fields studied by Raphael, I refer to Sect. 16.

15. Infinite dimensional Lie groups and Lie algebras

In the fall of '75 Raphael started working on representation theory of infinite dimensional groups of mappings, also called current groups or gauge groups. He pursued this work with Jean Marion, Daniel Testard, Bruno Torresani, Anatole Vershik and myself until his death. In fact Raphael attached great interest and importance to this work, and until the very end was seeking a general structure theorem.

It was all started as an attempt to extend the construction of generalized Markov fields to the case when they take values in non-commutative spaces, like Lie groups. By doing this, we discovered [42] what we called the energy representation of the groups of mappings from a d-dimensional Riemannian manifold M into a compact semisimple Lie group G. It is a unitary representation in a Hilbert space given essentially by exploiting the Maurer–Cartan cocycle and the scalar product given by the Killing form.

Our first paper on the subject was produced as a preprint at ZiF. Streater developed some of its points in connection with quantum fields [Str1], [Str3].

How big our surprise was when, while attending a conference on Information Theory in Repino, near Leningrad, June, 1976 (following an invitation by Dobrushin) we met Vershik and while chatting with him we discovered that he, Gelfand and Graev had also studied independently at the same time a representation which we guessed must have been the same. We got a preprint of [G-G-V1] from them and discovered that their representation was indeed isomorphic to ours! As a matter of fact, although their original motivation for the study was the development of a non-commutative distribution theory, both groups also had in mind applications to quantum fields. Later we discovered that Ismagilov [Ism] had also studied the same representation, for $G = SU(2)$, and that the construction can

be interpreted in terms of a general method of Parthasarathy and Schmidt [Pa-S].
During a stay in Marseille, Raphael realized that there had been an error in Gelfand, Graev and Vershik's proof [G-G-V1] of the irreducibility of the energy representation for $d > 3$, in fact their proof is only valid with modifications, for $d \geq 5$ (as corrected by the authors themselves in [G-G-V2]). Raphael, Testard and I started working on the problem and got a full proof [74] for the case $d \geq 3$ and $d = 2$ (the latter under an assumption on the length of the roots), essentially exploiting methods we had used for the construction of the exponential interaction model of quantum fields (cf. Sect. 4). For a recent proof see also [Wal]. An advantage of our approach in one dimension, already stressed in our first paper [42], was that we could show that the energy representation coincides with the regular representation given by left translation along the paths of a G-valued Brownian motion. We exploited this later on to give the decomposition into irreducible pieces of the energy representation for $d = 1$ [88], [109], [155], [Tes]. We also determined the type, in the case $M = I\!R$, and at that time furthermore developed a study of quasi-invariance which became more than 10 years later a topic of central attention in work of P. Malliavin, M. Malliavin and their co-workers (see [Ma-Ma], and also [Gro6] for a recent study of logarithmic Sobolev inequalities for loop groups).

But let us briefly describe the energy representation, for M a general Riemann manifold and G a semisimple Lie group with compact Lie algebra. Let $\Omega \equiv \Omega(TM; g)$ be the space of smooth 1-forms on the tangent bundle space TM, with values in g and with scalar product $< \omega, \omega' >$ given by the Killing form. By closure we get from Ω a Hilbert space $\mathcal{H}$.
Let G^M be the group of C^1-mappings from M into G (with pointwise multiplication). Let V be the representation of G^M in $\mathcal{H}$ given by

$$(V(\psi)\omega)(x) \equiv \text{Ad}\, \psi(x)\omega(x), \quad \psi \in G^M, \quad \omega \in \mathcal{H}, \quad x \in TM.$$

Let β be the Maurer–Cartan form on G, then $\psi \to \beta(\psi)$ is a 1-cocycle for V. Let μ be the canonical Gaussian measure associated with the real part of $\mathcal{H}$ and let

$$(U(\psi)f)(\omega)) = e^{i<\beta(\psi),\omega>} f(V^{-1}(\psi)\omega),$$

for all $f \in L^2(\mu)$. Then U is a unitary representation of G^M in $L^2(\mu)$, called the energy representation in [39], [42].
As we mentioned for $d = 1$ the representation can be expressed by Brownian motion on G, and is in this sense "Markovian". We proved the factoriality of the representation (with Testard and Vershik) for $G = SU(2)$ in '83 [88], and with Testard in '84 for $G = SU(n)$ [109]. We also showed that the representation is of type III if $M = I\!R$.
In 1981, during our visit to the Academy of Sciences of the USSR (following an invitation by Dobrushin and Minlos), we came to know I.M. Gelfand, among others, who invited us to give a lecture at his seminar. Raphael did that, and I remember vividly his joy at realizing the interest Gelfand had taken in our work.
Also the energy representation of the group of equivariant loops ($M = S^1$) in G has been studied in details ([Tes], [151]). Let T be a maximal torus in G, then the representation space $L^2(\mu)$ has been shown to have a direct integral decomposition

$$L^2(\mu) = \int^{\oplus} L^{2,\alpha}(\mu) d\mu_T(\alpha),$$

with μ_T the standard Wiener measure on $C(M,T)$. $L^{2,\alpha}(\mu)$ is, for μ_T-a.e. $\alpha \in C(M,T)$, isomorphic to $L^2(C(M,G/T),\mu_0)$, μ_0 being the the canonical image of the Wiener measure μ on G by the quotient map from $C(M,G)$ onto $C(M,G/T)$. The energy representation U decomposes directly, $U = \int^{\oplus} U^\alpha d\mu_T(\alpha)$, with

$$(U^\alpha(\psi)f)(\omega) = \Big(\frac{d\mu(\psi^{-1}\omega)}{d\mu(\omega)}\Big)^{1/2} f(\psi^{-1}\omega)\exp(-i < \alpha^{-1}d\alpha, \hat{\omega}^{-1}d\psi\, \psi^{-1}\,\hat{\omega} >),$$

for any $f \in L^{2,\alpha}(\mu)$. $\hat{\omega}$ is a representative of the class $\omega \in C(M,G/T)$.
In the case of the representation of the group of equivariant loops all paths ω should be replaced by equivariant ones. (Similar results have been obtained in the case G is a Sobolev–Lie group of mappings with values in some compact Lie group [Mar2]). For $G = SU(n), n \geq 2$, U is a factor representation, μ_T-almost all U^α have been proven to be irreducible (see [88], [109], and for the equivariant case [Tes], [151]), and one has in particular

$$U^\alpha = \int^{\oplus} U^\phi(\alpha)\, d\mu_h(\phi),$$

with μ_h a probability measure whose image by the restriction of the canonical mapping from $C(M,T)$ onto $C(M,T\backslash G/T)$ is the standard Wiener measure on $C(M,T\backslash G/T)$.
Later on the study of the energy representation and various extensions has been pursued in particular by Marion and Testard, [Mar1], [Tes], [M-T], and constribution in this volume. Many of these developments have gone into a book [153] in which Raphael took a lot of interest.
Around '81 Raphael started working with some students, Johannesen and especially Terje Wahl in Oslo, and later on with Torresani in Marseille, on the classification, construction and representation of a generalization of both semisimple and affine Kac–Moody Lie algebras, called "quasisimple Lie algebras". They are characterized by the existence of a finite-dimensional Cartan subalgebra, a non degenerate symmetric invariant Killing form, and nilpotent rootspaces attached to non-isotropic roots. In particular, a classification theorem for the possible irreducible elliptic quasisimple root systems was discovered. Some of these quasisimple Lie algebras have realizations as algebras of mappings of a ν dimensional torus T^ν into the Lie algebra g of a semisimple compact Lie group. These current algebras thus generalize the affine loop algebras to the higher dimensional case. For these impressive results by Raphael and Torresani see Torresani's Ph.D. Thesis [To1], [To2], and [151]. This work has been continued by Torresani (see contribution in these volumes), who considered the highest weight representation theory of elliptic quasisimple Lie algebras, both for compact and non compact groups G. In the case $G = SU(n,1)$ the connection is made with continuous tensor product representations studied before by Gelfand, Graev, Vershik and Delorme, see e.g. [155].

16. Group-valued homogeneous Markov fields and quantum fields

The background for this line of research is the essential role played by geometry in modern particle physics. Typically, the physical objects are sections of certain principal

or vector bundles.
In quantum theories, this leads via the Euclidean version of Feynman's formalism, to constructing random fields, which are in general indexed by geometrical objects such as curves, with values in Lie algebras or groups and with prescribed invariance properties. At the Lie algebra level the models can be expressed as follows. Start from a "classical" partial differential equation of the form $LA = F$, where L is a first order (pseudo) differential operator, and F is given. Now, let F be given by a probability measure with appropriate invariance, and solve for A to get a quantum field.
The material in Sect. 16.1 corresponds to the (nonlinear) Yang-Mills equation, so that $LA = D_A A = dA + \frac{1}{2}[A, A]$, the covariant derivative, and the construction is carried out in the Lie group.
The quaternionic model described in Sect. 16.2, one of Raphael's profound affections, is based on the Weyl equation. It is one half of the (mass zero) Dirac equations and it contains one half of the anti self dual Maxwell equations.
As Raphael often stressed, these models are very explicit. He was also aware of and enthusiastic about, the possibilities of using Poisson measures for the source F. In fact, in these models complicated constructions can be carried out [145] without renormalization; also, the use of Poisson measure leads to insight in the confinement problem [Tam].
The research was initiated in '83 by Raphael and myself and continued up to the present with other co-workers, in particular Helge Holden, Koichiro Iwata and Torbjörn Kolsrud.

16.1 Markov cosurfaces, multiplicative stochastic integrals

The problem lies in constructing and studying homogeneous (i.e. stationary) generalized Markov random fields associated with certain 1–codimensional hypersurfaces, instead of points. In the case where the underlying manifold is two dimensional and a special distribution is chosen, these fields can be identified with Euclidean gauge fields (hence fields which, after analytic continuation, yield relativistic quantized Yang–Mills fields), described by probability measures P on a space of connection 1–forms A on the Riemannian manifold M with values in the Lie algebra g of some compact Lie group G, of the heuristic form

$$dP(A) = Z^{-1} \exp\left(-\int_M F \wedge {}^*F dx\right) \prod_{x \in M} \prod_{\mu=1}^{d} dA_\mu(x) \,, \tag{16.1}$$

with $F = dA + \frac{1}{2}[A, A]$, the curvature 2–form associated with A, dx being the Riemann–Lebesgue volume form, * the Hodge dual and $\wedge$ the exterior product, Z the normalization. (A, P) is thus a heuristic stochastic connection to be constructed. Having A one can in principle construct a G–valued stochastic holonomy operator $X(\gamma)$ determined by $\int_\gamma A$, associated with oriented Jordan curves γ, obtaining in this way a stochastic process (random field) indexed by curves γ (in the abelian case we have $X(\gamma) = \exp\left(i \int_\gamma A\right)$). $X(\gamma)$ is a realization of a Markov cosurface in the case $\dim M = 2$ [108].
Let us now briefly describe the theory of Markov cosurfaces. Let M be an oriented Riemannian manifold. Let H_M be the family of $(d-1)$-dimensional oriented hypersurfaces

contained in M (piecewise smooth, connected and simply connected, closed, without self-intersections). It is possible to define a natural composition $\gamma_1 \circ \gamma_2$ for γ_1, γ_2 in a certain set containing H_M, and the inverse γ^{-1} of γ. (For $d = 2$, $\gamma_1 \circ \gamma_2$ is essentially defined as the point set $\gamma_1 \cup \gamma_2$ with orientation generated by γ_1, γ_2).

If G is a group, a stochastic cosurface $X(\gamma)(\omega)$ is for fixed $\gamma \in H_M$ a measurable map from H_M into G on a certain probability space.

One can extend $X(\cdot)(\omega)$ in a natural way to a certain class of (ordered) complexes $(\gamma_1, \ldots, \gamma_n)$, $\gamma_i \in H_M$. Suppose now for $d > 2$ that G is abelian. In the case $d = 2$, G is supposed to be compact and separable. Then it is possible to introduce a projective system of probability measures on a subfamily of complexes, the so-called saturated regular complexes. It is defined in terms of a Markov semigroup of probability measures on G and the normalized Haar measure on G. The coordinate process associated with the projective limit is a stochastic cosurface having the global Markov property. This is invariant (in law) under orientation preserving global diffeomorphisms of M leaving invariant the volume measure on M.

A partial converse can also be given, and continuity properties of the map $\gamma \to X(\gamma)(\cdot)$ are exhibited once a suitable topology is introduced in the space of all γ, see e.g. [108], [117], [121], [122], [124], [133], [135], [140], [Kau]. It has been shown that the Markov cosurfaces described above can be obtained as limits of cosurfaces associated with lattices. For $d = 2$ a stochastic cosurface is, by definition, a G-valued stochastic curve integral. In this case, if $\gamma = \partial B$ is a simple oriented loop enclosing a region B, we have that $X(\gamma)(\cdot) \equiv \eta(B)$ can be identified with a stochastic G-valued multiplicative measure. More generally a Markov cosurface $X(\gamma)$, for γ the boundary of a region B, can be identified with a stochastic G-valued measure $\eta(B)$. Stochastic G-valued measures are a particular case of stochastic multiplicative measures or integrals or "noise" as studied in [140]. If η is such an object, then $\eta(B_1)$ is independent of $\eta(B_2)$ if $B_1 \cap B_2 = \emptyset$, and $\eta(B_1 \cup B_2)$ is identically in law with $\eta(\mathcal{B}_1) \cdot \eta(\mathcal{B}_2)$. The law gives a "generalized semigroup of probability measures" p on G, indexed by M. η and p are then related by a Kolmogorov–type theorem established in [140]. η and p have been classified and constructed in [135], [140], starting from corresponding objects in the Lie algebra (Lévy–Khinchine representation). We refer to these references for further discussions of these concepts and ideas. If g is the Lie algebra of a Lie group G, then G-valued curve integrals X are in 1-1 correspondence with g-valued curve integrals χ, by $X \to \chi_{_X}(\gamma) \equiv \int_0^1 X(\tilde{\gamma}(s))^{-1} dX(\tilde{\gamma}(s))$, with $\tilde{\gamma}(s)$ the curve on M $\{\gamma(s),\ s \in [0,1]\}$ described until $\gamma(s)$.

If $\chi(\gamma) \equiv \int_\gamma A$ for some g-valued 1–form A on M, then X_χ (the integral determined by χ from the above correspondence) is the holonomy operator given by A. In case A is distributed according to (16.1) (which corresponds to p being given by $p_{|B|}$, with $|B|$ the volume of $B \subset M$, and p_t the heat semigroup on G) wee can look at X_χ as a realization of (usual) gauge fields.

For $d > 2$ and G abelian, $X_\chi(\gamma)$ can also be defined for $(d-1)$ dimensional hypersurfaces γ and is in 1-1 correspondence with G-valued stochastic measures and generalized Markov semigroups.

There is a procedure "by saturation" described in [108] and [Kau], by which given a multiplicative stochastic measure η, one can extend the corresponding object X, first defined

only on certain boundaries (loops for $d = 2$), to a Markov cosurface.
For a continuation of our work with applications to the study of quantized gauge fields on manifolds, see [G-K-S], [Driv], and [Sen].
One can also use Markov cosurfaces and stochastic multiplicative measures to construct models of Higgs-like fields in the continuum, at least for $d = 2$. These are probability measures heuristically given as follows.
Let G be a compact Lie group and V a finite dimensional complex Hilbert space carrying a unitary representation ρ of G. Let X be a Markov cosurface over $M = \mathbb{R}^2$, with values in G, obtained first on boundaries ∂B of smooth open sets in $\mathbb{R}^2$ by defining $X(\partial B) \equiv \eta(B)$, with η a stochastic G-valued measure obtained by taking over from the Lie algebra g to G a (not necessarily Gaussian) white noise ξ with Lévy measure invariant under the adjoint representation, so that η is inner invariant. X is a multiplicative stochastic curve integral and can be extended by saturation to general curves. Let A be such that $X(\partial B) = \int_{\partial B} A$. The measures we are interested in are given heuristically then, in the case where ξ is Gaussian, by

$$d\mu(A, \varphi) = Z^{-1} \exp(-W_\rho(A, \varphi)) d\varphi dA$$

with Z a normalization constant and

$$W_\rho(A, \varphi) = \frac{\lambda_1}{2} \int_M |D_A \varphi(x)|^2 dx + m_0^2 \int_M |\varphi(x)|^2 dx + \lambda_2 \int_M F \wedge^* F dx,$$

with $F \equiv dA + \frac{1}{2}[A, A]$, $D_A \varphi \equiv d\varphi + \rho(A)\varphi$ $\lambda_1, \lambda_2 \geq 0, m_0 > 0$ constants.
The non Gaussian case amounts to replacing the distribution given heuristically by "$\exp(-\lambda_2 \int F \wedge^* F dx) dA$" by one defined in terms of a Markov cosurface X. Quantities of interest are e.g. the correlation functions

$$S(A; x, y, \gamma) \equiv Z_0^{-1} \int < \varphi(x), \rho(X(\gamma))\varphi(y) > \mu(A, d\varphi), \tag{16.2}$$

where the integration is with respect to φ, A being kept fixed, $<,>$ is the scalar product in V, Z_0 is a normalization, γ is a curve going from x to y, $x, y \in \mathbb{R}^2$.
In [142], [145] a representation of (16.2) in terms of Brownian bridges is given:

$$S(A, x, y, \gamma) = \int_0^\infty e^{-m_0^2 t} E_{xy}^t \; Tr \; \rho(X(b\gamma^{-1}))dt,$$

with b a Brownian bridge from x to y in time t, E_{xy}^t the expectation with respect to it, $b\gamma^{-1}$ being the curve obtained by following b and then going along γ from y to x.
Moreover "$2n$-points, n-curves" correlation functions $S(A, x_1, ..., x_n, y_1, ..., y_n, \gamma_1, ..., \gamma_n)$ can be expressed similarly and shown to yield a theory satisfying postulates similar to the ones of gauge theories in terms of gauge invariant quantities, as in e.g. [Sei]. This is based on the use of Poisson measures; the Gaussian case requires renormalization.

16.2 Complex and quaternionic–valued fields

The objects discussed in Sect. 16.1 for dimensions ≥ 3 of the underlying manifold are interesting as homogeneous Markov fields indexed by geometric objects, however the associated relativistic fields might be too simple to yield interesting physical objects (at least when the topology of the underlying manifold is trivial). For this reason we started around '84 to look for other ideas, still preserving the basic geometrical inspiration. It was around '86 that Raphael and I, through the course of many discussions, often held while walking around the little lake outside Oslo where the Høegh-Krohn's had their house, stumbled upon the fact that the isomorphism of $\mathbb{R}^4$ with the field of quaternions could help solve the problem (much in the same way as the isomorphism of $\mathbb{R}^2$ with $\mathbb{C}$ leads to interesting conformal fields – a fact that sparked Raphael's interest already around '76, soon after [27] – unfortunately without any writing up of the results; see however [Os2], [A-I-K2], [A-I-K3], [Hab1] for some implementation of related ideas). Raphael and I wrote several papers on the topics and also got the help of Iwata and Kolsrud in this project. This research was one of the main topics of interest for Raphael in the last period of his life. A slip concerning the reflection invariance of the constructed Euclidean fields, present in the early papers on the subject, was announced in [137], corrected and discussed further in [146], [A-I-K1]. To the best of my knowledge the models constitute at the moment of writing the only known examples of non trivial (non Gaussian, non independent at every point) Markov fields, attached to points and which are invariant under the proper Euclidean group. Let me describe briefly the models, in the case where the underlying manifold is $\mathbb{R}^4$ and the values are in $\mathbb{R}^4$. The fields are obtained by solving, in law, the covariant stochastic partial differential equation $\partial A = F$, where A, F are $\mathbb{H}$–valued generalized random fields, $\mathbb{H}$ being the (non-commutative, associative) field of quaternions, identifiable, as a vector space, with $\mathbb{R}^4$. ∂ is the quaternionic Cauchy–Riemann operator. F is a given generalized random field over $\mathbb{R}^4$, with values in $\mathbb{H} \cong \mathbb{R}^4$, determined by its characteristic function

$$E\left(e^{i<f,F>}\right) = \exp\left(-\int_{\mathbb{R}^4} h\left(f(x)\right) dx\right) ,$$

$f \in \mathcal{S}(\mathbb{R}^4; \mathbb{H})$, with h a negative definite function (of Lévy–Khinchine type, with suitable invariance properties under rotations, and behavior at the origin).
The law of A is given by

$$E\left(e^{i<f,A>}\right) = \exp\left(-\int_{\mathbb{R}^4} h\left(-\bar{S} * f(x)\right) dx\right)$$

where $\bar{S}$ is the fundamental solution of $\bar{\partial}$ (the natural conjugate to ∂), and $*$ denotes convolution using quaternionic multiplication.

h is the sum of a Gaussian h_G and a non Gaussian component h_{nG}. It had been conjectured in [126] and proven in Iwata's Thesis [Iw1], [Iw2] (with methods extending [Kus1]; see also [A-I-K3], [A-I-K4]) that A is a Markov field. If $h_{nG} = 0$, A can be identified with the free Euclidean electromagnetic potential field, and in this case restricting the class of test functions f, one can get a reflection positive field. If $h_{nG} \neq 0$, the field $< f, A >$ is neither

Gaussian nor independent at every point: its support properties have been studied as well as approximation properties [146], [A-I-K1], [A-I-K4], [A-I-K5], [Schmi]. Its invariance under a "time reflection" induced by $x_0 \to -x_0$ is known to hold iff supp $h_{nG} \subset \mathbb{R} \times \{0\}$. The analytic continuation of moment functions has been discussed in [A-I-K1] and shown to yield relativistic invariant functions. The physical interpretation of the relativistic fields is still under discussion, however the highly non trivial content of the Euclidean counterpart, e.g. as showing a phenomenon of confinement [Tam], has already been worked out. One can look at $F(B)$ as ${}^*\omega(\partial B)$, with ${}^*\omega$ a $U(2)$-valued Euclidean multiplicative stochastic integral, using the isomorphism with $\mathbb{H}$ and $u(2)$. Forming the tensor product with $U(2)$ and a compact group G, we can extend this construction to other groups, to obtain non abelian Euclidean Markovian random fields also in dimensions four, cf. [131]. Similar constructions yielding reflection positive conformal Markovian fields have been discussed for the case of certain complex 2-dimensional manifolds where the quaternions are replaced by complex numbers, see the contribution by Osipov in these volumes and [A-I-K2], [A-I-K3], [Be].

17. Mathematical string theory

We already mentioned in Sect. 4 that the work by Raphael and co-workers on the model of scalar fields interacting by an exponential-type interaction has also found extensions and applications to string theory. Let us shortly describe this work. It is written up in [125], [149] and in several contributions to Proceedings [148], [A-P-S], see also [Hoh] and the Ph.D.-Theses of S. Paycha (Paris) and S. Scarlatti (Roma).

The connection between string theory and the Høegh-Krohn model arises as follows: In Polyakov's formulation the construction of quantized bosonic strings amounts to constructing a probability measure $d\mu(X,g)$ given heuristically by

$$d\mu(X,g) = Z^{-1} \exp\left(-A(X,g)\right)\; D_g\, X\; Dg\ ,$$

with $A(X,g)$ an action functional, depending on the $\mathbb{R}^d$-valued field X over a 2-dimensional Riemann surface Λ ("world sheet"), with metric tensor $g = (g_{ab})$, $a, b = 1, 2$:

$$A(X,g) = \frac{1}{2}\int_\Lambda \sqrt{\det g}\, g^{ab}\partial_a X^\nu \partial_b X^\nu\, d\eta\ ,$$

with $d\eta$ the volume element on Λ, $\det g$ the determinant of g, g^{ab} the inverse matrix to g_{ab}, $\partial_a = \frac{\partial}{\partial \eta^a}$, η^a being the a-th coordinate of η, X^ν the ν-th coordinate of X ($\nu = 1, \ldots, d$). η is a parametrization of Λ. d should be looked upon as the space-time dimension. Dg resp. $D_g\ X$ are heuristic "flat measures" on the spaces $M(\Lambda)$ of metrics resp. $E(\Lambda)$ of embeddings. Z is a "normalizing constant".

The mathematical construction is discussed in [149] using infinite dimensional differential geometry combined with stochastic analysis. The basic idea consists in splitting, in the case of surfaces Λ of genus ≥ 2, the metric g into $f^* e^\varphi g_t$, with f^* the pullback of a

diffeomorphism (connected to the identify), a Weyl function φ and a chosen metric g_t parametrized by a point t in Teichmüller space T.
Heuristically, the conditional measure obtained after integrating with respect to $D_g X$, exploiting the above splitting of the matrix g, is of the form

$$d\mu[t,\phi] = Z^{-1} \left(\frac{\det(-\Delta_g)}{\int_\Lambda \sqrt{\det g} d\eta} \right)^{-d/2} (\det F_g^* F_g)^{1/2} D[t] D\phi,$$

with $D[t]$ and $D[\phi]$ formal "flat measures" on Teichmüller space T and on $C(\Lambda; I\!R)$, respectively. The determinants det of elliptic operators Δ_g (the Laplace–Beltrami operator on Λ) and F_g (the Faddeev–Popov operator tangent to the map $(f, \phi, t) \to g$) are obtained by regularizations involving a parameter $\epsilon > 0$, and heat kernel methods rigororously developed in [149]. By a renormalization procedure replacing A by $A_\epsilon + \lambda_\epsilon \int \sqrt{g} d\eta$ (A_ϵ being a regularization of A) with $\lambda_\epsilon \to \infty$ as $\epsilon \to 0$, to cancel divergences as the regularization ϵ is removed, one obtains for $d < 13$ that the ϕ-dependence of $d\mu[t,\phi]$ is given by a well-defined probability measure $d\nu_t(\phi)$ indexed by $t \in T$. ν_t is of the type of the probability measures which describe a scalar quantum field model with mass zero over Λ with exponential interaction (Liouville model). This is exactly the Høegh-Krohn model with mass zero (cf. Sect. 4) in the case of genus 0. The restriction $d < 13$ (which contains the physical case of space-time dimension 4!) corresponds to the restriction $|\alpha|^2 < 4\pi$ discussed in Sect. 4. Hence one sees here yet another instance of the importance and unity of Raphael's work. For a continuation of his work in string theory, see e.g. [A-P-S].

18. Epilogue

We have gone through Raphael's published work and made a few glimpses into his life. It has been very hard for me, and for all of us who loved him and miss him so much, to go through this journey, confronted at every step by the absurd evidence that he is no longer there to be questioned, no longer able to settle the problems which bother us by a stroke of his superb intelligence. Much remains to be said, and I hope will some day be said, of the continued presence of this extraordinary man, who has left us a life work of extreme beauty with a texture of rare complexity and yet rendered so simple by his very sincere deep search for unity, rationality and above all, beauty in nature.
He left us full of ideas, experiencing, as he pointed out to me himself in one of our last conversations, walking along the sides of the Volga during a stay at Dubna in the Fall of '87, a sort of explosion of activity. The forms of thoughts, the theories, the models Raphael uncovered for us, have been a beautiful reading of the "great book of nature". Raphael and his work are still a source of great inspiration and will continue to accompany us in life. He was a great, inspired, generous man. All his work has been done with a spreading enthusiasm, with the open, cheerful smile so characteristic to him.
Would that this book, written by friends, co-workers, and scientists close to him, could convey a little of the precious heritage he has left us, and alleviate a little the enormous loss we experienced since he left us ...

Acknowledgements

I am very grateful to Helge Holden for having helped me enormously throughout the preparation of this paper, with his steady moral support and direct engagement; in particular I acknowledge his help with Sect. 13, and in getting valid information from various scientists concerning other sections. Moreover I thank him for reading the manuscript carefully and for making many very useful comments which led to important improvements.
I am also grateful to Rodolfo Figari, Magnus Landstad and Tom Lindstrøm for their help with Sect. 5, resp. 9.4., resp. 10, and to Philippe Blanchard, Daniel Gandolfo, Fritz Gesztesy, Koichiro Iwata, Torbjörn Kolsrud, Jean Marion, Sylvie Paycha, Michael Röckner, Ludwig Streit, Sandro Teta, Bruno Torresani for reading the manuscript and making useful comments.
I am indebted to Dr. Alan Harvey, Cambridge University Press, for his great help in improving the language in this paper.
I would also like to thank B. Richter, U. Weber as well as H. Holden, M. Jarrath, R. Kirchhoff, A. Krawczyk, H. Niehrenheim, N. Rise, G. Schlitt, C. Welge, B. Zegarlinski for skilfull technical help with the setting of the manuscript.

References

[A-W] L. Accardi, W. von Waldenfels (eds.), *Quantum Probability and Applications V*, Lecture Notes in Mathematics, Volume **1442**, Springer-Verlag, Berlin, 1990.

[Al1] S. Albeverio, *An introduction to some mathematical aspects of scattering theory in models of quantum fields*, in: *Scattering Theory in Mathematical Physics*, (eds. J.A. Lavita, J.P. Marchard), D. Reidel, Dordrecht 1974, pp. 299-382.

[Al2] S. Albeverio, *Scattering theory in some models of quantum fields I*, J. Math. Phys. **14** (1973) 1800-1816; II, Helv. Phys. Acta **45** (1972) 303-321.

[Al3] S. Albeverio, *Some points of interaction between stochastic analysis and quantum theory*, in: *Stochastic Differential Systems*, (eds. N. Christopeit, K. Helmes, M. Kohlmann), Lecture Notes in Control and Information Sciences, Volume **78**, Springer-Verlag, New York, 1986, pp. 1-26.

[Al4] S. Albeverio, *Some recent developments and applications of path integrals*, in: *Path Integrals from meV to MeV*, (eds. M.C. Gutzwiller, A. Inomata, J.R. Klauder, L. Streit), World Scientific, Singapore, 1986, pp. 3-32.

[Al5] S. Albeverio, *Applications of nonstandard analysis in mathematical physics*, in: *Non-Standard Analysis and its Applications*, (ed. N. Cutland), Cambridge University Press, 1988, pp. 182-220.

[**A-A**] S. Albeverio, T. Arede, *The relation between quantum mechanics and classical mechanics: a survey of some mathematical aspects*, in: *Chaotic Behavior in Quantum Systems, Theory and Applications*, (ed. G. Casati), Plenum Press, New York, 1985, pp. 37-76.

[**A-A-F**] S. Albeverio, T. Arede, M. de Faria, *Remarks on non linear filtering problems: White noise representation and asymptotic expansions*, in: *Stochastic Processes, Physics, and Geometry. Locarno 1990*, (eds. S. Albeverio, G. Casati, U. Cattaneo, D. Merlini, R. Moresi), World Scientific, Singapore, 1990, pp. 77-86.

[**A-BB-L**] S. Albeverio, A. M. Berthier Boutet de Monvel, Z. Brzezniak, G. Lebeau, work in preparation.

[**A-B-C-R-S-S**] S. Albeverio, Ph. Blanchard, Ph. Combe, R. Rodriguez, M. Sirugue, M. Sirugue-Collin, *Trapping in stochastic mechanics and applications to covers of clouds and radiation belts*, in: *Quantum Probability and Applications II*, (eds. L. Accardi, W. Waldenfels), Lecture Notes in Mathematics, Volume **1136**, Springer-Verlag, Berlin 1985, pp. 24-39.

[**A-B-G-S**] S. Albeverio, Ph. Blanchard, F. Gesztesy, L. Streit, *Quantum mechanical low energy scattering in terms of diffusion processes*, in: *Stochastic Aspects of Classical and Quantum Systems*, (eds. S. Albeverio, Ph. Combe, M. Sirugue-Collin), Lecture Notes in Mathematics, Volume **1109**, Springer-Verlag, Berlin, 1984, pp. 207-227.

[**A-B-R**] S. Albeverio, J. Brasche, M. Röckner, *Dirichlet forms and generalized Schrödinger operators*, in: *Schrödinger Operators*, (eds. H. Holden, A. Jensen), Lecture Notes in Physics, Volume **345**, Springer-Verlag, Berlin, 1989, pp. 1-42.

[**A-B1**] S. Albeverio, Z. Brzezniak, *Finite dimensional approximations approach to oscillatory integrals in infinite dimensions*, SFB **237** - Preprint 1990.

[**A-B2**] S. Albeverio, Z. Brzezniak, *On the limit from quantum mechanics to classical mechanics through infinite dimensional oscillatory integrals*, in preparation.

[**A-C**] S. Albeverio, A.B. Cruzeiro, *Global flows with invariant (Gibbs) measures for Euler and Navier-Stokes two dimensional fluids*, Comm. Math. Phys. **129** (1990) 431-444.

[**A-F-H-M-R**] S. Albeverio, M. Fukushima, W. Hansen, Zh. Ma, M. Röckner, *Capacities on Wiener space: Tightness and invariance*, C.R. Acad. Sci. Paris, t. 314, Série I (1991) 931-935.

[**A-F-K-S**] S. Albeverio, M. Fukushima, W. Karwowski, L. Streit, *Capacity and quantum mechanical tunneling*, Comm. Math. Phys. **80** (1981) 301-342.

[**A-G-K-S**] S. Albeverio, F. Gesztesy, W. Karwowski, L. Streit, *On the connection between Schrödinger and Dirichlet forms*, J. Math. Phys. **26** (1985) 2546-2553.

[A-G-H-K-M] S. Albeverio, R. Gielerak, H. Holden, T. Kolsrud, M. Mebkhout, *Low temperature expansions around classical crystalline ground states*, in: *Stochastic Processes, Physics, and Geometry. Locarno 1991*, (eds. S. Albeverio, U. Cattaneo, D. Merlini, R. Tartini), World Scientific, Singapore, 1992, to appear.

[A-H-P-R-S1] S. Albeverio, T. Hida, J. Potthoff, M. Röckner, L. Streit, *Dirichlet forms in terms of white noise analysis I - Construction and QFT examples*, Rev. Math. Phys. **1** (1990) 291-312.

[A-H-P-R-S2] S. Albeverio, T. Hida, J. Potthoff, M. Röckner, L. Streit, *Dirichlet forms in terms of white noise analysis II - Closability and diffusion processes*, Rev. Math. Phys. **1** (1990) 313-323.

[A-I-K1] S. Albeverio, K. Iwata, T. Kolsrud, *Random fields as solutions of the inhomogeneous quaternionic Cauchy-Riemann equation I. Invariance and Analytic Continuation*, Comm. Math. Phys. **132** (1990) 555-580.

[A-I-K2] S. Albeverio, K. Iwata, T. Kolsrud, *Conformally invariant random fields and processes – old and new*, in: *Proceedings of Lisboa Stochastic Analysis and Applications Conference, Sept. 1989*, (eds. A. B. Cruzeiro, J. C. Zambrini), Birkhäuser, Basel, to appear.

[A-I-K3] S. Albeverio, K. Iwata, T. Kolsrud, *Conformally invariant and reflection positive random fields in two dimensions*, in: *Stochastic Analysis. In Honor of Moshe Zakai*, (eds. E. Mayer-Wolf, E. Merzbach, A. Shwartz), Academic Press, New York, 1991, pp. 1-14.

[A-I-K4] S. Albeverio, K. Iwata, T. Kolsrud, *Homogeneous Markov generalized vector fields and quantum fields over 4-dimensional space-time*, in: *Proceedings of Trento Conference on "Stochastic Partial Differential Equations and Applications. III", Jan. 1990*, (eds. G. De Prato, L. Tubaro), to appear.

[A-I-K5] S. Albeverio, K. Iwata, T. Kolsrud, *A model of four space-time dimensional gauge fields: reflection positivity for associated random currents*, in: *Proceedings of the Liblice Conference on Rigorous Results in Quantum Dynamics, June 1990*, (eds. J. Dittrich, P. Exner), World Scientific, Singapore, 1991.

[A-K-R-S] S. Albeverio, W. Karwowski, M. Röckner, L. Streit, *Capacity, Green functions and Schrödinger equation*, in: *Infinite Dimensional Analysis and Stochastic Processes*, (ed. S. Albeverio), Pitman, Research Notes in Mathematics **124**, Boston-London, 1985, pp. 197-215.

[A-K-R] S. Albeverio, S. Kusuoka, M. Röckner, *On partial integration in infinite dimensional space and applications to Dirichlet forms*, J. London Math. Soc. **42** (1990) 122-136.

[**A-K-S**] S. Albeverio, S. Kusuoka, L. Streit, *Convergence of Dirichlet forms and associated Schrödinger operators*, J. Func. Anal. **68** (1986) 130-148.

[**A-M**] S. Albeverio, Zh. Ma, *A general correspondence between Dirichlet forms and right processes*, Bull. Amer. Math. Soc., to appear.

[**A-M-R**] S. Albeverio, Zh. Ma, M. Röckner, *Dirichlet forms and Markov fields - A report on recent developments*, in: *Diffusion Processes and Related Problems in Analysis*, Volume 1, (ed. M. Pinsky), Birkhäuser, New York-Basel 1990, pp. 325-347.

[**A-P-S**] S. Albeverio, S. Paycha, S. Scarlatti, *A short overview of mathematical approaches to functional integration*, in: *Proc. Karpacz XXV Winter School, Functional Integration, Geometry and Strings*, (eds. Z. Haba, J. Sobczyk), Birkhäuser, Basel, 1989, pp. 230-276.

[**A-R1**] S. Albeverio, M. Röckner, *Classical Dirichlet forms on topological vector spaces - the construction of the associated diffusion process*, Prob. Th. Rel. Fields **83** (1989) 405-438.

[**A-R2**] S. Albeverio, M. Röckner, *New developments in theory and applications of Dirichlet forms*, in: *Stochastic Processes, Physics, and Geometry. Locarno 1990*, (eds. S. Albeverio, G. Casati, U. Cattaneo, D. Merlini, R. Moresi), World Scientific, Singapore, 1990, pp. 27-76.

[**A-R3**] S. Albeverio, M. Röckner, *Classical Dirichlet forms on topological vector spaces - closability and a Cameron-Martin formula*, J. Func. Anal. **88** (1990) 395-436.

[**A-R4**] S. Albeverio, M. Röckner, *Stochastic differential equations in infinite dimensions: solutions via Dirichlet forms*, Prob. Th. Rel. Fields **89** (1991) 347-386.

[**A-Sch**] S. Albeverio, J. Schäfer, *A mathematical model of abelian Chern–Simons theory*, in: *Stochastic Processes, Physics and Geometry. Locarno 1991*, (eds. S. Albeverio, U. Cattaneo, D. Merlini, R. Tartini), World Scientific, Singapore, 1992, to appear.

[**A-S**] S. Albeverio, P. Šeba, *Wave chaos in quantum systems with point interaction*, J. Stat. Phys. **64** (1991) 369-384.

[**A-Za**] S. Albeverio, J.C. Zambrini, *Euclidean quantum mechanics: analytical approach*, Ann. Inst. H. Poincaré **49A** (1989) 259-308.

[**A-Z1**] S. Albeverio, B. Zegarlinski, *Construction of convergent simplicial approximations of quantum fields on Riemannian manifolds*, Comm. Math. Phys. **132** (1990) 39-71.

[**A-Z2**] S. Albeverio, B. Zegarlinski, *Some stochastic techniques in quantization, new developments in Markov fields and quantum fields*, in: *Stochastic Quantization*, (eds. P.H. Damgaard, H. Hüffel, A. Rosenblum), Plenum Press, 1990, pp. 1–19.

[A-J-S] W.O. Amrein, J.M. Jauch, K.B. Sinha, *Scattering Theory in Quantum Mechanics*, W.A. Benjamin, Reading, 1977.

[Ar] T. Arak, *A class of Markov fields with finite range*, in: *Proceedings of the International Congress of Mathematics*, (ed. A. M. Gleason), Berkeley, USA, 1968, pp. 994-999.

[Ar-S] T. Arak, D. Surgailis, *Markov fields with polygonal realizations*, Prob. Th. Rel. Fields **80** (1989) 543-579.

[Ark] L. Arkeryd, *A non-standard approach to the Boltzmann equation*, Arch. Rat. Mech. Anal. **77** (1981) 1-10.

[B-W] H. Baumgärtel, M. Wollenberg, *Mathematical Scattering Theory*, Akademie-Verlag, Berlin 1983.

[Be] C. Becker, *Wilson loops als Weisses Rauschen*, Diplomarbeit, Bochum 1991.

[B-P-P] G. Benfatto, P. Picco, M. Pulvirenti, *On the invariant measures for the two-dimensional Euler flow*, J. Stat. Phys. **46** (1987) 729-742.

[B-C-D-D] E. Benoit, J.L. Callot, F. Diener, M. Diener, *Chasse au canard*, Collect. Math. **32** (1981) 77–97.

[Be-F] F. A. Berezin, L. D. Faddeev, *A remark on Schrödinger's equation with a singular potential*, Soviet. Math. Dokl. **2** (1961) 372-375.

[B-R] J. Bertrand, G. Rideau, *Stochastic jump processes in the phase space representation of quantum mechanics*, in: *Mathematical Problems in Theoretical Physics*, (eds. R. Schrader, R. Seiler, D.A. Uhlenbrock), Lecture Notes in Physics, Volume **153**, Springer-Verlag, Berlin 1982, pp. 276-277.

[Bi] B. Birkeland, *A singular Sturm-Liouville problem treated by nonstandard analysis*, Math. Scand. **47** (1980) 275–294.

[B-N] B. Birkeland, D. Normann, *A non-standard treatment of the equation* $y' = f(y,t)$, Preprint, University of Oslo 1980.

[Bi-W] J. Bisognano, E. Wichmann, *On the duality condition for a Hermitean scalar field*, J. Math. Phys. **16** (1975) 985-1007.

[B-C-S-SC] Ph. Blanchard, Ph. Combe, M. Sirugue, M. Sirugue-Collin, *Jump processes: an introduction and some applications in quantum theories*, Rend. Circ. Mat. Palermo Ser. II, no 17 (1987) 47-104.

[B-C-Z] Ph. Blanchard, Ph. Combe, W. Zheng, *Mathematical and Physical Aspects of Stochastic Mechanics*, Lecture Notes in Physics, Volume **281**, Springer-Verlag, Berlin 1987.

[B-G] Ph. Blanchard, S. Golin, *Diffusion processes with singular drift fields*, Comm. Math. Phys. **109** (1987) 421-435.

[B-L-T] N.N. Bogoliubov, A.A. Logunov, R.T. Todorov, *Introduction to Axiomatic Quantum Field Theory*, Benjamin, Reading 1975.

[B-F1] C. Boldrighini, S. Frigio, *Equilibrium states for the two dimensional incompressible Euler fluid*, Atti Sem. Mat. Fis. Univ. Modena **27** (1978) 106-125.

[B-F2] C. Boldrighini, S. Frigio, *Equilibrium states for a plane incompressible perfect fluid*, Comm. Math. Phys. **78** (1980) 55-76.

[B-H] N. Bouleau, F. Hirsh, *Formes de Dirichlet générales et densité de variables aléatoires sur l'espace de Wiener*, J. Func. Anal. **69** (1986) 229-259.

[Bou] D.G. Boulware, *Quantum field theory in Schwarzschild and Rindler spaces*, Phys. Rev. **11D** (1975) 1404-1423.

[Bra] M. De Brabanter, *The classification of rational rotation C^*-algebras*, Arch. Math. **43** (1984) 79-83.

[B-Z] M. De Brabanter, H.H. Zettl, *C^*-algebras associated with rotation groups and characters*, Manuscr. Math. **47** (1984) 153-174.

[Br1] J.F. Brasche, *Perturbations of Schrödinger Hamiltonians by measures - Selfadjointness and lower semiboundedness*, J. Math. Phys. **26** (1985) 621-626.

[Br2] J.F. Brasche, *Dirichlet forms and generalized Schrödinger operators*, in: *Schrödinger Operators, Standard and Non-Standard* (eds. P. Exner, P. Šeba), World Scientific, Singapore, 1989, pp. 43-58.

[Br3] J.F. Brasche, *An inverse problem in spectral analysis and the Efimov effect*, in: *Stochastic Processes, Physics, and Geometry. Locarno 1990*, (eds. S. Albeverio, G. Casati, U. Cattaneo, D. Merlini), World Scientific, Singapore, 1990, pp. 207-244.

[B-K] J.F. Brasche, W. Karwowski, *On boundary theory for Schrödinger operators and stochastic processes*, Operator Th: Adv. and Appl. **46** (1990) 199-208.

[C-M1] R. Calinon, D. Merlini, *Inhomogeneous stationary states of two dimensional magnetofluids*, J. Phys. of Plasma **30** (1983) 95-107.

[C-M2] R. Calinon, D. Merlini, *Equilibrium statistical mechanics treatment of a "modified" 2-dimensional guiding center plasma*, Phys. Fluids **26** (1983) 3508-3514.

[Ca-S] R. Cameron, D.A. Storvick, *A simple definition of the Feynman integral, with applications*, Mem. Amer. Math. Soc. **288** (1983) 1-46.

[Ca-C] M. Capinski, N. Cutland, *The Euler equation: a uniform nonstandard construction of a global flow, invariant measures, and statistical solutions*, Preprint, Hull 1991.

[C-G] S. Caprino, S. De Gregorio, *On the statistical solutions of the two-dimensional periodic Euler equation*, Math. Methods in the Appl. Sciences **7** (1985) 55-73.

[Car1] E. Carlen, *Conservative diffusion*, Comm. Math. Phys. **94** (1984) 293-315.

[Car2] E. Carlen, *Existence and sample path properties of the diffusions in Nelson's stochastic mechanics*, in: *Stochastic Processes – Mathematics and Physics*, (eds. S. Albeverio, Ph. Blanchard, L. Streit), Lecture Notes in Mathematics, Volume **1158**, Springer-Verlag, Berlin, 1986, pp. 25-51.

[Car-S] E. Carlen, D. Stroock, *An application of the Bakry-Emery criterian in infinite dimensional diffusions*, in: *Séminaire de Probabilités XX*, (eds. J. Azéma, M. Yor), Lecture Notes in Mathematics, Volume **1204**, Springer-Verlag, Berlin, 1985, pp. 341-348.

[Ca] R. Carmona, *Regularity properties of Schrödinger and Dirichlet semigroups*, J. Func. Anal. **33** (1979) 259-296.

[C-M] A.M. Chebotarev, V.P. Maslov, *Processus de sauts et leurs applications dans la mécanique quantique*, in: *Feynman Path Integrals*, (eds. S. Albeverio, Ph. Combe, R. Høegh-Krohn, G. Rideau, M. Sirugue-Collin, M. Sirugue, R. Stora), Lecture Notes in Physics, Volume **106**, Springer-Verlag, Berlin, 1979, pp. 58-72.

[Ch] S.E. Cheremshantsev, *Hamiltonians with zero-range interactions supported by a Brownian path*, LOMI-preprint, Leningrad, 1989, to appear in Ann. Inst. H. Poincaré, Section **A**.

[C-R-R-S] Ph. Combe, G. Rideau, R. Rodriguez, M. Sirugue-Collin, *On the cylindrical approximation of the Feynman path integral*, Rep. Math. Phys. **13** (1978) 279-294.

[C-S] C. Conway, J. Smoller, *Global solutions of the Cauchy problem for quasi-linear first order equations in several space dimensions*, Comm. Pure. Appl. Math. **6** (1966) 95-105.

[Cr] A. B. Cruzeiro, *Solutions et measures invariantes pour des équations d'evolution stochastiques du type Navier-Stokes*, Exp. Math. **7** (1989) 73-82.

[Da] C.M. Dafermos, *Polygonal approximation of solutions of the inital value problem for a conservation law*, J. Math. Anal. Appl. **38** (1972) 33-41.

[Da-L] E. B. Davies, J. M. Lindsay, *Non-commutative Markov semigroups*, London–Nottingham preprint, 1990.

[DeA-D-G] G. F. De Angelis, D. De Falco, G. Di Genova, *Quantum fields on a gravitational background from random fields on Riemannian manifolds*, in: *Stochastic Processes in Classical and Quantum Systems*, Proc. Ascona 1985, (eds. S. Albeverio, G. Casati, D. Merlini), Lecture Notes in Physics, Volume **262**, Springer-Verlag, Berlin, 1986, pp. 170-178.

[D-O] Yu. N. Demkov, V. N. Ostrovskii, *Zero-Range Potentials and Their Applications in Atomic Physics*, Plenum, New York 1988, (Russian original, Nauka, Moscow, 1975).

[Di] J. Dimock, *Algebras of local observables on a manifold*, Comm. Math. Phys. **77** (1980) 219-228

[D-R] S. Disney, I. Raeburn, *Homogeneous C^*-algebras whose spectra are tori*, J. Austral. Math. Soc. Ser. A **38** (1985) 9-39.

[D-M] R.L. Dobrushin, R.A. Minlos, *Investigation of the properties of generalized Gaussian random fields*, Sel. Math. Sov. **1** (1981) 215-263.

[Dri] V.G. Drinfeld, *Quantum groups*, in: *Proceedings of the International Congress of Mathematicians, Berkeley 1986*, pp. 798-820.

[Driv] B.K. Driver, : YM_2 : *continuum expectations, lattice convergence and lassos*, Comm. Math. Phys. **123** (1989) 575 – 616.

[D-W] N. G. Duffield, R. F. Werner, *Mean-field dynamical semigroups on C^*–algebras*, DIAS-STP-Preprint (1990).

[D-S] N. Dunford, J.T. Schwartz, *Linear Operators*, Vols. I-III, Wiley 1971.

[D-P] D. Dürr, M. Pulvirenti, *On the vortex flow in bounded domains*, Comm. Math. Phys. **83** (1983) 265-273.

[Dy1] E.B. Dynkin, *Markov processes and random fields*, Bull. Amer. Math. Soc. **3** (1980) 975-999.

[Dy2] E.B. Dynkin, *Green's and Dirichlet spaces associated with fine Markov processes*, J. Func. Anal. **47** (1982) 381-418.

[Dy3] E.B. Dynkin, *Gaussian and non Gaussian random fields associated with Markov processes*, J. Func. Anal. **55** (1984), 344–376.

[Eck] J.P. Eckmann, *Hypercontractivity for anharmonic oscillators*, J. Func. Anal. **16** (1974) 388-404.

[E-L] S.F. Edwards, A. Lenard, *Exact statistical mechanics of a one-dimensional system with Coulomb forces. II. The method of functional integration*, J. Math. Phys. **3** (1962) 778-792.

[E-T1] D. Elworthy, A. Truman, *A Cameron-Martin formula for Feynman integrals (the origin of Maslov indices)*, in: *Mathematical Problems in Theoretical Physics, Proceedings, Berlin 1981*, (eds. R. Schrader, R. Seiler, D. A. Uhlenbrock), Lecture Notes in Physics, Volume **153**, Springer-Verlag, Berlin, 1982, pp. 288-294.

[E-T2] D. Elworthy, A. Truman, *Feynman maps, Cameron-Martin formulae and anharmonic oscillators*, Ann. Inst. H. Poincaré **41A** (1984) 115-142.

[Es] T. Espeli, *Gaussiske Markovprossesser*, Cand. real. Thesis, Institute of Mathematics, University of Oslo, 1975.

[Ex] P. Exner, *Open Quantum Systems and Feynman Integrals*, Reidel, Dordrecht 1985.

[E-K-S] H. Ezawa, J.R. Klauder, L.A. Shepp, *Vestigial effects of singular potentials in diffusion theory and quantum mechanics*, J. Math. Phys. **16** (1975) 793-799.

[E-S1] P. Exner, P. Šeba, *A simple model of thin-film point contact in two and three dimensions*, Czech. J. Phys. **38** (1988) 1095-1110.

[E-S2] P. Exner, P. Šeba (eds.), *Applications of Self-Adjoint Extension in Quantum Mechanics*, Lecture Notes in Physics, Volume **324**, Springer-Verlag, Berlin, 1989.

[E-S3] P. Exner, P. Šeba (eds.), *Schrödinger Operators – Standard and Non-Standard*, World Scientific, Singapore, 1989.

[Far1] M. de Faria, *Mesures de Gibbs et équation d'Euler de l'hydrodynamique*, Thèse de IIIe Cycle, Université de Provence, Marseille 1979.

[Far2] M.M.C. Ribeiro de Faria, *Fluido de Euler bidimensional: construção de medidas estacionárias e fluxo estocástico*, Dissertation, Universidade do Minho 1986.

[F-P-S] M. de Faria, J. Potthof, L. Streit, *The Feynman integrand as a Hida distribution*, J. Math. Phys. **32** (1991) 2123-2127.

[F-H-T] R. Figari, H. Holden, A. Teta, *A law of large numbers and a central limit theorem for the Schrödinger operator with zero-range potentials*, J. Stat. Phys. **51** (1988) 206-214.

[F-T] R. Figari, A. Teta, *A boundary value problem of mixed type on perforated domains*, Preprint SFB 237, Bochum 1991.

[Fö1] H. Föllmer, *On the global Markov property*, in: *Quantum Fields – Algebras*, (ed. L. Streit), Springer-Verlag, Berlin 1975, pp. 293-302.

[Fö2] H. Föllmer, *Von der Brownschen Bewegung zum Brownschen Blatt: einige neuere Richtungen in der Theorie der stochastischen Prozesse*, in: *Perspective in Mathematics*, (eds. W. Jäger, J. Moser, R. Rempert), Birkhäuser, Basel 1984, pp. 159-190.

[Fr1] K.O. Friedrichs, *Mathematical Aspects of the Quantum Theory of Fields*, Interscience, New York 1953.

[Fr2] K.O. Friedrichs, *Perturbation of Spectra in Hilbert Space*, American Mathematical Society, Providence, 1965.

[F-S] K.O. Friedrichs, H.N. Shapiro, *Integration in Functional Spaces*, Courant Institute of Mathematical Sciences, New York 1955.

[Fri1] A. Frigerio, *Duality of completely positive quasi-free maps and a theorem of L. Accardi and C. Cecchini*, Bull. UMI **2B** (1983) 269-281.

[Fri2] A. Frigerio, *Stationary states of quantum dynamical semigroups*, Comm. Math. Phys. **63** (1978) 269-276.

[F-V] A. Frigerio, M. Verri, *Long-time asymptotic properties of dynamical semigroups on W^*-algebras*, Math. Z. **180** (1982) 275-286.

[Frö1] J. Fröhlich, *The reconstruction of quantum fields from Euclidean Green's functions at arbitrary temperatures in models of a self-interacting Bose field in two space-time dimensions*, Helv. Phys. Acta **48** (1975) 355-363.

[Frö2] J. Fröhlich, *Classical and quantum statistical mechanics in one and two dimensions: Two component Yukawa and Coulomb systems*, Comm. Math. Phys. **47** (1976) 233-268.

[F-P1] J. Fröhlich, Y.M. Park, *Remarks on exponential interactions and the quantum sine-Gordon equation in two space-time dimensions*, Helv. Phys. Acta **50** (1977) 315-329.

[F-P2] J. Fröhlich, Y.M. Park, *Correlation inequalities and thermodynamic limit for classical and quantum continuous systems*, Comm. Math. Phys. **59** (1978) 235-266.

[Fuj] D. Fujiwara, *The Feynman path integral as an improper integral over the Sobolev space*, Saint Jean de Monts 1990, Soc. Math. de France (1990) , XIV, pp. 1-15..

[Fu1] M. Fukushima, *Dirichlet Forms and Markov Processes*, Kodansha and North-Holland, 1980.

[Fu2] M. Fukushima, *On a stochastic calculus related to Dirichlet forms and distorted Brownian motion*, Phys. Rep. **77** (1981) 225-262.

[Fu3] M. Fukushima, *Energy forms and diffusion processes*, in: *Mathematics and Physics, Lectures on Recent Results* **1**, (ed. L. Streit), World Scientific, Singapore, 1985, pp. 65-97.

[Fu4] M. Fukushima, *On the generation of Markov processes by symmetric forms*, in: *Proceedings 2nd Japan-USSR Symposium on Probability Theory*, (eds. G. Maruyama, Yu. V. Prokhorov), Lecture Notes in Mathematics, Volume **330**, Springer-Verlag, Berlin 1973, pp. 46-79.

[Ful] S.A. Fulling, *Aspects of Quantum Field Theory in Curved Space-Time*, Cambridge University Press, Cambridge 1989.

[Fu-R] S.A. Fulling, S.N.M. Ruijsenaars, *Temperature, periodicity and horizons*, Phys. Rep. **152** (1987) 135-176.

[Gan] D. Gandolfo, *I–Equilibre et Perturbation d'un Plasma Toroidal avec Flot. II–Modèle de Dynamique Stochastique*, Thèse de Doctorat de Troisième Cycle, Marseille 1985.

[G-G-V1] I.M. Gelfand, M.I. Graev, A.M. Vershik, *Representations of a group of smooth maps of a manifold X into a compact Lie group*, Comp. Math. **35** (1977) 299-334.

[G-G-V2] I.M. Gelfand, M.I. Graev, A.M. Vershik, *Representations of the group of functions taking values in a compact Lie group*, Comp. Math. **42** (1981) 217-243.

[G-V] I. Gelfand, N. Vilenkin, *Generalized Functions*, Vol 4 (English Translation), Academic Press, New York 1964.

[Geo] H.O. Georgii, *Gibbs Measures and Phase Transitions*, De Gruyter, New York-Berlin 1988.

[Gie] R. Gielerak, *Verification of the global Markov property in some class of strongly coupled exponential interactions*, J. Math. Phys. **24** (1983) 347-355.

[Gl-J] J. Glimm, A. Jaffe, *Quantum Physics – A Functional Integral Point of View*, Springer-Verlag, New York, 1987.

[Gl-K] S. A. Globa, Yu. G. Kondrat'ev, *The construction of Gibbs states of quantum lattice systems*, Selecta Math. Sov. **9** (1990) 297-307.

[Gol] S. Goldstein, *Remarks on the global Markov property*, Comm. Math. Phys. **74** (1980) 223-234.

[G-S] R. Goodman, I. Segal, *Proceedings of the Conference on the Mathematical Theory of Elementary Particles*, (Dedham, 1965), MIT Press, Cambridge, MA 1966.

[Gro1] L. Gross, *Abstract Wiener spaces*, in:*Proceedings of the 5th Berkeley Symposium on Mathematical Statistics and Probability*, (eds. L. M. Le Cam, J. Neyman), University of California Press, Berkeley, 1967, Volume **2**, part 2, pp. 31-42

[Gro2] L. Gross, *Existence and uniqueness of physical ground states*, J. Func. Anal. **10** (1972) 52-109.

[Gro3] L. Gross, *Analytic vectors for representations of the canonical commutation relations and nondegeneracy of ground states*, J. Func. Anal. **17** (1974) 104-111.

[Gro4] L. Gross, *Logarithmic Sobolev inequalities*, Amer. Math. J. **97** (1975) 1061-1083.

[Gro5] L. Gross, *Logarithmic Sobolev inequalities on loop groups*, Preprint, Cornell 1991.

[G-K-S] L. Gross, C. King, A. Sengupta, *Two dimensional Yang-Mills theory via stochastic differential equations*, Ann. Phys. **194** (1989) 65-112.

[Guc] J. Guckenheimer, *Shocks and rarefactions in two space dimensions*, Arch. Rat. Mech. Anal. **59** (1975) 281-291.

[Gu1] F. Guerra, *Uniqueness of the vacuum energy density and van Hove phenomenon in the infinite volume limit for two-dimensional self-coupled Bose fields*, Phys. Rev. Lett. **28** (1972) 1213-1215.

[Gu2] F. Guerra, *Local algebras in Euclidean quantum field theory*, in: *Symp. Math. XX*, Academic Press, New York, 1976, pp. 13-26.

[Gu3] F. Guerra, *Quantum field theory and probability theory. Outlook on new possible developments*, in: *Trends and Developments in the Eighties*, (eds. S. Albeverio, Ph. Blanchard), World Scientific, Singapore, 1985, pp. 214-243.

[G-R-S1] F. Guerra, L. Rosen, B. Simon, *The $P(\phi)_2$ Euclidean quantum field theory as classical statistical mechanics*, Ann. Math. **101** (1975) 111-259.

[G-R-S2] F. Guerra, L. Rosen, B. Simon, *Boundary conditions in the $P(\phi)_2$ Euclidean field theory*, Ann. Inst. Henri Poincaré **15** (1976) 231-334.

[Gu] M. Gutzwiller, *Chaos in Classical and Quantum Mechanics*, Springer-Verlag, Berlin 1990.

[Hab1] Z. Haba, *Stochastic equations for some Euclidean fields*, in: *Stochastic Processes in Classical and Quantum Systems*, (eds. S. Albeverio, G. Casati, D. Merlini), Lecture Notes in Physics, Volume **262**, Springer-Verlag, 1986, pp. 315-328.

[Hab2] Z. Haba, *Ergodicity and invariant measures of some randomly perturbed classical fields*, Wroclaw preprint 1991.

[Haw] S. W. Hawking, *Particle creation by black holes*, Comm. Math. Phys. **43** (1975) 199-220.

[H-L] L.L. Helms, P.A. Loeb, *Applications of nonstandard analysis to spin models*, J. Math. Anal. Appl. **69** (1979), 341–352.

[H-K-P-S] T. Hida, H.H. Kuo, J. Potthoff, L. Streit, *White Noise: An Infinite Dimensional Calculus*, in preparation.

[**Hoh**] E. Hohler, *On the exponential interaction on a compact 2-dimensional manifold and bosonic strings*, Cand. scient. Thesis, Institute of Physics, University of Oslo, 1989.

[**Ho1**] H. Holden, *Konvergens mot punkt-interaksjoner*, Cand. real. Thesis, Institute of Mathematics, University of Oslo, 1981.

[**Ho2**] H. Holden, *Point Interactions. A Solvable Model in Quantum Mechanics and Its Approximation*, Dr. Philos. Thesis, University of Oslo, 1985.

[**Hu-M**] Y.Z. Hu, P.A. Meyer, *Chaos de Wiener et intégrale de Feynman*, in: *Séminaire de Probabilités XXII*, (eds. J. Azéma, M. Yor), Lecture Notes in Mathematics, Volume **1321**, Springer-Verlag, Berlin 1988, pp. 51-71.

[**Hur**] A.E. Hurd, *Nonstandard analysis and lattice statistical mechanics: a variational principle*, Trans. Amer. Math. Soc. **263** (1981) 89-110.

[**I-O**] N. Ikeda, Y. Ogura, *A degenerating sequence of Riemannian metrics on a manifold and their Brownian motion*, Osaka Preprint 1990.

[**Ish**] C.J. Isham, *Quantum field theory in curved space-times: A general mathematical framework*, in: *Differential Geometrical Methods in Mathematical Physics II*, (eds. K. Bleuler, H.R. Petry, A. Reetz), Lecture Notes in Mathematics, Volume **676**, Springer-Verlag, Berlin, pp. 459–512.

[**Ism**] R. Ismagilov, *On unitary representations of the group* $C_0^\infty(X, G)$, $G = SU(2)$, Math. Sbornik **29** (1976) 105-117.

[**Ito**] K. Ito, *Generalized uniform complex measures in the Hilbertian metric space with their applications to the Feynman path integral*, in: *Proc. of the 5th Berkeley Symposium on Mathematical Statistics and Probability*, Volume **II**, part 1, University of California Press, Berkeley, 1966, pp. 145-161.

[**Iw1**] K. Iwata, *On linear maps preserving Markov properties and applications to multi-component generalized random fields*, Ph.D. Thesis, Bochum 1990.

[**Iw2**] K. Iwata, *The inverse of a local operator preserves the Markov property*, Preprint SFB 237, Bochum 1991.

[**J-L**] G.W. Johnson, M.L. Lapidus, *Generalized Dyson Series, generalized Feynman integrals and Feynman's operational calculus*, Mem. Amer. Math. Soc. **62** (1986) 1-78.

[**JL**] G. Jona-Lasinio, *Stochastic processes and quantum mechanics*, Colloque en l'honneur de L. Schwartz, **2**, Astérisque **132** (1985) 203-216.

[J-M-S] G. Jona-Lasinio, F. Martinelli, E. Scoppola, *The semi classical limit of quantum mechanics: A qualitative theory via stochastic mechanics*, Phys. Rep. **77** (1981) 313-327.

[Jo] R. Jost (ed.), *Local Quantum Theory*, Academic Press, New York 1969.

[K-B] G. Kallianpur, C. Bromley, *Generalized Feynman integrals using analytic continuation in several complex variables*, in: *Stochastic Analysis and Applications*, (ed. M. Pinsky), Plenum, New York, 1984, pp. 217-267.

[K-K-K] G. Kallianpur, D. Kannan, R.L. Karandikar, *Analytic and sequential Feynman integrals on abstract Wiener and Hilbert spaces, and a Cameron-Martin formula*, Ann. Inst. H. Poincaré **21B** (1985) 323-361.

[Ka-M] W. Karwowski, J. Marion, *On the closability of some positive definite differential forms on* $C^\infty(\Omega)$, J. Func. Anal. **62** (1985) 266-275.

[Kas1] D. Kastler, *Introduction a l'Electrodynamique Quantique*, Dunod, Paris 1961.

[Kas2] D. Kastler, *Does ergodicity plus locality imply the Gibbs structure?*, Proc. Symp. Pure Math. **38** (1982) 467-489.

[Kas3] D. Kastler, *Introduction to non-commutative geometry and Yang-Mills model-building*, Preprint, Marseille 1990.

[Ka1] T. Kato, *Perturbation Theory for Linear Operators*, Springer-Verlag, Berlin 1966.

[Ka2] T. Kato, *Wave operators and similarity for some non-selfadjoint operators*, Math. Ann. **162** (1966) 258-279.

[Ka-M1] Y. Kato, N. Mugibayashi, *Regular perturbations and asymptotic limits of operators in quantum field theory*, Progr. Theor. Phys. **30** (1963) 103-133.

[Ka-M2] Y. Kato, N. Mugibayashi, *Asymptotic fields in model space theories I*, Progr. Theor. Phys. **45** (1971) 628–639

[Kau] A.G. Kaufman, *Stetigkeit von gruppenwertigen stochastischen Koflächen*, Diploma Thesis, Bochum, 1986.

[Ke1] Ch. Kessler, *Examples of extremal lattice fields without the global Markov property*, Publ. RIMS **21** (1985) 877-888.

[Ke2] Ch. Kessler, *Attractiveness of interactions for binary lattice systems and the global Markov property*, Stoch. Proc. Appl. **24** (1987) 309-313.

[Ke3] Ch. Kessler, *On hyperfinite representations of distributions*, Bull. London Math. Soc. **20** (1988), 139–144.

[Kla] M. Klaus, *Some remarks on double wells in one and three dimensions*, Ann. Inst. H. Poincaré **34A** (1981) 405-417.

[K-S] M. Klaus, B. Simon, *Binding of Schrödinger particles through conspiracy of potential wells*, Ann. Inst. H. Poincaré **30A** (1979) 83-87.

[K-L] A. Klein, L.J. Landau, *Stochastic processes associated with KMS states*, J. Func. Anal. **42** (1981) 368-428.

[Ko] T. Kolsrud, *On the Markov property for certain Gaussian random fields*, Prob. Th. Rel. Fields **74** (1986) 393-402.

[Ko1] J.G. Kondrat'ev, *Dirichlet operators and smoothness of the solutions of infinite-dimensional elliptic equations*, Sov. Math. Dokl. **31** (1985) 461-464.

[Ko2] J.G. Kondrat'ev, *Functional integrals which corresponds to the temperature states of quantum lattice systems*, Preprint SFB 237, Bochum, 1991.

[Ko-K] J.G. Kondrat'ev, V.D. Košmanenko, *The scattering problem for operators associated with Dirichlet forms*, Sov. Math. Dokl. **26** (1982) 585-589.

[K-M] R.H. Kraichnan, D. Montgomery, *Two-dimensional turbulence*, Rep. Progr. Phys. **43** (1980) 547-619.

[K-P] R de L. Kronig, W. G. Penney, *Quantum mechanics of electrons in crystal lattices*, Proc. Roy. Soc. (London) **130A** (1931) 499-513.

[Kr] S.N. Kruzkov, *First order quasilinear equations in several independent variables*, Mat. Sbornik **81** (1970) 217-243.

[Kum-M] B. Kümmerer, H. Maassen, *The essentially commutative dilations of the dynamical semigroup M_n*, Comm. Math. Phys. **109** (1987) 1-22.

[Ku] H. Kunita, *Stochastic Flows and Stochastic Differential Equations*, Cambridge University press, Cambridge, 1990.

[K-P-K-M-M-Y] Y.A. Kuperin, B.S. Pavlov, P.B. Kurasov, K.A. Makarov, Y.B. Melnikov, U.V. Yevstratov, *Scattering theory for self-adjoint extensions*, Preprint, University of Linköping, 1989.

[Kus1] S. Kusuoka, *Markov fields and local operators*, J. Fac. Science Tokyo, Ser. A **1** (1979) 199-212.

[Kus2] S. Kusuoka, *Dirichlet forms and diffusion processes on Banach space*, J. Fac. Science Univ. Tokyo, Sec. 1A **29** (1982) 79-95.

[**L-C**] R. Lavine, M. O'Carroll, *Ground state properties and lower bounds for energy levels of a particle in a uniform magnetic field and external potential*, J. Math. Phys. **18** (1977) 1908-1912.

[**Law**] G. F. Lawler, *A self-avoiding random walk*, Duke Math. J. **47** (1980), 655–693.

[**La**] P. D. Lax, *Hyperbolic systems of conservation laws. II*, Comm. Pure Appl. Math. **10** (1957) 537-566.

[**Lee**] T.D. Lee, *On some statistical properties of hydrodynamics and magnetohydrodynamical fluids*, Quart. J. Appl. Math. **10** (1952) 69-74.

[**Li**] T. Lindstrøm, *Hyperfinite stochastic integration I, II, III*, Math. Scand. **46** (1980) 265-333.

[**Li-S**] V.A. Liskevich, Yu.A. Semenov, *Dirichlet operators. A priori estimates. Uniqueness problem*, Kiev Preprint (1990).

[**Liu1**] T.P. Liu, *The Riemann problem for general* 2×2 *conservation laws*, Trans. Amer. Math. Soc. **199** (1974) 89-112.

[**Liu2**] T.P. Liu, *The Riemann problem for general systems of conservation laws*, J. Diff. Eq. **18** (1975) 218-234.

[**Mal**] P. Maldowney, *A General Theory of Integration in Functional Spaces, Including Wiener and Feynman Integration*, Longman, Essex, 1987.

[**Ma-Ma**] M.P. Malliavin, P. Malliavin, *Integration on loop groups. I. Quasi invariant measures*, J. Func. Anal. **93** (1990) 207-237.

[**Man**] V. Mandrekar, *Some remarks on various definitions of Feynman integral*, in: *Probability in Banach Spaces IV, Oberwolfach 1982*, (eds. A. Beck, K. Jacobs), Lecture Notes in Mathematics, Volume **990**, Springer-Verlag, Berlin, 1983, pp. 170-177.

[**M-P**] C. Marchioro, M. Pulvirenti, *Vortex Methods in Two-Dimensional Fluid Mechanics*, Lecture Notes in Physics, Volume **203**, Springer-Verlag, Berlin 1984.

[**Mar1**] J. Marion, *A survey on the unitary representations of gauge groups, and some remaining open questions*, in: *Trends and Developments in the Eighties*, (eds. S. Albeverio, Ph. Blanchard), World Scientific, Singapore, 1985, pp. 309-329.

[**Mar2**] J. Marion, *Outline of harmonic analysis on groups of paths with values in Sobolev gauge groups*, in: *Stochastic Processes, Physics and Geometry. Locarno 1990*, (eds. S. Albeverio, G. Casati, U. Cattaneo, D. Merlini, R. Moresi), World Scientific, Singapore, 1990, pp. 575-584.

[**M-T**] J. Marion, D. Testard, *Energy representations of gauge groups associated with Riemannian flags*, J. Func. Anal. **76** (1988) 160-175.

[M-Se] W.T. Martin, I. Segal, *Mathematical Theory of Elementary Particles*, (Dedham, 1963), MIT Press, Cambridge, MA 1966.

[Mey] P.A. Meyer, *Fock space and probability theory*, in: *Stochastic Processes in Mathematics and Physics II, Bielefeld 1985*, (eds. S. Albeverio, Ph. Blanchard, L. Streit), Lecture Notes in Mathematics, Volume **1250**, Springer-Verlag, Berlin 1985, pp. 160-170.

[Mo] M.I. Monastyrskii, *Appendix to F.J. Dyson's paper: Missed opportunities*, Russ. Math. Surv. **35:1** (1980) 199-208.

[Nag1] H. Nagasawa, *Segregation of a population in an environment*, J. Math. Biology **9** (1980) 213-235.

[Nag2] H. Nagasawa, *Can the Schrödinger equation be a Boltzmann equation?*, in: *Diffusion Processes and Related Problems in Analysis*, (ed. M. Pinsky), Vol. 1, Birkhäuser, New York-Basel, 1991, pp. 155-200.

[Nel1] E. Nelson, *Derivation of the Schrödinger equation from Newtonian mechanics*, Phys. Rev. **150** (1966) 1079-1085.

[Nel2] E. Nelson, *A quartic interaction in two dimensions*, in: *Mathematical Theory of Elementary Particles*, (eds. R. Goodman, I. Segal), MIT Press, Cambridge, MA 1966, pp. 69–73.

[Nel3] E. Nelson, *The construction of quantum fields from Markov fields*, J. Func. Anal. **12** (1973) 97-112.

[Nel4] E. Nelson, *The free Markov field*, J. Func. Anal. **12** (1973) 211-217.

[Nel5] E. Nelson, *Probability theory and Euclidean field theory*, in: *Constructive Quantum Field Theory* eds. G. Velo, A. Wightman), Springer-Verlag, Berlin 1973, pp. 94–124.

[Nel6] E. Nelson, *Internal set theory*, Bull. AMS **83** (1977) 1165–1193.

[Nel7] E. Nelson, *Quantum Fluctuations*, Princeton University Press, Princeton, NJ, 1985.

[Nel8] E. Nelson, *Field theory and the future of stochastic mechanics*, in: *Stochastic Processes in Classical and Quantum System*, (eds. S. Albeverio, G. Casati, D. Merlini), Lecture Notes in Physics, Volume **262**, Springer-Verlag, Berlin, 1986, pp. 438-469.

[New] C. Newman, *The construction of stationary two dimensional Markoff fields with an application to quantum field theory*, J. Func. Anal. **14** (1973) 44-61.

[Øg] S. Øgrim, *Anvendelse av* Γ*-metoden og resolventmetoden ved studiet av spredningsproblemer med singulære potensialer*, Cand. real. Thesis, Institute of Mathematics, University of Oslo, 1970.

[O-P-T] D. Olesen, G.K. Pedersen, M. Takesaki, *Ergodic actions of compact abelian groups*, J. Operator Theory **3** (1980) 237-269.

[On] L. Onsager, *Statistical hydrodynamics*, Nuovo Cimento Suppl. **6** (1949) 279-282.

[O-P-C] T. A. Osborn, L. Papiez, R. Corns, *Constructive representatives of propagators for quantum systems with electromagnetic fields*, J. Math. Phys. **28** (1987) 103-123.

[Osh] Y. Oshima, *Lectures on Dirichlet spaces*, Preprint, Erlangen 1988.

[Os1] E.P. Osipov, *On triviality of the* $:\exp \lambda\phi:_4$ *quantum field theory in a finite volume*, Rep. Math. Phys. **20** (1984) 111-116.

[Os2] E.P. Osipov, *Two-dimensional random fields as solutions of stochastic differential equations*, Bochum preprint 1989.

[Pa] P. Paclet, *Espaces de Dirichlet et capacités fonctionnelles sur triplet de Hilbert-Schmidt*, Séminaire P. Krée, 1979, Exp. 5, 36 pp.

[Pan] D. Pantić, *Stochastic calculus on distorted Brownian motion*, J. Math. Phys. **29** (1988) 207-209.

[Pap] G. Papanicolaou, *On the convergence of the Feynman path integrals for a certain class of potentials*, J. Math. Phys. **31** (1990) 342-347.

[Pa-S] K. Parthasarathy, M. Schmidt, *A new method of constructing factorisable representations for current groups and current algebras*, Comm. Math. Phys. **50** (1976), 167–175.

[Per] J. Persson, *Second order linear ordinary differential equations with measures as coefficients*, Matematiche **36** (1981) 151-171.

[Pru] E. Prugovecki, *Scattering theory in Fock space*, J. Math. Phys. **13** (1972) 969-976.

[Ra] E. A. Razafimananitena, *Construction of a general class of Dirichlet forms in terms of white noise analysis*, to appear in Stoch. Proc. Appl.

[R-Si] M. Reed, B. Simon, *Methods of Modern Mathematical Physics*, Vols. I, II, III, IV, Academic Press, New York, 1972-1979.

[Rez1] J. Rezende, *Stationary phase method on Hilbert space and semi-classical approximation in quantum mechanics*, Bielefeld preprint, Project no 2, 1984.

[Rez2] J. Rezende, *The method of stationary phase for oscillatory integrals on Hilbert space*, Comm. Math. Phys. **101** (1985) 187-206

[Rie1] M.A. Rieffel, *C^*-algebras associated with irrational rotations*, Pac. J. Math. **93** (1981) 415-429.

[Rie2] M.A. Rieffel, *Noncommutative tori - case study of non-commutative differentiable manifolds*, Contemp. Math. **105** (1990) 191-211.

[Ri] N.H. Risebro, *The partial differential equation* $u_t + \sum_{i=1}^n f_i(u)_{x_i} = 0$. *A numerical method*, Cand. scient. Thesis, Institute of Mathematics, University of Oslo 1987.

[Rö1] M. Röckner, *Dirichlet problem for distributions and specifications for random fields*, Mem. AMS **324** (1985), 76 pp.

[Rö2] M. Röckner, *Generalized Markov fields and Dirichlet forms*, Acta Appl. Math. **3** (1985) 285-311.

[Rö3] M. Röckner, *Specifications and Martin boundary for* $P(\phi)_2$ *random fields*, Comm. Math. Phys. **106** (1986) 105-135.

[R-Z] M. Röckner, B. Zegarlinski, *The Dirichlet problem for quasilinear partial differential operators with boundary data given by a distribution*, in: *Stochastic Processes and Their Applications in Mathematics and Physics*, (eds. S. Albeverio, Ph. Blanchard, L. Streit), Reidel, Dordrecht, pp. 301-326.

[R-Zh] M. Röckner, T.S. Zhang, *On uniqueness of generalized Schrödinger operators and applications*, J. Func. Anal., to appear

[Ro] Yu. A. Rozanov, *Boundary problems for stochastic partial differential equations*, in: *Stochastic Processes – Mathematics and Physics II*, (eds. S. Albeverio, Ph. Blanchard, L. Streit), Lecture Notes in Mathematics, Volume **1250**, Springer-Verlag, Berlin 1987, pp. 233-267.

[R-S] K. Rullkötter, U. Spönemann, *Dirichletformen und Diffusionsprozesse*, Diplomarbeid, Bielefeld 1983.

[Schmi] M. Schmidt, *Die Gitterapproximation nicht-linearer elektromagnetischen Felder in vier Dimensionen*, Diploma Thesis, Bochum 1991.

[Sch] B. Schmuland, *An alternative compactification for classical Dirichlet forms*, Stochastics **33** (1990) 75-90.

[Se] I.E. Segal, *Mathematical Problems of Relativistic Physics*, AMS, Providence, 1963.

[Sei] E. Seiler, *Gauge Theories as a Problem of Constructive Field Theory and Statistical Mechanics*, Lecture Notes in Physics, Volume **159**, Springer-Verlag, Berlin 1982.

[Sen] A. Sengupta, *The Yang–Mills measure for* S^2, J. Func. Anal., to appear.

[Sew] G. Sewell, *Quantum fields on manifolds: PCT and gravitationally reduced thermal states*, Ann. Phys. **141** (1982) 201-224.

[Sil] M.L. Silverstein, *Symmetric Markov Processes*, Lecture Notes in Mathematics, Volume **426**, Springer-Verlag, Berlin, 1974.

[Si1] B. Simon, *The $P(\phi)_2$ Euclidean (Quantum) Field Theory*, Princeton University Press, Princeton, NJ, 1974.

[Si2] B. Simon, *Schrödinger semigroups*, Bull. Am. Math. Soc. **7** (1982) 447-526.

[Sl] A. Sloan, *The polaron without cutoffs in two space dimensions*, J. Math. Phys. **15** (1974) 190-201.

[Sm] J. Smoller, *Shock Waves and Reaction-Diffusion Equations*, Springer-Verlag, New York 1983.

[Sm-K] O.G. Smoluanov, A.Yu. Khrennikov, *The central limit theorem for generalized measures of infinite dimensional spaces*, Sov. Math. Dokl. **31** (1985) 301-304.

[Sto] A. Stoll, *Invariance principles for Brownian intersection local time and polymer measure*, Math. Scan. **64** (1989) 133–160.

[Stø] E. Størmer, *Spectra of ergodic transformations*, J. Func. Anal. **15** (1972) 665-779.

[Str1] R.F. Streater, *Markovian representation of current algebras*, J. Phys. **A10** (1977) 261-266.

[Str2] R.F. Streater, *Euclidean quantum mechanics and stochastic integrals*, in: *Stochastic Integrals*, (ed. D. Williams), Lecture Notes in Mathematics, Volume **851**, Springer-Verlag, Berlin 1981, pp. 371-393.

[Str3] R.F. Streater, *Why should anyone want to axiomatize quantum field theory ?*, BiBoS preprint no 223, 1986.

[S-W] R.F. Streater, A. S. Wightman, *PCT, Spin and Statistics, and All That*, Benjamin, New York 1964.

[S-Wil] R.F. Streater, I.F. Wilde, *The time evolution of quantised fields with bounded quasi-local interaction density*, Comm. Math. Phys. **17** (1970) 21-32.

[S-Z] D. Stroock, B. Zegarlinski, *The logarithmic inequality for continuous spin systems on a lattice*, J. Func. Anal., to appear.

[Sym] K. Symanzik, *Euclidean quantum field theory*, in: *Local Quantum Theory*, (ed. R. Jost), Academic Press, New York 1969, pp. 152-226.

[Ta1] M. Takeda, *On the uniqueness of the Markovian self-adjoint extension*, in: *Stochastic Processes – Mathematics and Physics II*, (eds. S. Albeverio, Ph. Blanchard, L. Streit), Lecture Notes in Mathematics, Volume **1250**, Springer-Verlag, Berlin, 1987, pp. 319-325.

[Ta2] M. Takeda, *The maximum Markovian self-adjoint extensions of generalized Schrödinger operators*, Preprint 1990.

[Tam] H. Tamura, *On the possibility of confinement caused by nonlinear electromagnetic interaction*, J. Math. Phys. **32** (1991) 897-904.

[Tes] D. Testard, *Representation of the group of equivariant loops in SU(n)*, in: *Stochastic Processes – Mathematics and Physics II, Bielefeld*, (eds. S. Albeverio, Ph. Blanchard, L. Streit), Lecture Notes in Mathematics, Volume **1250**, Springer-Verlag, Berlin, 1987, pp. 326-341.

[Te] A. Teta, *Quadratic forms for singular perturbations of the Laplacian*, Publ. RIMS **26** (1990) 803–819.

[To1] B. Torresani, *Représentations projectives des groupes de transformations de jauges locales*, Thesis, Marseille (1986).

[To2] B. Torresani, *Unitary positive energy representations of the gauge group*, Lett. Math. Phys. **13** (1987) 7-15.

[Wal] N.R. Wallach, *On the irreducility and inequivalence of unitary representations of gauge groups*, Comp. Math. **64** (1987) 3-29.

[W-Z] E. Wang, M. Zakai, *Isotropic Gauss-Markov currents*, Prob. Th. Rel. Fields **82** (1989) 137-154.

[Was1] A. Wassermann, *Ergodic actions of compact groups on operator algebras: III. Classification for* $SU(2)$, Inv. Math. **93** (1988) 309-354.

[Was2] A. Wassermann, *Coactions and Yang-Baxter Equations for ergodic actions and subfactors*, in: *Operator Algebras and Applications*, (eds. D. Evans, M. Takesaki), Volume II, Cambridge University Press, London Math. Soc. Lecture Note Series, **135** (1988) 203-236.

[Wa] H. Watanabe, *Path integral for some systems of partial differential equations*, Proc. Japan Acad. Ser. A **60** (1984) 86-89

[Wat] Y. Watatani, *Toral automorphisms on irrational rotation algebras*, Math. Japan. **26** (1981) 479-484.

[Wel] B. Welz, *Stochastische Gleichgewichtsverteilungen eines 2-dimensionalen Superfluids*, Diploma Thesis, Bochum 1986.

[Wic] W.D. Wick, *On the absolute continuity of a convolution with an infinite dimensional measure*, Preprint, University of Washington, Seattle, 1979.

[Wie] N. Wielens, *The essential self-adjointness of generalized Schrödinger operators*, J. Func. Anal. **61** (1985) 98-115.

[Zam] J.C. Zambrini, *Stochastic mechanics according to E. Schrödinger*, Phys. Rev. **A33** (1986) 1532-1548.

[Za] T. Zastawniak, *Approximation of Feynman path integrals by integrals over finite dimensional spaces*, Bull. Pol. Acad. Sci. Math. **34** (1986) 355-372.

[Ze1] B. Zegarlinski, *Uniqueness and the global Markov property for Euclidean fields; the case of general exponential interactions*, Comm. Math. Phys. **96** (1984) 195-221.

[Ze2] B. Zegarlinski, *The Gibbs measures and partial differential equations I*, Comm. Math. Phys. **107** (1986) 411-429.

[Ze3] B. Zegarlinski, *Extremality and the global Markov property II. The global Markov property for non-FKG maximal measures*, J. Stat. Phys. **43** (1986) 687-705.

[Ze4] B. Zegarlinski, *Extremality and the global Markov property III. The Euclidean fields on a lattice*, J. Multiv. Anal. **21** (1987) 158-167.

[Z-M] W. Zheng, P.A. Meyer, *Sur la construction de certaines diffusions*, in: *Séminaire de Probabilités XX*, (eds. J. Azéma, M. Yor), Lecture Notes in Mathematics, Volume **1204**, Springer-Verlag, Berlin, 1986, pp. 334-337.

SOME EUCLIDEAN INTEGER-VALUED RANDOM FIELDS WITH MARKOV PROPERTIES

by

Sergio Albeverio [1], Raphael Høegh-Krohn [2,†], Donatas Surgailis [3]

[1] Fakültat für Mathematik, Ruhr-Universität Bochum, 4630 Bochum 1, Germany; SFB 237, BiBoS; CERFIM (Locarno).

[2] Institute of Mathematics, University of Oslo Blindern, Oslo, Norway.

† Deceased 24 January 1988.

[3] Institute of Mathematics and Informatics, 232600 Vilnius, Lithuania.

Abstract

We associate to the Gibbs states of classical mechanical systems with finite range interaction certain integer-valued random fields. These are homogeneous with respect to the Euclidean group, have the Markov property with respect to arbitrary closed separated sets, and in the dimensions ≥ 3, also the sharp global Markov property with respect to a large class of partitions of the space.

Introduction

A main interest in random fields in $\mathbb{R}^d$ which are homogeneous with respect to the Euclidean group and have the (global) Markov property is to be found in Quantum Field Theory see e. g. [Ne, AHKZ, AMR, AIK1,2, AZ, Gi]. However, Markov random fields are also of independent interest, as multi-parameter generalizations of the classical Markov processes with one-dimensional parameter (time).

There is a well known characterization of Markov fields on the lattice $\mathbb{Z}^d$ as Gibbs fields, see e.g.[A, Sp, AHK2, AHKO, Ge]. The situation is much more complicated in the case of continuous 'time'$\mathbb{R}^d$, where there is no basic Markov field with respect to which all other homogeneous fields are locally absolutely continuous. Thc most studied class of Markov fields with continuous parameter is Gaussian, see [Ne, Ko, Ro, Rö] and references therein. It is notable that, to the best of our knowledge , until the present paper there were not known any examples of homogeneous Markov fields in $I\!R^d (d \geq 3)$ taking a finite number of values. (The approach of [ArS], where a class of Markov fields with polygonal realizations was constructed, does not seem to go beyond the dimension $d = 2$.)

The aim of this paper is to construct homogeneous Markov fields in $I\!R^d (d \geq 2)$ with finite or countable state space, from the classical Gibbs states

$$\nu = \lim_{\Lambda \nearrow I\!R^d} \nu_\Lambda \tag{1.1}$$

in $I\!R^d$, where

$$d\nu_\Lambda(\gamma) = Z_\Lambda^{-1} z^{|\gamma|} e^{-\beta V(\gamma)} d\lambda(\gamma) \tag{1.2}$$

is the Gibbs measure on the set $\mathcal{B}_\Lambda$ of all configurations (finite subsets) $\gamma \subset \Lambda$, with the pair potential

$$V(\gamma) = \sum_{\{x_i, x_j\} \subset \gamma} \phi(x_i - x_j), \tag{1.3}$$

(with ϕ a real-valued function on $I\!R^d$), and the parameters $z > 0$ (the 'activity') and $\beta > 0$ (the 'inverse temperature') ; λ is the Lebesgue measure on $\mathcal{B}_\Lambda$ and Z_Λ the normalizing factor (the 'partition function'). The limit (1.1) exists if the potential is stable and the activity is sufficiently small, (see e. g. [Ru], and also Sect.2 of this paper, for details.)

The $Z\!\!Z_+$-valued random field $\xi(x), x \in I\!R^d$ associated with the Gibbs state (1.1) is defined as

$$\xi(x)(\gamma) = \sum_{y \in \gamma} \chi(|x - y| \leq r), \tag{1.4}$$

where $r > 0$ is a parameter, γ is a configuration in $I\!R^d$ and $\chi(\cdot)$ the indicator function. In particular if the potential (1.3) has hard core of radius $r_0 \leq r$, then $\xi(x)$ takes a.s. a finite number $m = m(r, r_0, d) < \infty$ of values; $m = 2$ for $r = r_0$. One of the main results of this paper (Theorem 4.1) says that if $r \in (0, \infty)$ is the range of the potential (1.3) (i. e. $\phi(x) = 0$ for $|x| > r$) which satisfies an additional conditional stability condition and the activity is sufficiently small, then the random field (1.4) has the global Markov property, in the sense that for any disjoint closed sets $\Lambda_1, \Lambda_2 \subset I\!R^d$ separated by $\Lambda_0 = I\!R^d - (\Lambda_1 \cup \Lambda_2)$, the σ-algebras $\mathcal{A}_{\Lambda_i} = \sigma\{\xi(x) | x \in \Lambda_i\}, i = 1, 2$ are conditionally independent given the σ-algebra $\mathcal{A}_{\Lambda_0} = \sigma\{\xi(x) | x \in \Lambda_0\}$. Theorem 4.1 is based on the result about the global Markov property of the Gibbs state ν, Theorem 3.2, under the same conditions on z and $V(\gamma)$, whose proof takes a large part of the paper and relies on a detailed analysis of conditional correlation functions and the corresponding Kirkwood-Salzburg equation.

It turns out that in dimensions $d \geq 3$ the random field (1.4) has the so-called sharp global Markov property for a large class of partitions Λ_1, Λ_2 of the space $I\!R^d$, which means that the σ-algebras $\mathcal{A}_{\Lambda_1}, \mathcal{A}_{\Lambda_2}$ are conditionally independent given the trace σ-algebra $\mathcal{A}_{\Lambda_0} = \sigma\{\xi(x) | x \in \Lambda_0\}$ on the $(d-1)$-dimensional hypersurface $\Lambda_0 = \partial\Lambda_i, i = 1, 2$. The condition on Λ_0 for such a Markov property is roughly that Λ_0 is not spherical or flat in any neighborhood of its points (we call such hypersurfaces spherically irregular) (unfortunately the case of Λ_0 being a hyperplane is excluded by this condition).

2. Gibbs states of classical particle systems

In this Section we present some known facts about the thermodynamic limit (1.1) of low density classical particle systems (see e. g. [Ru]).

Let $\{\Lambda\}$ (resp. $\{b\Lambda\}$) denote the class of all measurable subsets of $I\!R^d$ (resp. all bounded measurable subsets of $I\!R^d$). We shall write $\Lambda \to \infty$ if $\Lambda \in \{b\Lambda\}$ increases to $I\!R^d$ in such a way that eventually each compact set is covered.

Let a set $\Lambda \in \{b\Lambda\}$ be given. For any $n = 0, 1, 2, ..., \mathcal{B}_\Lambda^{(n)}$ is the collection of all subsets of Λ consisting of exactly n points. We shall call the elements in $\mathcal{B}_\Lambda^{(n)}$ "configurations of n points in Λ". $\mathcal{B}_\Lambda^{(n)}$ is naturally isomorphic with $\tilde{\Lambda}^n \setminus S_n$, where $\tilde{\Lambda}^n \subset \Lambda^n$ is the subset of Λ^n such that $(x_1, ..., x_n) \in \tilde{\Lambda}^n$ iff $x_i \neq x_j$ for $i \neq j$, and S_n is the symmetric group of order n operating naturally in $\tilde{\Lambda}^n$ by $\sigma(x_1, ..., x_n) = (x_{\sigma(1)}, ..., x_{\sigma(n)})$. $\mathcal{B}_\Lambda^{(0)}$ consists of only one element, namely the empty set. Let

$$\mathcal{B}_\Lambda = \bigcup_{n=0}^{\infty} \mathcal{B}_\Lambda^{(n)},$$

so that $\mathcal{B}_\Lambda$ is the collection of all finite subsets of Λ. $\mathcal{B}_\Lambda$ is a topological space in the natural topology given in the components $\mathcal{B}_\Lambda^{(n)}$ by their identification with $\tilde{\Lambda}^n \setminus S_n$, and in this topology $\mathcal{B}_\Lambda$ is locally compact. Let Σ_Λ denote the σ-algebra of Borel subsets of $\mathcal{B}_\Lambda$. The Lebesgue measure on Λ induces in a natural way a Lebesgue measure λ on $\mathcal{B}_\Lambda$ such that for any measurable set $A \subset \mathcal{B}_\Lambda^{(n)}$,

$$\lambda(A) = \frac{1}{n!} \int_{\tilde{A}} \cdots \int dx_1 ... dx_n, \tag{2.1}$$

where $\tilde{A}$ is the preimage of A under the mapping $\tilde{\Lambda}^n \to \tilde{\Lambda}^n \setminus S_n = \mathcal{B}_\Lambda^{(n)}$. Since S_n acts discretely on $\tilde{\Lambda}^n$, it has a fundamental domain D_n which is in fact a convex cone bounded by a finite number of hyperplanes in $\tilde{\Lambda}^n$, and up to some identification on the boundary, $\mathcal{B}_\Lambda^{(n)}$ may be represented by D_n, and (2.1) is just the Lebesgue measure on D_n. Note that

$$\lambda(\mathcal{B}_\Lambda) = \exp\{\text{ vol } \Lambda\}. \tag{2.2}$$

We define the real function $|\gamma|$ on $\mathcal{B}_\Lambda$ by $|\gamma| = n$ for $\gamma \in \mathcal{B}_\Lambda^{(n)}$. A $I\!R \cup \{+\infty\}$-valued function $V(\gamma)$ on $\mathcal{B}_\Lambda$ is called a <u>two-body potential</u> if for any $\gamma = \{x_1, ..., x_n\}$ we have

$$V(\gamma) = \sum_{i \neq j} \phi(x_i - x_j), \tag{2.3}$$

where $\phi(x) = \phi(-x)$ is a $I\!R \cup \{+\infty\}$-valued function on $I\!R^d$. The corresponding Gibbs measure for the grand canonical ensemble is given by

$$\nu_\Lambda(A) = \int_A p_\Lambda(\gamma) d\lambda(\gamma), \quad A \in \Sigma_\Lambda, \tag{2.4}$$

where

$$p_\Lambda(\gamma) = Z_\Lambda^{-1} z^{|\gamma|} e^{-\beta V(\gamma)}. \tag{2.5}$$

Z_Λ^{-1} is the normalization so that (2.4) is a probability measure, and z and β are positive constants called the <u>activity</u> and the <u>inverse temperature</u>. The normalization is only possible if the potential $V(\gamma)$ is <u>stable</u>, i. e. if there is some constant $c < \infty$ such that

$$V(\gamma) \geq -c|\gamma|. \tag{2.6}$$

We say that the potential $V(\gamma)$ has a <u>finite range</u> (resp., a <u>hard core</u>) if $\phi(x) = 0$ for all $|x| > r$ and some $r < \infty$ (resp., $\phi(x) = +\infty$ for all $|x| \leq r_0$ and some $r_0 > 0$). A particular case of potential with finite range and hard core is the potential of non-interacting hard spheres of radius r, corresponding to

$$\phi(x) = \begin{cases} +\infty & \text{for } |x| \leq 2r, \\ 0 & \text{for } |x| > 2r. \end{cases} \tag{2.7}$$

<u>The correlation function.</u> For any configuration $\gamma_0 \in \mathcal{B}_\Lambda$, the mapping $\gamma \to \gamma \cup \gamma_0$ is a measurable transformation of $\mathcal{B}_\Lambda$. Moreover, the image $\gamma_0 \cup \mathcal{B}_\Lambda$ consist exactly of those configurations that contain γ_0. The correlation function ρ_Λ is a real valued function on $\mathcal{B}_\Lambda$ defined by

$$\rho_\Lambda(\gamma_0) = \int_{\mathcal{B}_\Lambda} p_\Lambda(\gamma_0 \cup \gamma) d\lambda(\gamma) = \int_{\mathcal{B}_\Lambda} d\nu_\Lambda(\gamma_0 \cup \gamma). \tag{2.8}$$

By definition, $\rho_\Lambda(\emptyset) = 1$. Moreover by what is said above and the fact that $d\lambda(\gamma_0 \cup \gamma) = d\lambda(\gamma)$ we see that for any $A \in \Sigma_\Lambda^{(n)} = \{A \in \Sigma_\Lambda | A \subset \mathcal{B}_\Lambda^{(n)}\}$ we have

$$\int_A \rho_\Lambda(\gamma) d\lambda(\gamma) = \int_{\mathcal{B}_\Lambda} \binom{|\gamma \cap A|}{n} d\nu_\Lambda(\gamma). \tag{2.9}$$

The inversion of (2.8) gives the measure ν_Λ in terms of the correlation function ρ_Λ. Namely for any $\gamma_0 \in \mathcal{B}_\Lambda$ we have

$$p_\Lambda(\gamma_0) = \int_{\mathcal{B}_\Lambda} (-1)^{|\gamma|} \rho_\Lambda(\gamma_0 \cup \gamma) d\lambda(\gamma). \tag{2.10}$$

We know that $\rho(\gamma)$ satisfies the well-known systems of Kirkwood-Salzburg and Mayer-Montroll equations, see [Ru], and also Sect.3. Denote

$$H(\beta) = \int_{\mathbb{R}^d} |e^{-\beta\phi(x)} - 1| dx. \tag{2.11}$$

We assume $H(\beta) < \infty$ (i. e. ϕ is regular in the sense of [Ru]). For

$$|z| < z^* = e^{-2\beta c - 1} H(\beta)^{-1} \tag{2.12}$$

$\rho_\Lambda(\gamma)$ is an analytic function of z for all $\gamma \in \mathcal{B}_\Lambda$. For $0 \le z < z^*$ one has the bound uniform with respect to Λ :

$$0 \le \rho_\Lambda(\gamma) \le C(z)^{|\gamma|}, \tag{2.13}$$

where $C(z) < \infty$ is some constant. (Here and below we designate by $C, C(\cdot)$ different constants which depend on the quantities in brackets.) For non-negative potentials, other detailed upper and lower bounds, the so-called "alternating bounds", are available, see [Ru].

The thermodynamic limit. We discuss now the limit of ν_Λ as $\Lambda \to \mathbb{R}^d$ written as $(\Lambda \to \infty)$.

Let us introduce some notation. For any $\Lambda \in \{\Lambda\}$, let $\mathcal{B}_\Lambda$ be the collection of all subsets $\gamma \subset \Lambda$ such that

$$\pi_{\Lambda,\Lambda'}\gamma \equiv \gamma \cap \Lambda' \in \mathcal{B}_{\Lambda'} \tag{2.14}$$

for any $\Lambda' \in \{b\Lambda\}, \Lambda' \subset \Lambda$. Let Σ_Λ be the σ-algebra on $\mathcal{B}_\Lambda$ induced by the mappings $\pi_{\Lambda,\Lambda'} : \mathcal{B}_\Lambda \to \mathcal{B}_{\Lambda'}$, where $\Lambda' \in \{b\Lambda\}, \Lambda' \subset \Lambda$. It is clear that these definitions coincide with the earlier ones in the case of bounded sets Λ. Set $\mathcal{B}_{\Lambda,0} = \{\gamma \in \mathcal{B}_\Lambda \mid |\gamma| < \infty\}$.

Write $\mathcal{B} = \mathcal{B}_{\mathbb{R}^d}, \mathcal{B}_0 = \mathcal{B}_{\mathbb{R}^d,0}, \Sigma = \Sigma_{\mathbb{R}^d}, \pi_{\Lambda'} = \pi_{\mathbb{R}^d,\Lambda'}(\Lambda' \in \{b\Lambda\})$. Then $\mathcal{B}$ is the projective limit of the spaces $\mathcal{B}_{\Lambda'}, \Lambda' \in \{b\Lambda\}$ and $\mathcal{B} - \mathcal{B}_0$ is isomorphic with $\widetilde{(\mathbb{R}^d)^\infty} \setminus S_\infty$, where $\widetilde{(\mathbb{R}^d)^\infty} = \{x = (x_1, x_2, ...) | x_i \ne x_j (i \ne j), \{x_j\}$ has no limit points in $\mathbb{R}^d\}$, and S_∞ is the group of permutations of the natural numbers acting naturally on $(\mathbb{R}^d)^\infty$.

Let probability measures μ, μ'_Λ be given on $\mathcal{B}, \mathcal{B}_{\Lambda'}$, respectively. We define the restriction of μ to $\mathcal{B}_\Lambda, \Lambda \in \{\Lambda\}$ by

$$\mu^{(\Lambda)}(A) = \mu(\pi_\Lambda^{-1}A), A \in \Sigma_\Lambda ,$$

(with $\pi_\Lambda \equiv \pi_{\mathbb{R}^d,\Lambda}$).
Similarly, one can define the restriction of $\mu_{\Lambda'}, \Lambda' \supset \Lambda$ to $\mathcal{B}_\Lambda$ which we denote by $\mu_{\Lambda'}^{(\Lambda)}$.

Let $\mu_\Lambda, \Lambda \in \{b\Lambda\}$ be a family of probability measures μ_Λ on $\mathcal{B}_\Lambda$. We say that μ_Λ has a thermodynamic limit as $\Lambda \to \infty$ if for each $\Lambda \in \{b\Lambda\}$ and any $A \in \Sigma_\Lambda$ the limit

$$\lim_{\Lambda' \to \infty} \mu_{\Lambda'}^{(\Lambda)}(A) \tag{2.15}$$

exists. In such a case there is a unique probability measure μ on $\mathcal{B}$ denoted by $\mu = \lim_{\Lambda \to \infty} \mu_\Lambda$ such that, for each $\Lambda \in \{b\Lambda\}, \mu^{(\Lambda)}(A)$ coincides with (2.15).

Coming back to the thermodynamic limit of the Gibbs measures (2.4), consider the limit as $\Lambda \to \infty$ of the correlation function ρ_Λ. We extend naturally the definition of $\rho_\Lambda(\gamma)$ to all $\gamma \in \mathcal{B}_0$ by setting $\rho_\Lambda(\gamma)$ equal to zero if at least one of the points of γ is not in Λ. Then it is well-known, see [Ru], that for all $|z| < z^*$ the $\rho_\Lambda(\gamma)$ converge pointwise as $\Lambda \to \infty$. The convergence is uniform for $\gamma \in \mathcal{B}_\Lambda, \Lambda \in \{b\Lambda\}$ being any bounded set, and is exponentially fast if the potential $V(\gamma)$ has the finite range, in the sense that

$$|\rho_\Lambda(\gamma) - \rho(\gamma)| \le C_1^{|\gamma|} \exp\{-C_2 \text{ dist } (\gamma, (\Lambda')^c)\}, \tag{2.16}$$

where $C_i > 0, i = 1,2$ are some constants independent of $\gamma, \Lambda'; (\Lambda')^c = \mathbb{R}^d - \Lambda'$. The limit satisfies the Kirkwood-Salzburg equation for all $|z| < z^*$, hence in particular $\rho(\gamma)$ is an analytic function of z in the circle $|z| < z^*$. The $\rho(\gamma)$ are invariant under the Euclidean group in the case $\phi(x)$ depends on $|x|$ only. They have exponentially strong cluster properties with respect to translations in the sense that if $\gamma_1, \gamma_2 \in \mathcal{B}_0$, then

$$|\rho(\gamma_1 \cup (\gamma_2 + a)) - \rho(\gamma_1)\rho(\gamma_2)| \leq \tilde{C_1}^{|\gamma_1 \cup \gamma_2|} e^{-\tilde{C_2}|a|}, \tag{2.17}$$

for all $a \in \mathbb{R}^d$, where $\gamma + a$ is the configuration consisting of all the points of γ translated by a, and $\tilde{C_i} > 0, i = 1,2$ are constants, see e. g. [Ru], [BoPKh], [AHK2]. As a consequence of the results of these references we have

Theorem 2.1 Let ν_Λ be the Gibbs measure on $\mathcal{B}_\Lambda(\Lambda \in \{b\Lambda\})$ corresponding to a two-body stable potential $V(\gamma)$ (with ϕ regular). Then for $|z| < z^*$ there exists the thermodynamic limit

$$\nu = \lim_{\Lambda \to \infty} \nu_\Lambda \tag{2.18}$$

such that for any $\Lambda \in \{b\Lambda\}$ and $A \in \Sigma_\Lambda$

$$\nu^{(\Lambda)}(A) = \int_A p^{(\Lambda)}(\gamma) d\lambda(\gamma), \tag{2.19}$$

where

$$p^{(\Lambda)}(\gamma) = \int_{\mathcal{B}_\Lambda} (-1)^{|\gamma'|} \rho(\gamma \cup \gamma') d\lambda(\gamma') \tag{2.20}$$

and $\rho(\gamma), \gamma \in \mathcal{B}_0$ is the limiting correlation function. The measure ν is invariant with respect to translation in $\mathbb{R}^d$ and is invariant with respect to all Euclidean motions if $\phi(x)$ depends on $|x|$ only.

Let us briefly show how Th. 2.1 follows from the discussion above of the correlation function $\rho_\Lambda(\gamma)$ and its convergence as $\Lambda \to \infty$, as a similar argument will be employed in Sect. 3 to prove the convergence of conditional Gibbs distributions. According to (2.10) for any $\Lambda \subset \Lambda', \Lambda, \Lambda' \in \{b\Lambda\}$ we have

$$\nu_{\Lambda'}^{(\Lambda)}(A) = \int_A p_{\Lambda'}^{(\Lambda)}(\gamma) d\lambda(\gamma), \quad A \in \Sigma_\Lambda, \tag{2.21}$$

where

$$p_{\Lambda'}^{(\Lambda)}(\gamma) = \int_{\mathcal{B}_\Lambda} (-1)^{|\gamma'|} \rho_{\Lambda'}(\gamma \cup \gamma') d\lambda(\gamma'). \tag{2.22}$$

By (2.16), $\rho_{\Lambda'}(\gamma \cup \gamma') \to \rho(\gamma \cup \gamma')$ as $\Lambda' \to \infty$ for any $\gamma, \gamma' \in \mathcal{B}_\Lambda$. On the other hand from (2.13) and (2.2) we have

$$\int_{\mathcal{B}_\Lambda} \int_{\mathcal{B}_\Lambda} \rho_{\Lambda'}(\gamma \cup \gamma') d\lambda(\gamma) d\lambda(\gamma') \leq \int_{\mathcal{B}_\Lambda} \int_{\mathcal{B}_\Lambda} C^{|\gamma| + |\gamma'|} d\lambda(\gamma) d\lambda(\gamma') = e^{2C \text{ vol } \Lambda} < \infty.$$

Hence by the Lebesgue dominated convergence theorem the limit $\lim \nu_{\Lambda'}^{(\Lambda)}(A)(\Lambda' \to \infty)$ exists for each $A \in \Sigma_\Lambda$ and is given by (2.19), (2.20).

3. The Markov property of Gibbs states

In this Section we discuss the global Markov property of the Gibbs state (2.18), in the case when the potential has finite range.

Let $\Lambda_0 \subset \Lambda$, where $\Lambda_0, \Lambda \in \{b\Lambda\}$. Since $\Lambda_1 \equiv \Lambda - \Lambda_0$ and Λ_0 are disjoint we have that $\mathcal{B}_\Lambda \cong \mathcal{B}_{\Lambda_1} \times \mathcal{B}_{\Lambda_0}$, where the isomorphism is given by $\gamma = \gamma_1 \cup \gamma_0$, with $\gamma_i = \gamma \cap \Lambda_i, i = 0, 1$. Let ν_Λ be the Gibbs measure (2.4) and $\nu_\Lambda^{(\Lambda_0)}$ the measure induced by it on $\mathcal{B}_{\Lambda_0}$. We easily see that $d\nu_\Lambda^{(\Lambda_0)}(\gamma) = p_\Lambda^{(\Lambda_0)}(\gamma) d\lambda_{\Lambda_0}(\gamma)$, where

$$p_\Lambda^{(\Lambda_0)}(\gamma) = Z_\Lambda^{-1} \int_{\mathcal{B}_{\Lambda_1}} z^{|\gamma_0|+|\gamma_1|} e^{-\beta V(\gamma_0 \cup \gamma_1)} d\lambda_{\Lambda_1}(\gamma_1) \tag{3.1}$$

$d\lambda_{\Lambda_i}$ is the Lebesgue measure on $\mathcal{B}_{\Lambda_i}, i = 0, 1$, and we have used that $d\lambda_\Lambda = d\lambda_{\Lambda_1} \times d\lambda_{\Lambda_0}$ under the splitting $\mathcal{B}_\Lambda \cong \mathcal{B}_{\Lambda_1} \times \mathcal{B}_{\Lambda_0}$.

Consider now the conditional probability measure with respect to the second factor in the splitting $\mathcal{B}_\Lambda \cong \mathcal{B}_{\Lambda_1} \times \mathcal{B}_{\Lambda_0}$ and we find that it is given by

$$d\nu_\Lambda^{(\Lambda_1|\Lambda_0)}(\gamma_1|\gamma_0) = \frac{z^{|\gamma_1|} e^{-\beta V(\gamma_0 \cup \gamma_1)} d\lambda_{\Lambda_1}(\gamma_1)}{\int z^{|\gamma_1|} e^{-\beta V(\gamma_0 \cup \gamma_1)} d\lambda_{\Lambda_1}(\gamma_1)} \tag{3.2}$$

as a probability measure on $\mathcal{B}_{\Lambda_1} = \mathcal{B}_{\Lambda - \Lambda_0}$. Introducing

$$V(\gamma_1|\gamma_0) = V(\gamma_0 \cup \gamma_1) - V(\gamma_0) \tag{3.3}$$

and

$$Z_\Lambda(\gamma_0) = Z_\Lambda^{(\Lambda_1|\Lambda_0)}(\gamma_0) = \int_{\mathcal{B}_{\Lambda_1}} z^{|\gamma_1|} e^{-\beta V(\gamma_1|\gamma_0)} d\lambda_{\Lambda_1}(\gamma_1) \tag{3.4}$$

we may write (3.2) as

$$d\nu_\Lambda^{(\Lambda_1|\Lambda_0)}(\gamma_1|\gamma_0) = p_\Lambda^{(\Lambda_1|\Lambda_0)}(\gamma_1|\gamma_0) d\lambda_{\Lambda_1}(\gamma_1), \tag{3.5}$$

where

$$p_\Lambda^{(\Lambda_1|\Lambda_0)}(\gamma_1|\gamma_0) = Z_\Lambda(\gamma_0)^{-1} z^{|\gamma_1|} e^{-\beta V(\gamma_1|\gamma_0)}. \tag{3.6}$$

In particular, in the case of non-interacting hard spheres of radius r we have of course the corresponding expression (3.6) with $e^{-\beta V(\gamma_1|\gamma_0)} = 1$ if $|x - x'| > 2r$ for all $\{x, x'\} \subset \gamma_0 \cup \gamma_1$ such that $\{x, x'\} \cap \gamma_1 \neq \emptyset$, and zero if not.

We prove first the following Markov property of the Gibbs measure ν_Λ in a "finite volume" $\Lambda \in \{b\Lambda\}$ (cfr. [Pr], [AHKO], [MaM], [Ge], for related results in lattice systems).

Theorem 3.1. Assume that the potential $V(\gamma)$ has a finite range $2r < \infty$ (and is stable). Let $\Lambda_i, \Lambda \in \{b\Lambda\}, i = 0,1,2$ be such that $\Lambda_0 \cup \Lambda_1 \cup \Lambda_2 = \Lambda$ and dist $(\Lambda_1, \Lambda_2) \geq 2r$.

Then with respect to the natural splitting $\mathcal{B}_{\Lambda-\Lambda_0} \cong \mathcal{B}_{\Lambda_1} \times \mathcal{B}_{\Lambda_0}$ we have that

$$d\nu_\Lambda^{(\Lambda_1 \cup \Lambda_2 | \Lambda_0)}(\gamma|\gamma_0) = d\nu_\Lambda^{(\Lambda_1|\Lambda_0)}(\gamma_1|\gamma_0) \times d\nu_\Lambda^{(\Lambda_2|\Lambda_0)}(\gamma_2|\gamma_0), \tag{3.7}$$

where $\gamma = \gamma_1 \cup \gamma_2$.

Proof. Let $\gamma \in \mathcal{B}_{\Lambda-\Lambda_0}$ and $\gamma = \gamma_1 \cup \gamma_2$, where $\gamma_i = \gamma \cap \Lambda_i$, and $\gamma_0 \in \Lambda_0$. Then

$$z^{|\gamma|}e^{-\beta V(\gamma|\gamma_0)} = z^{|\gamma_1|}e^{-\beta V(\gamma_1|\gamma_0)} z^{|\gamma_2|}e^{-\beta V(\gamma_2|\gamma_0)}, \tag{3.8}$$

since

$$\begin{aligned} V(\gamma|\gamma_0) &= \sum_{\{x,x'\}\subset\gamma_0\cup\gamma} \phi(x-x') - \sum_{\{x,x'\}\subset\gamma_0} \phi(x-x') \\ &= \sum_{\{x,x'\}\subset\gamma} \phi(x-x') + \sum_{x\in\gamma, y\in\gamma_o} \phi(x-y) \\ &= \sum_{\{x,x'\}\subset\gamma_1} \phi(x-x') + \sum_{\{x,x'\}\subset\gamma_2} \phi(x-x') \\ &+ \sum_{x\in\gamma_1, y\in\gamma_0} \phi(x-y) + \sum_{x\in\gamma_2, y\in\gamma_0} \phi(x-y) \\ &= V(\gamma_1|\gamma_0) + V(\gamma_2|\gamma_0), \end{aligned} \tag{3.9}$$

if dist $(\gamma_1, \gamma_2) > 2r$. Hence we have proved that (3.8) holds for λ-almost all configurations $\gamma \in \mathcal{B}_{\Lambda-\Lambda_0}$. If we now recall that $d\lambda_{\Lambda-\Lambda_0} = d\lambda_{\Lambda_1} \times d\lambda_{\Lambda_2}$, the conclusion of the theorem follows from (3.8) and (3.4) -(3.6). □

Now we shall extend Th. 3.1 to the limiting Gibbs state $\nu = \lim_{\Lambda\to\infty} \nu_\Lambda$ and unbounded domains $\Lambda_i, i = 0,1,2$. To do this, we shall prove the existence of the limit as $\Lambda' \to \infty$ of the conditional Gibbs measures $d\nu_{\Lambda'}^{(\Lambda_1|\Lambda_0)}(\gamma_1|\gamma_0)$, under some additional conditions on the potential and the activity. The method of the proof is similar to that of Th. 2.1 and employs conditional correlation functions and the corresponding Kirkwood-Salzburg equations.

Given disjoint sets $\Lambda_0, \Lambda_1 \in \{b\Lambda\}$ and a constant $c_0 < \infty$, we define $\Gamma_{\Lambda_0,\Lambda_1,c_0}$ as the set of all $\gamma_0 \in \mathcal{B}_{\Lambda_0}$ such that the inequality

$$V(\gamma_1|\tilde{\gamma}_0) \geq -c_0|\gamma_1| \tag{3.10}$$

holds for every $\gamma_1 \in \mathcal{B}_{\Lambda_1}$ and every $\tilde{\gamma}_0 \subset \gamma_0$.

Definition 3.1. We say that the potential (2.3) is <u>conditionally stable</u> if there exists $c_0 < \infty$ such that for any disjoint sets $\Lambda_0, \Lambda_1 \in \{b\Lambda\}$

$$\nu_{\Lambda_0\cup\Lambda_1}^{(\Lambda_0)}(\Gamma_{\Lambda_0,\Lambda_1,c_0}) = 1. \tag{3.11}$$

Remark 3.1. $V(\gamma)$ being conditionally stable with $c_0 < \infty$ implies that it is stable with $c \le c_0$. Indeed from (3.11) we have

$$\nu^{(\Lambda_0)}_{\Lambda_0\cup\Lambda_1}\left(\Gamma_{\Lambda_0,\Lambda_1,c_0} \cap \mathcal{B}^{(0)}_{\Lambda_0}\right) = \nu^{(\Lambda_0)}_{\Lambda_0\cup\Lambda_1}\left(\mathcal{B}^{(0)}_{\Lambda_0}\right), \tag{3.12}$$

where $\mathcal{B}^{(0)}_{\Lambda_0}$ consists of the empty configuration in Λ_0. By the definition of $\Gamma_{\Lambda_0,\Lambda_1,c_0}$, the intersection $\Gamma_{\Lambda_0,\Lambda_1,c_0} \cap \mathcal{B}^{(0)}_{\Lambda_0}$ equals $\mathcal{B}^{(0)}_{\Lambda_0}$ if $V(\gamma_1) \ge -c_0|\gamma_1|$ for all $\gamma_1 \in \mathcal{B}_{\Lambda_1}$, and is empty if otherwise. Therefore (3.12) implies $V(\gamma_1) \ge -c_0|\gamma_1|$ for all $\gamma_1 \in \Lambda_1$. As $\Lambda_1 \in \{b\Lambda\}$ is arbitrary, so $V(\gamma)$ is stable with stability constant c_0.

Remark 3.2. Any non-negative two-body potential $V(\gamma)$ is conditionally stable with $c_0 = 0$, and in this case one can take $\Gamma_{\Lambda_0,\Lambda_1,c_0} = \mathcal{B}_{\Lambda_0}$. Another example of a conditionally stable potential is a potential with a hard core r_0 and a finite range $r \ge r_0$. Here, Def. 3.1 is satisfied with $c_0 = c + 2cm(r, r_0, d)$ and

$$\Gamma_{\Lambda_0,\Lambda_1,c_0} = \{\gamma_0 \in \mathcal{B}_{\Lambda_0} \,\Big|\, |x - x'| > 2r_0 \quad \forall\{x, x'\} \subset \gamma_0\},$$

where c comes from the stability condition (2.6) (which implies in particular that $\inf\{\phi(x)|x \in I\!R^r - \{0\}\} \ge -c$) and $m(r, r_0, d)$ is the maximal number of hard spheres of radius r_0 whose centers belong to a sphere of radius r.
<u>In the rest of the paper we assume that the potential</u> $V(\gamma)$ <u>is stable and conditionally stable with some constants</u> $c \le c_0$, <u>respectively, and that it has a finite range</u> $2r < \infty$.

Introduce the conditional correlation function (c. f. (2.8)):

$$\rho^{(\Lambda_1|\Lambda_0)}_{\Lambda'}(\gamma_1|\gamma_0) = \int_{\mathcal{B}_{\Lambda_1}} d\nu^{(\Lambda_1|\Lambda_0)}_{\Lambda'}(\gamma_1 \cup \gamma|\gamma_0), \tag{3.13}$$

where $\Lambda_i \subset \Lambda' \in \{b\Lambda\}, i = 0, 1$ are disjoint sets, and $\gamma_i \in \mathcal{B}_{\Lambda_i}$. Without loss of generality we shall assume that

$$d\nu^{(\Lambda_1|\Lambda_0)}_{\Lambda'}(\gamma_1|\gamma_0) = \rho^{(\Lambda_1|\Lambda_0)}_{\Lambda'}(\gamma_1|\gamma_0) = 0 \tag{3.14}$$

if $\gamma_0 \in \mathcal{B}_{\Lambda_0} - \Gamma_{\Lambda_1,\Lambda_0,c_0}$. Note that (3.13) does not depend on $\Lambda_1 \subset \Lambda' - \Lambda_0$ in the sense that

$$\rho^{(\Lambda_1|\Lambda_0)}_{\Lambda'}(\gamma_1|\gamma_0) = \rho^{(\Lambda'-\Lambda_0|\Lambda_0)}_{\Lambda'}(\gamma_1|\gamma_0) \tag{3.15}$$

for any $\gamma_1 \in \mathcal{B}_{\Lambda'-\Lambda_0}$ and any $\gamma_1 \subset \Lambda_1 \subset \Lambda' - \Lambda_0$. Therefore we shall suppress the superscripts in the notation (3.13), when the set Λ_0 is clear from the context.

From (3.13), (3.15) and (3.4) - (3.6) we have that

$$\rho_{\Lambda'}(\gamma_1|\gamma_0) = Z_{\Lambda'}(\gamma_0)^{-1} \int_{\mathcal{B}_{\Lambda'-\Lambda_0}} z^{|\gamma_1|+|\gamma|} e^{-\beta V(\gamma_1\cup\gamma|\gamma_0)} d\lambda(\gamma),$$

where $Z_{\Lambda'}(\gamma_0) = Z_{\Lambda'}^{(\Lambda'-\Lambda_0|\Lambda_0)}(\gamma_0)$ (see (3.4)). By (3.10), (3.14) and the conditional stability condition for any $\gamma_1 \in \mathcal{B}_{\Lambda_1}$ we have the bound

$$0 \leq \rho_{\Lambda'}(\gamma_1|\gamma_0) \leq C(\gamma_0, \Lambda')(ze^{\beta c_0})^{|\gamma_1|}, \tag{3.16}$$

where $C(\gamma_0, \Lambda') = Z_{\Lambda'}(\gamma_0)^{-1} \exp\{ze^{\beta c_0} \text{ vol } \Lambda'\} < \infty$. Similarly as in (2.22) we have that

$$p_{\Lambda'}^{(\Lambda_1|\Lambda_0)}(\gamma_1|\gamma_0) = \int_{\mathcal{B}_{\Lambda_1}} (-1)^{|\gamma|} \rho_{\Lambda'}(\gamma_1 \cup \gamma|\gamma_0) d\lambda(\gamma), \tag{3.17}$$

where

$$p_{\Lambda'}^{(\Lambda_1|\Lambda_0)}(\gamma_1|\gamma_0) = d\nu_{\Lambda'}^{(\Lambda_1|\Lambda_0)}(\gamma_1|\gamma_0)/d\lambda_{\Lambda_1}(\gamma_1) \tag{3.17'}$$

is the Radon-Nikodym derivative of the conditional Gibbs measure restricted to Λ_1.

Lemma 3.1. Let

$$|z| < z_0^* = e^{-\beta(c+c_0)} H(\beta)^{-1}, \tag{3.18}$$

where $H(\beta)$ is defined in (2.11). Then for any $\Lambda_0 \in \{\Lambda\}$ there exists a measurable set $\Gamma_{\Lambda_0} \subset \mathcal{B}_{\Lambda_0}$ such that

$$\nu^{\Lambda_0}(\Gamma_{\Lambda_0}) = 1 \tag{3.19}$$

and such that for any $\gamma_1 \in \mathcal{B}_{\Lambda_0^c,0}$ and any $\gamma_0 \in \Gamma_{\Lambda_0}$ the limit

$$\lim_{\Lambda' \to \infty} \rho_{\Lambda'}(\gamma_1|\gamma_0') \equiv \rho(\gamma_1|\gamma_0) \tag{3.20}$$

exists, where $\gamma_0' = \gamma_0 \cap \Lambda'$. Moreover, there are constants $C_i > 0, i = 1,2$ independent of $\Lambda', \Lambda_0, \gamma_i, i = 0,1$, and such that

$$|\rho_{\Lambda'}(\gamma_1|\gamma_0') - \rho(\gamma_1|\gamma_0)| \leq C_1^{|\gamma_1|+1} \exp\{-C_2 \text{ dist } (\gamma_1, (\Lambda')^c)\} \tag{3.21}$$

holds for any $\gamma_1 \in \mathcal{B}_{(\Lambda_0)^c,0}$.

Proof. Define

$$\Gamma_{\Lambda_0} = \bigcap_{\tilde{\Lambda}_0 \subset \Lambda_0, \Lambda_1 \subset \Lambda_0^c} \hat{\Gamma}_{\Lambda_1, \tilde{\Lambda}_0, c_0}, \tag{3.22}$$

where the intersection is taken over all $\tilde{\Lambda}_0 \subset \Lambda_0, \Lambda_1 \subset \Lambda_0^c = \mathbb{R}^d - \Lambda_0, \Lambda_0, \Lambda_1 \in \{b\Lambda\}$ and $\hat{\Gamma}_{\Lambda_1,\tilde{\Lambda}_0,c_0} = \pi_{\Lambda_0,\tilde{\Lambda}_0}^{-1} \Gamma_{\Lambda_1,\tilde{\Lambda}_0,c_0}$, where $\pi_{\Lambda_0,\tilde{\Lambda}_0}\gamma = \gamma \cap \tilde{\Lambda}_0$ is the projection from $\mathcal{B}_{\Lambda_0}$ onto $\mathcal{B}_{\tilde{\Lambda}_0}(\tilde{\Lambda}_0 \subset \Lambda_0)$. Note that $\hat{\Gamma}_{\Lambda_1,\tilde{\Lambda}_0,c_0}$ decreases when either Λ_1 or $\tilde{\Lambda}_0$ increase. Therefore Γ_{Λ_0} can be written as the countable intersection:

$$\Gamma_{\Lambda_0} = \bigcap_{k=1}^{\infty} \hat{\Gamma}_{\Lambda_1(k),\Lambda_0(k),c_0}, \tag{3.23}$$

where $\Lambda_1(k) = \Lambda_0^c \cap [-k,k]^d$, $\Lambda_0(k) = \Lambda_0 \cap [-k,k]^d$. Indeed, write Γ'_{Λ_0} for the right hand side of (3.23). Then $\Gamma'_{\Lambda_0} \supset \Gamma_{\Lambda_0}$ by definition. On the other hand, each $\hat{\Gamma}_{\Lambda_1,\tilde{\Lambda}_0,c_0} \supset \hat{\Gamma}_{\Lambda_1(k),\Lambda_0(k),c_0}$ for all $k \geq 1$ sufficiently large by the monotonicity, hence $\hat{\Gamma}_{\Lambda_1,\tilde{\Lambda}_0,c_0} \supset \Gamma'_{\Lambda_0}$ and $\Gamma_{\Lambda_0} \supset \Gamma'_{\Lambda_0}$, which proves (3.23).

Let us show (3.19). In view of (3.23), it suffices to show that for any $k \geq 1$

$$\nu^{(\Lambda_0)}\left(\hat{\Gamma}_{\Lambda_1(k),\Lambda_0(k),c_0}\right) = 1.$$

As $\hat{\Gamma}_{\Lambda_1(k),\Lambda_0(k),c_0}$ is a cylinder set measurable with respect to the σ-algebra $\hat{\Sigma}_{\Lambda_0,\Lambda_0(k)} = \pi^{-1}_{\Lambda_0,\Lambda_0(k)}\Sigma_{\Lambda_0(k)}$, so by Theorem 2.1

$$\nu^{\Lambda_0}\left(\hat{\Gamma}_{\Lambda_1(k),\Lambda_0(k),c_0}\right) = \nu^{(\Lambda_0(k))}\left(\Gamma_{\Lambda_1(k),\Lambda_0(k),c_0}\right) = \lim_{\Lambda' \to \infty} \nu_{\Lambda'}^{(\Lambda_0(k))}\left(\Gamma_{\Lambda'-\Lambda_0(k),\Lambda_0(k),c_0}\right) = 1.$$

Next, $\Gamma_{\Lambda_1(k),\Lambda_0(k),c_0} \supset \Gamma_{\Lambda'-\Lambda_0(k),\Lambda_0(k),c_0}$ if $\Lambda' \supset \Lambda_1(k)$, and $\nu_{\Lambda'}^{(\Lambda_0(k))}\left(\Gamma_{\Lambda'-\Lambda_0(k),\Lambda_0(k),c_0}\right) = 1$ because of the conditional stability; see Def. 3.1. Thus Γ_{Λ_0} satisfies (3.19).

For any $\xi > 0$ introduce the Banach space $E_\xi = E_\xi(\Lambda_0^c)$ of all measurable functions $\varphi : \mathcal{B}_{\Lambda_0^c,0} \to \mathbb{C}$ with the norm

$$\|\varphi\|_\xi = \sup_{\gamma \in \mathcal{B}_{\Lambda_0^c,0}} \xi^{-|\gamma|}|\varphi(\gamma)| < \infty. \tag{3.24}$$

Given $\gamma_0 \in \mathcal{B}_{\Lambda_0}$ and $\Lambda' \in \{b\Lambda\}$, we extend $\rho_{\Lambda'}(\gamma_1|\gamma_0')$ for $\gamma_1 \in \mathcal{B}_{\Lambda_0^c,0}$ by putting $\rho_{\Lambda'}(\gamma_1|\gamma_0') = 0$ if $\gamma_1 \not\subset \Lambda' - \Lambda_0, \gamma_1 \in \mathcal{B}_{\Lambda_0^c,0}$. Then from (3.16) we have $\rho_{\Lambda'}(\cdot|\gamma_0') \in E_\xi$ provided $\xi \geq ze^{\beta c_0}$.

Let us fix a configuration $\gamma_0 \in \Gamma_{\Lambda_0}$ and consider the expressions:

$$W^x_{\Lambda'}(\gamma|\gamma_0') = \sum_{x' \in \gamma - \{x\}} \phi(x - x') + \sum_{y \in \gamma_0'} \phi(x - y) \tag{3.25}$$

(we recall that $\gamma_0' = \gamma_0 \cap \Lambda'$) and

$$W^x(\gamma|\gamma_0) = \sum_{x' \in \gamma - \{x\}} \phi(x - x') + \sum_{y \in \gamma_0} \phi(x - y), \tag{3.26}$$

where $\gamma \in \mathcal{B}_{\Lambda'-\Lambda_0}$ and $x \in \gamma$. Note that all sums in (3.25), (3.26) are finite because such is the configuration γ and the potential has finite range. Then according to (2.10) and (3.10)

$$\sum_{x \in \gamma} W^x_{\Lambda'}(\gamma|\gamma_0') = V(\gamma) + V(\gamma|\gamma_0') \geq -(c + c_0)|\gamma| \tag{3.27}$$

and, similarly,

$$\sum_{x \in \gamma} W^x(\gamma|\gamma_0) \geq -(c + c_0)|\gamma|. \tag{3.28}$$

Hence there exist $x_* = x_*(\gamma, \gamma_0) \in \gamma$ and $x'_* = x'_*(\gamma, \gamma_0, \Lambda') \in \gamma$ such that

$$W_{\Lambda'}^{x'_*}(\gamma|\gamma'_0) \geq -c - c_0 \tag{3.29}$$

and

$$W^{x_*}(\gamma|\gamma_0) \geq -c - c_0. \tag{3.30}$$

Introduce the integral operator:

$$K_{\Lambda',\gamma_0}\varphi(\gamma) = \exp\{-\beta W_{\Lambda'}^{x'_*}(\gamma|\gamma'_0)\} \times \tag{3.31}$$

$$\times \left[\varphi(\gamma - \{x'_*\})\chi(|\gamma| > 1) + \int_{\mathcal{B}_{\Lambda_0^c}} K(x'_*, \tilde{\gamma})\varphi(\gamma \cup \tilde{\gamma} - \{x'_*\})d\lambda(\tilde{\gamma}) \right]$$

for $\gamma \in \mathcal{B}_{\Lambda_0^c,0}, |\gamma| > 0$, where $\chi(\cdot)$ is the indicator function and

$$K(x, \gamma) = \prod_{y \in \gamma} \left(e^{-\beta\phi(y-x)} - 1 \right)$$

for $|\gamma| > 0$; for $|\gamma| = 0$ we put $K_{\Lambda',\gamma_0}\varphi(\gamma)$ and $K(x, \gamma)$ both equal to zero. The operator K_{Λ',γ_0} acts linearly and boundedly in each of the spaces $E_\xi, \xi > 0$, which follows from the estimate

$$|K_{\Lambda',\gamma_0}\varphi(\gamma)| \leq \|\varphi\|_\xi e^{\beta(c+c_0)} \xi^{|\gamma|-1} e^{\xi H(\beta)},$$

see (3.29), (3.31), and the proof of the corresponding estimate in [Ru]. Consequently,

$$\|K_{\Lambda',\gamma_0}\|_\xi \leq \xi^{-1} e^{\beta(c+c_0)} e^{\xi H(\beta)}. \tag{3.32}$$

Similarly as in [Ru] one can verify that the conditional correlation function satisfies the following Kirkwood-Salzburg equation:

$$\rho_{\Lambda'}(\gamma|\gamma'_0) = z\chi_{\Lambda'_1}(\gamma)\alpha'(\gamma) + z\chi_{\Lambda'_1}(\gamma)K_{\Lambda',\gamma_0}\rho_{\Lambda'}(\gamma|\gamma'_0), \gamma \in \mathcal{B}_{\Lambda_0^c,0}, \tag{3.33}$$

where $\Lambda'_1 = \Lambda' \cap \Lambda_0^c, \chi_\Lambda(\gamma) = \chi(\gamma \subset \Lambda)$ and

$$\alpha'(\gamma) = \chi(|\gamma| = 1) \cdot e^{-\beta V(\gamma|\gamma'_0)}.$$

In view of (3.32), Eq. (3.33) has a unique solution in E_ξ, provided

$$|z| < e^{-\beta(c+c_0)}\xi e^{-\xi H(\beta)}. \tag{3.34}$$

The above condition implies $\rho_{\Lambda'}(\cdot|\gamma'_0) \in E_\xi$ by (3.16) and coincides with the condition $|z| < z_0^*$ of the lemma for $\xi = H(\beta)^{-1}$.

Define $\rho(\gamma|\gamma_0), \gamma \in \mathcal{B}_{\Lambda_0^c,0}$ as the solution of the limiting Kirkwood-Salzburg equation:

$$\rho(\gamma|\gamma_0) = z\alpha(\gamma) + zK_{\gamma_0}\rho(\gamma|\gamma_0), \tag{3.35}$$

where $\alpha(\gamma) = \chi(|\gamma| = 1)e^{-\beta V(\gamma|\gamma_0)}$ and

$$K_{\gamma_0}\phi(\gamma) = \exp\{-\beta W^{x_*}(\gamma|\gamma_0)\}\times$$

$$\times\left[\varphi(\gamma - \{x_*\})\chi(|\gamma| > 1) + \int_{\mathcal{B}_{\Lambda_0^c}} K(x_*, \tilde{\gamma})\varphi(\gamma \cup \tilde{\gamma} - \{x_*\})d\lambda(\tilde{\gamma})\right]$$

if $|\gamma| > 0$, $= 0$ if otherwise. Due to (3.30), Eq. (3.35) also has a unique solution in E_ξ under (3.34).

In order to prove the convergence of $\rho_{\Lambda'}(\gamma|\gamma_0')$ to $\rho(\gamma|\gamma_0)$ as $\Lambda' \to \infty$, write

$$\rho_{\Lambda'}(\gamma|\gamma_0') = \sum_{n=0}^{\infty}(z\chi_{\Lambda_1'}K_{\Lambda',\gamma_0})^n z\chi_{\Lambda_1'}\alpha', \tag{3.36}$$

$$\rho(\gamma|\gamma_0) = \sum_{n=0}^{\infty}(zK_{\gamma_0})^n z\alpha. \tag{3.37}$$

Because of the finite range $2r$ of the potential, for any measurable sets $\Lambda \subset \Lambda_0^c$ and $\Lambda' \supset \Lambda, \tilde{\Lambda} \supset \Lambda$ such that

$$\text{dist}\,(\Lambda, (\Lambda')^c) > 2r, \quad \text{dist}\,(\Lambda, \tilde{\Lambda}^c) > 2r \tag{3.38}$$

we have

$$\chi_\Lambda K_{\Lambda',\gamma_0}\chi_{\Lambda'\cap\Lambda_0^c} = \chi_\Lambda K_{\gamma_0}\chi_{\tilde{\Lambda}\cap\Lambda_0^c} \tag{3.39}$$

and

$$\chi_\Lambda K_{\gamma_0}\chi_{\Lambda'\cap\Lambda_0^c} = \chi_\Lambda K_{\gamma_0}\chi_{\tilde{\Lambda}\cap\Lambda_0^c}. \tag{3.40}$$

Indeed, if $\gamma \subset \Lambda \subset \Lambda_1' = \Lambda' \cap \Lambda_0^c$ and (3.38) holds, then $W^x_{\Lambda'}(\gamma|\gamma_0') = W^x(\gamma|\gamma_0)$ and therefore $\chi_\Lambda K_{\Lambda',\gamma_0} = \chi_\Lambda K_{\gamma_0}$. Moreover, $K(x_*, \tilde{\gamma}) = 0$ if $|x_* - y| > 2r$ for some $y \in \tilde{\gamma}$ which implies $\chi_\Lambda K_{\gamma_0} = \chi_\Lambda K_{\gamma_0}\chi_{\tilde{\Lambda}\cap\Lambda_0^c}$. Eq. (3.40) is proved analogously.

By using repeatedly (3.39) and (3.40) and the fact that $\chi_{\tilde{\Lambda}}\alpha' = \chi_{\tilde{\Lambda}}\alpha$ if dist $(\tilde{\Lambda}, (\Lambda')^c) >$ $2r$ we obtain

$$\chi_\Lambda(\chi_{\Lambda_1'}K_{\Lambda',\gamma_0})^k\chi_{\Lambda_1'}\alpha' = \chi_\Lambda(K_{\gamma_0})^k\alpha$$

provided $\Lambda \subset \Lambda_1' \equiv \Lambda' \cap \Lambda_0^c$ satisfies dist $(\Lambda, (\Lambda')^c) > 2kr (k = 1, 2, \ldots)$. Hence by (3.36), (3.37)

$$\chi_\Lambda(\rho_{\Lambda'}(\gamma|\gamma_0') - \rho(\gamma|\gamma_0)) = \sum_{n=[d/2r]}^{\infty} z^{n+1}\left\{\chi_\Lambda(\chi_{\Lambda_1'}K_{\Lambda',\gamma_0})^n\alpha' - \chi_\Lambda K_{\gamma_0}^n\alpha\right\}, \tag{3.41}$$

where $d =$ dist $(\Lambda, (\Lambda')^c)$ and $[\cdot]$ denotes the entire part. Finally, using (3.32) and a similar bound for $\|K_{\gamma_0}\|_\xi$ for z, ξ satisfying (3.34), together with $\|\alpha\|_\xi \le e^{\beta c_0}, \|\alpha'\|_\xi \le e^{\beta c_0}$ we obtain from (3.14) that

$$\|\chi_\Lambda(\rho_{\Lambda'}(\cdot|\gamma_0') - \rho(\cdot|\gamma_0))\|_\xi \le \sum_{n=[d/2r]}^{\infty} e^{\beta c_0}|z|^{n+1}\left\{\|K_{\Lambda',\gamma_0}\|_\xi^n + \|K_{\gamma_0}\|_\xi^n\right\}$$

$$\leq 2ze^{\beta c_0}\left(\xi^{-1}e^{\beta(c+c_0)}e^{\xi H(\beta)}\right)^{d/2r} / \left(1-\xi^{-1}e^{\beta(c+c_0)}e^{\xi H(\beta)}\right),$$

or the exponential rate of convergence in (3.21). □

With the help of the previous lemma we shall prove a similar statement about the convergence of the conditional Radon-Nikodym derivatives $p_{\Lambda'}^{(\Lambda_1|\Lambda_0)}(\gamma_1|\gamma_0')$ as $\Lambda' \to \infty$.

Lemma 3.2. Let $0 < z < z_0^*$. Then for any $\Lambda_0 \in \{\Lambda\}, \Lambda_1 \in \{b\Lambda\}, \Lambda_1 \subset \Lambda_0^c \equiv \mathbb{R}^d - \Lambda_0$ and any $\gamma_1 \in \mathcal{B}_{\Lambda_1}, \gamma_0 \in \mathcal{B}_{\Lambda_0}$ there exists the limit

$$\lim_{\Lambda' \to \infty} p_{\Lambda'}^{(\Lambda_1|\Lambda_0')}(\gamma_1|\gamma_0') \equiv p^{(\Lambda_1|\Lambda_0)}(\gamma_1|\gamma_0), \tag{3.42}$$

where $\Lambda_0' = \Lambda' \cap \Lambda_0, \gamma_0' = \gamma_0 \cap \Lambda'$, such that for any measurable sets $A_1 \subset \mathcal{B}_{\Lambda_1}, A_0 \subset \mathcal{B}_{\Lambda_0}$

$$\int_{A_0}\int_{A_1} p^{(\Lambda_1|\Lambda_0)}(\gamma_1|\gamma_0)d\lambda_{\Lambda_1}(\gamma_1)d\nu^{(\Lambda_0)}(\gamma_0) = \nu(\hat{A}_1 \cap \hat{A}_0), \tag{3.43}$$

where $\hat{A} = \pi_{\Lambda_i}^{-1}A_i \in \Sigma, i = 0,1$. The convergence (3.42) is uniform in $\gamma_0 \in \mathcal{B}_{\Lambda_0}$ and is exponentially fast so that there exist constants $\tilde{C}_i > 0, i = 1,2$ independent of $\Lambda', \gamma_0, \gamma_1$ and such that

$$\left|p_{\Lambda'}^{(\Lambda_1|\Lambda_0')}(\gamma_1|\gamma_0') - p^{(\Lambda_1|\Lambda_0)}(\gamma_1|\gamma_0)\right| \leq \tilde{C}_1^{|\gamma_1|+1}\exp\{-\tilde{C}_2 \text{ dist }(\Lambda_1,(\Lambda')^c)\}. \tag{3.44}$$

Proof. We'll show that the limit (3.42) exists and is given by

$$p^{(\Lambda_1|\Lambda_0)}(\gamma_1|\gamma_0) = \int_{\mathcal{B}_{\Lambda_1}}(-1)^{|\gamma|}\rho(\gamma_1 \cup \gamma|\gamma_0)d\lambda(\gamma), \tag{3.45}$$

where $\rho(\gamma_1|\gamma_0) = \rho^{(\Lambda_1|\Lambda_0)}(\gamma_1|\gamma_0)$ is the limiting conditional correlation function of Lemma 3.1 (for $\gamma_0 \in \mathcal{B}_{\Lambda_0} - \Gamma_{\Lambda_0}$ we put $\rho(\cdot|\gamma_0) = p^{(\Lambda_1|\Lambda_0)}(\cdot|\gamma_0) = 0$). As $0 \leq \rho(\gamma_1|\gamma_0) \leq C\xi^{-|\gamma_1|}$ with $\xi = H(\beta)^{-1}$ and $C < \infty$ independent of γ_1, γ_0; see the proof of Lemma 3.1; so (3.45) is well-defined and finite and from (3.17), (3.45) and Lemma 3.1 (3.21) we find that

$$|p_{\Lambda'}^{(\Lambda_1|\Lambda_0')}(\gamma_1|\gamma_0') - p^{(\Lambda_1|\Lambda_0)}(\gamma_1|\gamma_0)|$$

$$\leq \int_{\mathcal{B}_{\Lambda_1}} |\rho_{\Lambda'}(\gamma_1 \cup \gamma|\gamma_0') - \rho(\gamma_1 \cup \gamma|\gamma_0)|d\lambda(\gamma)$$

$$\leq \exp\{-C_2\tilde{d}\}C_1^{|\gamma_1|+1}\int_{\mathcal{B}_{\Lambda_1}} C_1^{|\gamma|}d\lambda(\gamma) \leq \tilde{C}_1^{|\gamma_1|+1}e^{-C_2\tilde{d}},$$

where $\tilde{d} =$ dist $(\Lambda_1,(\Lambda')^c)$ and $\tilde{C}_1 = C_1e^{C_1 \text{ vol } \Lambda_1}$.

Let us prove (3.43). Note that it suffices to prove it for cylinder sets $A_0 \in \Sigma_{\tilde{\Lambda}_0}, \tilde{\Lambda}_0 \in \{b\Lambda\}, \tilde{\Lambda}_0 \subset \Lambda_0$ only. If, moreover, $\tilde{\Lambda}_0$ and Λ_1 are subsets of $\Lambda' \in \{b\Lambda\}$ and $\Lambda_0' = \Lambda_0 \cap \Lambda'$, then by definition of conditional probability measure we have

$$\int_{A_0} \left\{ \int_{A_1} p_{\Lambda'}^{(\Lambda_1|\Lambda_0')}(\gamma_1|\gamma_0) d\lambda(\gamma_1) \right\} d\nu_{\Lambda'}^{(\Lambda_0')}(\gamma_0) = \nu_{\Lambda'}^{(\Lambda_1 \cup \tilde{\Lambda}_0)}(A_1 \cap A_0). \qquad (3.46)$$

By Th. 2.1, the right hand side of (3.46) tends to the right hand side of (3.43) as $\Lambda' \to \infty$. Write $\Delta_{\Lambda'}$ for the difference between the corresponding left hand sides, and set

$$f_{\Lambda'}(\gamma_0) = \int_{A_1} p_{\Lambda'}^{(\Lambda_1|\Lambda_0')}(\gamma_1|\gamma_0) d\lambda(\gamma_1),$$
$$f(\gamma_0) = \int_{A_1} p^{(\Lambda_1|\Lambda_0)}(\gamma_1|\gamma_0) d\lambda(\gamma_1)$$

By (3.44), for any $\varepsilon > 0$ there exists $\Lambda'' \in \{b\Lambda\}$ such that for all $\Lambda' \supset \Lambda'', \Lambda' \in \{b\Lambda\}$ and any $\gamma_0 \in \mathcal{B}_{\Lambda_0}$

$$|f_{\Lambda'}(\gamma_0') - f_{\Lambda''}(\gamma_0'')| < \varepsilon, \qquad (3.47)$$
$$|f(\gamma_0) - f_{\Lambda''}(\gamma_0'')| < \varepsilon, \qquad (3.48)$$

where $\gamma_0' = \gamma_0 \cap \Lambda', \gamma_0'' = \gamma_0 \cap \Lambda''$. Write $\Delta_{\Lambda'}$ as

$$\begin{aligned} \Delta_{\Lambda'} = &\int_{A_0} (f_{\Lambda'}(\gamma_0) - f_{\Lambda''}(\gamma_0'')) d\nu_{\Lambda'}^{(\Lambda_0')}(\gamma_0) \\ &+ \int_{A_0} (f_{\Lambda''}(\gamma_0'') - f(\gamma_0)) d\nu^{(\Lambda_0)}(\gamma_0) \\ &+ \int \chi_{A_0} f_{\Lambda''}[d\nu_{\Lambda'}^{(\Lambda_0')} - d\nu^{(\Lambda_0)}] \equiv \Delta_{\Lambda'}^1 + \Delta_{\Lambda''}^2 + \Delta_{\Lambda''}^3 \end{aligned}$$

Here $|\Delta_{\Lambda'}^i|, i = 1, 2$ do not exceed ε by (3.47), (3.48), while $\Delta_{\Lambda'}^3$ vanishes as $\Lambda' \to \infty$ by Th. 2.1, as $\chi_{A_0} f_{\Lambda''}$ is a bounded cylindric function. This proves (3.43) and the lemma as well. □

Now we come to the main result of this Section.

Theorem 3.2. Assume, as above, that the potential $V(\gamma)$ is stable and conditionally stable with constants c and c_0, respectively, and that it has finite range $2r$. Let $0 < z < z_0^*$, where z_0^* is defined in (3.18), and let $\Lambda_i \in \{\Lambda\}, i = 0, 1, 2$ be pairwise disjoint sets such that dist $(\Lambda_1, \Lambda_2) \geq 2r$. Then with respect to the natural splitting $\mathcal{B}_{\Lambda_1 \cup \Lambda_2} \cong \mathcal{B}_{\Lambda_1} \times \mathcal{B}_{\Lambda_2}$ we have that

$$d\nu^{(\Lambda_1 \cup \Lambda_2 | \Lambda_0)}(\gamma|\gamma_0) = d\nu^{(\Lambda_1|\Lambda_0)}(\gamma_1|\gamma_0) \times d\nu^{(\Lambda_2|\Lambda_1)}(\gamma_2|\gamma_0), \tag{3.49}$$

where $\gamma = \gamma_1 \cup \gamma_2$.

Proof. It suffices to prove the theorem for bounded sets $\Lambda_i, i = 1, 2$, as the σ-algebra Σ_{Λ_i} is generated by cylinder sets $\hat{A}_i = \pi^{-1}_{\Lambda_i, \Lambda} A, A \in \Sigma_\Lambda, \Lambda \in \{b\Lambda\}$.

Let $\Lambda_i \subset \Lambda' \in \{b\Lambda\}, i = 1, 2; \Lambda_0' = \Lambda_0 \cap \Lambda'$. Then by Th. 3.1 the equality

$$p_{\Lambda'}^{(\Lambda_1 \cup \Lambda_2|\Lambda_0')}(\gamma|\gamma_0') = p_{\Lambda'}^{(\Lambda_1|\Lambda_0')}(\gamma_1|\gamma_0') \times p_{\Lambda'}^{(\Lambda_2|\Lambda_0')}(\gamma_2|\gamma_0') \tag{3.50}$$

holds for any $\gamma \in \mathcal{B}_{\Lambda_1 \cup \Lambda_2}$ and $\gamma_0 \in \mathcal{B}_{\Lambda_0}$, where $\gamma_i = \gamma \cap \Lambda_i, i = 1, 2$ and $\gamma_0' = \gamma_0 \cap \Lambda'$. Lemma 3.2 enables now to pass to the limit as $\Lambda' \to \infty$ in this equality, and guarantees that the limiting expressions coincide with the Radon-Nikodym derivatives of the corresponding conditional Gibbs distributions in (3.49). □

4. The random field associated with the Gibbs state

We associate with the Gibbs state ν a random field $\xi(x), x \in \mathbb{R}^d$ defined by

$$\xi(x)(\gamma) = \sum_{y \in \gamma} \chi(|x - y| \leq r), \tag{4.1}$$

where $2r > 0$ is the range of the potential. The sum (4.1) is well-defined and finite for any $(x, \gamma) \in \mathbb{R}^d \times \mathcal{B}$, and is jointly measurable as a function from $\mathbb{R}^d \times \mathcal{B}$ to $\mathbb{Z}_+ = \{0, 1, 2, ...\}$. Hence $\xi(x), x \in \mathbb{R}^d$ is a $\mathbb{Z}_+$-valued random field defined on the probability space $(\mathcal{B}, \Sigma, \nu)$. For each $\Lambda \in \{\Lambda\}$ introduce the σ-algebra

$$\mathcal{A}_\Lambda = \sigma\{\xi(x) | x \in \Lambda\}. \tag{4.2}$$

Put $\Lambda_r = \{y \in \mathbb{R}^d | \exists x \in \Lambda \text{ s.t. } |x - y| \leq r\}, S(x, a) = \{y \in \mathbb{R}^d || x - y| \leq a\}, \partial S(x, a) = \{y \in \mathbb{R}^d || x - y| = a\}$.

Lemma 4.1. (i) For any $\Lambda \in \{\Lambda\}$ we have $\mathcal{A}_\Lambda \subset \Sigma_{\Lambda_r}$.
(ii) Let Λ be an open connected set such that $\operatorname{diam} \Lambda = \sup\{|x-y| \mid x, y \in \Lambda\} > 2r$, then $\mathcal{A}_\Lambda = \Sigma_{\Lambda_r}$.

Proof. (i) For each $x \in \Lambda$ we have $\xi(x)(\gamma) = \xi(x)(\gamma \cap \Lambda_r)$ hence $\xi(x)$ is Σ_{Λ_r}-measurable, and $\mathcal{A}_\Lambda \subset \Sigma_{\Lambda_r}$.
(ii) Let $\gamma \cap \Lambda_r = \{y_1, y_2, ...\}$. We will show that each y_i is uniquely determined by the values $\xi(x)(\gamma \cap \Lambda_r), x \in \Lambda$, i.e. that $\mathcal{A}_\Lambda \supset \Sigma_{\Lambda_r}$. As $y_i \in \Lambda_r$, so $S(y_i, r) \cap \Lambda \neq \emptyset$. Moreover, $\partial S(y_i, r) \cap \Lambda \neq \emptyset$ as $\operatorname{diam} \Lambda > 2r$ and Λ is connected. Indeed, as $\operatorname{diam} \Lambda > 2r$, there are $x_1 \in S(y_i, r) \cap \Lambda$ and $x_2 \in \Lambda - S(y_i, r)$. As Λ is connected, there is a continuous curve lying in Λ and connecting the points x_1 and x_2. This curve must intersect $\partial S(y_i, r)$ at some point $y \in \partial S(y_i, r) \cap \Lambda$.

As Λ is open, there is $\varepsilon > 0$ such that $\tilde{S}_i \equiv S(y, \varepsilon) \cap \partial S(y_i, r) \subset \Lambda$. It is clear that $\tilde{S}_i$ determines y_i completely. In turn, $\tilde{S}_i$ is completely determined by $\xi(x)(\gamma \cap \Lambda_r), x \in \Lambda$ as it belongs to the surface on which this function has a unit jump. □

Theorem 4.1. Let Λ_1, Λ_2 be closed sets such that $\Lambda_1 \cap \Lambda_2 = \emptyset$ and such that $\Lambda_0 \equiv \mathbb{R}^d - (\Lambda_1 \cup \Lambda_2)$ is connected and $\operatorname{diam} \Lambda_0 > 2r$. Let f_i be $\mathcal{A}_{\Lambda_i}$-measurable and bounded and E_{Λ_0} be the conditional expectation with respect to the σ-algebra $\mathcal{A}_{\Lambda_0}$. Then

$$E_{\Lambda_0}(f_1 \cdot f_2) = E_{\Lambda_0}(f_1) \cdot E_{\Lambda_0}(f_2). \tag{4.3}$$

Proof. By Lemma 4.1 (i), $\mathcal{A}_{\Lambda_i} \subset \Sigma_{(\Lambda_i)_r} \subset \sigma(\Sigma_{\Lambda_i - (\Lambda_0)_r}, \Sigma_{(\Lambda_0)_r}), i = 1, 2$, while $\mathcal{A}_{\Lambda_0} = \Sigma_{(\Lambda_0)_r}$.
The statement of the theorem follows now from Th.3.2, according to which the σ-algebras $\Sigma_{\Lambda_i - (\Lambda_0)_r}, i = 1, 2$ are conditionally independent given the σ-algebra $\Sigma_{(\Lambda_0)_r}$. □

Remark 4.1. One can associate with the Gibbs state ν also the $\{0, 1\}$-valued random field

$$\tilde{\xi}(x)(\gamma) = \xi(x)(\gamma) \wedge 1 = \chi(\text{ dist } (x, \gamma) \leq r), x \in \mathbb{R}^d. \tag{4.4}$$

It is easy to understand, however, that for the random field $\tilde{\xi}(x), x \in \mathbb{R}^d$ the Markov property as in Th. 4.1 is not true in general. An exception in this respect is the case of non-interacting hard spheres of the radius r, as in this case $\tilde{\xi}(x)(\gamma) = \xi(x)(\gamma)$ ν-a.s. for each $x \in \mathbb{R}^d$, i.e. $\tilde{\xi}(x)$ is a version of the random field (4.1) and as such satisfies the Markov property of Th.4.1.

What can be said about the Markov property of $\xi(x)$ with respect to partitions of $\mathbb{R}^d$ formed by $(d-1)$-dimensional hypersurfaces rather than open sets Λ_0 ?

Let $\{\Lambda\}_0$ denote the class of all $(d-1)$ dimensional continuous manifolds Λ_0 in $\mathbb{R}^d$ such that $\mathbb{R}^d - \Lambda_0 = \Lambda_1 \cup \Lambda_2$, where $\partial \Lambda_i = \Lambda_0, i = 1, 2$ and $\Lambda_i, i = 1, 2$ are disjoint and connected. For $\Lambda_0 \in \{\Lambda\}_0$ we introduce the σ-algebra.

$$\bar{\mathcal{A}}_{\Lambda_0} = \bigcap_{\varepsilon > 0} \mathcal{A}_{(\Lambda_0)_{\varepsilon^-}}, \tag{4.5}$$

where $(\Lambda_0)_{\varepsilon^-} = \{x \in \mathbb{R}^d | \exists y \in \Lambda_0 \text{ s.t. } |x-y| < \varepsilon\}$. Clearly $\bar{\mathcal{A}}_{\Lambda_0} \supset \mathcal{A}_{\Lambda_0}$, where $\mathcal{A}_{\Lambda_0} = \sigma\{\xi(x) \mid x \in \Lambda_0\}$.

Theorem 4.2. let $\Lambda_0 \in \{\Lambda\}_0$ be a $(d-1)$-dimensional manifold which separates $\Lambda_i, i = 1,2$, and diam $\Lambda_0 > 2r$. Let f_i be bounded and $\mathcal{A}_{\Lambda_i}$- measurable, and $\bar{E}_{\Lambda_0}$ be the conditional expectation with respect to $\bar{\mathcal{A}}_{\Lambda_0}$. Then

$$\bar{E}_{\Lambda_0}(f_1 \cdot f_2) = \bar{E}_{\Lambda_0}(f_1) \cdot \bar{E}_{\Lambda_0}(f_2). \tag{4.6}$$

Proof. Let f_i be $\mathcal{A}_{\Lambda_i - (\Lambda_0)_{\varepsilon^-}}$-measurable for some $\varepsilon > 0$. Then by Th.4.1

$$E_{(\Lambda_0)_{\varepsilon^-}}(f_1 \cdot f_2) = E_{(\Lambda_0)_{\varepsilon^-}}(f_1) \cdot E_{(\Lambda_0)_{\varepsilon^-}}(f_2). \tag{4.7}$$

As $\varepsilon \downarrow 0$, the conditional expectations in (4.7) converge a.s. to the corresponding conditional expectations in (4.6). It remains to use the fact that the σ-algebra $\mathcal{A}_{\Lambda_i}$ is generated by $\mathcal{A}_{\Lambda_i - (\Lambda_0)_{\varepsilon^-}}, \varepsilon > 0 (i = 1,2)$. □

It is remarkable that in dimensions $d \geq 3$ the Markov property (4.6) holds also for the conditional expectation E_{Λ_0} with respect to the trace σ-algebra $\mathcal{A}_{\Lambda_0} = \sigma\{\xi(x) \mid x \in \Lambda_0\}$ instead of the expectation $\bar{E}_{\Lambda_0}$ with respect to the larger σ- algebra $\bar{\mathcal{A}}_{\Lambda_0}$, for a large class of $(d-1)$-dimensional hypersurfaces Λ_0.

Definition 4.1. Let $d \geq 3$. A manifold $\Lambda_0 \in \{\Lambda\}_0$ is called r-spherically irregular if for any sphere $S(x,r)$ intersecting Λ_0 and $|\partial S(x,r) \cap \Lambda_0| > 1$, the intersection $\partial S(x,r) \cap \Lambda_0$ does not lie in a $(d-1)$-dimensional hyperplane.

Theorem 4.3. let $\Lambda_0, \Lambda_i, f_i, i = 1,2$ be the same as in Th.4.2. Assume in addition that $d \geq 3$ and Λ_0 is r-spherically irregular. Then

$$E_{\Lambda_0}(f_1 \cdot f_2) = E_{\Lambda_0}(f_1) \cdot E_{\Lambda_0}(f_2). \tag{4.8}$$

Proof. By Lemma 4.1 (i), $\mathcal{A}_{\Lambda_0} \subset \Sigma_{(\Lambda_0)_r}$. We will show below that under the conditions of the theorem,

$$\mathcal{A}_{\Lambda_0} = \Sigma_{(\Lambda_0)_r} \quad \nu - \text{ a.s.} \tag{4.9}$$

After that (4.8) follows from Th.3.2.

To prove (4.9) we will use the following elementary fact from geometry : each sphere $S(x,r) \subset \mathbb{R}^d$ is completely determined by $d+1$ points $x_i \in \partial S(x,r), i = 1, ..., d+1$ which do not belong to a single hyperplane. At first, we shall exclude from the consideration those configurations which have a point exactly at the distance r from Λ_0. Put

$$A_{\Lambda_0} = \{\gamma \in \mathcal{B} \mid \exists x \in \gamma \text{ s.t. dist } (x, \Lambda_0) = r\}.$$

Then $\nu(A_{\Lambda_0}) = 0$ as the measure ν is locally absolutely continuous with respect to the Lebesgue measure λ on $\mathcal{B}$, and $\lambda(A_{\Lambda_0}) = 0$. Now let $\gamma \notin A_{\Lambda_0}$, then for each $x \in \gamma \cap (\Lambda_0)_r$ the intersection $\partial S(x,r) \cap \Lambda_0$ contains infinitely many points which do not lie in a single hyperplane by the assumption of the theorem, and therefore this intersection determines completely $S(x,r)$ and x (see above). In turn, the function $x \to \xi(x)(\gamma) = \xi(x)(\gamma \cap (\Lambda_0)_r), x \in \Lambda_0$ determines completely $\gamma \cap (\Lambda_0)_r$. This proves (4.9). □

Let us discuss some examples of spherically irregular surfaces. Note first that any hypersurface $\Lambda_0 \in \{\Lambda\}_0$ which is either spherical (i.e. the surface of a sphere in $\mathbb{R}^d$) or a hyperplane, or coincides in some neighborhood with these two types, is not r-spherically irregular for any $r > 0$ by definition.

Example 1. A cylinder $C_\alpha(a) = \{x = (x^{(1)}, ...x^{(d)}) \in \mathbb{R}^d \mid (x^{(1)} - a^{(1)})^2 + (x^{(2)} - a^{(2)})^2 = \alpha^2\}$ is r-spherically irregular iff $\alpha > r$. A simple proof of this fact in the dimension $d = 3$ can be given as follows. Assume there are $b \in \mathbb{R}^3$ and a plane $L \subset R^3$ such that $\varphi \equiv \partial S(b,r) \cap C_\alpha(a) \subset L$ and dist $(b, C_\alpha(a)) < r$. Then φ is a continuous closed curve lying in L which is the intersection of $\varphi_1 = \partial S(b,r) \cap L$, which is a circle, and $\varphi_2 = C_\alpha(a) \cap L$, which is an ellipse (or consists of two lines). Of course this is possible only if φ_2 is a circle itself in which case L is perpendicular to the axis of the cylinder. But this implies in turn that b belongs to this axis and $r \geq \alpha$, which is a contradiction. The proof can be generalized to $d > 3$ with appropriate modifications.

Example 2. The "harmonic" hypersurface

$$H(\tau, \lambda, \theta) = \{x \in \mathbb{R}^d \mid x^{(1)} = \tau + \prod_{i=2}^{d} \cos(\lambda^{(i)} x^{(i)} + \theta^{(i)})\},$$

$$\tau \in \mathbb{R}, 0 \neq \lambda = (\lambda^{(2)}, ..., \lambda^{(d)}) \in \mathbb{R}^{d-1}, \quad \theta = (\theta^{(2)}, ..., \theta^{(d)}) \in \mathbb{R}^{d-1}$$

is r-spherically irregular for any $r > 0$. This follows from the observation that for any $(d-1)$-dimensional hyperplane L, the intersection $L \cap H(\tau, \lambda, \theta)$ is either a $(d-2)$ - dimensional hyperplane, or a $(d-2)$-dimensional hypersurface which is not algebraic, and therefore cannot coincide locally which the $(d-2)$-dimensional hypersphere $L \cap \partial S(x,r)$ for any $x \in \mathbb{R}^d$ and $r > 0$ such that dist $(x, L) < r$.

We shall now see that we may use Th. 4.3 to form a Markov process. Let $\Lambda_0 \in \{\Lambda\}_0$ be a r-spherically irregular hypersurface such that $T_t\Lambda_0 \cap \Lambda_0 = \emptyset$ for any $t \in \mathbb{R} - \{0\}$, where $T_t x = T_t(x^{(1)}, ..., x^{(d)}) = (x^{(1)} + t, x^{(2)}, ..., x^{(d)})$ is the translation in the $x^{(1)}$-direction.

Let E_t be the conditional expectation with respect to the σ-algebra $\Sigma_t = \mathcal{A}_{T_t\Lambda_0}$ generated by the $\mathbb{Z}_+$-valued random field (4.1) on $T_t\Lambda_0$. From Th. 4.3, for $d \geq 3$, we have that for any $t \in \mathbb{R}$ and any bounded functions f_1, f_2, measurable with respect to $\bigcup_{s \leq t} \Sigma_s$ and $\bigcup_{s \geq t} \Sigma_s$, respectively,

$$E_t(f_1 \cdot f_2) = E_t(f_1) \cdot E_t(f_2). \tag{4.10}$$

For $f \in L^\infty(\Sigma_0)$ we define

$$P_t f = E_0 T_t f,$$

where T_t is the transformation of functions corresponding to the translation T_t in $I\!R^d$. It follows from (4.10) that $P_{t+s} = P_t P_s$ so that P_t are the transition probabilities of a Markov process with the restriction to Σ_0 of the Gibbs measure ν in $I\!R^d$ as the invariant measure. We have thus proved the following theorem.

Theorem 4.4. Let $d \geq 3$, and $\Lambda_0 \in \{\Lambda\}_0$ be a r-spherically irregular $(d-1)$-dimensional hypersurface such that $T_t\Lambda_0 \cap \Lambda_0 = \emptyset$ for each $t \neq 0$, where T_t is the translation in the $x^{(1)}$-direction.

If Σ_t is the σ-algebra generated by $\xi(x)$, as given by (4.1) for $x \in T_t\Lambda_0$, and E_t is the conditional expectation with respect to Σ_t and for $f \in L^\infty(\Sigma_0)$

$$P_t = E_0 T_t f,$$

then on $L^\infty(\Sigma_0)$

$$P_t P_s = P_{t+s}$$

for any $t \geq 0, s \geq 0$, and $P_t 1 = 1$.

Remark 4.2. P_t is not symmetric because Λ_0 is not invariant with respect to $(x^{(1)}, ..., x^{(d)}) \to (-x^{(1)}, x^{(2)}, ..., x^{(d)})$, under the conditions of Th. 4.4.

Note

Work on this paper was begun in the middle of the seventies, when the first named author was in Oslo with Raphael Høegh-Krohn, but was not completed at that time. New developments stimulated by the work [ArS] encouraged the completion of the work, which took place after the premature departure of Raphael Høegh-Krohn. The first and third author remember Raphael with great gratitude and in deep sorrows mourn his passing away.

Acknowledgements

The support of the European Science Twinning Project "Stochastic Processes in Analysis" and the SFB 237 (Essen–Bochum–Düsseldorf) during the stay of the third author in Bochum is gratefully acknowledged.

References

[AFHKL] S. Albeverio, J. E. Fenstad, R. Høegh-Krohn, T. Lindstrøm, Non standard methods in stochastic analysis and mathematical physics, Academic Press (1986).

[AHK1] S. Albeverio, R. Høegh-Krohn, Uniqueness and the global Markov property for Euclidean fields, The case of trigonometric interactions, Comm. Math. Phys. 68, 95-128 (1979).

[AHK2] S. Albeverio, R. Høegh-Krohn, Homogeneous random fields and statistical mechanics, J. Funct. Anal., 19, 242-272 (1975).

[AHKO] S. Albeverio, R. Høegh-Krohn, G. Olsen, The global Markov property for lattice fields, J. Multiv. Anal. 11, 599-607 (1981).

[AHKZ] S. Albeverio, R. Høegh-Krohn, B. Zegarlinski, Uniqueness and global Markov property for Euclidean fields: the case of general polynomial interactions, Commun. Math. Phys. 123, 377-424 (1989).

[AIK1] S. Albeverio, K. Iwata, T. Kolsrud, Homogenous Markov generalized vector fields and quantum fields over 4-dimensional space-time, SFB 237 - Preprint, to appear in Proc. Trento, G. Da Prato, L. Tubaro Edts. (1991).

[AIK2] S. Albeverio, K. Iwata, T. Kolsrud, Conformally invariant and reflection positive random fields in two dimensions, Bochum–Stokholm Preprint (1990), to appear in Volume dedicated to M. Zakai, Edt. E. Merzbach, Acad. Press, New York.

[AMR] S. Albeverio, Z. M. Ma, M. Röckner, Dirichlet form and Markov fields - A report on recent developments, to appear in M. Pinsky Edt; "Diffusion Processes and Related Problems in Analysis", Birkhäuser, New York (1990).

[ArS] T. Arak, D. Surgailis, Markov fields with polygonal realizations, Prob. Th. Rel. Fields 80, 543-579 (1989).

[Av] M. B. Averintsev, On a method of describing discrete parameter random fields, Prob. Per. Inf., 6, 100-109 (1970).

[AZ] S. Albeverio, B. Zegarlinski, Global Markov property: results and open problems, this volume.

[BaM] A. Baddeley, J. Møller, Nearest-neighbor Markov point processes and random sets, Int. Stat. J. 57, 89-121 (1989).

[BeHK] J. Bellissard, R Høegh-Krohn, Compactness and the maximal Gibbs state for random Gibbs fields on a lattice, Commun. Math. Phys. 84, 297-327 (1982).

[BoPKh] N.N. Bogolinbon, D.Ya. Petrina, B.I. Khatset, Mathematical description of the equilibrum state of classical systems on the basis of the canonical ensemble formalism, Theor. and Math. Phys. 1, 194 (1969)

[Fö] H. Föllmer, On the global Markov property, in L. Streit Ed. Quantum Fields - Algebras, Processes, Springer, Berlin (1975).

[Ge] H. O. Georgii, Gibbs measures and phase transitions, De Gruyter, Berlin (1988).

[Gi] R. Gielerak, Verification of the global Markov property for strongly coupled trigonometric interactions, these volume.

[I] K. Iwata, On linear maps preserving Markov properties and applications to multicomponent generalized random fields, Ph. D. Thesis, Bochum (1990).

[Ko] T. Kolsrud, Gaussian random fields, infinite dimensional Ornstein-Uhlenbeck processes, and symmetric Markov processes, Acta Appl. Math. 12, 237-263 (1988).

[MaM] V. A. Malyshev, R. A. Minlos, Gibbs random fields. The method of cluster expansions, (in russ.), Nauka, Moscou (1985).

[Ne] E. Nelson, The free Markov field, J. Funct. Anal. 12, 211-217 (1978).

[Os] E. P. Osipov, Markov properties of solutions of stochastic partial differential in a finite volume, this volume.

[Pr] Ch. Preston, Random fields, Lect. Notes Maths. 534, Springer, (1976).

[RiK] B. D. Ripley, F. P. Kelly, Markov point processes, J. London Math. Soc. 15, 188-192 (1877).

[Ro1] Yu. Rozanov, Markov random fields, Springer, Berlin (1982).

[Ro2] Yu. Rozanov, Boundary problems for stochastic partial differential equations, pp. 233-268 in "Stochastic Processes–Mathematics and Physics II", Edts. S. Albeverio , Ph. Blanchard, L. Streit. Lect. Notes Maths.1250, Springer, Berlin (1987).

[Rö] M. Röckner , Specification and Markov boundaries for $P(\varphi)_2$ - random fields, Commun. Math. Phys. 106, 105-135 (1986).

[Ru] D. Ruelle, Statistical mechanics: Rigorous results, Benjamin, New York (1969).

[Sp] F. Spitzer, Introduction au processus de Markov à paramètre dans $\mathbb{Z}_\nu$, Lect. Notes Maths. 390 (1973).

[Su] D. Surgailis, On covariant stochastic differential equations and Markov property of their solutions, Prepr. 144, Univ. Roma (1979).

[Ze] B. Zegarlinski, Extremality and global Markow property I: The Euclidean fields on lattice, J. Multiv. Anal. 21, 158-167 (1987).

A Beurling–Deny type structure theorem for Dirichlet forms on general state spaces

by

Sergio Albeverio*, Zhi-Ming Ma and Michael Röckner*****

* Fakultät für Mathematik
Ruhr-Universität
D–4630 Bochum 1 (Germany)
SFB 237 – Essen – Bochum – Düsseldorf
BiBoS – Bielefeld – Bochum
CERFIM – Locarno

** Institute of Applied Mathematics
Academia Sinica
P.O.Box 2734
Beijing 100080 (China)

*** Institut für angewandte Mathematik
Universität Bonn
Wegelerstraße 6
D–5300 Bonn 1 (Germany)

1. Introduction and main result

The purpose of this paper is to extend the Beurling – Deny structure theorem for symmetric Dirichlet forms (cf.[F80, §2.2, §4.5]) to general (not necessarily locally compact) state spaces. Apart from theoretical interest our work was motivated by applications of the following kind. One is interested in a limit of a sequence of explicitely given Dirichlet forms e.g. on an infinite dimensional (linear) state space. This limit form is in general quite hard to investigate (see e.g. [AHKS80][AKS86] for a special case in the locally compact setting), but a Beurling – Deny type structure theorem can give useful a priori information about it. We start with describing our framework.

Let X be a Hausdorff topological space such that each compact subset is metrizable and such that its Borel σ–algebra $\mathcal{B}(X)$ is countably generated (which both holds if e.g. X is a *Souslin space* i.e. the continuous image of a complete separable metric space). Let m be a positive measure on $\mathcal{B}(X)$ and let $(\mathcal{E}, \mathcal{F})$ be a (symmetric) *Dirichlet form* on (real) $L^2(X; m)$, i.e. $\mathcal{F}$ is a dense linear subspace of $L^2(X; m)$ and $\mathcal{E} : \mathcal{F} \times \mathcal{F} \to I\!R$ is a positive definite symmetric bilinear form satisfiying the following properties:

$\mathcal{F}$ equipped with $\mathcal{E}_1^{\frac{1}{2}}$ (i.e. the norm given by $\mathcal{E}_1$) is complete, where (1.1)

$\mathcal{E}_1 := \mathcal{E} + (\ ,\ )_m$ and $(\ ,\ )_m$ is the usual inner product in $L^2(E;m)$;

for all $u \in \mathcal{F}$, $u^\sharp := (u \vee 0) \wedge 1 \in \mathcal{F}$ and $\mathcal{E}(u^\sharp, u^\sharp) \leq \mathcal{E}(u,u)$. (1.2)

Recall the definition of the capacity associated with $(\mathcal{E},\mathcal{F})$ which is defined for $U \subset X$, U open, by

$$\mathrm{Cap}(U) := \inf\{\mathcal{E}_1(u,u) | u \in \mathcal{F}\,,\ u \geq 1 \quad m \text{ a.e. on } U\} \tag{1.3}$$

and for any $A \subset X$,

$$\mathrm{Cap}(A) := \inf\{\mathrm{Cap}(U) | A \subset U \subset X\,,\ U \text{ open}\}. \tag{1.4}$$

We say a property holds *quasi-everywhere* (abbreviated *q.e.*) if it holds for all $x \in X \backslash N$ where $N \subset X$ with $\mathrm{Cap}(N) = 0$. Note that since $\mathrm{Cap}(A) \geq m(A)$ for all $A \in \mathcal{B}(X)$, any Borel capacity zero set has m-measure zero. A q.e. defined function u on E is called *quasi-continuous* if for each $\varepsilon > 0$ there exists $F \subset X$, F closed, such that $u_{|F}$ (i.e. the restriction of u to F) is (everywhere defined and) continuous and $\mathrm{Cap}(X \backslash F) < \varepsilon$. For details about the above notions in the locally compact case we refer to [F80].
Suppose, in addition, that

$$1 \in \mathcal{F} \tag{1.5}$$

(or equivalently $\mathrm{Cap}(X) < \infty$) which implies $m(X) < \infty$. Then we call $(\mathcal{E},\mathcal{F})$ *quasi-regular*, if

(i) Cap is *tight*, i.e. $\lim \mathrm{Cap}(E \backslash K_n) = 0$ for some compact sets $K_n \subset X$, $n \in I\!N$. (1.6)

(ii) There is an $\mathcal{E}_1^{\frac{1}{2}}$-dense subset $\mathcal{F}_0 \subset \mathcal{F}$ such that each $u \in \mathcal{F}_0$ has a quasi-continuous m-version.

(iii) There exists a set $N \in \mathcal{B}(X)$ of zero capacity and quasi-continuous functions $u_n \in \mathcal{F}$, $n \in I\!N$, such that $\sigma\{u_n | n \in I\!N\} \supset \mathcal{B}(X) \cap (X \backslash N)$ (with $\sigma\{u_n | n \in I\!N\}$ the σ-algebra generated by u_n, $n \in I\!N$)

Remark 1.0. By (1.6) (ii) one proves as in [F80, §3.1] that each $u \in \mathcal{F}$ has a quasi-continuous (m-) version $\tilde{u}$ (cf. [AM90, Propositions 3.2, 3.3] for details) and that $\tilde{u} \geq 0$ q.e. on U if $u \geq 0$ m-q.e. on U where $U \subset X$, U open.

Now we can state the extension of the Beurling–Deny structure theorem:

Theorem 1.1. Let $(\mathcal{E},\mathcal{F})$ be a symmetric Dirichlet form on $L^2(X;m)$ with $1 \in \mathcal{F}$ which is quasi-regular. Then there exists unique $\mathcal{E}^c$, J and k satisfying

(i) $(\mathcal{E}^c,\mathcal{F})$ is a positive definite symmetric bilinear form which is *strongly local*, i.e.

$$\mathcal{E}^c(u,v) = 0 \text{ if } u,v \in \mathcal{F} \text{ with } u = \text{const},\ m\text{-a.e. on a neighbourhood of } \mathrm{supp}[|v| \cdot m]$$

(ii) J is a σ-finite symmetric positive measure on $\mathcal{B}(X \times X \backslash d)$ (where $d :=$ diagonal) such that J does not charge any part of $X \times X \backslash d$ whose projection on the factor X has capacity zero.

(iii) k is a finite positive measure on X which charges no set of capacity zero.

(iv) For all $u, v \in \mathcal{F}$

$$\mathcal{E}(u,v) = \mathcal{E}^c(u,v) + \int_{X \times X \setminus d} (\tilde{u}(x) - \tilde{u}(y))(\tilde{v}(x) - \tilde{v}(y))J(dxdy) + \int \tilde{u}\tilde{v}dk.$$

where $\tilde{u}, \tilde{v}$ are the quasi–continuous (m–) versions of u, v respectively.

We shall prove Theorem 1.1 in the next section.

Remark 1.2. (i) More information about $\mathcal{E}^c$ can be obtained if more structure on X is given (cf. e.g. [F80, Theorem 2.2.2] for the case $X \subset \mathbb{R}^d$, X open.). For corresponding results for more general X we refer to [Sp].

(ii) If X is a locally compact separable metric space and $(\mathcal{E}, \mathcal{F})$ is *regular* in the sense of [F80], then by Proposition 1.3 below $(\mathcal{E}, \mathcal{F})$ is quasi–regular. Hence Theorem 1.1 is indeed an extension of the classical Beurling–Deny theorem.

One can reformulate the notion of "quasi–regularity" without using capacity as follows. For $F \subset X$, F closed, define

$$\mathcal{F}_F := \{u \in \mathcal{F} | u = 0 \ m\text{–a.e. on } X \setminus F\} \tag{1.7}$$

An increasing sequence of closed sets $F_k \subset X$, $k \in \mathbb{N}$, is called *$\mathcal{E}$–nest* if $\bigcup_{k \geq 1} \mathcal{F}_{F_k}$ is $\mathcal{E}_1^{\frac{1}{2}}$–dense in $\mathcal{F}$. A set $N \subset X$ is called *$\mathcal{E}$–exceptional* if there exists an $\mathcal{E}$–nest $(F_k)_{k \in \mathbb{N}}$ such that $N \subset \bigcap_{k \geq 1}(X \setminus F_k)$. A function $u : X \to \mathbb{R}$ is called *$\mathcal{E}$–quasi–continuous* if there exists an $\mathcal{E}$–nest $(F_k)_{k \in \mathbb{N}}$ such that $u_{|F_k}$ is continuous for each $k \in \mathbb{N}$. $(\mathcal{E}, \mathcal{F})$ is called *quasi–regular* if

(i)' There exists an $\mathcal{E}$–nest consisting of compact subsets of X.

(ii)' There exists an $\mathcal{E}_1$–dense subset $\mathcal{F}_0 \subset \mathcal{F}$ such that each $u \in \mathcal{F}_0$ has an $\mathcal{E}$–quasi–continuous version.

(iii)' There exists an $\mathcal{E}$–exceptional set $N \subset X$ and quasi–continuous functions $u_n \in \mathcal{F}$, $n \in \mathbb{N}$, such that

$$\sigma\{u_n | n \in \mathbb{N}\} \supset \mathcal{B}(X) \cap (X \setminus N).$$

We shall prove in section 2 the following

Proposition 1.3. Assume $1 \in \mathcal{F}$ then (i)—(iii) is equivalent with (i)'—(iii)'.

This notion of quasi–regularity i.e. (i)'—(iii)', is natural since by [AM90] it is necessary and sufficient for the existence of a special standard process associated with $(\mathcal{E}, \mathcal{F})$.

Remark 1.4. If $1 \notin \mathcal{F}$ we can always consider the Dirichlet form $(\mathcal{E}_h, \mathcal{F}_h)$ on $L^2(X; h^2 \cdot m)$ obtained by h–transformation for some q.e. strictly positive 1–excessive function $h \in \mathcal{F}$ i.e.

$$\mathcal{F}_h := \{u \in L^2(X; h^2 \cdot m) | \, u \cdot h \in \mathcal{F}\}$$

$$\mathcal{E}_h(u,v) = \mathcal{E}(uh, vh) \quad ; \ u, v \in \mathcal{F}_h.$$

(for earlier use of h–transformations in classical potential theory on locally compact spaces see e.g. [Do84]). Then $1 \in \mathcal{F}_h$ and $(\mathcal{E}_h, \mathcal{F}_h)$ is quasi–regular, provided $(\mathcal{E}, \mathcal{F})$ satisfies (i)'—(iii)', and we can apply Theorem 1.1 to $(\mathcal{E}_h, \mathcal{F}_h)$. Thus we also obtain structural information about $(\mathcal{E}, \mathcal{F})$.

In the course of the proof of Theorem 1.1 we develop a compactification method extending those in [F71],[AHK77],[K82],[AR89a,b,90a] and [T90]. It allows to transfer all results on regular Dirichlet forms on locally compact spaces (cf.[F80]) to the quasi–regular Dirichlet form $(\mathcal{E}, \mathcal{F})$ on $L^2(X; m)$ as above. In particular, the fundamental correspondence between *additive functionals* of the process associated with $(\mathcal{E}, \mathcal{F})$ and *smooth measures* of $(\mathcal{E}, \mathcal{F})$ still holds.

2. Proofs

Proof of Theorem 1.1.
First we prove existence. We do this in three steps.

Step 1. Construction of a suitable compactification.
Since $L^2(X; m)$ is separable, $\mathcal{F}$ equipped with $\mathcal{E}_1^{\frac{1}{2}}$ is separable. Thus by (1.6)(iii) and Remark 1.0 there exists a countable $\mathbb{Q}$-algebra (i.e. an algebra over the rationals) $\mathcal{D} := \{g_n | n \in \mathbb{N}\} \subset \mathcal{F}$ containing 1 which is $\mathcal{E}_1^{\frac{1}{2}}$-dense in $\mathcal{F}$ such that each g_n is bounded and quasi–continuous and such that for some $N \in \mathcal{B}(X)$ of capacity zero,

$$\sigma\{g_n | n \in \mathbb{N}\} \supset \mathcal{B}(X) \cap (X \backslash N). \tag{2.1}$$

By (1.6)(i) and a diagonal argument (cf. [F80,Theorem 3.1.2]) we can construct an increasing sequence of compact sets $F_k \subset X \backslash N$, $k \in \mathbb{N}$, such that $g_{n|F_k}$ is continuous for all $n, k \in \mathbb{N}$ and $\mathrm{Cap}(X \backslash F_k) < \frac{1}{k}$. Set

$$Y := \bigcup_{k \geq 1} F_k \tag{2.2}$$

and

$$f_n := \frac{2}{\pi} \arctan g_n, \; n \in \mathbb{N},$$

Since Y is a Lusin space (cf.[Sch73]) we easily obtain by (2.1) that f_n, $n \in \mathbb{N}$, separate the points of Y. Hence we can define a metric

$$\rho(x, y) := \sum_{n=1}^{\infty} \frac{1}{2^n} |f_n(x) - f_n(y)| \; ; \; x, y \in Y. \tag{2.3}$$

Observe that since (Y, ρ) is isometric to a subspace of $[-1, 1]^{\mathbb{N}}$, the completion $(\hat{Y}, \rho)$ is a compact metric space, i.e $\hat{Y}$ is a *compactification* of Y. Each g_n has a unique continuous extension $\hat{g}_n$ to $\hat{Y}$ and,

for each compact subset of Y the trace topologies inherited from $Y, \hat{Y}$ respectively, coincide. (2.4)

Furthermore,

$$\mathcal{B}(Y) = \{A \in \mathcal{B}(\hat{Y}) | A \subset Y\} \tag{2.5}$$

since Y is a Lusin space.

Step 2. Image Dirichlet form.

Let $\hat{m}$ be the image measure on $\mathcal{B}(\hat{Y})$ of m under the embedding map $i : Y \to \hat{Y}$. Define the image Dirichlet form $(\hat{\mathcal{E}}, \hat{\mathcal{F}})$ (under i) as follows

$$\hat{\mathcal{F}} := \{\hat{u} \in L^2(\hat{Y}; \hat{m}) | \hat{u} \circ i \in \mathcal{F}\}$$

and

$$\hat{\mathcal{E}}(\hat{u}, \hat{v}) := \mathcal{E}(\hat{u} \circ i, \hat{v} \circ i) ; \hat{u}, \hat{v} \in \hat{\mathcal{F}},$$

where we consider each function u on Y as a function on X by setting $u \equiv 0$ on $X \backslash Y$. Note that for a function $u \in \mathcal{F}$ its values on $X \backslash Y$ are irrelevant since $X \backslash Y$ is of capacity zero. Obviously, $(\hat{\mathcal{E}}, \hat{\mathcal{F}})$ is a Dirichlet form on $L^2(\hat{Y}; \hat{m})$ and $\hat{D} := \{\hat{g}_n | n \in \mathbb{N}\}$ is $\hat{\mathcal{E}}_1^{\frac{1}{2}}$-dense in $\hat{\mathcal{F}}$; here $\hat{\mathcal{E}}_1 := \hat{\mathcal{E}} + (\,,\,)_{\hat{m}}$. Since $\hat{D}$ is a $\mathbb{Q}$-algebra of continuous functions on $\hat{Y}$ containing 1 and separating the points of $\hat{Y}$, $\hat{D}$ is dense in the space of all continuous functions $C(\hat{Y})$ on $\hat{Y}$ w.r.t. uniform norm. This means that $(\hat{\mathcal{E}}, \hat{\mathcal{F}})$ is a regular Dirichlet form on $L^2(\hat{Y}; \hat{m})$ in the sense of [F80]. Denote the associated capacity (analogously defined as Cap above) by Câp. By [AFHMR90] (see also [AR90b]) it follows (since Cap is tight) that

$$\widehat{\mathrm{Cap}}(A) = \mathrm{Cap}(A \cap Y) \text{ for all } A \subset \hat{Y}. \tag{2.6}$$

In particular,

$$\widehat{\mathrm{Cap}}(\hat{Y} \backslash Y) = 0. \tag{2.7}$$

For notational convenience from now on we consider any function u on Y as a function on $\hat{Y}$ by setting it equal to zero on $\hat{Y} \backslash Y$ (as we do it with X as mentioned above). Then since Cap is tight and by (2.4), (2.6)

the notion of quasicontinuity w.r.t. Cap, Câp respectively, coincide. (2.8)

Step 3. Existence of $\mathcal{E}^c$, J and k.

By the classical Beurling–Deny formula for regular Dirichlet forms on locally compact state space (cf.[F80, Theorem 2.2.1, Lemma 4.5.4]) there exists $\hat{\mathcal{E}}^c$, $\hat{J}$ and $\hat{k}$ satisfying (i)—(iv) in Theorem 1.1 with $(\hat{\mathcal{E}}, \hat{\mathcal{F}}), \hat{Y}$ replacing $(\mathcal{E}, \mathcal{F}), X$ respectively. J is even finite on every compact subset of $\hat{Y} \times \hat{Y} \backslash d$. (Note that the assumption supp $\hat{m} = \hat{Y}$ made in [F80] is superfluous and can be dropped). Define

$$J(A) := \hat{J}(A \cap (Y \times Y)) \qquad , A \in \mathcal{B}(X \times X \backslash d) \tag{2.9}$$

$$k(A) := \hat{k}(A \cap Y) \qquad , A \in \mathcal{B}(X) \tag{2.10}$$

$$\mathcal{E}^c(u, v) := \mathcal{E}(u, v) - \int_{X \times X \backslash d} (\tilde{u}(x) - \tilde{u}(y))(\tilde{v}(x) - \tilde{v}(y)) J(dxdy)$$
$$- \int_X \tilde{u}(x)\tilde{v}(x)k(dx) \qquad u, v \in \mathcal{F}, \tag{2.11}$$

where $\tilde{u}, \tilde{v}$ are the quasi–continuous versions of u, v respectively. Note that by construction the integrals in (2.11) are well–defined. So are also the right hand sides of (2.9), (2.10) since $\mathcal{B}(X)$ is countably generated and Y is a Lusin space, hence $\mathcal{B}(X) \cap Y = \mathcal{B}(Y)$. Then from (2.6), (2.7) we see that (ii), (iii) in Theorem 1.1 hold for J, k. (iv) is satisfied by the definition of $\mathcal{E}^c$. So, it remains to prove that $\mathcal{E}^c$ satisfies (i) in Theorem 1.1. It follows by the definition of $\mathcal{E}^c$ and by (2.7) that

$$\mathcal{E}^c(u,v) = \hat{\mathcal{E}}^c(u,v) \quad \text{for all } u, v \in \mathcal{F}, \tag{2.12}$$

in particular $(\mathcal{E}^c, \mathcal{F})$ is a positive definite symmetric bilinear form on $L^2(X; m)$. To check strong locality for $\mathcal{E}^c$, note that if $u, v \in \mathcal{F}$ with u equals a constant c m–a.e. on a neighbourhood of supp$[|v| \cdot m]$ then

$$\begin{aligned} \mathcal{E}^c(u,v) &= \mathcal{E}^c(u-c, v) + \hat{\mathcal{E}}^c(c, v) \\ &= \mathcal{E}^c(u-c, v). \end{aligned}$$

So, we only need to consider the case $u = 0$ m–a.e. on a neighbourhood of supp$[|v| \cdot m]$. We may also assume that u is bounded.

Let h_k be the $\mathcal{E}_1$–orthogonal projection of $1 \in \mathcal{F}$ onto $\mathcal{F}_{F_k}$, (cf.(1.7), (2.2)), $n \in \mathbb{N}$. Since $h_k \to 1$ w.r.t. $\mathcal{E}_1$ and $|h_k| \le 1$, $k \in \mathbb{N}$, it follows from [F80, Theorem 1.4.2 (ii)] that $\sup_{k \in \mathbb{N}} \mathcal{E}_1(uh_k, uh_k) < \infty$. Hence we can find a subsequence $(k_n)_{n \in \mathbb{N}}$ such that

$$u_j := \left(\frac{1}{j} \sum_{n=1}^{j} h_{k_n} \right) \cdot u \qquad , j \in \mathbb{N}, \tag{2.13}$$

converges to u w.r.t. $\mathcal{E}_1$ and hence w.r.t. $\mathcal{E}^c$.
Consequently,

$$\lim_{j \to \infty} \mathcal{E}^c(u_j, v) = \mathcal{E}^c(u, v). \tag{2.14}$$

Setting $\tilde{F}_j := \bigcup_{n=1}^{j} F_{k_n}$ we know that $u_j = 0$ $\hat{m}$–a.e. on $\hat{Y} \backslash \tilde{F}_j$. Since the two relative topologies (induced by Y and $\hat{Y}$) coincide on $\hat{F}_j$, we thus see that $u_j = 0$ $\hat{m}$–a.e. on a $\hat{Y}$–neighbourhood of $\widehat{\text{supp}}[|v| \cdot \hat{m}]$ where $\widehat{\text{supp}}$ is w.r.t. the topology on $\hat{Y}$. Consequently,

$$\mathcal{E}^c(u_j, v) = \hat{\mathcal{E}}^c(u_j, v) = 0 \text{ for all } j \in \mathbb{N}$$

and hence by (2.14)

$$\mathcal{E}^c(u,v) = 0. \tag{2.15}$$

This shows strong locality for $\mathcal{E}^c$ and completes the existence proof.

Now we prove uniqueness. So, let $\bar{\mathcal{E}}^c$, $\bar{J}$, $\bar{k}$ be such that they also fulfill (i)—(iv) in Theorem 1.1. Since $\mathcal{E}(u,1) = \int \tilde{u} dk = \int \tilde{u} d\bar{k}$ for all $u \in \mathcal{F}$ by (iv) and strong locality, we can conclude by (2.1) and the monotone class theorem that $k = \bar{k}$ on $\mathcal{B}(X) \cap (X \backslash N)$. But $k(N) = \bar{k}(N) = 0$ hence $k = \bar{k}$ on $\mathcal{B}(X)$. To prove $J = \bar{J}$, let $u, v \in \mathcal{F}$, u, v bounded, such that $\text{supp}[|u| \cdot m] \cap \text{supp}[|v| \cdot m] = \emptyset$. Then by Remark 1.0, $\tilde{u} \cdot \tilde{v} = 0$ q.e., hence by (iv) in Theorem 1.1

$$\int_{Y \times Y \backslash d} \tilde{u}(x)\tilde{v}(y)J(dxdy) = \int_{Y \times Y \backslash d} \tilde{u}(x)\tilde{v}(y)\bar{J}(dxdy). \tag{2.16}$$

Since (2.16) in particular holds for all $u, v \in \hat{\mathcal{F}} \cap C(\hat{Y})$ with $\widehat{\text{supp}}\, u \cap \widehat{\text{supp}}\, v = \emptyset$ which form a dense subset of $C(\hat{Y})$ w.r.t. uniform norm, it follows that $J = \bar{J}$ on $\mathcal{B}(Y \times Y \backslash d)$, hence $J = \bar{J}$ by property (ii) in Theorem 1.1. Consequently by property (iv), $\mathcal{E}^c = \bar{\mathcal{E}}^c$ and we have shown uniqueness. ■

We now turn to the proof of Proposition 1.3 which is in fact an obvious consequence of the following Lemma, in whose proof we use the terminology of [F80].

Lemma 2.1. Suppose that $1 \in \mathcal{F}$. Then an increasing sequence $(F_k)_{k \in \mathbb{N}}$ of closed sets in X, is an $\mathcal{E}$-nest if and only if $lim_{k \to \infty}$ $\text{Cap}(X \backslash F_k) = 0$.

Proof. Let F_k, $k \in \mathbb{N}$, be an increasing sequence of closed sets. Let e_k be the *1-equilibrium potential* of $X \backslash F_k$, i.e. e_k is the unique 1-excessive function in $\mathcal{F}$ such that $e_k = 1$ m–a.e. on $X \backslash F_k$ and

$$\mathcal{E}_1(e_k, u) = 0\,, \quad \forall \quad u \in \mathcal{F}_{F_k}. \tag{2.17}$$

It is easy to check from (2.17) that e_k, $k \in \mathbb{N}$, is an $\mathcal{E}_1$–Cauchy sequence whose limit we denote by e_∞. It is clear that

$$\mathcal{E}_1(e_\infty, u) = 0\,, \quad \forall \quad u \in \bigcup_{k \geq 1} \mathcal{F}_{F_k}. \tag{2.18}$$

Suppose now that $(F_k)_{k \in \mathbb{N}}$ is an $\mathcal{E}$–nest. Then (2.18) implies that $\mathcal{E}_1(e_\infty, e_\infty) = 0$ and consequently

$$\lim_{k \to \infty} \text{Cap}(X \backslash F_k) = \lim_{k \to \infty} \mathcal{E}_1(e_k, e_k) = \mathcal{E}_1(e_\infty, e_\infty) = 0.$$

Conversely, suppose that $\lim_{k \to \infty} \text{Cap}(X \backslash F_k) = 0$. Let h be an arbitrary 1–excessive function in $\mathcal{F}$ such that $h \leq 1$ m–a.e.. Let h_k be the 1–*reduced function* of h on $X \backslash F_k$. i.e., h_k is the unique 1–excessive function in $\mathcal{F}$ with the properties that $h_k = h$ m–a.e. on $X \backslash F_k$ and

$$\mathcal{E}_1(h_k, u) = 0\,, \quad \forall \quad u \in \mathcal{F}_{F_k}. \tag{2.19}$$

We have, since $h_k \leq e_k$ m–a.e. and both of them are 1–excessive,

$$0 \leq \mathcal{E}_1(h_k, h_k) \leq \mathcal{E}_1(e_k, e_k) = \text{Cap}(X \backslash F_k) \to 0 \quad \text{as } k \to \infty.$$

Hence $(h - h_k)_{k \in \mathbb{N}}$ converges to h w.r.t $\mathcal{E}_1^{\frac{1}{2}}$, i.e. h is an $\mathcal{E}_1^{\frac{1}{2}}$–limit of elements of $\bigcup_{k \geq 1} \mathcal{F}_{F_k}$, which in turn implies that $\bigcup_{k \geq 1} \mathcal{F}_{F_k}$ is $\mathcal{E}_1^{\frac{1}{2}}$–dense in $\mathcal{F}$. ■

Remark 2.2. There is a more general version of Lemma 2.1 which also applies if $1 \notin \mathcal{F}$ (see [AM90, Proposition 2.5]).

Acknowledgement. Kind invitations by Professor T.Hida, M.Fukushima, S.Watanabe and H.Watanabe to stays in Japan which greatly stimulated this work are gratefully acknowledged. In particular we thank M.Takeda for very interesting discussions related to this work. We also like to thank M.Fukushima and T.S.Zhang for their steady interest during the preparation of this paper.

References

[AFHKL86] Albeverio, S., Fenstad, J.E., Høegh-Krohn, R., Lindstrøm, T.: Nonstandard Methods in Stochastic Analysis and Mathematical Physics, Academic Press, Orlando (1986)

[AFHMR90] Albeverio, S., Fukushima, M., Hansen, W., Ma, Z.M., Röckner, M.: In preparation

[AHK77] Albeverio, S., Høegh-Krohn, R.: Hunt processes and analytic potential theory on rigged Hilbert spaces, Ann. Inst. Henri Poincaré B13, 269–291 (1977)

[AHKS80] Albeverio, S., Høegh-Krohn, R., Streit, L.: Regularization of Hamiltonians and processes, J. Math. Phys. 21, 1636–1642 (1980)

[AKS86] Albeverio, S., Kusuoka, S., Streit, L.: Convergence of Dirichlet forms and associated Schrödinger operators, J. Funct. An. 68, 130–148 (1986)

[AM90] Albeverio, S., Ma, Z.M.: Necessary and sufficient conditions for the existence of m-perfect processes associated with Dirichlet forms. BiBoS-preprint 427 (1990).

[AR88] Albeverio, S., Röckner, M.: New developments in theory and applications of Dirichlet forms, Proc. "Ascona July 1988", Edts. S.Albeverio, G.Casati, U.Cattaneo, D.Merlini, R. Moresi, World Scient. Singapore (1990)

[AR89a] Albeverio, S., Röckner, M.: Classical Dirichlet forms on topological vector spaces — construction of an associated diffusion process. Prob. Th. Rel. Fields 83, 405–434 (1989).

[AR89b] Albeverio, S., Röckner, M.: Stochastic differential equations in infinite dimensions: solutions via Dirichlet forms. Preprint Edinburgh 1989. Publication in preparation.

[AR89c] Albeverio, S., Röckner, M.: Dirichlet forms, quantum fields and stochastic quantization. Stochastic analysis, path integration and dynamics. Research Notes in Math. Vol. 200, 1–21. Editors: K.D. Elworthy, J.C.Zambrini. Harlow: Longman 1989.

[AR90a] Albeverio, S., Röckner, M.: Classical Dirichlet forms on topological vector spaces — closability and a Cameron–Martin formula. J. Funct. Anal. 88, 395–436 (1990).

[AR90b] Albeverio, S., Röckner, M.: Infinite dimensional diffusions connected with positive generalized white noise functionals. pp 1–21 in: White noise analysis. Editors: T.Hida, H.H.Kuo, J.Potthoff, L.Streit, World Scient. Singapore (1990).

[Do84] Doob, J.L.: Classical Potential Theory and Its Probabilistic Counterpart, Springer New York (1984).

[F71] Fukushima, M.: Regular representations of Dirichlet forms. Trans. Amer. Math. Soc. 155, 455–473 (1971).

[F80] Fukushima, M.: Dirichlet forms and Markov processes. Amsterdam–Oxford–New York: North Holland 1980.

[K82] Kusuoka, S.: Dirichlet forms and diffusion processes on Banach spaces, J. Fac. Sci. Univ. Tokyo Sect. IA 29, 79–95 (1982).

[Sch73] Schwartz, L.: Radon measures on arbitrary topological spaces and cylindrical measures, Oxford Univ. Press, London (1973).

[Sp] Spönemann, U.: Dissertation. Publication in preparation.
[T90] Takeda, M.: The maximum Markovian self–adjoint extensions of generalized Schrödinger operators. Preprint (1990).

Log-Concavity of Radial Schrödinger Wave Functions and Convergence of Planetesimal Diffusions

by

Richard M. Durran and Aubrey Truman
Department of Mathematics and Computer Science
University College of Swansea
Singleton Park
Swansea,   SA2 8PP

Introduction

In this paper we give a mathematical model for the condensation of planets out of a protosolar nebula. We discuss models in the setting of diffusion theory and, in particular, Nelson's stochastic mechanics [13], [14] i.e. the stochastic description of quantum mechanics. Albeverio, Hoegh-Krohn and Blanchard have already discussed this problem in the context of the stochastic mechanics of the hydrogen atom, giving a possible mechanism for explaining the origin of the Titius-Bode law [1]. In [5] and [6] we also addressed this problem in detail when the interacting potential between the cloud nebula and the planetesimal corresponds to a point mass at the origin. There a model was constructed (from the $\psi_{n,\, n-1,\, n-1}$ orbital of the hydrogen atom) in which in the infinite time limit the planets condense out of the nebula describing coplanar circular orbits (with centre at the origin) at a rate consistent with Kepler's third law of planetary motion.

Our aim in this present work is to generalize those results to include a wider class of potentials corresponding to different cloud densities. For example the radial harmonic oscillator potential corresponds to a cloud with uniform density. We try to model the collisions of dust particles of the nebula with the planetesimal by using a diffusion process. Due to these collisions the planetesimal accrues mass and exchanges energy and angular momentum with the nebula. The statistics of the collisions and the accretion of mass in our model is governed by ε, the diffusion coefficient in our equations. We require our diffusion models to obey certain dynamical laws which are the natural generalizations of the classical laws of dynamics. In the next section we discuss the types of diffusions which we must consider in this setting. We shall see that the Schrödinger equation plays a very important role. Indeed we establish some results of independent interest on the log concavity of solutions of the Schrödinger equation in section 4.

1. Planetesimal Diffusions

Kinematics of Diffusions

Let $\underset{\sim}{X}$ be the position of a planetesimal in $\mathbb{R}^3$ which we assume satisfies an Itô stochastic differential equation of the form

$$d\underset{\sim}{X}(t) = \underset{\sim}{b}(\underset{\sim}{X}(t),t)\,dt + \varepsilon\, d\underset{\sim}{B}(t) \quad , \quad t > s \quad , \quad \underset{\sim}{X}(s) = \underset{\sim}{x} \quad , \tag{1}$$

where $\underset{\sim}{B}$ is a BM($\mathbb{R}^3$) process and ε is for the time being a positive constant. We assume that the process has a smooth density $\rho\,(\underset{\sim}{y}, t)$ satisfying the forward Kolmogorov equation

$$\frac{\partial}{\partial t}\,\rho(\underset{\sim}{y}, t) = \operatorname{div}\left\{\frac{\varepsilon^2}{2}\underset{\sim}{\nabla}\,\rho(\underset{\sim}{y}, t) - \underset{\sim}{b}(\underset{\sim}{y}, t)\,\rho(\underset{\sim}{y}, t)\right\} . \tag{2}$$

As is well known, the sample paths of $\underset{\sim}{X}$ are almost surely nowhere differentiable in the usual sense. In order to describe the kinematics of the diffusion we must therefore give a reasonable notion of derivative. Following Nelson we define the mean forward and backward derivatives, $D_{\pm}$, by

$$D_{\pm}\, f(\underset{\sim}{X}(t),t) = \lim_{h\downarrow o} \mathbb{E}\left\{\frac{f(\underset{\sim}{X}(t \pm h), t \pm h) - f(\underset{\sim}{X}(t),t)}{\pm h} \,\middle|\, \underset{\sim}{X}(t)\right\} . \tag{3}$$

Given mild regularity conditions on f and $\underset{\sim}{b}$, a straight-forward application of Itô's formula gives the explicit expression

$$D_{+}\, f(\underset{\sim}{X}(t),t) = \left\{\frac{\partial}{\partial t} + \underset{\sim}{b}(\underset{\sim}{X}(t),t)\,.\underset{\sim}{\nabla} + \frac{\varepsilon^2}{2}\Delta\right\} f(\underset{\sim}{X}(t),t) \quad , \tag{4}$$

from which one easily obtains $D_{+}\underset{\sim}{X}(t) = \underset{\sim}{b}(\underset{\sim}{X}(t),t) = \underset{\sim}{b}_{+}(\underset{\sim}{X}(t),t)$, the **forward drift.**

A similar expression to (4) exists for $D_- f$. We omit the details save to say that using the identity (see e.g. [14])

$$\frac{d}{dt}\mathbb{E}\left\{ g(\underset{\sim}{X}(t),t)\ h(\underset{\sim}{X}(t),t)\right\} = \mathbb{E}\left\{ g(\underset{\sim}{X}(t),t)\ D_-\ h(\underset{\sim}{X}(t),t)\right\} + \mathbb{E}\left\{ h(\underset{\sim}{X}(t),t)\ D_+\ g(\underset{\sim}{X}(t),t)\right\} \tag{5}$$

and integration by parts we deduce

$$D_- f(\underset{\sim}{X}(t),t) = \left\{\frac{\partial}{\partial t} + [\underset{\sim}{b}_+(\underset{\sim}{X}(t),t) - \varepsilon^2\ \underset{\sim}{\nabla}\ \log \rho(\underset{\sim}{X}(t),t)].\underset{\sim}{\nabla} - \frac{\varepsilon^2}{2}\Delta\right\} f(\underset{\sim}{X}(t),t)\ . \tag{6}$$

In particular, setting $f(\underset{\sim}{X}(t),t) = \underset{\sim}{X}(t)$, we obtain

$$D_- \underset{\sim}{X}(t) = \underset{\sim}{b}_+(\underset{\sim}{X}(t),t) - \varepsilon^2\ \underset{\sim}{\nabla}\ \log \rho(\underset{\sim}{X}(t),t) = \underset{\sim}{b}_-(\underset{\sim}{X}(t),t), \tag{7}$$

the **backward drift**.

$\underset{\sim}{b}_{\pm}$ embody the kinematics of the diffusion. For further details see [7].

2. Dynamics of Diffusions

Following Nelson, we define the stochastic acceleration, $\underset{\sim}{a}(\underset{\sim}{X}(t),t)$, by

$$\underset{\sim}{a}(\underset{\sim}{X}(t),t) = \frac{1}{2}(D_+ D_- + D_- D_+)\ \underset{\sim}{X}(t)\ . \tag{8}$$

Simple manipulations using (4) and (6) lead to the expression

$$\underset{\sim}{a}(\underset{\sim}{X}(t),t) = \left\{\frac{\partial}{\partial t}\underset{\sim}{v} + (\underset{\sim}{v}.\underset{\sim}{\nabla})\ \underset{\sim}{v} - (\underset{\sim}{u}.\underset{\sim}{\nabla})\ \underset{\sim}{u} - \frac{\varepsilon^2}{2}\Delta\underset{\sim}{u}\right\}(\underset{\sim}{X}(t),t)\ , \tag{9}$$

where $\underset{\sim}{v}$ and $\underset{\sim}{u}$ are respectively the current and osmotic velocities defined in terms of $\underset{\sim}{b}_{\pm}$ by

$$\underset{\sim}{v}(\underset{\sim}{X}(t),t) = \frac{1}{2}(\underset{\sim}{b}_+ + \underset{\sim}{b}_-)(\underset{\sim}{X}(t),t) = \underset{\sim}{b}_+(\underset{\sim}{X}(t),t) - \frac{\varepsilon^2}{2}\underset{\sim}{\nabla}\log\rho(\underset{\sim}{X}(t),t) \quad , \tag{10}$$

$$\underset{\sim}{u}(\underset{\sim}{X}(t),t) = \frac{1}{2}(\underset{\sim}{b}_+ - \underset{\sim}{b}_-)(\underset{\sim}{X}(t),t) = \frac{\varepsilon^2}{2}\underset{\sim}{\nabla}\log\rho(\underset{\sim}{X}(t),t) \quad . \tag{11}$$

The motivation for this definition arises from its connection with the Schrödinger equation.

To see this suppose that the density ρ is positive so that for some real R, $\rho(\underset{\sim}{X},t) = \exp\{2R(\underset{\sim}{X},t)\}$. Assume also that $\underset{\sim}{b}$ is a gradient so that for some real S, $\underset{\sim}{b}(\underset{\sim}{X},t) = \varepsilon^2\,\underset{\sim}{\nabla}\,(R+S)(\underset{\sim}{X},t)$. Under these assumptions and the further requirement that Newton's second law,

$$\text{Force} = \text{Mass x Acceleration}, \tag{12}$$

be valid in mean, equations (2) and (8) are equivalent to the coupled p.d.e.'s

$$\varepsilon^2\frac{\partial R}{\partial t} = -\frac{1}{2}\varepsilon^4\,\Delta S - \varepsilon^4\,\underset{\sim}{\nabla}R.\underset{\sim}{\nabla}S \tag{13}$$

$$m\left\{\varepsilon^2\frac{\partial \underset{\sim}{\nabla}S}{\partial t} + \frac{1}{2}\varepsilon^4\underset{\sim}{\nabla}(|\underset{\sim}{\nabla}S|^2 - |\underset{\sim}{\nabla}R|^2 - \Delta R)\right\} = -\underset{\sim}{\nabla}V \quad , \tag{14}$$

where the force is assumed to be minus the gradient of some potential V. Now define the complex valued function, Ψ , by $\Psi(\underset{\sim}{x},t) = \exp\{R(\underset{\sim}{x},t) + iS(\underset{\sim}{x},t)\}$. Then it is not too difficult to show using (13) and (14) that Ψ satisfies

$$i\varepsilon^2\frac{\partial\Psi}{\partial t}(\underset{\sim}{x},t) = -\frac{1}{2}\varepsilon^4\,\Delta\Psi(\underset{\sim}{x},t) + \frac{V}{m}\Psi(\underset{\sim}{x},t) \quad . \tag{15}$$

If we set $\varepsilon^2 = h/m$ this is of course precisely the Schrödinger equation as obtained by Nelson in his formulation of stochastic mechanics. (Here h is the reduced Planck constant.)

Remarks

The density of the diffusion corresponds exactly with the quantum mechanical particle density, i.e. $\rho(\underset{\sim}{X},t) = |\Psi(\underset{\sim}{X},t)|^2$.

To model the situation in which the planetesimal condenses out of the nebula we assume that ε, m and other parameters, such as angular momentum are functions of the time, with $\varepsilon(t) \to 0$ as $t \uparrow \infty$. It is possible to make this assumption at the outset. If in the above we assume $\varepsilon = \varepsilon(t)$, $m = m(t)$, to be deterministic functions of the time, and assume the dynamical law for time-dependent mass:

$$m(t)\frac{1}{2}(D_+ D_- + D_- D_+)\underset{\sim}{X}(t) + \dot{m}(t)\ \frac{1}{2}(D_+ + D_-)\underset{\sim}{X}(t) = -\underset{\sim}{\nabla}\ V(\underset{\sim}{X}(t),t)\ , \qquad (16)$$

everything goes through except that we obtain the following equation for Ψ:

$$i\varepsilon^2\frac{\partial\Psi}{\partial t}(\underset{\sim}{x},t) = -\frac{1}{2}\varepsilon^4(t)\ \Delta\Psi(\underset{\sim}{x},t) + \frac{V(\underset{\sim}{x},t)}{m(t)}\ \Psi(\underset{\sim}{x},t)$$

$$+\frac{i}{2m(t)}\frac{d}{dt}(m(t)\ \varepsilon^2(t))\ \Psi(\underset{\sim}{x},t)\log\frac{\Psi^*(\underset{\sim}{x},t)}{\Psi(\underset{\sim}{x},t)}\ , \qquad (17)$$

where * denotes complex conjugation. This non-linear equation is similar to the "Schrödinger equation" for dissipative forces, except for the time-dependent coefficients. (See for example [11] and [16].) Naturally it reduces to (15) when ε and m are constants. It is however very difficult to solve for time-dependent ε and m.

We shall assume that the time-dependent parameters ε and m are deterministic, continuous, **piecewise constant** functions of the time. When these parameters are all constant, we think of the collisions between the particles making up the nebula and the planetesimal as being non-violent. Interspersed with these non-violent collision phases there will be violent collision phases during which the parameters are rapidly varying functions of time. We shall assume however that the variation is such that in the non-violent phases the density (of the diffusion with time-dependent parameters) reduces to the appropriate quantum mechanical density, for which the parameters are constant. In these non-violent phases during which $\dot{m} = \dot{\varepsilon} = 0$ the analysis below shows that the Nelson-Newton law is valid. An illustrative example follows.

Example

Consider the orbital stationary state solution of the 3-dimensional harmonic oscillator corresponding to the potential, $V(\underset{\sim}{x}) = 2^{-1}\, m\, |\underset{\sim}{x}|^2$. The corresponding radial diffusion on $(0, \infty)$ satisfies the stochastic differential equation

$$dR(t) = \left(\frac{\alpha\varepsilon^2}{R(t)} - R(t)\right)dt + \varepsilon dB(t) \quad , \quad t > s \quad , \quad R(s) = x \quad , \tag{18}$$

where B is a BM($\mathbb{R}$) process, $\varepsilon(= h/m)$ is the diffusion coefficient and α is related to the quantum number n by $\alpha = n-1$ $(n \geq 1)$. (Here we may choose it to be any positive constant though.) The appropriate quantum mechanical density is

$$\rho(x,t) = \frac{2}{\Gamma(\alpha+\frac{1}{2})}\left(\frac{1}{\varepsilon^2}\right)^{\alpha+\frac{1}{2}} x^{2\alpha} \exp\left\{-\frac{x^2}{\varepsilon^2}\right\} . \tag{19}$$

Now assume ε is time-dependent of the form

$$\varepsilon(t) = \begin{cases} \varepsilon_{2n}, \text{ a constant, for } t_{2n} \leq t \leq t_{2n+1} , \\ \varepsilon_{2n+1}(t), \text{ for } t_{2n+1} \leq t \leq t_{2n+2}, \text{ with} \end{cases}$$

$$\varepsilon_{2n+1}(t_{2n+1}) = \varepsilon_{2n} , \quad \varepsilon_{2n+1}(t_{2n+2}) = \varepsilon_{2n+2} \quad , \tag{20}$$

where $s = t_0 < t_1 < t_2 < \ldots.$ Here the non-violent phases are (t_{2n}, t_{2n+1}) and violent collision phases (t_{2n+1}, t_{2n+2}). The transition density for this process is, for each $t > s$,

$$\rho(x,s;y,t) = \frac{1}{T}\left(\frac{y}{x}\right)^{\alpha+\frac{1}{2}} \exp\left\{(\alpha - \tfrac{1}{2})(t-s) - \frac{1}{2T}(y^2 + x^2 e^{-2(t-s)})\right\} I_{\alpha-\frac{1}{2}}\left(xy\frac{e^{-(t-s)}}{T}\right) , \tag{21}$$

where $T = e^{-2t} \int_s^t \varepsilon^2(u)\, e^{2u}\, du$ and $I_{\alpha-\frac{1}{2}}$ is the modified Bessel function of order $\alpha - \frac{1}{2}$.

For the initial density at time s, ρ_{ε_0}, defined by

$$\rho_{\varepsilon_0}(x) = \frac{2}{\Gamma(\alpha+\frac{1}{2})}\left(\frac{1}{\varepsilon_0^2}\right)^{\alpha+\frac{1}{2}} x^{2\alpha} \exp\left\{-\frac{x^2}{\varepsilon_0^2}\right\} , \tag{22}$$

with $\varepsilon(s) = \varepsilon_0$, the density at time $t > s$ is $\rho(y,t)$ where

$$\rho(y,t) = \frac{2}{\Gamma(\alpha+\frac{1}{2})}\left\{2\, e^{-2t} \int_s^t \varepsilon^2(u) e^{2u}\, du + \varepsilon_0^2\, e^{-2(t-s)}\right\}^{-\alpha-\frac{1}{2}}$$

$$y^{2\alpha} \exp\left\{\frac{-y^2}{2\, e^{-2t} \int_s^t \varepsilon^2(u)\, e^{2u}\, du + \varepsilon_0^2\, e^{-2(t-s)}}\right\} . \tag{23}$$

This density reduces to the corresponding quantum mechanical density on $[t_{2n}, t_{2n+1}]$, where ε is constant, if and only if,

$$2\, e^{-2t} \int_s^t \varepsilon^2(u)\, e^{2u}\, du + \varepsilon_0^2\, e^{-2(t-s)} = \varepsilon_{2n}^2 \;, \text{ for each } t \in [t_{2n}, t_{2n+1}] \;.$$

It can be shown that this condition is satisfied as long as ε satisfies, for each $n \geq 0$,

$$\varepsilon_{2n+2} > \varepsilon_{2n}\, e^{-(t_{2n+2}-t_{2n+1})} \;, \quad \int_{t_{2n+1}}^{t_{2n+2}} \dot{\varepsilon}_{2n+1}(u)\, \varepsilon_{2n+1}(u)\, e^{2u}\, du = 0 \;. \tag{24}$$

Evidently there are infinitely many ε's satisfying the last two conditions. For these ε it is not difficult to show that, for each $t \in [t_{2n}, t_{2n+1}]$, the Nelson-Newton law is valid in the form

$$\frac{1}{2} m (D_+ D_- + D_- D_+) R(t) = -\frac{d}{dR}\left\{\frac{m\,\alpha(\alpha+1)}{2R^2}\varepsilon_{2n}^2 + \frac{1}{2} m R^2\right\}, \tag{25}$$

each $n \geq 0$. i.e. the appropriate Nelson-Newton law is valid. More examples and a full justification are given in [7].

In the next section we discuss the types of solutions of the Schrödinger equation with which we model planetesimal diffusions.

3. Planetesimal Diffusions and the Schrödinger Equation

Although it is possible to construct models corresponding to potentials with angular dependence, e.g. the Hartmann potential $V(r,\theta) = Ar^{-1} - B(r^2 \sin^2\theta)^{-1}$ ([10] gives explicit stationary state solutions of the Schrödinger equation), we content ourselves by treating only spherically symmetric potentials $V(\underset{\sim}{x}) = V(|\underset{\sim}{x}|)$, $|\underset{\sim}{x}| = r$ being the Euclidean norm in $\mathbb{R}^3$. To see how the potential relates to the density of the cloud in our model we note that the Nelson-Newton law of motion for a planetesimal of mass m moving in a cloud of density $\rho(r)$ reads

$$\frac{1}{2} m (D_+ D_- + D_- D_+) \underset{\sim}{X}(t) = -\left\{\frac{4}{3}\pi \int_0^{|\underset{\sim}{X}|} \rho(u)\, u^2\, du\right\} m \frac{\underset{\sim}{X}}{|\underset{\sim}{X}|^3}, \tag{26}$$

the term in brackets on the right-hand side being the mass of the cloud interacting with the planetesimal at $\underset{\sim}{X}$. Hence, if $V(r)$ is the potential, then

$$V'(r) = m r^{-2}\left\{\frac{4}{3}\pi \int_0^r \rho(u)\, u^2\, du\right\}. \tag{27}$$

This equation can be used to determine ρ given the potential V:

$$\rho(r) = \frac{3}{4\pi m}\left\{V''(r) + \frac{2}{r} V'(r)\right\}. \tag{28}$$

Of course the Coulomb potential is the special case corresponding to ρ equal to the Dirac delta-function at the origin. As expected this expresses the fact that the density of the cloud is concentrated at the origin.

Naturally, models are only physically realizable when the potential leads to a **positive** mass density. The results presented below include potentials for which such a positive density does not exist.

We study solutions of the Schrödinger equation

$$i\,\varepsilon^2 \frac{\partial \Psi}{\partial t}(\underset{\sim}{x}, t) = -\frac{1}{2}\varepsilon^4 \Delta \Psi(\underset{\sim}{x}, t) + \frac{V(|\underset{\sim}{x}|)}{m}\Psi(\underset{\sim}{x}, t) \quad , \tag{29}$$

$\underset{\sim}{x}$ in $\mathbb{R}^3$, $r = |\underset{\sim}{x}|$ being the radial coordinate. We look for stationary state solutions, in polar coordinates (r, θ, φ), of the form

$$\Psi(\underset{\sim}{x}, t) = e^{-iEt/\varepsilon^2} r^{-1} u(r)(\sin\theta)^{-1+\lambda/\varepsilon^2} e^{i(-1+\lambda/\varepsilon^2)\phi} \quad , \tag{30}$$

where u(r) satisfies the radial Schrödinger equation

$$-\frac{\varepsilon^4}{2}\frac{d^2u(r)}{dr^2} + \left\{\frac{\lambda(\lambda-\varepsilon^2)}{2r^2} + \frac{V(r)}{m}\right\} u(r) = E\,u(r) \quad . \tag{31}$$

By simple applications of Itô's formula it can be shown that the corresponding diffusion, $\underset{\sim}{X}(t)$, has spherical polar coordinates (R, θ, Φ) satisfying:

$$dR(t) = -F(R(t), \alpha)\,dt + \varepsilon\, d B_r(t) \quad , \tag{32}$$

$$d\theta(t) = (\lambda - \frac{\varepsilon^2}{2})\frac{\cot\theta(t)}{R^2(t)}\,dt + \frac{\varepsilon}{R(t)} dB_\theta(t) \quad , \tag{33}$$

$$d\Phi(t) = (\lambda - \varepsilon^2)\frac{\operatorname{cosec}^2\theta(t)}{R^2(t)} + \frac{\operatorname{cosec}\theta(t)}{R(t)}\varepsilon\, d B_\phi(t) \quad , \tag{34}$$

where (B_r, B_θ, B_ϕ) are independent BM($\mathbb{R}$) processes. The function F is related to u by $F(r,\alpha) = -\varepsilon^2 \frac{d}{dr} \log u(r, \alpha)$, where we have noted that in general u will depend on parameters such as ε, λ, m and those appearing in the potential. We use α as a collective label for such parameters.

It is worth remarking that the radial process $R(t) = |\underset{\sim}{X}(t)|$ also satisfies a Nelson-Newton law in the form

$$\frac{1}{2} m (D_+ D_- + D_- D_+) |\underset{\sim}{X}(t)| = -\frac{d}{d|\underset{\sim}{X}|} \left\{ \frac{m\lambda(\lambda - \varepsilon^2)}{2|\underset{\sim}{X}|^2} + V(|\underset{\sim}{X}|) \right\} , \tag{35}$$

in the state with quantum mechanical density $|\Psi|^2$. The physical significance of E and λ will now be explained.

Setting

$$H(\underset{\sim}{X}(t)) = \frac{1}{2} m \left\{ \frac{\underset{\sim}{b}_+^2(\underset{\sim}{X}(t)) + \underset{\sim}{b}_-^2(\underset{\sim}{X}(t))}{2} \right\} + V(|\underset{\sim}{X}(t)|) , \tag{36}$$

the total energy of the stochastic process, and

$$\underset{\sim}{L}(\underset{\sim}{X}(t)) = \underset{\sim}{X}(t) \wedge m \frac{1}{2} \{ \underset{\sim}{b}_+(\underset{\sim}{X}(t)) + \underset{\sim}{b}_-(\underset{\sim}{X}(t)) \} , \tag{37}$$

the stochastic angular momentum, we see that in the state with density $|\Psi|^2$

$$\mathbb{E}\{H(\underset{\sim}{X}(t))\} = mE(\alpha) \quad \text{and} \quad \mathbb{E}\{L_3(\underset{\sim}{X}(t))\} = m(\lambda - \varepsilon^2) . \tag{38}$$

These are the stochastic analogues of the usual quantum mechanical eigenvalue relations. The last identity reveals the reason for choosing such a simple angular dependence for the solutions. Moreover, it can be seen from equation (34) that the third component of forward angular momentum, $L_3^+(\underset{\sim}{X}(t)) = \{\underset{\sim}{X} \wedge m\, \underset{\sim}{b}_+(\underset{\sim}{X}(t))\}_3 = m(\lambda - \varepsilon^2)$, with probability one, irrespective of the initial conditions.

In the next section we concentrate our attention on the radial equation (32) and in particular on its dependence on u. We shall discuss the convergence to (Keplerian) circular orbits of the solutions to equation (32), when we allow the parameters α to be continuous piecewise constant functions of the time, with $\varepsilon \to 0$ as $t \uparrow \infty$. We discuss some sufficient conditions on the potential V to give the desired convergence. It is there that we present some new results, of independent interest, on log-concavity of radial Schrödinger wave functions.

4. Radial Planetesimal Diffusions

In this section we consider the radial equation with time-dependent parameters.

$$dR(t) = -F(R(t),t)dt + \varepsilon(t)\, dB(t) \ , \tag{39}$$

where $F(R(t),t) \equiv F(R(t), \alpha(t)) \equiv F(R(t), \lambda(t), m(t), \varepsilon(t), \ldots)$. We shall assume the limiting conditions

(a) $\lambda(t) \to \lambda\ (>0)$, $m(t) \to m\ (>0)$, ... as $t \uparrow \infty$,

and (b) $\varepsilon(t) \to 0$ as $t \uparrow \infty$, with $\int_0^\infty \varepsilon^2(u)\, du < \infty$.

The reason for the L^2 condition on ε will become clear later.

We define

$$V_e(r,t) = \left\{\frac{m(t)\,\lambda(t)\,(\lambda(t) - \varepsilon^2(t))}{2r^2} + V(r,t)\right\} \tag{40}$$

to be the time-dependent effective potential. $V(r,t)$ denotes the potential with time-dependent parameters. e.g. $V(r,t) = -m(t)\,\mu(t)\, r^{-1}$ is the potential corresponding to a point mass.

Since we are looking for circular orbits we require the radial coordinate to converge in the infinite time limit, to some fixed value r_0, say. We expect from classical considerations to have to assume that r_0 is a minimizer of

$$U(r) = \lim_{t\uparrow\infty} V_e(r,t) = \left\{\frac{m\lambda^2}{2r^2} + V(r)\right\} , \tag{41}$$

with $U(r_0) = E$, the energy. Also that $G'(r) = \lim_{t\uparrow\infty} F'(r,t)$ be positive.

We shall make the following basic assumptions:-

(i) $\lim_{t\uparrow\infty} E(\alpha(t)) = E$ exists and $E = U(r_0)$, where r_0 is the unique minimizer of U. Assume also that $U(r) \sim \alpha r^{-2} - \beta r^{-1} + \gamma r^2$, as $r \sim 0$ or ∞ , for possibly different $\alpha > 0$ and $\beta, \gamma \geq 0$.

(ii) U(r) is smooth save at $r = 0$ and $U'(r)$ has a unique zero at r_0.

(iii) $\lim_{t\uparrow\infty} F(r,t) = G(r)$ exists. (It will follow from the above that $G'(r) > M_k > 0$ for $r \in K$, each compact $K \subset \mathbb{R}$.)

(iv) F(r,t) and F′(r,t) are jointly continuous in $K \times (0, \infty)$, where K is any compact not containing the origin. In this case $F(r,t) \to G(r)$ as $t \uparrow \infty$ uniformly in K.

(v) $F(r,t) \to G(r)$ as $t \uparrow \infty$ in such a way that

$$\int_0^\infty | F(r(t), t) - G(r(t)) | \, dt < \infty ,$$

where r(t) is the solution of $\dot{r}(t) = -G(r(t))$.

(vi) It will turn out that, because of the nature of solutions, r_0 will be the unique positive zero of G(r). Thus we shall assume that here.

To see that our results are not vacuous we give two basic examples where all that above assumptions are satisfied, for appropriate ε, λ, m, A and B.

Example 1

$$V(r) = \frac{mA}{r^2} - \frac{mB}{r} \ ; \ E(\lambda, \varepsilon, A, B) = -2B^2 \left\{\varepsilon^2 + \sqrt{\{(2\lambda - \varepsilon^2)^2 + 8A\}}\right\}^{-2} ,$$

$$-F(r,\lambda,\varepsilon,A,B) = \frac{\left\{\varepsilon^2 + \sqrt{\{(2\lambda - \varepsilon^2)^2 + 8A\}}\right\}}{2r} - \frac{2B}{\left\{\varepsilon^2 + \sqrt{\{(2\lambda - \varepsilon^2)^2 + 8A\}}\right\}}$$

Here $u(r,\alpha) = \rho^{s+1} e^{-\frac{\rho}{2}}$ for $\rho = \frac{2}{\varepsilon^2}\sqrt{\{-2E(\alpha)\}}r$ and s the positive solution of $\varepsilon^4 s(s+1) = \lambda(\lambda-\varepsilon^2) + 2A$.

Example 2

$$V(r) = \frac{mA}{r^2} + \frac{1}{2}m B^2 r^2 \ ; \ E(\lambda, \varepsilon, A, B) = \frac{B}{2}\left\{2\varepsilon^2 + \sqrt{\{(2\lambda - \varepsilon^2)^2 + 8A\}}\right\} ,$$

$$-F(r,\lambda,\varepsilon,A,B) = \frac{\left\{\varepsilon^2 + \sqrt{\{(2\lambda - \varepsilon^2)^2 + 8A\}}\right\}}{2r} - Br .$$

Here $u(r,\alpha) = \xi^{s+\frac{1}{2}} e^{-\frac{\xi}{2}}$ for $\xi = \frac{Br^2}{\varepsilon^2}$ and s the positive solution of $2\varepsilon^4 s(s+1) = \lambda(\lambda-\varepsilon^2) + 2A$.

See for example [12] for the complete set of stationary state solutions for the above potentials.

We now discuss the convergence of the solutions of equation (32). We begin with an elementary lemma.

Lemma 4.1

Given the above assumptions the solution, r, of

$$\frac{dr(t)}{dt} = -G(r(t)), \quad \text{with non zero } r(0) \neq r_0 ,$$

is such that $\lim_{t\uparrow\infty} r(t) = r_0$. Further r(t) is monotonic increasing in t if $r(0) < r_0$ and monotonic decreasing if $r(0) > r_0$.

Proof

The last assertion is a trivial consequence of the fact that G is strictly increasing because $G' > 0$. Since $G(r_0) = 0$ by assumption, we may write our equation as

$$\frac{d}{dt}(r(t) - r_0) = - G(r(t)) + G(r_0) \tag{42}$$
$$= - (r(t) - r_0) \int_0^1 G'(r_0 + \theta(r(t) - r_0))\, d\theta \ .$$

$$\therefore \quad (r(t) - r_0) = (r(0) - r_0) \exp\left\{- \int_0^t \{ \int_0^1 G'(r_0 + \theta(r(u) - r_0))\, d\theta\}\, du \right\} \ . \tag{43}$$

Since $G' > 0$, we see that $|r(t) - r_0| < |r(0) - r_0|$ so $(r_0 + \theta(r(t) - r_0))$ lies in a compact subset $K \subseteq \mathbb{R}$ for $0 \le \theta \le 1$. Hence, by assumption (iii) ,

$$|r(t) - r_0| < |r(0) - r_0| \exp(-M_k t) \to 0 \quad \text{as } t \uparrow \infty \ . \qquad //$$

We now aim to extend this last result to the equation

$$\frac{ds(t)}{dt} = - F(s(t),t) \ , \text{ with } s(0) = r(0) \ , \tag{44}$$

and to the corresponding stochastic differential equation (32). We need some preliminary results which are of some independent interest. The first of these is a new result on the log-concavity of Schrödinger wave-functions. We begin with

Lemma 4.2

Let u(r) be a radial Schrödinger wave function i.e. a C^1 solution of

$$\frac{1}{2}\frac{d^2u(r)}{dr^2} + (E - V_e(r))\, u(r) = 0 \ , \tag{45}$$

with u(r), u′(r) and $V_e(r)\, u^2(r) \to 0$ as $r \uparrow \infty$, where $V_e(r) = \dfrac{\ell(\ell+1)}{2r^2} + V(r)$.

Then

$$\frac{d}{dr}\left(- \frac{u'(r)}{u(r)}\right) = \frac{2}{u^2(r)} \int_r^\infty V_e'(x)\, u^2(x)\, dx \ , \quad u(r) \ne 0 \ . \tag{46}$$

Proof

$$2^{-1}u''(r)\,u'(r) + (E - V_e(r))\,u(r)\,u'(r) = 0 \quad .$$

Integrating gives

$$\frac{u'^2(a)}{4} - \frac{u'^2(r)}{4} + \int_r^a (E - V_e(x))\,u(x)\,u'(x)\,dx = 0 \quad . \tag{47}$$

Integrating by parts and letting $a \uparrow \infty$ then gives

$$u'^2(r) = 4 \lim_{a\uparrow\infty}\left\{[(E - V_e(x))\frac{u^2(x)}{2}]_r^a - \int_r^a (E - V_e(x))'\frac{u^2(x)}{2}\,dx\right\} \quad . \tag{48}$$

i.e. $$u'^2(r) = -2\,(E - V_e(r))\,u^2(r) + 2\int_r^\infty V_e'(x)\,u^2(x)\,dx \tag{49}$$

$$= u(r)\,u''(r) + 2\int_r^\infty V_e'(x)\,u^2(x)\,dx \quad , \text{ by } (45) \quad . \tag{50}$$

Therefore, for $u(r) \neq 0$, rearranging gives

$$\frac{d}{dr}\left(-\frac{u'(r)}{u(r)}\right) = \frac{2}{u^2(r)}\int_r^\infty V_e(x)\,u^2(x)\,dx \quad . \qquad //$$

We also need:

Lemma 4.3

Let $u(r)$ be a C^1 solution of

$$\frac{1}{2}\frac{d^2u(r)}{dr^2} + (E - V_e(r))\,u(r) = 0 \quad , \tag{51}$$

with $u(r)$, $u'(r)$ and $V_e(r)\,u^2(r) \to 0$ as $r \downarrow 0$. Then

$$\frac{d}{dr}\left(-\frac{u'(r)}{u(r)}\right) = -\frac{2}{u^2(r)}\int_0^r V_e'(x)\,u^2(x)\,dx \quad , \quad u(r) \neq 0 \quad . \tag{52}$$

Proof

As in previous lemma. //

Corollary 4.4

If u satisfies the hypotheses of the last two lemmas then

$$<V_e'> = \int_0^\infty V_e'(r)\, u^2(r)\, dr = 0 \ . \tag{53}$$

i.e. average outward forces and average inward forces balance exactly!

Proof

Simply equate (46) and (52). //

Corollary 4.5

With the above hypotheses on the potential V_e, $\lim_{r\uparrow\infty} \frac{d}{dr}\left(-\frac{u'(r)}{u(r)}\right) \geq 0$.

Proof

Because of the last corollary and the fact that $u(r) \to 0$ as $r \uparrow \infty$ we may apply L'Hopital's rule to (46) to obtain

$$\lim_{r\uparrow\infty} \frac{d}{dr}\left(-\frac{u'(r)}{u(r)}\right) = \lim_{r\uparrow\infty} -\frac{V_e'(r)\, u(r)}{u'(r)} \ . \tag{54}$$

The result follows easily for all $\alpha > 0$, $\beta, \gamma \geq 0$ by using the explicit solutions of examples 1 and 2. //

We now come to our new result on log-concavity of radial Schrödinger wave-functions.

Proposition 4.6

Let $V_e(r) \sim \alpha r^{-2} - \beta r^{-1} + \gamma r^2$, as $r \sim 0$ or ∞, with $\alpha > 0$, $\beta, \gamma \geq 0$. Let u(r) be a solution of

$$\frac{1}{2}\frac{d^2u(r)}{dr^2} + (E - V_e(r))\, u(r) = 0 \tag{55}$$

satisfying the hypotheses of the last two lemmas and corollaries. Then a sufficient condition for u to be log-concave for $r > 0$ is that either V_e' has a single zero ξ which is a local minimum of V_e, or V_e' has several zeros ξ and for each ξ_{max}, a zero corresponding to a local maximum of V_e,

$$\int_0^{\xi_{max}} V_e'(x)\, u^2(x)\, dx < 0 \quad . \tag{56}$$

<u>Proof</u>

The first case is trivial since, for $0 < r < \xi$, $V_e'(r) < 0$ and, for $\xi < r < \infty$, $V_e'(r) > 0$. The result then follows from the last two lemmas. In the second case it is also easy to see that the stated condition is sufficient to ensure that

$$\frac{d}{dr}\left(-\frac{u'(r)}{u(r)}\right) = -\frac{2}{u^2(r)}\int_0^r V_e'(x)\, u^2(x)\, dx \tag{57}$$

is positive definite for finite $r > 0$. The limit as $r \uparrow \infty$ has been dealt with in corollary 4.5.

//

Reinstating ε and recasting in terms of F, the last result shows that if the above assumptions on the effective potential $V_e = 2^{-1}\lambda(\lambda - \varepsilon^2)\, r^{-2} + m^{-1}\, V(r)$ are satisfied, then $F'(r, \alpha) > 0$. This strict monotone property of $F(r, \alpha)$ will be crucial in what follows. Recall $F(r, t) \equiv F(r, \alpha(t))$ where α is the collective label for the finite number of variable parameters.

Now by Heine's theorem and assumption (iv), given $\eta > 0$ there exists a t_o such that for $r \in K$

$$|F'(r, t) - G'(r)| < \eta \ , \ \text{for } t > t_0 \ , \ \text{independent of } r \quad . \tag{58}$$

By assumption (iii) and the left hand inequality of (58) we see that if we choose $\eta = M_k/2$ then for $t > t_0$

$$F'(r, t) > M_k/2 \quad . \tag{59}$$

We can now extend the result in lemma 4.1 to equation (44) . We have

Proposition 4.7

Let s(t) satisfy

$$\frac{ds(t)}{dt} = - F(s(t), t) \quad , \text{ with } \quad s(o) = r(o) \quad . \tag{60}$$

Then given the above assumptions on V_e and F it follows that

$$\lim_{t\uparrow\infty} s(t) = r_0 \quad ,$$

the unique minimizer of U(r) .

Proof

We compare s(t) with r(t) , the solution in lemma 4.1. We have

$$\frac{ds(t)}{dt} - \frac{dr(t)}{dt} = - F(s(t), t) + G(r(t)) \tag{61}$$

$$= - F(s(t), t) + F(r(t), t) - F(r(t), t) + G(r(t)) \quad . \tag{62}$$

This can be written

$$\frac{d}{dt}(s(t) - r(t)) = - (s(t) - r(t)) \int_0^1 F'(r(t) + \theta(s(t) - r(t)), t)\, d\theta \; - \; [F(r(t), t) - G(r(t))] \quad . \tag{63}$$

Rearranging and integrating gives

$$(s(t) - r(t)) = - \int_o^t \exp\left\{- \int_u^t \int_0^1 F'(r(t) + \theta(s(v) - r(v)), v)\, d\theta\, dv\right\} (F(r(u),u) - G(r(u)))\, du \quad . \tag{64}$$

This gives, since $F' > 0$,

$$|s(t) - r(t)| \leq \int_0^\infty |F(r(u), u) - G(r(u))|\, du \tag{65}$$

$$< \infty \;, \text{ by assumption (v)} .$$

It follows that $[r(v) + \theta(s(v) - r(v))] \in K$, some compact subset of $\mathbb{R}$, for $0 \le \theta \le 1$ and $v \in (0, \infty)$. If K does not contain 0 then for sufficiently large t

$$F'(r(t) + \theta(s(t) - r(t)), t) > M_k/2 > 0 \quad . \tag{66}$$

We now exploit this condition. Integrating (63) for $a < b$ gives

$$\begin{aligned}(s(b) - r(b)) &= (s(a) - r(a)) \exp\left\{-\int_a^b \int_0^1 F'(r(u) + \theta(s(u) - r(u)), u)\, d\theta\, du\right\} \\ &+ \int_a^b \exp\left\{-\int_v^b \int_0^1 F'(r(u) + \theta(s(u) - r(u)), u)\, d\theta\, du\right\} (G(r(v)) - F(r(v),v))\, dv \quad .\end{aligned} \tag{67}$$

Choosing a sufficiently large, if $[r(u) + \theta(s(u) - r(a))] \in K$ for $u > a$, K not containing 0, $F'(r(u) + \theta(s(u) - r(u)), u) > M_k/2$ and letting $b \uparrow \infty$ the first term of (67) tends to zero. For the second term we apply L'Hopital's rule to obtain

$$\lim_{b\uparrow\infty} |s(b) - r(b)| \le \lim_{b\uparrow\infty} \frac{|G(r(b)) - F(r(b), b)|}{\int_0^1 F'(r(b) + \theta(s(b) - r(b), b))\, d\theta} \tag{68}$$

$$< \lim_{b\uparrow\infty} \frac{2}{M_k} |G(r(b)) - F(r(b), b)| \tag{69}$$

$$= 0 \quad .$$

It only remains to show that K can be chosen so as not to contain the origin. Because of the monotonicity of r the only problem lies with s. Define ξ_t by $F(\xi_t, t) = 0$. Then, because $F'(., t) > 0$, $s(t)$ is moving toward ξ_t. But clearly $\xi_t \to r_0$ as $t \uparrow \infty$ so we can choose t_0 so that $\xi_t > r_0/2$ for $t > t_o$. So if $s(a) \neq 0$ then $s(u)$ is bounded away from zero for $u > a \geq t_0$. Clearly, therefore,

$$[s(u) + \theta(s(u) - r(u))] \in K, \text{ some compact set not containing the origin.}$$

We finally arrive at our result for the stochastic differential equation (32).

<u>Proposition 4.8</u>

Let $\varepsilon(t) \to 0$ as $t \uparrow \infty$ with $\int_0^\infty \varepsilon^2(u)\,du < \infty$. Define $R(t)$ by

$$dR(t) = -F(R(t), t)\,dt + \varepsilon(t)dB_r(t), \quad R(o) = s(o) = r(o), \tag{70}$$

B_r being a BM($\mathbb{R}$) process. Then, almost surely, $\lim_{t\uparrow\infty} R(t) = r_0$, the unique minimizer of $U(r)$.

<u>Proof</u>

We simply compare R with s, the solution in the last proposition. Define

$$M(t) = (R(t) - s(t))^2 + \int_t^\infty \varepsilon^2(u)\,du \quad (\geq 0). \tag{71}$$

Ito's formula gives

$$dM(t) = -2(R(t) - s(t))^2 \int_0^1 F'(s(t) + \theta(R(t) - s(t), t)\,d\theta + 2(R(t) - s(t))\,\varepsilon(t)\,dB_r(t). \tag{72}$$

The first term is negative semi-definite so that $M(t)$ is a local supermartingale. Since $M \geq 0$ then M is a true supermartingale. (see e.g. [15]). This gives

$$\mathbb{E}\,(R(t) - s(t))^2 \leq \int_0^t \varepsilon^2(u)\,du < \infty, \quad \text{by hypothesis}. \tag{73}$$

By Doob's convergence theorem it follows that $\lim_{t\uparrow\infty} (R(t) - s(t))^2$ exists a.s. But $\lim_{t\uparrow\infty} s(t)$ exists ($=r_0$) so $\lim_{t\uparrow\infty} R(t)$ exists a.s. However, for W a BM($\mathbb{R}$) process

$$R(t) - R(0) = -\int_0^t F(R(u), u)\, du + W\left\{\int_0^t \varepsilon^2(u)\, du\right\} . \tag{74}$$

Since $\int_0^\infty \varepsilon^2(u)\, du < \infty$, letting $t \uparrow \infty$ it follows that, $\lim_{t\uparrow\infty} \int_0^t F(R(u), u)\, du$ exits a.s. Since $\lim_{t\uparrow\infty} F(.,t) = G(.)$ exists with $G(r_0) = 0$ and $\lim_{t\uparrow\infty} R(t)$ exists a.s. the almost sure convergence of the last integral shows that $\lim_{t\uparrow\infty} R(t) = r_0$ a.s. //

This last result shows that our planetesimal diffusion eventually lies on a sphere of radius r_0 . In the final section we show that the diffusion converges to the plane $z = 0$. The convergence of the Φ process will then follow easily.

5. Convergence to Coplanar Circular Orbits

Using equations (32) and (33) a simple application of Itô's formula shows that the Z cartesian coordinate process satisfies

$$dZ(t) = -\left\{\frac{F(R(t),t)}{R(t)} + \frac{\lambda(t)}{R^2(t)}\right\} Z(t)\, dt + \varepsilon(t)\, dB_z(t) , \tag{75}$$

where B_z is a BM($\mathbb{R}$) process. R(t) is the above radial process.

We define the underlying deterministic process z(t) by

$$\dot{z}(t) = -\left\{\frac{G(r(t))}{r(t)} + \frac{\lambda}{r^2(t)}\right\} z(t) , \tag{76}$$

where r(t) is the solution of $\dot{r}(t) = -G(r(t))$ discussed in lemma 4.1 . We have the following:

<u>Lemma 5.1</u>

Let z(t) be the solution of the above equation (76) with $z(t{=}0) = z(0)$. Then $z(t) \to 0$ as $t \uparrow \infty$.

<u>Proof</u>

It is easy to show that

$$z(t) = z_0 \exp\left\{ - \int_0^t \frac{G(r(s))}{r(s)} ds \right\} \exp\left\{ - \int_0^t \frac{\lambda}{r^2(s)} ds \right\} , \tag{77}$$

and also that

$$\frac{r(t)}{r(o)} = \exp\left\{ - \int_0^t \frac{G(r(s))}{r(s)} ds \right\} , \quad r(o) \neq 0 . \tag{78}$$

Hence,

$$z(t) = \frac{z(o)}{r(o)} r(t) \exp\left\{ - \int_0^t \frac{\lambda}{r^2(s)} ds \right\} \to 0 \text{ as } t \uparrow \infty , \tag{79}$$

since $\int_0^t \frac{\lambda}{r^2(s)} ds \to \infty$ as $t \uparrow \infty$ by lemma 4.1. //

This lemma is necessary when we consider the Φ process. For the Z process we have

<u>Proposition 5.2</u>

Let Z(t) be the solution of

$$dZ(t) = - \left\{ \frac{F(R(t),t)}{R(t)} + \frac{\lambda(t)}{R^2(t)} \right\} Z(t)\, dt + \varepsilon(t)\, dB_z(t) , \quad Z(o) = z(o) , \tag{80}$$

where R(t) is the solution of (39). Then, if $\int_0^\infty \varepsilon^2(u)\, du < \infty$, $Z(t) \to 0$ as $t \uparrow \infty$ almost surely.

<u>Proof</u> (Outline)

Using Itô's formula and equation (32) it can be shown that Z has the 'explicit' solution

$$Z(t) = R(t) \left\{ \frac{z(o)}{r(o)} \exp\{- A(t)\} + \exp\{- A(t)\} \int_0^t \exp\{A(s)\} \frac{\varepsilon(s)}{R(s)} dB_z(s) \right\} , \tag{81}$$

where

$$A(t) = \int_0^t \frac{\lambda(s)}{R^2(s)} ds - \frac{1}{2}\int_0^t \frac{\varepsilon^2(s)}{R^2(s)} ds + W_r\left\{\int_0^t \frac{\varepsilon^2(s)}{R^2(s)} ds\right\} , \tag{82}$$

for W_r a BM(ℝ) process. Since $\int_0^\infty \varepsilon^2(s)\,ds < \infty$, and $R(t) \to r_0$ as $t \uparrow \infty$ a.s., the last two terms of $A(t)$ are almost surely finite. Hence,

$$A(t) \sim \int_0^t \frac{\lambda(s)}{R^2(s)} ds \uparrow \infty \quad \text{as } t \uparrow \infty \quad \text{a.s.,} \tag{83}$$

since $\lambda(t) \to \lambda(>0)$ as $t \uparrow \infty$. It follows, therefore, that the first term on the right hand side of (81) vanishes almost surely. As for the second, we have for W a BM(ℝ) process

$$e^{-A(t)} \int_0^t e^{A(s)} \frac{\varepsilon(s)}{R(s)} dB_z(s) = e^{-A(t)}\, W\left\{\int_0^t e^{2A(s)} \frac{\varepsilon^2(s)}{R^2(s)} ds\right\} \tag{84}$$

$$= O\left\{e^{-2A(t)}\left(\int_0^t e^{2A(s)} \frac{\varepsilon^2(s)}{R^2(s)} ds\right) \log\log\left[\int_0^t e^{2A(s)} \frac{\varepsilon^2(s)}{R^2(s)} ds\right]\right\}^{\frac{1}{2}} \quad \text{as } t \sim \infty , \tag{85}$$

where we have used the law of the iterated logarithm. By L'Hopital's rule

$$e^{-2A(t)} \int_0^t e^{2A(s)} \frac{\varepsilon^2(s)}{R^2(s)} ds = 0\,(\varepsilon^2(t)/2\lambda(t)) . \tag{86}$$

Hence,

$$e^{-A(t)} \int_0^t e^{A(s)} \frac{\varepsilon(s)}{R(s)} dB_z(s) = O\left\{\frac{\varepsilon^2(t)}{2\lambda(t)} \log\log\left[\int_0^t \frac{\varepsilon^2(s)}{R^2(s)} \exp\left\{2\int_0^s \frac{\lambda(u)}{R^2(u)} du\right\} ds\right]\right\})^{\frac{1}{2}} \tag{87}$$

$$\to 0 \text{ as } t \uparrow \infty \text{ a.s.,}$$

since $\lim_{t\uparrow\infty} \varepsilon(t) = 0$, $\lim_{t\uparrow\infty} \lambda(t) = \lambda(>0)$ and $\lim_{t\uparrow\infty} R(t) = r_0$ a.s. //

The desired result for the Φ process now follows easily.

Proposition 5.3

Let Φ be the solution of

$$d\Phi(t) = \frac{(\lambda(t) - \varepsilon^2(t))}{(R^2(t) - Z^2(t))} dt + \frac{\varepsilon(t)}{(R^2(t) - Z^2(t))^{\frac{1}{2}}} dB_\phi(t) , \tag{88}$$

where B_ϕ is a BM($\mathbb{R}$) process. Then, if $\int_0^\infty \varepsilon^2(u)\, du < \infty$,

$$\left| \Phi(t+h) - \Phi(t) - \int_t^{t+h} \frac{\lambda(s)}{(r^2(s) - z^2(s))} \right| \to 0 \quad \text{a.s. as } t \uparrow \infty , \text{ for fixed } h, \tag{89}$$

where r and z are the corresponding deterministic solutions of lemmas 4.1 and 5.1.

Proof

Firstly we note that equation (34) is equivalent to equation (88) since $(r \sin\theta)^2 = r^2 - z^2$. Now integration gives, for $\tilde{W}$ a BM($\mathbb{R}$) process,

$$\Phi(t+h) - \Phi(t) - \int_t^{t+h} \frac{\lambda(s)}{r^2(s) - z^2(s)} ds = \int_t^{t+h} \left\{ \frac{1}{R^2(s) - Z^2(s)} - \frac{1}{r^2(s) - z^2(s)} \right\} \lambda(s)\, ds$$
$$- \int_t^{t+h} \frac{\varepsilon^2(s)}{R^2(s) - Z^2(s)} ds + \tilde{W}\left\{ \int_t^{t+h} \frac{\varepsilon^2(s)}{R^2(s) - Z^2(s)} ds \right\} . \tag{90}$$

Since $R^2(s) - Z^2(s) \to r^2(s) - z^2(s) \to r_0^2$ a.s. as $s \uparrow \infty$ and $\int_0^\infty \varepsilon^2(u)\, du < \infty$ and $\lambda(s) \to \lambda(>0)$ as $s \uparrow \infty$, given $\delta > 0 \ \exists \ T(\delta)$ such that for all $t > T(\delta)$,

$$\left| \int_t^{t+h} \left\{ \frac{1}{R^2(s) - Z^2(s)} - \frac{1}{r^2(s) - z^2(s)} \right\} \lambda(s)\, ds \right| < \frac{\delta}{2} \text{ and } \left| \int_t^{t+h} \frac{\varepsilon^2(s)}{R^2(s) - Z^2(s)} ds \right| < \frac{\delta}{2} \text{ a.s.} \tag{91}$$

Hence, for $t > T$,

$$\left| \Phi(t+h) - \Phi(t) - \int_t^{t+h} \frac{\lambda(s)}{R^2(s) - Z^2(s)} ds \right| \leq \delta + \sup_{0 \leq s \leq \frac{\delta}{2}} |\tilde{W}(s)| \quad \text{a.s.} \tag{92}$$

and the result follows. //

The last result is equivalent to saying that the angular velocity of the process, after infinite time, is given by $\lim_{t\uparrow\infty} \lambda(t)\,(r^2(t) - z^2(t))^{-1} = \lambda/r_0^2$.

Remarks

The L^2 condition on ε is sufficient to ensure almost sure convergence of our diffusions but by no means necessary. For "best possible" results in this direction we refer to the work of Williams and Chan in [3] and [4]. There they prove for certain classes of diffusion that almost sure convergence is equivalent to the condition $\varepsilon^2(t)\log(t) \to 0$ as $t \uparrow \infty$.

Acknowlegement

It is a pleasure to thank Professor David Williams for helpful conversations concerning some of this work.

References

[1] Albeverio, S., Blanchard, Ph., Hoegh-Krohn, R.: A Stochastic Model for the Orbits of Planets and Satellites: an Interpretation of the Titius-Bode Law. Expositiones Mathematicae 1, 365-373, (1984).

[2] Albeverio, S., Blanchard, Ph., Hoegh-Krohn, R.: Newtonian Diffusions and Planets, with a Remark on Non-Standard Dirichlet Forms and Polymers. In 'Stochastic Analysis and Applications', Proceedings, Swansea 1983, editors A. Truman and D. Williams, 1-25. Lecture Notes in Maths. 1095, Springer-Verlag.

[3] Chan, T. and Williams, D.: An Excursion Approach to an Annealing Problem, Math. Proc. Camb. Phil. Soc. Vol.105, 169-176, (1989).

[4] Chan, T.: On Multi-Dimensional Annealing Problems, Math. Proc. Camb. Phil. Soc. Vol.150, 177-184, (1989).

[5] Durran, R.M. and Truman, A.: Planetesimal Diffusions. In 'Stochastic Mechanics and Stochastic Processes', Proceedings, Swansea 1986, editors I.M. Davies and A. Truman, 76-88. Lecture Notes in Maths. 1325, Springer-Verlag.

[6] Durran, R.M. and Truman, A.: Planetesimal Diffusions in Stochastic Mechanics. In 'Stochastic analysis, path integration and dynamics', editors K.D. Elworthy and J-C. Zambrini, 197-214. Pitman Research Notes in Mathematics, Series 200. Longman Scientific and Technical.

[7] Durran, R.M. Ph.D. thesis in preparation.

[8] Gihman, I.I., and Skorohod, A.V. (1972): Stochastic Differential Equations. Ergebnisse der Mathematik, Berlin: Springer-Verlag.

[9] Guba, A. and Mukherjee, S.: Exact Solutions of the Schrödinger Equation with Noncentral Parabolic Potentials, J.M.P., Vol.28, No.4, 840-843, (1987).

[10] Kan, K.K. and Griffin, J.J.: Quantized Friction and the Correspondence Principle: Single Particle with Friction, Phys. Lett., Vol.50B, No.2, 241-243, (1974).

[11] Landau, L.D., and Lifshitz, E.M. (1965): Quantum Mechanics: Non-Relativistic Theory, Second Edition. Pergamon Press.

[12] Nelson, E. (1967): Dynamical Theories of Brownian Motion. Mathematical Notes, Princeton: Princeton University Press.

[13] Nelson, E. (1985): Quantum Fluctuations, Princeton: Princeton University Press.

[14] Rogers, L.C.G. and Williams, D. (1987): Diffusions, Markov Processes, and Martingales: Itô Calculus. Chichester: Wiley.

[15] Yasue, K.: Quantization of Dissipative Dynamical Systems, Phys. Lett., Vol.64B, No.3, 239-241, (1976).

Dirichlet forms, diffusion processes and spectral dimensions for nested fractals

MASATOSHI FUKUSHIMA
Department of Mathematical Science
Faculty of Engineering Science
Osaka University, Toyonaka Osaka

1. Introduction

The (regular local)Dirichlet form for the Sierpinski gasket in the Euclidean k-space R^k has been introduced in Fukushima-Shima[5] as a basis to formulate the spectral analysis for the gasket. The associated self-adjoint operator coincides with the Laplacian introduced previously by Kigami[6]. The associated diffusion recovers the Brownian motion constructed by Kusuoka[9] and Barlow- Parkins[1].

The Dirichlet form in [5] has two special properties ; the first is that the Dirichlet norm is simply obtained as the limit of an increasing sequence of finite sums evaluated on the successive pre-gaskets (see §2), and the second is that the Dirichlet space is continuously embedded into the space of continuous functions. These properties are never shared by the ordinary Sobolev space H^1 on R^k except for the case that $k = 1$. They seem to be due to the following geometrical property of the gasket called the finite ramifiedness:if we try to connect any distinct two points of the gasket by a continuous curve on it,the curve should cross at least one of the finite number of specific points.

Indeed Dirichlet forms with these properties are nicely extended to general classes of finitely ramified fractals by Kusuoka[10] (for Lindstrom's nested fractals) and Kigami[7] (for post critically finite (PCF) self similar sets), yielding a quickest way to construct diffusion processes on the respective fractal sets. Accordingly the diffusion associated with the Dirichlet form of [10] recovers the Brownian motion on the nested fractal already constructed by Lindstrom[11].

In this paper, we consider the Dirichlet form of [10] and observe a simple scaling property it exhibits when the size of the underlying finite nested fractal is expanded. We shall present two straightforward applications of this scaling property. The first is to prove the point recurrence of the Brownian motion on the infinite nested fractal. The second is to identify the spectral dimension (which is strictly smaller than 2)with the exponent of polynomial growth of eigenvalues for the finite nested fractal. This identification was established in [11] by a different method. But we shall simultaneously derive similar tail behaviors of the integrated density of states (IDS) for the infinite nested fractal.

Recently Kumagaya[8] employs a method similar to ours to identify the spectral dimension of Kigami's PCF self similar set. On the other hand, Shima[16] uses the Dirichlet-Neumann bracketing method in terms of the Dirichlet form in getting the Lifschitz tail behavior(which also involves the spectral dimension in an exponential decay rate) of the

IDS for the infinite nested fractal under the presence of random Poisson obstacles or random Poisson noise potentials. The Lifschitz tail of the IDS for the Sierpinski gasket in R^2 with Poisson obstacles has been derived by Katarzyna Pietuska-Paluba[13]using a different method.

Thus the knowledge of the Dirichlet form and its scaling property is enough to extract the notion of the spectral dimension from the behaviors of eigenvalues for the fairly general finitely ramified fractal set. However more subtle spectral properties were studied so far only for the Sierpinski gasket. For instance, the IDS for the Sierpinski gasket has been shown in [5] to be purely discontinuous.[1] It is interesting to know if such wild spectral phenomena are common among the finitely ramified fractal sets.

2. The Dirichlet form for the nested fractal

In this section, we describe those notions and relations in Lindstrom[11] and Kusuoka[10] which we shall use later on.

For $\alpha > 1$, a mapping Ψ from R^k to R^k is said to be an *α-similitude* if $\Psi x = \alpha^{-1}Ux + \beta, x \in R^k$, for some unitary map U and $\beta \in R^k$. Given a collection $\Psi = \{\Psi_1, \Psi_2, \ldots, \Psi_N\}$ of α-similitudes,we let $\Psi(A) = \bigcup_{i=1}^N \Psi_i(A), A \subset R^k$. There exists then a unique compact set $E \subset R^k$ such that $\Psi(E) = E$. The pair (Ψ, E) is called a *self similar fractal.*

For $A \subset R^k$ and integer $n \geq 1$, we let

$$A_{i_1 \ldots i_n} = \Psi_{i_1} \ldots \Psi_{i_n}(A), \qquad 1 \leq i_1, \ldots, i_n \leq N$$

$$A^{(n)} = \Psi^{(n)}(A) = \bigcup_{1 \leq i_1, \ldots, i_n \leq N} A_{i_1 \ldots i_n}, \qquad A^{(0)} = A$$

. We denote by F the set of all essential fixed points of Ψ([11]). $\sharp F \leq N$. Lindstrom[11] calls a self similar fractal (Ψ, E) a *nested fractal* if three axioms (axioms of connectivity, symmetry and nesting) and the open set condition are fulfilled and $\sharp F \geq 2$. We refer the readers to [11] for details but we note that the nesting axiom requires

$$E_{i_1 \ldots i_n} \cap E_{j_1 \ldots j_n} = F_{i_1 \ldots i_n} \cap F_{j_1 \ldots j_n} \qquad (i_1, \ldots, i_n) \neq (j_1, \ldots, j_n),$$

which expresses the finite ramifiedness mentioned in §1.

Given a nested fractal (Ψ, E), the Hausdorff dimension of E is known to be equal to $\frac{\log N}{\log \alpha}$. The normalized Hausdorff measure on E is denoted by $\mu : \mu(E) = 1$. The sequence $\{F^{(n)}\}$ of finite sets is increasing and $\cup_{n=0}^{\infty} F^{(n)}$ is denoted by $F^{(\infty)}$, which may be called the pre-nested fractal since $\overline{F^{(\infty)}} = E$.

A typical example of nested fractals is the Sierpinski gasket in R^k. We now explane how to introduce a natural Dirichlet form on the Siepinski gasket in R^2. Let $F = \{p_1, p_2, p_3\}$ be the vertices of the regular triangle of side length 1 in R^2. Let Ψ_i be the 2-similitude on R^2 without rotation $(U = I)$ making the point p_i fixed $(i = 1, 2, 3)$. The self-similar fractal$E$ determined by $\Psi = (\Psi_1, \Psi_2, \Psi_3)$ is the finite Sierpinski gasket,which is easily seen to be a nested fractal.

[1] See also [14] in this connection

For any real valued function f on the pre-gasket $F^{(\infty)}$, we set

(2.1)

$$\mathcal{E}^{(n)}(f,f) = c\left(\frac{5}{3}\right)^n \sum_{1\le i_1,\dots,i_n\le 3}\sum_{\xi,\eta\in F}(f(\Psi_{i_1}\dots\Psi_{i_n}\xi)-f(\Psi_{i_1}\dots\Psi_{i_n}\eta))^2, \quad n=0,1,2,\dots$$

c being a positive constant. It is easy to see that $\mathcal{E}^{(n)}(f,f)$ is increasing in n. In fact $\mathcal{E}^{(0)}(f,f)\le\mathcal{E}^{(1)}(f,f)$ reduces to the elementary absolute inequality

$$\begin{aligned}(2.2)\quad (A_1-A_2)^2+(A_2-A_3)^2+(A_3-A_1)\\ \le\frac{5}{3}\left\{(a_1-a_2)^2+(a_2-a_3)^2+(a_3-a_1)^2\right\}\\ +\frac{5}{3}\left\{(A_1-a_2)^2+(A_1-a_3)^2+(A_2-a_1)^2+(A_2-a_3)^2+(A_3-a_1)^2+(A_3-a_2)^2\right\}\end{aligned}$$

holding for any real numbers $A_i, a_i, i=1,2,3$. This inequality is honestly inherited to the next step inequality $\mathcal{E}^{(1)}(f,f)\le\mathcal{E}^{(2)}(f,f)$ and so on.

Hence it seems natural to introduce the space

$$(2.3)\qquad \begin{aligned}\mathcal{F}&=\left\{f: functiononF^{(\infty)}, \sup_n \mathcal{E}^{(n)}(f,f)<\infty\right\}\\ \mathcal{E}(f,g)&=\lim_{n\to\infty}\mathcal{E}^{(n)}(f,g), \qquad f,g\in\mathcal{F}.\end{aligned}$$

It is actually proven in [5] that any function in $\mathcal{F}$ is extended to be a continuous function on the gasket $\mathcal{E}=\overline{F^{(\infty)}}$ and that $\mathcal{E}$ on $\mathcal{F}$ is a local regular Dirichlet form on $L^2(E;\mu)$. It is clear that behind this analytical approach is the sequence of random walks on $F^{(n)}, n=0,1,2,\dots$,moving to the nearest neighbours with equal probability. The renormalizing factor $\frac{5}{3}$ should have to do with these random walks. In fact,the equality is attained in (2.2) when each a_i is the harmonic average of A_1,A_2,A_3 with respect to the random walk on $F^{(1)}$.

As was proved in Kusuoka[10], the above mentioned situation for the Sierpinski gasket is totally unchanged for the general nested fractal if one takes, as the random walks on $F^{(n)}$'s, those governed by Lindstrom's invariant probability[11].

In what follows, we work with a fixed nested fractal (Ψ,E) in R^k. By a random walk on F, we mean a Markov chain on F whose transition probability from $x\in F$ to $y\in F, x\ne y$, depends only on the distance $|x-y|$ and decreases strictly if the value $|x-y|$ increases. Thus we let

$$\{|x-y|: x,y\in F, x\ne y\}=\{\ell_1,\dots,\ell_r\},\quad 0<\ell_1<\cdots<\ell_r$$

$$m_s=\sharp\{y\in F: |x-y|=\ell_s\}\quad \text{for a fixed } x\in F, 1\le s\le r$$

$$\mathcal{P}=\{\mathbf{P}=(p_1,\dots,p_r): p_1>p_2>\cdots>p_r>0\}.$$

m_s is independent of $x\in F$ because of the axiom of symmetry.

Each element $\mathbf{P} \in \mathcal{P}$ not only decides a random walk on F, but also determines a Markov chain on $F^{(1)}$ with transition probability

$$p(x,y,\mathbf{P}) = \begin{cases} \frac{\rho(x,y)}{\rho(x)} p_s & \rho(x,y) \geq 1, |x-y| = \alpha^{-1}\ell_s \\ 0 & \text{otherwise,} \end{cases}$$

where $\rho(x) = \sharp\{i : x \in F_i\}, \rho(x,y) = \sharp\{i : x,y \in F_i\}$. This Markov chain on $F^{(1)}$ in turn induces a Markov chain on F by letting the sample path start at $x \in F$ and observing the first time the path hits $F \setminus \{x\}$ and so on. The invariant probability is an element of $\mathcal{P}$ such that the induced Markov chain on F in the above sense coincides in law with the original walk on F. Lindstrom[11] proved its existence and Barlow[2] recently shows its uniqueness for a certain class of nested fractals including Lindstrom's snow flake..

Let us denote the invariant probability by $\mathbf{P}_0 = (p_1, \ldots, p_r)$. Let $(X(n), P_x)$ be the Markov chain on $F^{(1)}$ with trasition probability $p(x,y,\mathbf{P}_0), x,y \in F^{(1)}$ and let

$$c = P_x(X_{\tilde{\sigma}} = x),\ x \in F, \text{ where } \tilde{\sigma} = \{n > 0 : X(n) \in F\}.$$

By virtue of the axiom of symmetry,c is independent of $x \in F$ and evidently $0 < c < 1$. In view of the above description of the invariance property of $\mathbf{P}_0$, we may expect the quanitity c to play an intrinsically important role.

For $\xi, \eta \in F, \xi \neq \eta$, we let $\pi_{\xi\eta} = p_s$ if $|\xi - \eta| = \ell_s$. Then $\sum_{\eta \in F} \pi_{\xi\eta} = 1$ and $\pi_{\xi\eta} = \pi_{\eta\xi}$. We denote by $\mathcal{D}$ the set of all real valued functions on $F^{(\infty)} = \cup_{n=0}^{\infty} F^{(n)}$. For $f, g \in \mathcal{D}$, we define $\mathcal{E}^{(n)}(f,g)$ by

$$(2.4)\quad \mathcal{E}^{(n)}(f,g) = \frac{1}{2}(1-c)^{-n} \sum_{1 \leq k_1,\ldots,k_n \leq N} \sum_{\substack{\xi,\eta \in F \\ \xi \neq \eta}} (f(\Psi_{k_1}\ldots\Psi_{k_n}\xi) - f(\Psi_{k_1}\ldots\Psi_{k_n}\eta)) \\ (g(\Psi_{k_1}\ldots\Psi_{k_n}\xi) - g(\Psi_{k_1}\ldots\Psi_{k_n}\eta))\pi_{\xi\eta}$$

In the case of the Sierpinski gasket in R^2, $c = \frac{2}{5}, \pi_{\xi\eta} = \frac{1}{2}, \xi \neq \eta$, and hence (2.4) reduces to (2.1).

THEOREM 2.1(KUSUOKA[10]).

(1) *For any $f \in \mathcal{D}, \mathcal{E}^{(n)}(f,f)$ defined by (2.4) is non-decreasing in n. Hence $(\mathcal{F}, \mathcal{E})$ is well defined by (2.3).*

(2) *Any function of $\mathcal{F}$ can be uniquely extended to a continuous function on $E = \overline{F^{(\infty)}}$.*

(3) *$(\mathcal{F}, \mathcal{E})$ is a regular local Dirichlet form on $L^2(E;\mu)$.*[2]

Since $\pi_{\xi\eta}/p_r \geq 1, \xi, \eta \in F$, we have for any function f on F

$$(2.5)\qquad \max\{|f(\xi) - f(\eta)| : \xi, \eta \in F\} \leq \frac{1}{\sqrt{p_r}} \left\{ \frac{1}{2} \sum_{\xi,\eta \in F} (f(\xi) - f(\eta))^2 \pi_{\xi\eta} \right\}^{1/2}$$

This elementary esimate leads us to the next theorem. Actually Theorem 2.1(2) is a corollary of the much stronger assertion Theorem 2.2(1).

[2] μ can be replaced by any everywhere dense positive Radon measure on E.

THEOREM 2.2(KUSUOKA[10]).

(1) *Let B be the set of those functions f on $F^{(\infty)}$ with $\sup_n \mathcal{E}^{(n)}(f,f) \leq 1$. Then*

$$\lim_{\varepsilon\to 0} \sup\left\{|f(x)-f(y)| : f \in B, x, y \in F^{(\infty)}, |x-y| < \varepsilon\right\} = 0.$$

(2)

$$\sup\{|f(x)-f(y)| : x, y \in E\} \leq 4N\left(\frac{N}{p_r}\right)^{1/2} \frac{1-c}{c}\mathcal{E}_1(f,f)^{1/2},\ f \in \mathcal{F}.$$

(3)

$$\sup\{|f(x)| : x \in E\} \leq \sqrt{2}\left(\frac{N}{p_r}\right)^{1/2} \frac{1-c}{c}\mathcal{E}_1(f,f)^{1/2},\ f \in \mathcal{F},$$

where $\mathcal{E}_\alpha(f,g) = \mathcal{E}(f,g) + \alpha(f,g)_{L^2(\mu)}$, $\alpha > 0$.

We can easily draw several conclusions from the above two theorems.

THEOREM 2.3.

(1) *For each $\alpha > 0$, the Hilbert space $(\mathcal{F}, \mathcal{E}_\alpha)$ admits a reproducing kernel $g_\alpha(x,y)$: for each $y \in E$ there exists $g_\alpha(\cdot,y) \in \mathcal{F}$ such that*

$$\mathcal{E}_\alpha(g_\alpha(\cdot,y), v) = v(y), \quad v \in \mathcal{F}. \tag{2.6}$$

(2) *$g_\alpha(x,y)$ is positive continuous symmetric on $E \times E$.*

(3) *Each one point set has a positive capacity:*

$$Cap(\{y\}) = \frac{1}{g_1(y,y)}, \qquad y \in E.$$

(4) *The associated diffusion on E is point recurrent:*

$$P_x\left(\sigma_{\{y\}} < \infty\right) = 1 \text{ for any } x, y \in E,$$

$\sigma_{\{y\}}$ being the first hitting time for $\{y\}$.

PROOF: (1) follows from Theorem 2.2(3) which implies that the map sending $f \in \mathcal{F}$ to $f(y) \in R$ is bounded.
(2). Symmetry is obvious from (2.6). From (2.6), we also have

$$\mathcal{E}_\alpha(g_\alpha(\cdot,y), g_\alpha(\cdot,y)) = g_\alpha(y,y)$$

which is positive because otherwise $v(y)$ vanishes for any $v \in \mathcal{F}$. Further, by Theorem 2.2(3)

$$\sup_{y\in E}\sqrt{g_\alpha(y,y)} \le \sup_{y\in E}\frac{\sup_{x\in E} g_\alpha(x,y)}{\sqrt{g_\alpha(y,y)}} \le \sup_{f\in\mathcal{F}}\frac{\sup_{x\in E}|f(x)|}{\sqrt{\mathcal{E}_\alpha(f,f)}} < \infty.$$

Therefore the family of functions $\{g_\alpha(\cdot,y); y\in E\}$ is equi uniformly continuous on E by Theorem 2.2(1) and we can get the continuity of $g_\alpha(x,y)$ in x,y from

$$|g_\alpha(x,y)-g_\alpha(x\prime,y\prime)| \le |g_\alpha(x,y)-g_\alpha(x\prime,y)| + |g_\alpha(y,x\prime)-g_\alpha(y\prime,x\prime)|.$$

(3).If we let $p_1^y(x)=\frac{g_1(x,y)}{g_1(y,y)}$, then $p_1^y\in\mathcal{F}, p_1^y(y)=1$ and $\mathcal{E}_1(p_1^y,v)\ge 0$ for any $v\in\mathcal{F}$ with $v(y)\ge 0$. Therefore p_1^y is the 1-equilibrium potential of the one point set {y} and

$$Cap(\{y\}) = \mathcal{E}_1(p_1^y,p_1^y) = \frac{1}{g_1(y,y)}.$$

(4). $(\mathcal{F},\mathcal{E})$ is irreducible because otherwise $\mathcal{F}$ must contain a discontinuous function([4]) contradicting to Theorem 2.1(2). Since $1\in\mathcal{F}$ and $\mathcal{E}(1,1)=0, (\mathcal{F},\mathcal{E})$ is recurrent([4]). Becuase of (3), the associated diffusion is point recurrent. In particular,

$$g_\alpha(x,y) = E_x\left(e^{-\alpha\sigma_{\{y\}}}\right) g_\alpha(y,y) > 0, \quad x,y\in E.$$

We denote by Δ the self-adjoint operator on $L^2(E;\mu)$ associated with $(\mathcal{F},\mathcal{E})$:

$$\mathcal{D}(\Delta)\subset\mathcal{F} \qquad \mathcal{E}(f,g) = -(\Delta f,g), f\in\mathcal{D}, g\in\mathcal{F}.$$

By virtue of Theorem 2.3 (2), $-\Delta$ is of compact resolvent and Mercer's theorem leads to the absolutely uniformly convergent series expansion :

$$g_\alpha(x,y) = \sum_{i=1}^{\infty}\frac{1}{\alpha+\lambda_i}\varphi_i(x)\varphi_i(y), \tag{2.7}$$

where

$$0\le\lambda_1\le\lambda_2\le\cdots\le\lambda_n\le\ldots$$

are eigenvalues of $-\Delta$ and $\{\varphi_n\}$ are the corresponding normalized eigenfunctions. Notice that κ is an eigenvalue of $-\Delta$ with an eigenfunction f if

$$f\in\mathcal{F} \quad \text{and} \quad \mathcal{E}(f,g) = \kappa(f,g)_{L^2(E;\mu)} \quad \text{for any} \quad g\in\mathcal{F}. \tag{2.8}$$

3. Dirichlet forms on expanded nested fractals

As in the preceding section, we fix a nested fractal (Ψ, E) in R^k. Without loss of generality, we assume that $0 \in F$ and $\Psi_1 x = \alpha^{-1}x, x \in R^k$. We let

$$(3.1) \qquad E^{\langle \ell \rangle} = \alpha^\ell E, \quad \ell = 0, 1, \ldots, \quad E^{\langle \infty \rangle} = \bigcup_{\ell=0}^{\infty} E^{\langle \ell \rangle}.$$

We call $E^{\langle \ell \rangle}$ (resp.$E^{\langle 0 \rangle} = E$)the *infinite* (resp.*unit*) nested fractal. $E^{\langle \ell \rangle}$ is the union of N^ℓ -number of sets congruent to E :

$$E^{\langle \ell \rangle} = \bigcup_{1 \le i_1, \ldots, i_\ell \le N} E^{\langle \ell \rangle}_{i_1 \ldots i\ell} \quad \text{and} \quad E^{\langle \ell \rangle}_{i_1 \ldots i_\ell} = \Phi_{i_1 \ldots i_\ell} E \quad \text{with} \quad \Phi_{i_1 \ldots i_\ell} = \alpha^\ell \Psi_{i_1} \ldots \Psi_{i_\ell}.$$

We define the mapping σ_ℓ by

$$(3.2) \qquad (\sigma_\ell f)(x) = f(\alpha^\ell x) \qquad x \in E,$$

which maps a function f on $E^{\langle \ell \rangle}$ to a function $\sigma_\ell f$ on the unit nested fractal E.

Recall that we have a Hausdorff measure μ on E with $\mu(E) = 1$ and a Dirichlet form $(\mathcal{E}, \mathcal{F})$ on $L^2(E; \mu)$ introduced in Theorem 2.1. We extend μ to $E^{\langle \infty \rangle}$ by defining its value on B to be $\mu\left(\Phi^{-1}_{i_1 \ldots i_\ell} B)\right)$ if $B \subset E^{\langle \ell \rangle}_{i_1 \ldots i_\ell}$ (which does not depend on the choice of ℓ). We also define a Dirichlet form $(\mathcal{F}_{E^{\langle \ell \rangle}}, \mathcal{E}_{E^{\langle \ell \rangle}})$ on $L^2(E^{\langle \ell \rangle}; \mu)$ by

$$(3.3) \quad \mathcal{F}_{E^{\langle \ell \rangle}} = \sigma_\ell^{-1} \mathcal{F}$$

$$\mathcal{E}_{E^{\langle \ell \rangle}}(f, g) = \sum_{1 \le i_1, \ldots, i_\ell \le N} \mathcal{E}\left(f\left(\Phi_{i_1 \ldots i_\ell} \cdot\right), g\left(\Phi_{i_1 \ldots i_\ell} \cdot\right)\right), \quad f, g \in \mathcal{F}^{\langle \ell \rangle}, \ell = 1, 2, \ldots.$$

LEMMA 3.1(SCALING PROPERTIES).

(1) *For a function F on $E^{\langle \ell \rangle}$,*

$$\int_{E^{\langle \ell \rangle}} f \, d\mu = N^\ell \int_E (\sigma_\ell f) \, d\mu.$$

(2) *For $f \in \mathcal{F}_{E^{\langle \ell \rangle}}$,*

$$\mathcal{E}_{E^{\langle \ell \rangle}}(f, f) = (1 - c)^\ell \, \mathcal{E}(\sigma_\ell f, \sigma_\ell f).$$

PROOF: (1) In the case that $f = I_B$ for $B \subset E^{\langle \ell \rangle}_{i_1 \ldots i_\ell}$, the right hand side equals

$$N^\ell \mu\left(\alpha^{-\ell} B\right) = N^\ell \mu\left(\alpha^{-\ell} \Phi^{-1}_{i_1 \ldots i_\ell} B\right) = \mu\left(\Phi^{-1}_{i_1 \ldots i_\ell} B\right)$$

which coincides with the left hand side.

(2) By (3.3) and (2.4),

$$\mathcal{E}_{E^{\langle \ell \rangle}}(f,f) = \sum_{1\le i_1,\dots,i_\ell \le N} \mathcal{E}\left\{\sigma_\ell f\left(\Psi_{i_1\dots i_\ell}\cdot\right), \sigma_\ell f\left(\Psi_{i_1\dots i_\ell}\cdot\right)\right\}$$

$$= \lim_{n\to\infty} \frac{1}{2}(1-c)^{-n} \sum_{1\le i_1,\dots,i_{\ell+n}\le N} \sum_{\substack{\xi,\eta\in F\\ \xi\neq\eta}} \left\{\sigma_\ell f\left(\Psi_{i_1}\dots\Psi_{i_{\ell+n}}\xi\right) - \sigma_\ell f\left(\Psi_{i_1}\dots\Psi_{i_{\ell+n}}\eta\right)\right\}^2 \pi_{\xi\eta}$$

$$= (1-c)^{\ell}\mathcal{E}\left(\sigma_\ell f, \sigma_\ell f\right).$$

At the end of §2,we introduced the self-adjoint operator Δ on $L^2(E;\mu)$ associated with $(\mathcal{F},\mathcal{E})$. Analogously we consider the self-adjoint operator $\Delta^{\langle \ell \rangle}$ on $L^2(E^{\langle \ell \rangle};\mu)$ associated with $(\mathcal{F}_{E^{\langle \ell \rangle}}, \mathcal{E}_{E^{\langle \ell \rangle}})$. From the above lemma and (2.8), we get

COROLLARY 3.2. *κ is an eigenvalue of $-\Delta$ iff $\left(\frac{1-c}{N}\right)^{\ell}\kappa$ is an eigenvalue of $-\Delta^{\langle \ell \rangle}$.*

It is clear from definition (3.3) that

$$\mathcal{E}_{E^{\langle \ell \rangle}}\left(f|_{E^{\langle \ell \rangle}}, f|_{E^{\langle \ell \rangle}}\right) \le \mathcal{E}_{E^{\langle m \rangle}}(f,f), \qquad \ell < m, \qquad f \in \mathcal{F}_{E^{\langle m \rangle}},$$

the equality holding when f vanishes on $E^{\langle m \rangle} - \overline{E^{\langle \ell \rangle}}$. We define the space $\mathcal{F}_{E^{\langle\infty\rangle}}$ of functions f on the infinite nested fractal $E^{\langle\infty\rangle}$ by

(3.4)

$$\mathcal{F}_{E^{\langle\infty\rangle}} = \left\{ f : f|_{E^{\langle \ell \rangle}} \in \mathcal{F}_{E^{\langle \ell \rangle}} \text{ for each } \ell \text{ and } \lim_{\ell\to\infty} \mathcal{E}_{E^{\langle \ell \rangle}}\left(f|_{E^{\langle \ell \rangle}}, f|_{E^{\langle \ell \rangle}}\right) < \infty \right\} \cap L^2(E^{\langle\infty\rangle};\mu).$$

We also let

$$\mathcal{E}_{E^{\langle\infty\rangle}}(f,g) = \lim_{\ell\to\infty} \mathcal{E}_{E^{\langle \ell \rangle}}\left(f|_{E^{\langle \ell \rangle}}, g|_{E^{\langle \ell \rangle}}\right), \qquad f,g \in \mathcal{F}_{E^{\langle\infty\rangle}}. \tag{3.5}$$

Denote by $C(E^{\langle\infty\rangle})$ (resp.$C_0(E^{\langle\infty\rangle})$) the space of continuous functions(resp.continuous functions with compact suport) on $E^{\langle\infty\rangle}$. We have $\mathcal{F}_{E^{\langle\infty\rangle}} \subset C(E^{\langle\infty\rangle})$. If $f \in C_0(E^{\langle\infty\rangle})$, supp$f \subset E^{\langle m \rangle}$ and $f \in \mathcal{F}_{E^{\langle m \rangle}}$, then $f \in \mathcal{F}_{E^{\langle\infty\rangle}}$ and

$$\mathcal{E}_{E^{\langle\infty\rangle}}(f,f) = \mathcal{E}_{E^{\langle m \rangle}}(f,f). \tag{3.6}$$

LEMMA 3.3. *There exist functions $\phi_\ell \in \mathcal{F}_{E^{\langle\infty\rangle}} \cup C_0\left(E^{\langle\infty\rangle}\right)$ such that*

$$\phi_\ell = 1 \text{ on } E^{\langle \ell-1 \rangle},\ \phi_\ell = 0 \text{ on } E^{\langle\infty\rangle} - E^{\langle \ell \rangle},\ 0 \le \phi_\ell \le 1 \text{ and } \lim_{\ell\to\infty} \mathcal{E}_{E^{\langle\infty\rangle}}\left(\phi_\ell, \phi_\ell\right) = 0.$$

PROOF: Take the function ψ on E with the properties that $\psi \in \mathcal{F}, \psi = 1$ on $F_1(= \Psi_1 F), \psi = 0$ on $F^{(1)} \setminus F_1$ and ψ is $\mathcal{E}$-harmonic on $E \setminus F^{(1)}$. Then $\psi = 1$ on $E^{(1)} = \psi_1 E$ and $0 \le \psi \le 1$ on E. It suffices to let ϕ_ℓ be $\sigma_\ell^{-1}\psi$ on $E^{\langle \ell \rangle}$ and 0 on $E^{\langle\infty\rangle} \setminus E^{\langle \ell \rangle}$. We get from Lemma 3.1 and (3.6)

$$\mathcal{E}_{E^{\langle\infty\rangle}}(\phi_\ell, \phi_\ell) = \mathcal{E}_{E^{\langle \ell \rangle}}(\phi_\ell, \phi_\ell) = (1-c)^{\ell}\mathcal{E}(\psi,\psi),$$

which tends to zero as $\ell \to \infty$.

THEOREM 3.4. *$(\mathcal{F}_{E^{\langle\infty\rangle}}, \mathcal{E}_{E^{\langle\infty\rangle}})$ is a regular local Dirichlet form on $L^2\left(E^{\langle\infty\rangle};\mu\right)$. For each $\alpha > 0, \mathcal{E}_{E^{\langle\infty\rangle},\alpha}$ admits a positive symmetric continuous reproducing kernel $g_\alpha(x,y)$ on $E^{\langle\infty\rangle} \times E^{\langle\infty\rangle}$ and*

$$Cap(\{y\}) = \frac{1}{g_\alpha(y,y)}(>0), \qquad y \in E^{\langle\infty\rangle}.$$

The associated diffusion on $E^{\langle\infty\rangle}$ is point recurent.

PROOF: The preceding lemma implies that $(\mathcal{F}_{E^{\langle\infty\rangle}}, \mathcal{E}_{E^{\langle\infty\rangle}})$ is non-transient([4]). Other assertions except for the regularity can be proven similarly as the proof of theorems of §2. To prove the regularity of $(\mathcal{F}_{E^{\langle\infty\rangle}}, \mathcal{E}_{E^{\langle\infty\rangle}})$, take any bounded function $f \in \mathcal{F}_{E^{\langle\infty\rangle}}$ and set $f_\ell = f \cdot \phi_\ell, \ell = 1, 2, \dots$, for ϕ_ℓ of Lemma 3.3. We write for $g \in \mathcal{F}_{E^{\langle\infty\rangle}}$

$$\mathcal{E}_{E^{\langle\infty\rangle}\setminus E^{\langle\ell\rangle}}(g,g) = \mathcal{E}_{E^{\langle\infty\rangle}}(g,g) - \mathcal{E}_{E^{\langle\ell\rangle}}\left(g|_{E^{\langle\ell\rangle}}, g|_{E^{\langle\ell\rangle}}\right).$$

We have then

$$\begin{aligned}\mathcal{E}_{E^{\langle\infty\rangle}}(f - f_\ell, f - f_\ell) &= \mathcal{E}_{E^{\langle\infty\rangle}\setminus E^{\langle\ell-1\rangle}}(f(1-\phi_\ell), f(1-\phi_\ell)) \\ &\le \|f\|_\infty^2 \mathcal{E}_{E^{\langle\infty\rangle}}(\phi_\ell, \phi_\ell) + \mathcal{E}_{E^{\langle\infty\rangle}\setminus E^{\langle\ell-1\rangle}}(f,f) \ \rightarrow\ \infty,\ \ell \to \infty.\end{aligned}$$

Since $f_\ell \in \mathcal{F}_{E^{\langle\infty\rangle}} \cap C_0\left(E^{\langle\infty\rangle}\right), \mathcal{F}_{E^{\langle\infty\rangle}} \cap C_0\left(E^{\langle\infty\rangle}\right)$ is $\mathcal{E}_{E^{\langle\infty\rangle},1}$-dense in $\mathcal{F}_{E^{\langle\infty\rangle}}$.

4. Asymptotics of the eigenvalue distribution and the integrated density of states

For Δ and $\Delta^{\langle\ell\rangle}$ considered in §3, we let

$$\rho(\lambda) = \sharp\{\text{eigenvalues of } -\Delta \le \lambda\}$$

$$k_\ell(\lambda) = \sharp\left\{\text{eigenvalues of} - \Delta^{\langle\ell\rangle} \le \lambda\right\}.$$

Together with $k_\ell(\lambda)$, we consider

$$k_\ell^0(\lambda) = \sharp\left\{\text{eigenvalues of} - \Delta_0^{\langle\ell\rangle} \le \lambda\right\},$$

where $\Delta_0^{\langle\ell\rangle}$ is the self-adjoint operator on $L^2(E^{\langle\ell\rangle};\mu)$ corresponding to the Dirichlet space $\left(\mathcal{F}^0_{E^{\langle\ell\rangle}}, \mathcal{E}_{E^{\langle\ell\rangle}}\right)$ with

$$\mathcal{F}^0_{E^{\langle\ell\rangle}} = \left\{f \in \mathcal{F}_{E^{\langle\ell\rangle}} : f(p) = 0 \quad p \in \alpha^\ell F\right\}.$$

We have then the inequality $k_\ell^0(\lambda) \le k_\ell(\lambda)$. Further we see that $\frac{k_\ell(\lambda)}{N^\ell}$ is non-increasing in ℓ. To see this, consider the space

$$\widetilde{\mathcal{F}}_{E^{\langle\ell+1\rangle}} = \left\{f : \text{function on } E^{\langle\ell+1\rangle},\ f(\alpha\Psi_k\ \cdot) \in \mathcal{F}_{E^{\langle\ell\rangle}},\quad 1 \le k \le N\right\}$$

and the self-adjoint operator $\tilde{\Delta}^{\langle\ell+1\rangle}$ on $L^2(E^{\langle\ell+1\rangle},\mu)$ associated with $\left(\widetilde{\mathcal{F}}_{E^{\langle\ell+1\rangle}}, \mathcal{E}_{E^{\langle\ell+1\rangle}}\right)$. Then

$$\sharp\left\{\text{eigenvalues of } -\tilde{\Delta}^{\langle\ell+1\rangle} \le \lambda\right\} = N\, k_\ell(\lambda).$$

But the left hand side is not smaller than $k_{\ell+1}(\lambda)$ because $\tilde{\mathcal{F}}_{E^{(\ell+1)}} \supset \mathcal{F}_{E^{(\ell+1)}}$. In the same way, we can see that $\frac{k_\ell^0(\lambda)}{N^\ell}$ is non-decreasing in ℓ. Therefore there exists a non-trivial right continuous non-decreasing function $\mathcal{N}(\lambda), \lambda \geq 0$, such that

$$\lim_{\ell\to\infty} \frac{k_\ell(\lambda)}{N^\ell} = \mathcal{N}(\lambda) \tag{4.1}$$

at each continuity point λ of $\mathcal{N}(\lambda)$. $\mathcal{N}(\lambda)$ is called the *integrated density of states* .

On the other hand, we have from Corollary 3.2 that

$$k_\ell(\lambda) = \rho\left(\left(\frac{N}{1-c}\right)^\ell \lambda\right). \tag{4.2}$$

Fix a $\lambda_0 > 0$ such that $\mathcal{N}(\lambda_0) > 0$. Let

$$d_s = 2\,\frac{\log N}{\log N - \log(1-c)}. \tag{4.3}$$

For $x \in \left[\left(\frac{N}{1-c}\right)^{\ell-1}\lambda_0, \left(\frac{N}{1-c}\right)^{\ell}\lambda_0\right]$, we have $N^{\ell-1}\lambda_0^{d_s/2} \leq x^{d_s/2} \leq N^\ell \lambda_0^{d_s/2}$ and , by (4.2)

$$\frac{k_{\ell-1}(\lambda_0)}{N^{\ell-1}}\frac{\lambda_0^{-d_s/2}}{N} \leq \frac{\rho(x)}{x^{d_s/2}} \leq \frac{k_\ell(\lambda_0)}{N^\ell} N\,\lambda_0^{-d_s/2}. \tag{4.4}$$

THEOREM 4.1.

(1)

$$0 < \underline{\lim}_{x\to\infty} \frac{\rho(x)}{x^{d_s/2}} \leq \overline{\lim}_{x\to\infty} \frac{\rho(x)}{x^{d_s/2}} < \infty.$$

(2) $\mathcal{N}\left(\left(\frac{N}{1-c}\right)\lambda\right) = N\cdot\mathcal{N}(\lambda)$ *for any* λ.

$$0 < \underline{\lim}_{x\to\infty} \frac{\mathcal{N}(x)}{x^{d_s/2}} \leq \overline{\lim}_{x\to\infty} \frac{\mathcal{N}(x)}{x^{d_s/2}} < \infty,$$

$$0 < \underline{\lim}_{x\downarrow 0} \frac{\mathcal{N}(x)}{x^{d_s/2}} \leq \overline{\lim}_{x\downarrow 0} \frac{\mathcal{N}(x)}{x^{d_s/2}} < \infty.$$

PROOF: By letting $x \to \infty$ in (4.4), we get (1) with lower bound $\mathcal{N}(\lambda_0)\frac{\lambda_0^{-d_s/2}}{N}$ and upper bound $\mathcal{N}(\lambda_0)N\lambda_0^{-d_s/2}$. (4.1) and(4.2) lead us to the above scaling property of $\mathcal{N}$ which in turn implies the above asymptotics of $\mathcal{N}$ at 0 and ∞.

In the case of the Sirpinski gasket,ρ and $\mathcal{N}$ are so wild that no equality holds in the middle of each of three inequalities in the above theorem([5]).

REFERENCES

[1] M.T.Barlow and E.A.Perkins, *Brownian motion on the Sierpinski gasket* , Prob.Theo.Rel.Fields **79** (1988), 543-624.

[2] M.T.Barlow, *Random walks, electrical resistence, and nested fractals*, preprint.

[3] M.Fukushima, "Dirichlet forms and Markov processes," Kodansha and NorthHolland, 1980.

[4] M.Fukushima, *On recurence criteria in the Dirichlet space theory*, in "From local times to global geometry, control and physics," ed. by K.D.Elworthy, Pitman Research Notes in Math. 150, Longman, 1986, pp. 100-110.

[5] M.Fukushima and T.Shima, *On a spectral analysis for the Sierpinski gasket*, J. of Potential Analysis (to appear).

[6] J.Kigami, *A harmonic calculus on the Sierpinski spaces*, Japan J.Appl.Math. **6** (1989), 259-290.

[7] J.Kigami, *Harmonic calculus on P.C.F. self-similar sets*, Trans.AMS (to appear).

[8] T.Kumagaya, *Regularity and spectral dimensions of Kigami's self-similar sets*, preprint.

[9] S.Kusuoka, *A diffusion process on a fractal*, in "Probabilistic methods on Mathematical Physics," ed. by K.Ito and N.Ikeda, Kinokuniya and North-Holland, 1987, pp. 251-274.

[10] S.Kusuoka, *Diffusion processes on nested fractals*, Lecture Note at Nankai University, 1989.

[11] T.Lindstrom, "Brownian motion on nested fractals," Mem.Amer.Math.Soc. 420, 1989.

[12] Katarzyna Pietruska-Paluba, *The Lifschitz singularity for the density of states on the Sierpinski gasket*, Preprint.

[13] R.Rammal and G.Toulouse, *Random walks on fractal sturctures and percolation clusters*, J.Physique Lett. **43** (1982), 13-22.

[14] R.Rammal, *Spectrum of harmonic excitations on fractals*, J.Physique **45** (1984), 191-206.

[15] T.Shima, *On eigenvalue problems for random walks on the Sierpinski pregaskets*, Japan J. appl.Math. (to appear).

[16] T.Shima, *On Lifschitz tails for the density of states on nested fractals*, Preprint.

Large Deviations for Weak Solutions of Stochastic Differential Equations

G.Jona-Lasinio

Dipartimento di Fisica - Università "La Sapienza", Roma

Abstract

We give an informal discussion of Freidlin-Ventzell type estimates for ordinary stochastic differential equations with measurable drift and constant diffusion.

1. In a recent paper[1] in collaboration with P.K.Mitter we established some large deviation estimates of the Freidlin-Ventzell type[2] for a stochastic parabolic P.D.E. in two space dimensions. This is an extreme case of an equation which has only weak solutions and in addition the equation itself is purely formal in so far as infinite constants appear among its coefficients. The problem of large deviations in this case required overcoming various special difficulties which are not present in simpler cases like, for example, ordinary differential equations with measurable coefficients. The general strategy remained therefore obscured to a certain extent by the technicalities involved.

In this paper we want to explain the basic idea, which is quite simple, by applying it to the situation of an ordinary SDE with measurable drift $\underline{b}$ and constant diffusion ε, that is to the equation

$$d\underline{x}_t = \underline{b}(\underline{x}_t)dt + \varepsilon\, d\underline{w}_r \tag{1}$$

Dimensionality does not play any role in our considerations so that we may assume without loss of generality that $x \in R^1$ and b is a scalar function. For our purpose the most convenient definition of a weak solution is via the Cameron-Martin-Girsanov formula, that is we define the evolution semigroup

$$E_x\,(f(x_t)) \overset{\text{def}}{=} E_x\,(f(w_t^\varepsilon)e^{\xi_T}) \tag{2}$$

where T>t and

$$\xi_T = \frac{1}{\varepsilon} \int_0^T b(w_s^\varepsilon)\, dw_s - \frac{1}{2\,\varepsilon^2} \int_0^T ds\, b^2(w_s^\varepsilon)\, ds \tag{3}$$

E_x is the expectation with respect to the Wiener process starting at x and w_t^ε is the process εw_t. In order that our definition (2) be meaningful it is sufficient that $E_x(e^{p\xi_T}) < \infty$ and $E_x\,(|f(w_t^\varepsilon)|^q) < \infty$, $\frac{1}{p} + \frac{1}{q} = 1$. We shall assume that this conditions are satisfied for all finite p, q $\geq$ 1 uniformly in ε. We further assume that b is locally integrable. We want to establish Freidlin-Ventzell (F-V) type estimates.

2. The first F-V estimate gives a lower bound to the probability $P_x\,(\sup_{0\leq t\leq T} |x_t - \varphi_t| < \delta)$ where φ_t is a preassigned trajectory starting at the same point x as the process. When eq.(1) has strong solutions the standard way to study this probability[2] is to reduce the problem to the case of the process εw_t using the continuity of the map connecting x_t to w_t. This method cannot be applied directly in our case. Let us introduce then a regularization of our drift $b^{(n)}(x)$ by taking for example the convolution of b(x) with smooth functions in such a way that $b^{(n)}(x) \underset{n\to\infty}{\to} b(x)$ almost everywhere. The equations

$$dx_t^{(n)} = b^{(n)}(x_t^{(n)})\, dt + \varepsilon\, dw_t \tag{4}$$

have strong solutions and for each n the usual F-V large deviation theory applies. On the other hand we may write using (2)

$$P_x\,(\sup_{0\leq t\leq T} |x_t - \varphi_t| < \delta) = E_x(\chi\,(\sup_{0\leq t\leq T} |w_t^\varepsilon - \varphi_t| < \delta)\, e^{\xi_T})$$

$$\geq E_x(\chi(A)\,\chi(|\xi_T - \xi_T^{(n)}| < \alpha)\, e^{\xi_T}) \geq$$

$$\geq E_x(\chi(A)\,\chi(B)\, e^{\xi_T^{(n)}})\, e^{-\alpha} \geq \tag{5}$$

$$\geq E_x(\chi(A)\, e^{\xi_T^{(n)}})\, e^{-\alpha} - E_x(\chi(\overline{B})\, e^{\xi_T^{(n)}})$$

$$\geq E_x(\chi(A)\, e^{\xi_T^{(n)}})\, e^{-\alpha} - P_x^{1/2}(\overline{B})\, E_x^{1/2}(e^{2\xi_T^{(n)}})$$

$\xi_T^{(n)}$ is obtained from ξ_T by substituting b with $b^{(n)}$, the χ's are characteristic functions, A and B have obvious meaning and $\overline{B}$ is the complementary event, $\alpha>0$ is a number to be chosen.

The first term in the last line, apart from multiplication by $e^{-\alpha}$, represents $P_x(\sup_{0\leq t\leq T} |x_t^{(n)} - \varphi_t| < \delta)$ which can be estimated à la F-V. For this purpose let us introduce the action functional

$$I_T^{(n)}(\varphi) = \int_0^T |\dot{\varphi}_s - b^{(n)}(\varphi_s)|^2\, ds \tag{6}$$

Then given some small number h we have

$$P_x(\sup_{0\leq t\leq T} |x_t^{(n)} - \varphi_t| < \delta) \geq e^{-(I_T^{(n)}(\varphi)+h)/2\varepsilon^2} \tag{7}$$

for $\varepsilon<\varepsilon_0(n,h,\delta)$. We assume $I_T^{(n)}(\varphi) \leq K$ for $\forall$ n which is a condition on φ. It is clear from (5) that we can obtain a reasonable estimate if we can choose α in such a way that the right hand side of (7) is not substantially altered by multiplication with $e^{-\alpha}$ and n so large that the second term in the last line of (5) becomes negligible with respect to the first. The first requirement is satisfied for example with $\alpha = h/\varepsilon^2$. To satisfy the second requirement the natural condition is, as much of the experience with large deviations suggests, the superexponential convergence

$$\lim_{n\to\infty}\ \lim_{\varepsilon\to 0} \varepsilon^2 \ln P_x(\overline{B}) = \lim_{n\to\infty}\ \lim_{\varepsilon\to 0} \varepsilon^2 \ln P_x (|\xi_T - \xi_T^{(n)}| > \frac{h}{\varepsilon^2}) = -\infty \tag{8}$$

(8) holds if, as it is often the case, we can prove an estimate of the form

$$P_x (|\xi_T - \xi_T^{(n)}| > \frac{h}{\varepsilon^2}) \le e^{-C(n)/\varepsilon^2} \tag{9}$$

for sufficiently large n with $C(n) \underset{n\to\infty}{\longrightarrow} \infty$. From (5) and (9) we obtain immediately for large n

$$\lim_{\varepsilon\to 0} \varepsilon^2 \ln P_x (\sup_{0\le t\le T} |x_t - \varphi_t| < \delta) \ge -\frac{1}{2} I_T^{(n)}(\varphi)$$

Since the first term of this inequality does not depend on n we can take the limit $n\to\infty$

$$\lim_{\varepsilon\to 0} \varepsilon^2 \ln P_x (\sup_{0\le t\le T} |x_t - \varphi_t| < \delta) \ge -\frac{1}{2} I_T(\varphi) \tag{10}$$

Under our assumptions we can take the limit under the integral sign.

3. We now consider the second F-V estimate which is more subtle. We define

$$\phi^{(n)}(s) = \{\varphi_t : I_T^{(n)}(\varphi) \le s\} \tag{11}$$

$\varphi_t \in C(0,T)$, $\varphi_0 = x$. We want to study the probability of the event

$$A = \{x_t : \rho(x, \phi^{(n)}(s)) > \delta\} \tag{12}$$

where ρ is the distance in the sup norm of x_t from the set $\phi^{(n)}(s)$. We have from (2), and using the same notations of the previous section

$$P_x(A) = E_x(\chi(\rho(w^\varepsilon, \phi^{(n)}(s)) > \delta)\, e^{\xi T})$$

$$= E_x(\chi(A)\,\chi(B)\, e^{\xi T}) + E_x(\chi(A)\,\chi(\bar{B})\, e^{\xi T})$$

$$\leq E_x(\chi(A)\, e^{\xi T^{(n)}})\, e^{\alpha} + E_x(\chi(\bar{B})\, e^{\xi T}) \tag{13}$$

$$\leq E_x(\chi(A)\, e^{\xi T^{(n)}})\, e^{\alpha} + P_x^{1/2}(\bar{B})\, E^{1/2}(e^{2\xi T})$$

The expectation in the first term of the last line of (13) is $P_x(\rho(x^{(n)}, \phi^{(n)}(s)) > \delta)$.

Applying the second F-V estimate to $x^{(n)}$

$$P_x(\rho(x^{(n)}, \phi^{(n)}(s)) > \delta) \leq e^{-(s-h)/2\varepsilon^2} \tag{14}$$

valid for $\varepsilon<\varepsilon_0(\delta,n,h)$.
Assuming again $\alpha=h/\varepsilon^2$, and the superexponential estimate (9) we obtain

$$\lim_{\varepsilon\to 0} \varepsilon^2 \ln P_x(\rho(x, \phi^{(n)}(s)) > \delta) \leq -s$$

Taking the limit $n\to\infty$, we have finally

$$\lim_{n\to\infty} \lim_{\varepsilon\to 0} \varepsilon^2 \ln P_x(\rho(x, \phi^{(n)}(s)) > \delta) \leq -s \tag{15}$$

(15) is clearly weaker than the usual F-V second estimate.

Acknowledgements

This note is based on my previous work with P.K.Mitter whom I wish to thank for a stimulating collaboration over many years. I am also very grateful to C.Landim for illuminating discussions and for pointing out an error in a first version of this paper.

References

[1] G.Jona-Lasinio, P.K.Mitter: "Large Deviation Estimates in the Stochastic Quantization of ϕ_2^4". Comm. Math. Phys. 130, 111 (1990).

[2] M.I. Freidlin, A.D. Ventzell: "Random Perturbations of Dynamical Systems" Springer 1984.

On a Class of Probabilistic Integrodifferential Equations

Torbjörn Kolsrud

Introduction. This note is inspired by a series of articles of Schneider and Wyss (see [7,8] and references therein) on the *fractional diffusion equation*

$$(*)\qquad u(t,x) = u_0(x) + \int_0^t \frac{(t-s)^{\alpha-1}}{\Gamma(\alpha)} \quad \frac{\partial^2 u}{\partial x^2}(x,s)ds,$$

and certain special functions and distributions related to (*). Our purpose is to show how to interprete (*) and generalisations of it probabilistically. The main ingredient is the local time of a one-dimensional diffusion, and the main idea is to understand the calculations on the Mittag-Leffler function performed in [3, pp. 453-4] from that point of view.

Consider the integrodifferential equation

$$(1)\qquad u(t,x) = u_0(x) + \int_0^t \frac{\partial^2 u}{\partial x^2}(x,t-s)\rho(ds),$$

where $\rho \geq 0$ is a positive measure on $\mathbf{R}_+ = [0,\infty)$. To avoid trivialities we assume $\rho \not\equiv 0$. Denote by $\hat{u}_0(\xi)$ the Fourier transform of u_0, with the convention

$$(2)\qquad \hat{u}_0(\xi) = \int_{-\infty}^{\infty} e^{-ix\xi} u_0(x)dx, \quad \xi \in \mathbf{R}.$$

Denote by $\mathcal{L}\rho(\tau)$ the Laplace transform of ρ (which we assume satisfies appropriate growth conditions at infinity):

$$(3)\qquad \mathcal{L}\rho(\tau) = \int_0^{\infty} e^{-t\tau}\rho(dt) \equiv \int_{[0,\infty)} e^{-t\tau}\rho(dt), \quad \tau > 0.$$

We denote by $v(\tau,\xi)$ the Fourier-Laplace transform of u :

$$(4)\qquad v(\tau,\xi) = \mathcal{L}\hat{u}(\tau,\xi) = \int_0^{\infty}\int_{-\infty}^{\infty} e^{-t\tau}e^{-ix\xi}u(x,t)dxdt.$$

Supported in part by grants from the Swedish Natural Science Research Council, NFR, and the Royal Swedish Academy of Sciences.

Then (1) becomes

$$v(\tau,\xi) = \frac{1}{\tau(1+\xi^2\mathcal{L}\rho(\tau))}\hat{u}_0(\xi), \tag{5}$$

which by inversion gives an expression for the kernel of the map $u_0(x) \to u(t,x)$ defined by (1). (We shall not be concerned here with questions on which regularity requirements u_0 must satisfy.)

We now put conditions on ρ that, as we shall see, allows for a probabilistic interpretation of (1). Let

$$\varphi(\tau) = \frac{1}{\mathcal{L}\rho(\tau)}\ , \tag{6}$$

and assume that φ is a *Bernstein function* ([1]), i.e. $(-1)^p\varphi^{(p)}(\tau) \leq 0\ , p = 1,2,\ldots$ Instead of (5) we get

$$v(\tau,\xi) = \frac{\varphi(\tau)}{\tau(\varphi(\tau)+\xi^2)}\hat{u}_0(\xi). \tag{7}$$

This latter multiplier, $\varphi(\tau)/\tau(\varphi(\tau)+\xi^2)$, can be described probabilistically, and it is best to start with its close relative

$$\frac{\varphi(\tau)}{\tau(\varphi(\tau)+\lambda)}, \quad \lambda \geq 0. \tag{8}$$

Our construction is based on local times for one-dimensional diffusions. It is well known ([2, 4]) that the inverses of local times are infinitely divisible. Denote by α a local time and by β its inverse, so that

$$\alpha_t \leq x \quad \Longleftrightarrow \quad \beta_x \geq t, \quad x,t \geq 0. \tag{9}$$

We assume that β is càdlàg.

Write

$$\begin{cases} F(t,x) = Pr(\beta_x \leq t) \\ G(t,x) = Pr(\alpha_t \leq x) \end{cases} \tag{10}$$

so that

$$G(t,x) = 1 - F(t,x). \tag{11}$$

For notational convenience we assume that F and G have densities:

$$\begin{cases} f(t,x) = \dfrac{\partial F}{\partial t}(t,x) \\ g(t,x) = \dfrac{\partial G}{\partial x}(t,x). \end{cases} \tag{12}$$

It is further assumed that f and φ are related by

$$\mathcal{L}f(\cdot,x)(\tau) = e^{-x\varphi(\tau)}\ , \tag{13}$$

which in fact is a demand on φ (hence on ρ); see [4], p. 217. We shall assume, finally, that

$$F(x,0) \equiv 0. \tag{14}$$

Denote the bivariate Laplace transform of g by

$$\mathcal{L}_2 g(\tau,\lambda) \equiv \int_0^\infty \int_0^\infty g(t,x)e^{-t\tau}e^{-x\lambda}dt\ dx. \tag{15}$$

Then we have

$$\mathcal{L}_2 g(\tau,\lambda) = \frac{\varphi(\tau)}{\tau(\varphi(\tau)+\lambda)}, \tag{16}$$

i.e. the bivariate Laplace transform of g is the function defined by (8).

PROOF: By (11) and (12)

$$g = -\frac{\partial F}{\partial x}, \tag{17}$$

so

$$\mathcal{L}_2 g = -\mathcal{L}_2 \frac{\partial F}{\partial x}. \tag{18}$$

Integrate (13) partially and use (14) to get

$$\mathcal{L}F(\cdot,x)(\tau) = \frac{1}{\tau}e^{-x\varphi(\tau)}. \tag{19}$$

Then

$$\mathcal{L}\left(-\frac{\partial F}{\partial x}(\cdot,x)\right)(\tau) = \frac{\varphi(\tau)}{\tau}e^{-x\varphi(\tau)}, \tag{20}$$

hence

$$\mathcal{L}_2\left(-\frac{\partial F}{\partial x}\right)(\tau,\lambda) = \frac{\varphi(\tau)}{\tau}\int_0^\infty e^{-x\varphi(\tau)}e^{-x\lambda}dx = \frac{\varphi(\tau)}{\tau(\varphi(\tau)+\lambda)}, \tag{21}$$

which together with (17) yields (16). ∎

Let now

$$p(t,x) = \int_0^\infty \frac{e^{-x^2/4s}}{\sqrt{4\pi s}}g(t,s)ds, \quad x \in \mathbf{R}. \tag{22}$$

Then, by Fubini's theorem,

$$
\begin{aligned}
&\int_0^\infty \int_{-\infty}^\infty p(t,x)e^{-t\tau}e^{-ix\xi}dt\,dx \\
&= \int_0^\infty \int_0^\infty g(t,s)e^{-t\tau} \int_{-\infty}^\infty \frac{e^{-x^2/4s}}{\sqrt{4\pi s}} e^{-ix\xi}dx\;dt\;ds \\
&= \int_0^\infty \int_0^\infty g(t,s)e^{-t\tau}e^{-s\xi^2}\;dt\;ds \\
&= \mathcal{L}_2 g(\tau,\xi^2) = \frac{\varphi(\tau)}{\tau(\varphi(\tau)+\xi^2)}.
\end{aligned}
\tag{23}
$$

It follows that (1) has the solution

$$
u(t,x) = \int_{-\infty}^\infty p(t,x-y)u_0(y)dy. \tag{24}
$$

To get a probabilistic interpretation of this formula, let $B = (B_s)$ be a Brownian motion, independent of α and starting at $x = 0$, and set

$$
X_t = B_{\alpha_t}. \tag{25}
$$

It is easily verified that then

$$
\int_A p(t,x)dx = Pr\Big(X_t \in A\Big) \tag{26}
$$

for any Borel set A in $\mathbf{R}$, so (24) is simply

$$
u(t,x) = \mathbf{E}\Big[u_0(x+X_t)\Big]. \tag{27}
$$

REMARKS: (a) It is easy to generalise this result to higher (space-) dimension. Also, we may replace ξ^2 by more general so-called negative definite ([1]) functions $\psi(\xi)$. This means that the integral in (1) should be replaced by $\int_0^t(-Au)(x,t-s)\rho(ds)$, where ψ is the Fourier symbol of $-A$. In fact translation invariance in space is not really needed (cf. [5]), but we must have a generalised subordination formula such as (25), though.

(b) The fractional diffusion equation (*) is associated with the local times of (fractional) Bessel processes. See Molchanov - Ostrovskii [6].

References:

[1] Berg, C., Forst, G., Potential Theory on Locally Compact Groups. Springer 1975.
[2] Blumenthal, R.M., Getoor, R.K., Markov Processes and Potential Theory. Academic Press 1968.
[3] Feller, W., An Introduction to Probability Theory and its Applications, vol. II, 2nd edition. Wiley 1971.
[4] Itô, K., Mc Kean, H.P., Diffusion Processes and their Sample Paths. Springer 1965.
[5] Kolsrud, T., Traces of Harmonic Functions, Capacities, and Traces of Symmetric Markov Processes. Preprint TRITA - MAT - 1988-11. To appear in J. Theor. Probability.
[6] Molchanov. S.A., Ostrovskii, E. Symmetric stable processes as traces of degenerate diffusion processes. Th. Prob. Appl. **14** (1969), 128-31.
[7] Schneider, W.R., Grey Noise. In this volume.
[8] ——, Wyss, W., Fractional diffusion and wave equations. J. Math. Phys 30 (1989), 134-144.

Department of Mathematics
The Royal Institute of Technology
S-100 44 Stockholm
Sweden

OPERATOR INTEGRALS AND MARTINGALE INTEGRALS WITH GENERAL PARAMETER

TORBJÖRN KOLSRUD

KTH, Stockholm, Sweden

Summary. Aiming at general martingale integration, we sketch an integration theory based on selecting commuting subfamilies of two spectral resolutions. In the several parameter case, neither explicit mention of the co-ordinates, nor assumptions on any kind of conditional independence are made.

Introduction.

The article Cairoli-Walsh [2] meant a breakthrough for martingales with several-dimensional (in practice the dimension is two, and the index set $\mathbf{R}^2$) time, and since then many papers on integration w.r.t. such martingales have appeared. –See the references in [9], especially P.A. Meyer's article (p. 1 ff.).

Generally speaking this theory is based on some kind of assumption on independence (e.g. condition (F4) in [2]) between the σ-algebras corresponding to the coordinate axes. In the present note we consider martingale integrals without such assumptions. In fact the basic idea is to develop martingale integration without the probabilistic frame, and emphasis is put instead on pure Hilbert space theory (projections, spectral measures,...) which is really what it is all about, as far as integration is concerned.

This is, of course, not the case when it comes to the truly special features of martingale theory: a.s. convergence, existence of quadratic variation etc, on which we have very little to say. –See however the articles [2, 3, 5] about the special features connected with several dimensional parameter martingales.

Basics.

1. The notions of projective systems and projective limits are essential when discussing martingales, and we need to define these.

Suppose we have a partially ordered set $(I, \leq)$, and to each $i \in I$ a space $X_i = (X_i, \mathcal{A}_i)$ which will be either topological or measurable. The system $(X_i, i \in I)$ is *projective* if there are maps

$$\pi_{ij}: X_i \to X_j, \quad j \leq i,$$

satisfying

$$\pi_{ik} = \pi_{ij} \circ \pi_{jk}, \quad k \leq j \leq i.$$

Supported in part by the Swedish Natural Science Research Council, NFR, and the Royal Swedish Academy of Sciences.

It is assumed that these maps are continuous if the spaces are topological, and measurable if they are measurable.

The projective limit is the space $X = \operatorname{proj\,lim}_{i\in I} X_i$ consisting of coherent systems:

$$X = \{(x_i)_{i\in I} \colon x_j = \pi_{ij}(x_i),\ j \leq i\},$$

equipped with the smallest topological/measurable structure that makes all the π_{ij} continuous/measurable.

When each X_i is a measurable space we may talk of a projective system of measures. This is a system $(\mu_i)_{i\in I}$, where for each $i \in I$, μ_i is a measure on $(X_i, \mathcal{A}_i)$, and

$$\mu_j = \mu_i \circ (\pi_{ij})^{-1},\ j \leq i.$$

2. Example. Let now $(\Omega, \mathcal{F}, \mathbf{P})$ be a probability space, and let $\mathcal{F}_i$, $i \in I$, be an increasing family of sub σ-algebras of $\mathcal{F}$. Define

$$X_i \equiv L^1(\Omega, \mathcal{F}_i, \mathbf{P}),\ i \in I,$$

and let π_{ij} be the maps obtained by taking conditional expectations, i.e.

$$\xi \to \pi_{ij}\xi \equiv \mathbf{E}[\xi|\mathcal{F}_j] \in X_j, \quad \xi \in X_i,\ j \leq i.$$

We see that a martingale $(\xi_i)_{i\in I}$ is nothing but an element in the projective limit of the X_i. We may of course also replace L^1 by L^p, $p > 1$, here. In what follows we shall only consider L^2-martingales.

Usually martingales are indexed by an ordered set on which the random variables 'live', such as $\mathbf{Z}$, $\mathbf{R}$, $\mathbf{R}^n$ or subsets thereof. The order relation then refers to this underlying set. It is however more natural, we think, and anyway less restrictive, to look upon the index set from other points of view. Here is the first basic, rather general, one. It was suggested by my colleague Lars Svensson.

3. Let $(\Omega, \mathcal{F}, \mathbf{P})$ be as above and define the *spectrum* of $\mathcal{F}$, to be denoted $\operatorname{spec}\mathcal{F}$, as all sub σ-algebras $\mathcal{G}$ of $\mathcal{F}$.

For each $\mathcal{G} \in \operatorname{spec}\mathcal{F}$, define the space

$$X_{\mathcal{G}} \equiv L^2(\Omega, \mathcal{G}, \mathbf{P}).$$

$\operatorname{Spec}\mathcal{F}$ is naturally ordered by inclusion, and this gives us another way of looking at martingales, simply as elements in the projective limit of $X_{\mathcal{G}}$ along $\operatorname{spec}\mathcal{F}$. (Strictly speaking one should have the martingales indexed by some subset of $\operatorname{spec}\mathcal{F}$. See further comments on this in section 5 below.)

It is suggestive to look upon the martingales as elements in a bundle, or sheaf. We have a base space, $\operatorname{spec}\mathcal{F}$, and over each point $\mathcal{G}$ in the base space we have a fibre $X_{\mathcal{G}}$. These are linked together by conditional projections.

Although we will not use it, we remark that $\operatorname{spec}\mathcal{F}$ can be topologised. The *Zariski topology* is given by the base of sets

$$U_F \equiv \{\mathcal{G} \in \operatorname{spec}\mathcal{F} \colon F \notin \mathcal{G}\}, \quad F \in \mathcal{F}.$$

In an ordinary filtration $(\mathcal{F}_t,\, t \in \mathbf{R})$ the Zariski topology will usually not make any difference between say $(-\infty, t)$ and $(-\infty, t]$. Of course this has to do with how one by convention makes the σ-algebras right-continuous. We will encounter this again in the next section.

It is also easy to define stopping times. A map $T \colon \Omega \to \operatorname{spec}\mathcal{F}$ is a *stopping time* provided

$$T^{-1}(\operatorname{spec}\mathcal{G}) \subset \mathcal{G}, \quad \mathcal{G} \in \operatorname{spec}\mathcal{F}.$$

4. Let us now turn to filtrations. Consider first a martingale $\xi = (\xi_t)$indexed by $\mathbf{R}$. Then the martingale property implies that ξ has *orthogonal increments*:

$$\mathbf{E}\big[(\xi_u - \xi_t)(\xi_t - \xi_s)^*\big] = 0, \quad s \le t \le u.$$

We can therefore find an increasing function f such that

$$\mathbf{E}\big[|\xi_t - \xi_s|^2\big] = f(t) - f(s), \quad s \le t,$$

and we may assume that f is right-continuous. By standard measure theory we can now find a Radon measure μ on $\mathbf{R}$ such that

$$\mu(s,t] = f(t) - f(s), \quad s \le t,$$

and we may view ξ as a set-indexed random field $(\xi(A),\, A \in \mathcal{B}(\mathbf{R}))$, satisfying

$$\mathbf{E}[\xi(A)\xi(B)^*] = \mu(A \cap B).$$

(Here $\mathcal{B}(\mathbf{R})$) denotes the Borel sets on $\mathbf{R}$.) The 'old' martingale is recovered by the relation $\xi_t = \xi((-\infty, t])$.

Now the measure μ depends on ξ. It is a general fact in operator theory that given two martingales ξ and η, which we look upon as random fields indexed by $\mathcal{B}(\mathbf{R})$), we can find a measure $\mu_{\xi,\eta}$ such that

$$\mathbf{E}[\xi(A)\eta(B)^*] = \mu_{\xi,\eta}(A \cap B).$$

The family of measures $\mu_{\xi,\eta}$ represents $L^2(\Omega, \mathcal{F}, \mathbf{P})$ as a subspace in a projective limit, and it can be associated with a spectral measure $\mathbf{Q} = (Q(A),\, A \in \mathcal{B}(\mathbf{R}))$ on $L^2(\Omega, \mathcal{F}, \mathbf{P})$. Then $\xi(A) = Q(A)\xi$, and

$$\mathbf{E}[Q(A)\xi\,(Q(B)\eta)^*] = \mu_{\xi,\eta}(A \cap B).$$

Strictly speaking this relation holds only for, say, relatively compact Borel sets. The measure $\mu = \mu_{\xi,\xi}$ above is finite if and only if our original martingale is closed, i.e. $\xi(\mathbf{R}) \in L^2(\Omega, \mathcal{F}, \mathbf{P})$.

Summing up, we have seen that ordinary, i.e. classical, martingales can be associated with spectral measures on the underlying space $L^2(\Omega, \mathcal{F}, \mathbf{P})$. The reasoning above is valid in $\mathbf{R}^n$ as well. We shall therefore in what follows consider filtrations as spectral measures (indexed by a general measurable space $(\mathbf{A}, \mathcal{A})$), simply. This will be our starting point for martingale integration below.

Note that given such a martingale we may consider an increasing one-parameter subfamily (A_t) of Borel sets. Then $\xi(A_t)$ will be a classical one-parameter martingale. If $(\mathbf{A}, \mathcal{A})$ is the underlying measurable space, we may view this as a free choice of time on $\mathbf{A}$, i.e. given a 'time' we have an ordinary martingale. This idea is in fact implicit in the theory of Gaussian Markov random fields (indexed by duals of Dirichlet spaces). See Kolsrud [7, 8], for instance.

If we are given a martingale in the formulation of section 3, we may consider continuous sections in the bundle. These may have various dimension, which means that they may be indexed by (subclasses of) $\mathbf{R}^n$ (for various n) or, more generally, a σ-algebra $\mathcal{A}$, where $(\mathbf{A}, \mathcal{A})$ is a measurable space. Thus a section will be a map from $\mathcal{A}$ into spec $\mathcal{F}$ which is additive and continuous, i.e. a spectral measure.

Integration Theory.

One of the motivations behind this article has been to separate from martingale theory what does not depend on the very special structure we actually have, viz. the maps (projections) defining our projective limit are in fact conditional expectations. It turns out that the basic ideas in martingale integration are really of a much more general nature; accordingly we will now present a more general machinery.

5. We start with a Hilbert space $\mathbf{H}$, with inner product $(\cdot, \cdot)$ and norm $\|\cdot\|$, and denote by $\mathcal{P}$ all orthogonal projections on $\mathbf{H}$. We can now look upon $\mathcal{P}$ as the base space, and the total space, i.e. the (Hilbert) bundle, is given by all pairs $(P, P\mathbf{H})$, $P \in \mathcal{P}$. When $\mathbf{H} = L^2(\Omega, \mathcal{F}, \mathbf{P})$, our old bundle (section 3) is a subbundle, since spec $\mathcal{F}$ can be seen as a subset of $\mathcal{P}$.

To obtain an "ordinary" martingale we really need some subclass $\mathcal{P}_0$ of $\mathcal{P}$, e.g. all projections P such that $(1-P)\mathbf{H}$ is infinite dimensional. A martingale is then an element in the subbundle with base space $\mathcal{P}_0$, and as such it may be looked upon as an element in the projective limit.

When it comes to martingale integration the full projective limit structure is not really needed since we can always replace an element ξ in $\operatorname{proj\,lim}_{\mathcal{P}} P\mathbf{H}$ by $P_0\xi$, where $P_0 \in \mathcal{P}_0$. We will therefore only consider elements ξ in $\mathbf{H}$. The martingale is then the system $(P\xi,\ P \in \mathcal{P})$

To define integrals with respect to a martingale, we need, as mentioned in section 4, a measurable space $(\mathbf{A}, \mathcal{A})$ and a map $Q : \mathcal{A} \to \mathcal{P}$ such that $\mathbf{Q} \equiv (Q(A),\ A \in \mathcal{A})$ is a spectral measure on $\mathbf{H}$. $\mathbf{Q}$ is our time, or filtration.

To each pair ξ, η in $\mathbf{H}$ we have a measure $\mu_{\xi,\eta}$, such that

$$(Q(A)\xi, Q(B)\eta) = \mu_{\xi,\eta}(A \cap B), \quad A, B \in \mathcal{A}.$$

We can integrate $\mathcal{A}$-measurable functions with respect to $\mu_{\xi,\eta}$ also. An alternative way to this, is to extend ξ and η from set-indexed random fields to random fields indexed by functions. For a function $\phi \in L^2(\mathbf{A}, \mathcal{A}, \mu_{\xi,\xi})$, we define $\xi(\phi)$ so that

$$\|\xi(\phi)\|^2 = \int \|\phi\|^2 \, d\mu_{\xi,\xi}.$$

This means that

$$\xi(\phi) = \int \phi(a) \, Q(da)(\xi),$$

in operator notation. More generally, we have

$$(\xi(\phi), \eta(\psi)) = \int \phi\psi^* \, d\mu_{\xi,\eta}.$$

If the underlying space is $\mathbf{R}^n$ or some manifold, we may let differential operators act on the martingales. If L is a differential operator, then

$$L\xi(\phi) \equiv \xi(L^t\phi) = \int L^t\phi(a) \, Q(da)(\xi),$$

where L^t is the formal adjoint of L. In this case

$$(L\xi(\phi), M\eta(\psi)) = \int L^t\phi(M^t\psi)^* \, d\mu_{\xi,\eta}.$$

The function-indexed random field is ξ. The classical martingale Ξ, which should be the integral of the noise ξ, is then given by the stochastic differential equation

$$\xi = \partial_1 \ldots \partial_n \Xi,$$

assuming the underlying space is $\mathbf{R}^n$, and $\partial_i = \partial / \partial x^i$. Consequently, the classical martingale, which is then indexed by points, may be defined by the relation

$$\Xi = (\partial_1 \ldots \partial_n)^{-1} \, \xi.$$

This means that

$$\Xi(\phi) = \int \Big[\int_{t_j \leq a_j} \phi(t_1, \ldots, t_n) \, dt_1 \ldots dt_n \Big] Q(da),$$

and it merely reflects the relation between distribution functions and measures, as in section 4.

From the stochastic integral point of view the integral we have defined above goes back to Paley and Wiener in the 1930s, when the underlying martingale was Brownian motion. (The general integral for one-dimensional processes with orthogonal increments is treated in Doob [5], p. 426 f.) We look upon it as an integral with deterministic integrand.

Let us now turn to integration with respect to more general (stochastic, if our bundle is based on spec $\mathcal{F}$) integrands.

6. To introduce new integrands we need one more filtration. Suppose that we also have a spectral measure $\mathbf{P} = (P_F,\ F \in \mathcal{F})$, where $(\mathbf{F}, \mathcal{F})$ is a measurable space. The reason for choosing the letter $\mathcal{F}$ is that, associated with spec $\mathcal{F}$ are the obvious projections

$$P_F\xi = 1_F \cdot \xi,$$

which are used when constructing classical martingale integrals (as in [4], for instance).

In general $\mathbf{P} \otimes \mathbf{Q}$ does not define a spectral measure on $\mathbf{H}$, because the P_F and Q_A need not commute. To remedy this, we select, considering $\mathbf{Q}$ as fixed, an appropriate subclass of $\mathbf{P}$ which will commute. This way we obtain a sub σ-algebra of $\mathcal{F} \otimes \mathcal{A}$, having $\mathbf{P}$ and $\mathbf{Q}$ as 'marginals.'

For $A \in \mathcal{A}$, define by $\gamma(A)$ those $F \in \mathcal{F}$ for which P_F and Q_A commute:

$$\gamma(A) \equiv \{F \in \mathcal{F} \colon [P_F, Q_A] = 0\}.$$

It is then easily seen that each $\gamma(A)$ is a σ-algebra, and that the 'rectangles'

$$E = F \times A, \quad F \in \gamma(A),\ A \in \mathcal{A},$$

can be associated with projections

$$\pi_E \equiv P_F Q_A,$$

on $\mathbf{H}$. We can now build a σ-algebra $\mathcal{E} \subset \mathcal{F} \otimes \mathcal{A}$ from such rectangles. In this section however, we shall only consider a finitely based situation, and discuss more general cases in §7. Accordingly, we let

$$\tau \colon \mathbf{A} = A_1 \cup \cdots \cup A_n$$

be a finite (it would make no difference in the following if τ were countably infinite) partitioning of $\mathbf{A}$ into disjoint and measurable sets. Then, with obvious notation,

$$\mathcal{E}^\tau \equiv (\gamma(A_1) \times A_1) \oplus \cdots \oplus (\gamma(A_n) \times A_n) \subset \mathcal{F} \otimes \mathcal{A},$$

is a σ-algebra, and each set

$$E = \bigcup E_j = \bigcup F_j \times A_j, \quad F_j \in \gamma(A_j),\ j = 1, 2, \ldots, n,$$

in $\mathcal{E}^\tau$ can be associated with the projection

$$\pi_E = \bigvee \pi(E_j) = \sum \pi(E_j).$$

Here the second equality follows from the $\pi(E_n)$ being mutually orthogonal:

$$\pi(E_n)\pi(E_m) = P(F_n)Q(A_n)P(F_m)Q(A_m) = P(F_n)Q(A_n)Q(A_m)P(F_m)$$

$$= P(F_n)Q(A_n \cap A_m)P(F_m) = P(F_n)\delta_{nm}Q(A_n)P(F_m) = \delta_{nm}\pi(E_n).$$

Hence

$$\boldsymbol{\pi} = \boldsymbol{\pi}^\tau \equiv (\pi_E,\ E \in \mathcal{E}^\tau)$$

is a spectral family on $\mathbf{H}$, indexed by the measurable space $(\mathbf{E}, \mathcal{E}^\tau)$, where

$$\mathbf{E} \equiv \mathbf{F} \times \mathbf{A}.$$

Consequently we have a measure representation of $\mathbf{H}$ associated with $\boldsymbol{\pi}$. To each $\xi,\ \eta \in \mathbf{H}$

$$\nu_{\xi,\eta}(E) \equiv (\pi_E \xi, \eta) = (\xi, \pi_E \eta), \quad E \in \mathcal{E}^\tau,$$

defines a measure on $(\mathbf{E}, \mathcal{E}^\tau)$. We can now use the $\nu_{\xi,\eta}$ to integrate $\mathcal{E}^\tau$-measurable functions, of course. If Φ satisfies

$$\Phi \in L^2(\mathbf{E}, \mathcal{E}^\tau, \nu_{\xi,\xi}),$$

we define

$$\int_{\mathbf{E}} \Phi(t)\pi(dt)(\xi) \equiv \int \Phi\, d\pi(\xi),$$

as the (in general only densely defined) map

$$\eta \to \int_{\mathbf{E}} \Phi(t)\nu_{\xi,\eta}(dt).$$

We have then the isomorphism formula

$$\int \|\Phi\, d\pi(\xi)\|^2 = \int \|\Phi\|^2 d\nu_{\xi,\xi},$$

and, more generally,

$$\int (\Phi\, d\pi(\xi), \Psi\, d\pi(\eta)) = \int \Phi\Psi^*\, d\nu_{\xi,\eta}.$$

REMARK: We do not need to use the whole family $\gamma(A)$ for each A. We could of course just as well perform the analogous construction w.r.t. a system $(\beta(A),\ A \in \mathcal{A})$, where each $\beta(A)$ is a sub σ-algebra of $\gamma(A)$.

7. The construction above carries over by 'transfer' to non-standard analysis, and one may use the Loob construction to obtain an internal measure corresponding to the continuous case. See [1], Chapter 3. For those who do not like non-standard analysis, we sketch an alternative procedure. The 'continuous case' is treated as a limit over the net $\mathcal{T}$ of finite partitions τ of $\mathbf{A}$. To make such an approach possible, we must demand more. First, we assume that $\mathbf{F}$ is Hausdorff. Secondly, we assume that $\mathbf{A}$ is a nice topological space, and for our purposes compact and Hausdorff is convenient. We also assume that $\gamma(A)$ increases with A, i.e.

(7.1) $$A, B \in \mathcal{A},\ A \subset B \Rightarrow \gamma(A) \subset \gamma(B)$$

For each $\tau \in \mathcal{T}$ we have a sub σ-algebra $\mathcal{E}^\tau$ of $\mathcal{F} \otimes \mathcal{A}$ and a corresponding measure representation ν^τ of the spectral family

$$\boldsymbol{\pi}^\tau \equiv (\pi^\tau(E),\ E \in \mathcal{E}^\tau),$$

i.e.

$$(\pi^\tau(E)\xi, \eta) = \nu^\tau_{\xi,\eta}(E), \quad \xi,\ \eta \in \mathbf{H},\ E \in \mathcal{E}^\tau.$$

Let now $\mathcal{E}_0$ be the *algebra*

$$\mathcal{E}_0 \equiv \sum\{\mathcal{E}^\tau,\ \tau \in \mathcal{T}\},$$

and denote by $\mathcal{E}$ the corresponding σ-algebra

$$\mathcal{E} \equiv \bigvee\{\mathcal{E}^\tau,\ \tau \in \mathcal{T}\} = \sigma(\mathcal{E}_0).$$

Since $\gamma(\cdot)$ is increasing, we immediately see that for fixed ξ and η in $\mathbf{H}$, $(\nu^\tau_{\xi,\,\eta},\ \tau \in \mathcal{T})$ is a projective system on $(\mathbf{E}, \mathcal{E})$.

Fix ξ and consider the corresponding system

$$(\nu^\tau,\ \tau \in \mathcal{T}) \equiv (\nu^\tau_{\xi,\,\xi},\ \tau \in \mathcal{T})$$

Obviously ν^τ is countably additive on $(\mathbf{E}, \mathcal{E}^\tau)$ for each $\tau \in \mathcal{T}$, and therefore finitely additive on $\mathcal{E}_0$. We want to have conditions such that the limit of ν^τ along $\mathcal{T}$ is countably additive on $(\mathbf{E}, \mathcal{E})$. We therefore assume that $\mathbf{P}$ is inner regular w.r.t. $\gamma(\cdot)$ in that for each $F \in \gamma(A)$

$$\|P_F\xi\|^2 = \sup\{\|P_K\xi\|^2 \colon K \text{ compact},\ K \subset F,\ K \in \gamma(A)\}.$$

Then we may use Henry's Theorem ([1], p. 358). According to this, there is some σ-algebra $\mathcal{S} \supset \mathcal{E}_0$ to which $\nu = \lim_{\tau\in\mathcal{T}} \nu^\tau$ has a countably additive extension. But by definition $\mathcal{E} \subset \mathcal{S}$, so ν is a countably additive measure on $(\mathbf{E}, \mathcal{E})$.

If the integrand Φ above is differentiable in the second variable $a \in \mathbf{A}$, we can define $L\xi(\Phi)$, as in section 5. For instance, assuming that $\mathbf{A}$ is a manifold, Φ of the form $\Phi = \sum f_i\phi_i$, where $\phi_i \in Q(A_i)\mathbf{H}$, the A_i are disjoint, and $f_i \in P(F_i)\mathbf{H}$, $F_i \in \gamma(A_i)$, then

$$L\xi(\Phi) \equiv \xi(L^t\Phi) = \xi\left(\sum f_i L^t\phi_i\right)$$

(where L is a differential operator), can be defined by

$$L\xi(\Phi) = \int \sum f_i(x) L^t\phi_i(a)\, P(dx) Q(da).$$

8. The integral we considered in the last two sections was completely 'unordered' and could of course be nothing else, since we did not assume $\mathbf{A}$ to be partially ordered. We will now introduce a kind of Itô, or previsible, integral. It is worth noting that no properties that are special for conditional expectations are needed. Everything works fine in general Hilbert spaces.

It should be mentioned here that in Norberg [10] a purely probabilistic approach to martingale integration with $\mathbf{R}^n$ as parameter space is presented. See also Walsh [11], Chapter 2.

We let $\mathbf{A} = \mathbf{R}^n$ with its Borel σ-algebra, and the lexiographic order: $s \leq t$ means $s_i \leq t_i$, $1 \leq i \leq n$, and similarly for $s < t$. We will use the natural notation for intervals in $\mathbf{R}^n$. $\mathbf{P}$ and $\mathbf{Q}$ are spectral measures on $\mathbf{H}$, as before. Also, condition (7.1) is assumed to hold.

Consider a half-open interval $A = (s,t]$, and instead of $\gamma(A)$, the class

$$\Gamma_s \equiv \{F\colon [P_F, Q_{(-\infty,s]}] = 0\} = \gamma((-\infty,s]),\ s \in \mathbf{R}^n.$$

By the definition of the commutator it is then clear that

$$F \in \Gamma_s \Rightarrow P_F Q_{(-\infty,s]}\mathbf{H} \subset Q_{(-\infty,s]}\mathbf{H}.$$

Since $\mathbf{Q}$ is a spectral measure $Q_{(-\infty,s]}\mathbf{H}$ increases with s. Consequently, for $u \geq s$ and $F \in \Gamma_s$, $P_F Q_{(-\infty,s]}\mathbf{H} \subset Q_{(-\infty,u]}\mathbf{H}$. Now $Q_{(s,t]}$ may be written as a sum of the type

$$Q_{(s,t]} = \sum a_j Q_{(-\infty,u_j]},$$

where the 2^n points u_j all satisfy $s \leq u_j$. It follows from (7.1) that

$$[P_F, Q_{(s,t]}] = 0, \quad F \in \Gamma_s,\ s < t.$$

We can therefore study a new sub σ-algebra $\mathcal{P} \subset \mathcal{F} \otimes \mathcal{A}$, the *previsible* σ-algebra, constructed from disjoint unions av rectangles of the type

$$E = F \times (s,t], \quad F \in \Gamma_s,\ s < t.$$

Again, under suitable tightness conditions, or with an internal construction, we obtain a spectral measure $\boldsymbol{\pi}$, with respect to which we may integrate $\mathcal{P}$-measurable functions. This then gives a spectral measure on $\mathbf{H}$, and the construction may be continued inductively to produce a spectral measure on the tensor algebra over $\mathbf{H}$, as in the Ito-Wiener construction.

References.

[1] S. Albeverio. J.E. Fenstad, R. Høegh-Krohn, T. Lindstrøm. Nonstandard methods in stochastic analysis and mathematical physics. Academic Press 1986.

[2] R. Cairoli, J. Walsh. Stochastic integrals in the plane. Acta. Math. **134** (1974), 111-83.

[3] S.D. Chatterji, Martingale theory: An analytic formulation with some applications in analysis. Pp. 109-67 in "Probability and analysis, Varenna (Como) 1985." G. Letta, M. Pratteli (Eds). Lect. Notes in Math. **1206**, Springer 1986.

[4] K. L. Chung, R. J. Williams. Introduction to stochastic integration. Birkhäuser 1983.

[5] J.L. Doob. Stochastic processes. Wiley 1953.

[6] P. Immerkeller. The structure of two-parameter martingales and their quadratic variation. Thesis, Munich 1987.

[7] T. Kolsrud. On the Markov property for certain Gaussian random fields. Prob. Th. Rel. Fields **74**, 393-402, 1987.

[8] T. Kolsrud. Gaussian random fields, symmetric Markov processes, and infinite dimensional Ornstein-Uhlenbeck processes. Acta Appl. Math. **12** (1988), 237-263.

[9] H. Korezlioglu, G. Mazziotto, J. Szpirglas (Eds). "Processus aleatoires a deux indices." Lect. Notes in Math. **863**, Springer 1981.

[10] T. Norberg, A co-ordinate free description of multiparameter stochastic integration. Preprint, Chalmers, Gothenburg 1988.

[11] J.B. Walsh, An introduction to stochastic partial differential equations. In 'École d'Été de Probabilités de Saint-Flour XIV-1984', Ed. P.L. Hennequin. Springer Lecture Notes in Math. 1180, 1986.

Department of Mathematics, Royal Institute of Technology S-100 44 Stockholm, SWEDEN

WICK MULTIPLICATION AND ITO-SKOROHOD STOCHASTIC DIFFERENTIAL EQUATIONS

Tom Lindstrøm, Bernt Øksendal and Jan Ubøe

Department of Mathematics, University of Oslo
Box 1053, Blindern, N-0316 Oslo 3, NORWAY.

Abstract

We show that there is a close connection between deterministic differential equations of the form

$$\frac{d\xi_t}{dt} = b(\xi_t) + \sigma(\xi_t) \cdot \sum_k \zeta_k(t) z_k$$

(where $z_k = x_k + iy_k$ are complex parameters) and Ito-Skorohod stochastic differential equations of the form

$$dX_t = b^\diamond(X_t)dt + \sigma^\diamond(X_t)\delta B_t,$$

where $b^\diamond, \sigma^\diamond$ denote the *Wick versions* of the functions b, σ.

The connection is provided by the Hermite transform $\mathcal{H}$, which maps L^2 stochastic processes X_t into (deterministic) analytic functions $\mathcal{H}(X_t)(z_1, z_2, \cdots)$ on

$$\mathbf{C}_0^{\mathbf{N}} = \{(z_1, z_2, \cdots); z_k \in \mathbf{C} \text{ and } \exists M \text{ with } z_j = 0 \text{ for } j > M\},$$

and by its inverse $\mathcal{H}^{-1}$, which can be given an explicit form.

As an application we consider a model for population growth in a crowded, stochastic environment.

§1. Introduction.

The purpose of this paper is to establish a link between deterministic differential equations and Ito-Skorohod stochastic differential equations. If the coefficients are analytic functions the connection becomes particularly simple. The key to the link is to replace ordinary products in the deterministic equation by Wick products $\diamond$ in the corresponding Ito-Skorohod equation. More generally, the given ordinary functions f should be replaced by their *Wick versions* $f^{\diamond}$.

The proof of this connection is provided by the use of the *Hermite transform* $\mathcal{H}$ and its (left) inverse $\mathcal{H}^{-1}$. The Hermite transform associates to a given (generalized) stochastic process X_t on the white noise probability space $(\mathcal{S}', \mathcal{F}, \mu)$ (see definition in §2) an analytic function $\mathcal{H}(X_t)(z_1, z_2, \cdots) = \tilde{X}_t(z_1, z_2, \cdots)$ on $\mathbf{C}_0^{\mathbf{N}}$. This particular transform was introduced by us in [LØU], but the general idea of associating analytic functions to functions on $\mathcal{S}'$ is much older. See [H], [HKPS] and the survey [MY] and the references there. An important property of $\mathcal{H}$ is that it transforms Wick products into ordinary complex products and this explains its role in the link above.

Another crucial property of $\mathcal{H}$ is that it has a (left) inverse $\mathcal{H}^{-1}$ which can be computed explicitly as an integral with respect to the infinite product of the normalized Gaussian measures on $\mathbf{R}$. This gives a useful method for solving Ito-Skorohod stochastic (possibly anticipating) differential equations involving Wick versions.

A key result (Theorem 3.3) states that if $\int \cdot \delta B_t$ denotes Skorohod integral (B_t is Brownian motion) then

$$\int_0^T Y_t \delta B_t = \int_0^T Y_t \diamond W_t dt$$

for all Skorohod integrable processes Y_t, where W_t denotes the white noise (generalized) process. Thus Wick multiplication appears naturally when Ito or Skorohod stochastic differential equations are used to model dynamical systems with noise. Ordinary and Wick multiplication coincide for deterministic processes, but for stochastic processes the products differ and we would like to stress that it is not obvious what type of product one should use to get the best model. As an example - and an illustration of our main result - we discuss (§4) the following model for population growth in a crowded, random environment:

$$dX_t = rX_t \diamond (N - X_t)dt + \alpha X_t \diamond (N - X_t)\delta B_t$$

§2. Some preliminaries.

Since white noise is so fundamental for our construction, we recall some basic facts about this generalized (i.e. distribution valued) process:

For $n = 1, 2, \cdots$ let $\mathcal{S}(\mathbf{R}^n)$ be the Schwartz space of all rapidly decreasing smooth (C^∞)

functions on $\mathbf{R}^n$. Then $\mathcal{S}(\mathbf{R}^n)$ is a Fréchet space under the family of seminorms

$$\|f\|_{N,\alpha} = \sup_{x\in\mathbf{R}^n}(1+|x|^N)|\partial^\alpha f(x)|,$$

where $N \geq 0$ is an integer and $\alpha = (\alpha_1,\cdots,\alpha_k)$ is a multi-index of non-negative integers α_j. The space of *tempered distributions* is the dual $\mathcal{S}'(\mathbf{R}^n)$ of $\mathcal{S}(\mathbf{R}^n)$, equipped with the weak star topology.

Now let $n = 1$ for the rest of this section and put $\mathcal{S}' = \mathcal{S}'(\mathbf{R})$. By the Bochner-Minlos theorem (see e.g. [GV]) there exists a probability measure μ on $(\mathcal{S}',\mathcal{F})$ (where $\mathcal{F}$ denotes the Borel subsets of $\mathcal{S}'$) such that

(2.1) $$E^\mu[e^{i<\cdot,\phi>}] := \int_{\mathcal{S}'} e^{i<\omega,\phi>}d\mu(\omega) = e^{-\frac{1}{2}\|\phi\|^2} \text{ for all } \phi\in\mathcal{S},$$

where $\|\phi\|^2 = \|\phi\|^2_{L^2(\mathbf{R})}$ and $<\omega,\phi>=\omega(\phi)$ for $\omega\in\mathcal{S}'$. We call $(\mathcal{S}',\mathcal{F},\mu)$ the *white noise probability space.*

It follows from (2.1) that

(2.2) $$\int_{\mathcal{S}'} f(<\omega,\phi>)d\mu(\omega) = (2\pi\|\phi\|^2)^{-\frac{1}{2}}\int_{\mathbf{R}} f(t)e^{-\frac{t^2}{2\|\phi\|^2}}dt;\ \phi\in\mathcal{S},$$

for all f such that the integral on the right converges. (It suffices to prove (2.2) for $f\in C_0^\infty$, i.e. f smooth with compact support. Such a function f is the inverse Fourier transform of its Fourier transform $\hat{f}$ and we obtain (2.2) by (2.1) and the Fubini theorem). In particular, if we choose $f(t) = t^2$ we get from (2.2)

(2.3) $$E^\mu[<\omega,\phi>^2] = \|\phi\|^2;\ \phi\in\mathcal{S}.$$

This allows us to extend the definition of $<\omega,\phi>$ from $\phi\in\mathcal{S}$ to $\phi\in L^2(\mathbf{R})$ for a.a. $\omega\in\mathcal{S}'$, as follows:

(2.4) $$<\omega,\phi>:=\lim_{k\to\infty}<\omega,\phi_k> \text{ for } \phi\in L^2(\mathbf{R}),$$

where ϕ_k is any sequence in $\mathcal{S}$ such that $\phi_k\to\phi$ in $L^2(\mathbf{R})$ and the limit in (2.4) is in $L^2(\mathcal{S}',\mu)$.

In particular, if we define

(2.5) $$\tilde{B}_t(\omega) :=<\omega,\chi_{[0,t]}>$$

then we see that $(\tilde{B}_t,\mathcal{S}',\mu)$ becomes a Gaussian process with mean 0 and covariance

$$E^\mu[\tilde{B}_t(\omega)\tilde{B}_s(\omega)] = \int_{\mathcal{S}'}<\omega,\chi_{[0,t]}>\cdot<\omega,\chi_{[0,s]}>d\mu(\omega)$$

$$= \int_{\mathbf{R}}\chi_{[0,t]}(x)\cdot\chi_{[0,s]}(x)dx = \min(s,t), \text{ using (2.3)}.$$

Therefore $\tilde{B}_t$ is essentially a Brownian motion, in the sense that there exists a t-continuous version B_t of $\tilde{B}_t$:

$$\mu(\{\omega; B_t(\omega) = \tilde{B}_t(\omega)\}) = 1 \text{ for all } t.$$

If $u \in L^2(\mathbf{R})$ we define, using (2.4)

$$\int_{-\infty}^{\infty} \phi(t) dB_t(\omega) = <\omega, \phi>,$$

which coincides with the classical Ito integral if supp $\phi \subset [0, \infty)$.

If we define the *white noise process* W_ϕ by

$$W_\phi(\omega) = <\omega, \phi> \text{ for } \phi \in \mathcal{S}, \omega \in \mathcal{S}' \tag{2.6}$$

then the *white noise process* W_ϕ may be regarded as the distributional derivative of B_t, in the sense that, if $\phi \in \mathcal{S}$

$$\begin{aligned} <\frac{d}{dt}B_t(\omega), \phi> &= -\int_{-\infty}^{\infty} \phi'(t)B_t(\omega)dt = \int_{-\infty}^{\infty} \phi(t)dB_t(\omega) \\ &= \lim_{\Delta t_j \to 0} \sum_j \phi(t_j)(B_{t_{j+1}} - B_{t_j}) = \lim_{\Delta t_j \to 0} \sum_j \phi(t_j) < \omega, \chi_{(t_j, t_{j+1}]} > \\ &= \lim_{\Delta t_j \to 0} < \omega, \sum_j \phi(t_j)\chi_{(t_j,t_{j+1}]} > = < \omega, \phi > = W_\phi(\omega), \end{aligned}$$

where the second identity is based on integration by parts for Ito integrals.

By the Wiener-Ito chaos theorem (see e.g. [I] and [HKPS]), we can write any function $f \in L^2(\mu)(= L^2(\mathcal{S}', \mu))$ on the form

$$f = \sum_{n=0}^{\infty} \int f_n dB^{\otimes n}, \tag{2.7}$$

where

$$f_n \in \hat{L^2}(\mathbf{R}^n, dx), \tag{2.8}$$

i.e. $f_n \in L^2(\mathbf{R}^n, dx)$ and f_n is *symmetric* (in the sense that $f_n(x_{\sigma_1}, x_{\sigma_2}, \cdots, x_{\sigma_n}) = f(x_1, \cdots, x_n)$ for all permutations σ of $(1, 2, \cdots, n)$) and

$$\begin{aligned} \int f_n dB^{\otimes n} &= \int_{\mathbf{R}^n} f_n(u) dB_u^{\otimes n} \\ &= n! \int_{-\infty}^{\infty} \Big(\int_{-\infty}^{u_n} \cdots \int_{-\infty}^{u_3} \Big(\int_{-\infty}^{u_2} f_n(u_1, \cdots, u_n) dB_{u_1} \Big) dB_{u_2} \cdots dB_{u_{n-1}} \Big) dB_{u_n} \end{aligned} \tag{2.9}$$

for $n \geq 1$, while $n = 0$ term in (3.1) is just a constant f_0.

For a general (non-symmetric) $f \in L^2(\mathbf{R}^n)$ we define

$$\int f dB^{\otimes n} := \int \hat{f} dB^{\otimes n} \tag{2.10}$$

where $\hat{f}$ is the symmetrization of f, defined by

$$\hat{f}(u_1, \cdots, u_n) = \frac{1}{n!} \sum_{\sigma} f(u_{\sigma_1}, \cdots, u_{\sigma_n}), \tag{2.11}$$

the sum being taken over all permutations σ of $(1, 2, \cdots, n)$.
With f, f_n as in (2.7) we have

$$\|f\|^2_{L^2(\mu)} = \sum_{n=0}^{\infty} n! \|f_n\|^2_{L^2(\mathbf{R}^n)} \tag{2.12}$$

Note that (2.12) follows from (2.7) and (2.9) by the Ito isometry, since

$$E[(\int_{\mathbf{R}^n} f_n dB^{\otimes n})(\int_{\mathbf{R}^m} f_m dB^{\otimes m})] = 0 \text{ for } n \neq m$$

and

$$E[(\int_{\mathbf{R}^n} f_n dB^{\otimes n})^2] = (n!)^2 E[(\int_{-\infty}^{\infty} \cdots (\int_{-\infty}^{u_2} f_n(u_1, \cdots, u_n) dB_1) \cdots dB_{u_n})^2]$$

$$= (n!)^2 \cdot \int_{-\infty}^{\infty} \cdots (\int_{-\infty}^{u_2} f_n^2(u_1, \cdots, u_n) du_1) \cdots du_n = n! \int_{\mathbf{R}^n} f_n^2(u) du$$

Here $B_t(\omega); t \geq 0, \omega \in \mathcal{S}'$ is the 1-dimensional Brownian motion associated with the white noise probability space $(\mathcal{S}', \mu)$ as explained above.

Now suppose that $X_t = X(t, \omega) : \mathbf{R} \times \mathcal{S}' \to \mathbf{R}$ is an $\mathcal{B} \times \mathcal{F}$-measurable stochastic process such that $E[X_t^2] < \infty$ for all t. (Here $\mathcal{B}$ denotes the Borel σ-algebra on $\mathbf{R}$). Then by the above there exist $f_n(t, \cdot) \in \hat{L}^2(\mathbf{R}^n)$ such that

$$X_t(\omega) = \sum_{n=0}^{\infty} \int_{\mathbf{R}^n} f_n(t, u_1, \cdots, u_n) dB_u^{\otimes n}(\omega) \tag{2.13}$$

Moreover, each f_n can be chosen measurable in all its variables (see [NZ]). Fix $T > 0$. Let $\hat{f}_n$ denote the symmetrization of $f_n \cdot \chi_{0 \leq t \leq T}$ with respect to its $n+1$ variables. Suppose

$$E[\int_0^T X_t^2 dt] + \sum_{n=0}^{\infty} (n+1)! \|\hat{f}_n\|^2_{L^2(\mathbf{R}^{n+1})} < \infty \tag{2.14}$$

Then *the Skorohod integral* of X_t, denoted by $\int_0^T X_t \delta B_t$, is defined by

(2.15) $$\int_0^T X_t \delta B_t = \sum_{n=0}^{\infty} \int_{\mathbf{R}^{n+1}} \hat{f}_n(t,u) dB^{\otimes(n+1)}$$

The Skorohod integral is an extension of the Ito integral in the following sense:

(2.16) If Y_t is adapted and $E[\int_0^T Y_t^2 dt] < \infty$ then the Skorohod integral of Y exists and

$$\int_0^T Y_t \delta B_t = \int_0^T Y_t dB_t.$$

(See [NZ]).

As is customary we let $H^s = H^s(\mathbf{R}^n)$ denote the Sobolev space

$$H^s(\mathbf{R}^n) = \{\psi \in \mathcal{S}'(\mathbf{R}^n); \|\psi\|^2_{H^s(\mathbf{R}^n)} := \int_{\mathbf{R}^n} |\hat{\psi}(y)|^2 (1 + |y|^2)^s dy < \infty\},$$

where $\hat{\psi}$ denotes the Fourier transform of ψ and $s \in \mathbf{R}$. Then the dual of H^s is H^{-s} for all $s \in \mathbf{R}$. For notational simplicity we put

$$H^{-\infty} = \bigcup_{k=1}^{\infty} H^{-k},$$

so that if $F \in H^{-\infty}$ then $F \in H^{-k}$ for some k.

We now recall the definition of *functional processes*, which were introduced in [LØU]:

DEFINITION 2.1. A *functional process* $\{X(\cdot,\omega)\}_{\omega \in \mathcal{S}'}$ is a sum of distribution valued processes of the form

(2.17) $$X_\phi(\omega) = X(\phi,\omega) = \sum_{n=0}^{\infty} \int_{\mathbf{R}^n} F^{(n)}(\phi^{\otimes n}) dB^{\otimes n}(\omega); \phi \in \mathcal{S}, \omega \in \mathcal{S}'$$

where

$$F^{(n)}(\cdot) \in H^{-\infty}(\mathbf{R}^n; L^2(\mathbf{R}^n)) \text{ for all } n \geq 1$$

and

$$F^{(0)}(\cdot) \in H^{-\infty}(\mathbf{R}).$$

Moreover, we assume that

$$E[|X(\phi,\omega)|^2] = \sum_{n=0}^{\infty} n! \int_{\mathbf{R}^n} < F^{(n)}, \phi^{\otimes n} >^2 (u)du < \infty \tag{2.18}$$

for all $\phi \in \mathcal{S}$ with $\|\phi\|_{L^2}$ sufficiently small.

To make the notation more suggestive we often write the functional process $X(\phi,\omega)$ on the form

$$X_t(\omega) = \sum_{n=0}^{\infty} \int_{\mathbf{R}^n} F^{(n)}_{t,\cdots,t}(u)dB_u^{\otimes n}(\omega) = \sum_{n=0}^{\infty} \int_{\mathbf{R}^n} F_t^{(n)} dB^{\otimes n}, \tag{2.19}$$

where each $F_t^{(n)}(u)$ is really an L^2-valued distribution in the t-variable, $t = (t_1, \cdots, t_n)$. The distributional derivative of X_t with respect to t is then defined by

$$\frac{dX_t}{dt}(\omega) = \sum_{n=0}^{\infty} \int_{\mathbf{R}^n} \frac{d}{dt} F^{(n)}_{t,\cdots,t}(u)dB_u^{\otimes n}(\omega) \tag{2.20}$$

where

$$\frac{d}{dt}F^{(n)}_{t,t,\cdots,t} = (\sum_{j=1}^{n} \frac{\partial F^{(n)}}{\partial x_j})_{x=(t,\cdots,t)},$$

$\frac{\partial}{\partial x_j}$ denoting the usual distributional derivative with respect to x_j, i.e.

$$< \frac{\partial F^{(n)}}{\partial x_j}, \psi >= - < F^{(n)}, \frac{\partial \psi}{\partial x_j} > \text{ for } \psi = \psi(x_1, \cdots, x_n) \in \mathcal{S}(\mathbf{R}^n).$$

EXAMPLE 2.2. The *white noise process* W_t can be represented as a functional process as follows:

$$W_t = \int_{-\infty}^{\infty} \delta_t(u)dB_u \tag{2.21}$$

where $\delta_t(u)$ is the usual Dirac measure, i.e.

$$< \delta_t(u), \phi(t) >= \phi(u)$$

To see this note that, according to the definition above, (2.21) means that

$$W_\phi(\omega) = \int \phi(u)dB_u(\omega) \tag{2.22}$$

which is just a reformulation of (2.6).

If $X_\phi = \int\limits_{\mathbf{R}^n} F_\phi^{(n)}(u)dB_u^{\otimes n}$ and $Y_\phi = \int\limits_{\mathbf{R}^m} G_\phi^{(m)}(v)dB_v^{\otimes m}$ are two functional processes with just one term we define *the Wick product* of X and Y by

(2.23) $$(X \diamond Y)_\phi = \int\limits_{\mathbf{R}^{n+m}} F_\phi^{(n)} \hat{\otimes} G_\phi^{(m)} dB^{\otimes(n+m)},$$

where $F_\phi^{(n)} \hat{\otimes} G_\phi^{(m)}$ denotes the symmetrized tensor product of $F_\phi^{(n)}(u)$ and $G_\phi^{(n)}(v)$.

We extend this by linearity to the case when X_ϕ and Y_ϕ are finite sums of such terms. Under certain growth conditions (which we will return to later) this definition extends to infinite sums, i.e. to functional processes.

For an explanation of the relation between this multiplication and the classical Wick product as defined e.g. in [Si], we refer to [LØU, §5].

EXAMPLE 2.3. By the previous example together with (3.15) we can represent *the square of white noise* as follows:

(2.24) $$W_t^{\diamond 2} = W_t \diamond W_t = \int\limits_{\mathbf{R}^2} \delta_t(u)\delta_t(v)dB_u dB_v,$$

This means that as a distribution valued process $W_t^{\diamond 2}$ can be written

$$W^{\diamond 2}(\phi \otimes \psi, \omega) = \int\limits_{\mathbf{R}^2} \phi(u)\psi(v)dB_u dB_v(\omega); \phi, \psi \in \mathcal{S}$$

In particular,

$$W^{\diamond 2}(\phi \otimes \phi, \omega) = \int\limits_{\mathbf{R}^2} \phi(u)\phi(v)dB_u dB_v(\omega); \phi \in \mathcal{S}$$

so a more correct notation for $W_t^{\diamond 2}$ would be $W_{t,t}^{\diamond 2}$.

Finally we recall the definition of *the Hermite transform*, introduced in [LØU]:
Fix an orthonormal family $\{\zeta_k\}_{k=1}^\infty$ in $L^2(\mathbf{R})$. Let

$$X_\phi = \sum_{n=0}^\infty \int F_\phi^{(n)}(u)dB^{\otimes n}$$

be a functional process, and keep $\phi \in \mathcal{S}$ fixed. Using multi-index notation each function $F_\phi^{(n)}(u_1, \cdots, u_n)$ may be (uniquely) written

$$F_\phi^{(n)}(u_1, \cdots, u_n) = \sum_{|\alpha|=n} c_\alpha^{(n)} \zeta^{\otimes\alpha}(u_1, \cdots, u_n)$$

for suitable constants $c_\alpha^{(n)} = c_\alpha^{(n)}(\phi)$, where $\alpha = (\alpha_1, \cdots, \alpha_m), |\alpha| = \alpha_1 + \cdots + \alpha_m$ and

$$\zeta^{\otimes\alpha} = \zeta_1^{\otimes\alpha_1} \otimes \zeta_2^{\otimes\alpha_2} \otimes \cdots \otimes \zeta_m^{\otimes\alpha_m}$$

This gives the (unique) representation

$$X_\phi = \sum_{n=0}^{\infty} \sum_{|\alpha|=n} c_\alpha^{(n)} \int \zeta^{\otimes\alpha} dB^{\otimes n} = \sum_\alpha c_\alpha \int \zeta^{\otimes\alpha} dB^{\otimes|\alpha|} \tag{2.25}$$

where $c_\alpha^{(n)}(\cdot) \in H^{-\infty}(\mathbf{R}^n)$ for $n \geq 1, c_\alpha^{(0)}(\cdot) \in H^{-\infty}(\mathbf{R})$.

By symmetrization of $\zeta^{\otimes\alpha}$ and use of the Ito isometry we see that with such a representation we have

$$E[X_\phi^2] = \sum_\alpha \alpha! c_\alpha^2 \tag{2.26}$$

where $\alpha! = \alpha_1!\alpha_2!\cdots\alpha_m!$ if $\alpha = (\alpha_1, \alpha_2, \cdots, \alpha_m)$.

DEFINITION 2.4. Let X_ϕ be the functional process with the representation (2.25). Then the *Hermite transform* (or $\mathcal{H}$-transform) of X_ϕ is the formal power series in infinitely many complex variables $z_1, z_2, \cdots$ given by

$$\mathcal{H}(X_\phi)(z) = \tilde{X}_\phi(z) = \sum_{n=0}^{\infty} \sum_{|\alpha|=n} c_\alpha z^\alpha = \sum_\alpha c_\alpha z^\alpha, \tag{2.27}$$

where $z = (z_1, z_2, \cdots), z^\alpha = z_1^{\alpha_1} \cdot z_2^{\alpha_2} \cdots z_m^{\alpha_m}$ if $\alpha = (\alpha_1, \cdots, \alpha_m)$.

The main properties of the Hermite transform can be summarized as follows:

THEOREM 2.5 [LØU].
(a) For each integer N put

$$z^{(N)} := (z_1, \cdots, z_N, 0, 0, \cdots) \text{ if } z = (z_1, \cdots, z_N, z_{N+1}, \cdots) \in \mathbf{C}^{\mathbf{N}}$$

and define

$$\tilde{X}_\phi^{(N)}(z) = \tilde{X}_\phi(z^{(N)})$$

Then the power series for $\tilde{X}_\phi^{(N)}$ converges uniformly on compacts in $\mathbf{C}^N$ and hence represents an analytic function in $\mathbf{C}^N$, for each N.

(b) (Inverse $\mathcal{H}$-transform). Define the measure λ on the product σ-algebra on $\mathbf{R}^{\mathbf{N}}$ by
(2.28)

$$\int f(y)d\lambda(y) = \int_{-\infty}^{\infty} \cdots (\int_{-\infty}^{\infty} (\int_{-\infty}^{\infty} f(y_1, \cdots, y_n) e^{-\frac{1}{2}y_1^2} \frac{dy_1}{\sqrt{2\pi}}) e^{-\frac{1}{2}y_2^2} \frac{dy_2}{\sqrt{2\pi}}) \cdots e^{-\frac{1}{2}y_n^2} \frac{dy_n}{\sqrt{2\pi}}$$

if $f : \mathbf{R}^{\mathbf{N}} \to \mathbf{R}$ is a bounded function depending only on finitely many variables $y_1, \cdots, y_n$. (This defines λ as a premeasure on the algebra generated by finite products of sets in $\mathbf{R}$ and so λ extends uniquely to a measure on the product σ-algebra of $\mathbf{R}^{\mathbf{N}}$). Then

$$X_\phi = \mathcal{H}^{-1}(\tilde{X}_\phi) := \left[\int \tilde{X}_\phi(x+iy)d\lambda(y)\right]_{x=\int \zeta dB} \tag{2.29}$$

where $x+iy = (x_1+iy_1, x_2+iy_2, \cdots)$ and "$x = \int \zeta dB$" is a short-hand notation for the substitution $x_k = \int \zeta_k dB, k = 1, 2, \cdots$.

(c) Suppose $p, q \geq 1$ is such that $\frac{1}{p} + \frac{1}{q} = 1$. Let X_ϕ, Y_ϕ be functional processes such that

$$\tilde{X}_\phi \in L^{2p}(\lambda \times \lambda), \tilde{Y}_\phi \in L^{2q}(\lambda \times \lambda)$$

for all $\phi \in \mathcal{S}$ sufficiently small. Then $X_\phi \diamond Y_\phi$ is a functional process and

$$\mathcal{H}(X_\phi \diamond Y_\phi) = \mathcal{H}(X_\phi) \cdot \mathcal{H}(Y_\phi),$$

where the product on the right is the usual complex product in the complex variables z_j (and a tensor product in the coefficients (as functions of ϕ)).

§3. Wick multiplication and Ito-Skorohod stochastic differential equations.

In this section we establish a connection between deterministic and Ito-Skorohod differential equations. The key to this connection is that ordinary multiplication should be replaced by Wick multiplication or, more generally, given functions should be replaced by their *Wick versions*. This will be explained more closely below. Heuristically, our main result could be formulated as follows:
Ito-Skorohod calculus using usual multiplication is equivalent to usual calculus using Wick multiplication.

First we make some remarks about the Hermite transform and its inverse, explained in §2:

DEFINITION 3.1. Let $u_t(z) = u(t; z_1, z_2, \cdots) : [0, \infty) \times \mathbf{C}_0^{\mathbf{N}} \to \mathbf{C}$ be measurable and satisfy the conditions

(3.1) (Antisymmetry) $\qquad u_t(\overline{z}) = \overline{u_t(z)}$ for all t, z,

where $\overline{}$ denotes complex conjugation, and

$$\int_0^T \int \int |u_t(z)|^2 d\lambda(x) d\lambda(y) dt < \infty \tag{3.2}$$

for all $T < \infty$, where $z = x + iy$ as before.

Then we say that $u_t(t)$ is a *generalized Hermite transform*. The family of such functions is denoted by $\mathcal{G}$. If - in addition - $u_t(z)$ satisfies the requirement

$$u_t(\cdot) \quad \text{is analytic in each } z_k \in \mathbf{C}, k = 1, 2, \cdots \tag{3.3}$$

we call $u_t(z)$ an *analytic Hermite transform*. The family of such functions is denoted by $\mathcal{A}$.

Note that if $u \in \mathcal{A}$ we can write

$$u_t(z) = \sum_\alpha c_\alpha(t) z^\alpha \quad \text{for } z \in \mathbf{C}_0^{\mathbf{N}}$$

and we see that

$$u_t(z) = \tilde{X}_t(z)$$

where

$$X_t = \sum_\alpha c_\alpha(t) \int \zeta^{\otimes\alpha} dB^{\otimes|\alpha|} = \mathcal{H}^{-1}(u_t).$$

is the inverse Hermite transform of $u_t(\cdot)$. So u is indeed the Hermite transform of a functional process X.

However, if we start with a general $v_t(z) \in \mathcal{G}$ and apply the inverse Hermite transform

$$Y_t := \mathcal{H}^{-1}(v_t) = [\int v_t(z) d\lambda(y)]_{x=\int \zeta dB}$$

we get a functional process Y_t whose Hermite transform $\tilde{Y}_t = \mathcal{H}(Y_t)$ does not necessarily coincide with v_t. For example, if

$$v_t(z) = c(t)|z_1|^2 \quad z = (z_1, z_2, \cdots)$$

then

$$Y_t = [\int c(t)[x_1^2 + y_1^2] d\lambda(y_1)]_{x_1=\int \zeta_1 dB} = c(t)[(\int \zeta_1 dB)^2 + 1].$$

This can be written in canonical form

$$Y_t = c(t)[\int \zeta_1^{\otimes 2} dB^{\otimes 2} + 2],$$

from which we see that

$$\tilde{Y}_t(z) = c(t)[z_1^2 + 2].$$

But this argument shows that to any given $v_t \in \mathcal{G}$ we can always find a (unique) *analytic* $\hat{v}_t = \mathcal{H}(\mathcal{H}^{-1}(v_t)) \in \mathcal{A}$ with the same inverse Hermite transform as that of v_t. We call $\hat{v}_t$ the *analytic representative* of v_t. Note that if v_t depends only on $z_1 \in \mathbf{C}$ then $\hat{v}_t$ is also a function of z_1 only.

Suppose we are given a function $g : \mathbf{R} \to \mathbf{R}$ such that $g \in L^2(\lambda)$. Regard g as a function on $\mathbf{C}$ by putting

$$g(z) = g(x) \quad \text{when} \quad z = x + iy \in \mathbf{C}.$$

Let $\hat{g}$ be the analytic representative of g. Then define $\breve{g} : \mathbf{C} \to \mathbf{C}$ by

$$\breve{g}(z) = g(x) + i \operatorname{Im} \hat{g}(z) \quad ; \; z = x + iy \in \mathbf{C}. \tag{3.4}$$

We call $\breve{g}$ the *canonical complex extension* of g. Note that

$$\breve{g} = g \text{ on } \mathbf{R}, \breve{g}(\overline{z}) = \overline{\breve{g}(z)} \quad \text{and} \quad \mathcal{H}^{-1}(g) = \mathcal{H}^{-1}(\breve{g}).$$

We will need the following result:

LEMMA 3.2. Let $g : \mathbf{R} \to \mathbf{R}$ belong to $L^2(\lambda)$. Let $X \in L^2(\mu)$. Then

$$\int (g(\tilde{X}))^\wedge \cdot z_k d\lambda(y) = \int \breve{g}(\tilde{X}) \cdot z_k d\lambda(y)$$

for all $k = 1, 2, \cdots$, provided the integrals exist.

Proof. Since $\mathcal{H}^{-1}((g(\tilde{X}))^\wedge) = \mathcal{H}^{-1}(g(\tilde{X})) = \mathcal{H}^{-1}(\breve{g}(\tilde{X}))$ we see that it suffices to prove that

$$\int \operatorname{Im}((g(\tilde{X}))^\wedge) y_k d\lambda = \int \operatorname{Im}(\breve{g}(\tilde{X})) y_k d\lambda$$

and this follows from (3.4) plus the fact that $(g(\tilde{X}))^\wedge = \hat{g}(\tilde{X})$ (both sides are analytic and they have the same inverse Hermite transform).

We proceed to prove the following basic relation between Ito/Skorohod integrals and white noise calculus: (As usual we let $W_t = \int \delta_t(u) dB_u$ denote the white noise functional process and

$$\tilde{W}_t(z) = \sum_{j=1}^{\infty} \zeta_j(t) z_j$$

its Hermite transform)

THEOREM 3.3. Let $T > 0$ and let Y_t be a stochastic process such that

$$\int_0^T \left(\int \int |\tilde{Y}_t(t) \cdot \tilde{W}_t(z)|^2 d\lambda(x) d\lambda(y) \right) dt < \infty.$$

Then its Skorohod integral $\int_0^T Y_t(\omega) \delta B_t$ exists and

$$\int_0^T Y_t(\omega) \delta B_t = \int_0^T \left[\int \tilde{Y}_t(z) \cdot \tilde{W}_t(z) d\lambda \right]_{x = \int \zeta dB} dt \quad (z = x + iy) \tag{3.5}$$

In particular, if $\{Y_t\}$ is $\{\mathcal{F}_t\}$-adapted, then

$$\int_0^T Y_t(\omega)dB_t = \int_0^T [\int \tilde{Y}_t(z)\cdot \tilde{W}_t(z)d\lambda]_{x=\int \zeta dB} dt \tag{3.6}$$

Proof. If we put $Y_t = 0$ for $t \notin [0,T]$ and write

$$Y_t = \sum_\alpha c_\alpha(t) \int \zeta^{\otimes\alpha} dB^{\otimes|\alpha|} = \sum_{a,k} (c_\alpha, \zeta_k)\zeta_k(t) \int \zeta^{\otimes\alpha} dB^{\otimes|\alpha|},$$

then we get (all $d\lambda$-integrals are evaluated at $x = \int \zeta dB$)

$$\begin{aligned}
\int_{-\infty}^{\infty} Y_t \delta B_t &= \sum_{\alpha,k} (c_\alpha, \zeta_k) \int \zeta_k \otimes \zeta^{\otimes\alpha} dB^{\otimes|\alpha|+1} \\
&= \sum_{\alpha,k} (c_\alpha, \zeta_k) \int z_k z^\alpha d\lambda \\
&= \sum_{\alpha,k,j} (c_\alpha, \zeta_k) \int z_j z^\alpha d\lambda \cdot \int_{-\infty}^{\infty} \zeta_k(t)\zeta_j(t)dt \\
&= \int_{-\infty}^{\infty} (\int \sum_{\alpha,k} (c_\alpha, \zeta_k)\zeta_k(t) z^\alpha (\sum_j \zeta_j(t) z_j) d\lambda) dt \\
&= \int_{-\infty}^{\infty} (\int \tilde{Y}_t(z) \cdot (\sum_j \zeta_j(t) z_j) d\lambda) dt = \int_{-\infty}^{\infty} (\int \tilde{Y}_t(z) \tilde{W}_t(z) d\lambda) dt,
\end{aligned}$$

as claimed.

A more striking way of stating Theorem 3.3 is the following:

COROLLARY 3.4. Let Y_t be as in Theorem 3.3. Then

$$\int_0^T Y_t(\omega)\delta B_t = \int_0^T Y_t \diamond W_t dt \tag{3.7}$$

If, in addition, Y_t is $\mathcal{F}_t$-adapted then

$$\int_0^T Y_t(\omega)dB_t = \int_0^T Y_t \diamond W_t dt \tag{3.8}$$

In other words: *Ito integration is equivalent to Wick multiplication by white noise followed by usual (Lebesgue) integration.*

Remark: Using (3.8) repeatedly we see that the Wiener-Ito chaos formula (2.7) may be written

$$f = \sum_{n=0}^{\infty} \int_{\mathbf{R}^n} f_n \diamond W_t^{\diamond n} dt^{\otimes n},$$

which is strikingly similar to the Taylor expansion of a real analytic function. See [St] for a discussion about this.

From (3.8) we see that if we model a white noise differential equation

$$\frac{dX_t}{dt} = b(X_t) + \sigma(X_t) \cdot \text{ "white noise"}$$

by the Ito stochastic differential equation

$$dX_t = b(X_t)dt + \sigma(X_t)dB_t,$$

we are actually interpreting the product

$$\sigma(X_t) \cdot \text{ "white noise" } \quad \text{as the Wick product} \quad \sigma(X_t) \diamond W_t.$$

This raises the question whether it may be more appropriate to interpret other nonlinear terms in the equation in the "Wick sense" as well. For example, as a model for population growth in a crowded - and random - environment we could use the equation

$$\frac{dX_t}{dt} = (r + \alpha W_t)X_t(N - X_t)$$

in the "traditional" sense, i.e.

$$dX_t = rX_t(N - X_t)dt + \alpha X_t(N - X_t)dB_t \tag{3.9}$$

or we could use the Wick version of the products:

$$dX_t = rX_t \diamond (N - X_t)dt + \alpha X_t \diamond (N - X_t)dB_t. \tag{3.10}$$

Which model gives the best description of the situation? We will examine this example more closely in §4. (We remark that it follows from the main result in [LØU] that if $0 \leq X_s \leq N$ for $s \leq t$ then $X_s \diamond (N - X_s) \geq 0$ for $s \leq t$.)

First we introduce the general concept of *the Wick version* $f^\diamond$ of a given real function f:

DEFINITION 3.5. Let $f : \mathbf{C} \to \mathbf{C}$ be measurable. If X_t is a stochastic process in $L^2(\mu)$ such that

$$\int\int |f(\tilde{X}_t(x + iy))|^2 d\lambda(x) d\lambda(y) < \infty$$

then

$$(3.11) \qquad Y_t := [\int f(\tilde{X}_t(x+iy))d\lambda(y)]_{x=\int \varsigma dB} = \mathcal{H}^{-1}(f(\mathcal{H}(X_t)))$$

defines a stochastic process in $L^2(\mu)$. This process Y_t is denoted by $f^\diamond(X_t)$ and called *the Wick version* of $f(X_t)$.

EXAMPLE 3.6. If $f(z) = \sum_{k=0}^{n} a_k z^k$ is a complex polynomial then

$$f^\diamond(X_t) = \sum_{k=0}^{n} a_k X_t^{\diamond k},$$

i.e. the $\diamond$-polynomial obtained by replacing the usual powers by Wick powers (assuming the latter exist).

THEOREM 3.7. Let $b : \mathbf{R} \to \mathbf{R}$ and $\sigma : \mathbf{R} \to \mathbf{R}$ be measurable functions. Extend b to $\mathbf{C}$ such that $b(\overline{z}) = \overline{b(z)}$ and assume that σ is extended to $\mathbf{C}$ in the canonical way (3.4). Suppose there exists $T = T(x_1, x_2, \cdots) > 0$ such that for all $k \in \mathbf{N}$ and all $(z_1, \cdots, z_k) = (x_1 + iy_1, \cdots, x_k + iy_k)$ there is a unique solution $\xi_t^{(k)} \in L^2(\chi_{[0,T]}dt \times d\lambda \times d\lambda)$ of the (deterministic) differential equation

$$(3.12) \qquad \begin{aligned} \frac{d\xi_t^{(k)}}{dt} &= b(\mathcal{H}(\mathcal{H}^{-1}(\xi_t^{(k)}))) + \sigma(\mathcal{H}(\mathcal{H}^{-1}(\xi_t^{(k)}))) \cdot \sum_{j=1}^{k} \varsigma_j(t) z_j; \\ \xi_0^{(k)} &= \xi_0 \in L^2(\lambda \times \lambda) \text{ given.} \end{aligned}$$

Define $\xi_t(z)$ for $z \in \mathbf{C}_0^{\mathbf{N}}$ by putting

$$(3.13) \qquad \xi_t(z) = \xi_t^{(k)}(z) \text{ if } z = (z_1, \cdots, z_k, 0, \cdots).$$

Assume that $b(\mathcal{H}(\mathcal{H}^{-1}(\xi_t)))$ and $\sigma(\mathcal{H}(\mathcal{H}^{-1}(\xi_t))) \cdot \tilde{W}_t$ belong to $L^2(\chi_{[0,T]}dt \times d\lambda \times d\lambda)$.

Define

$$(3.14) \qquad X_t(\omega) = \mathcal{H}^{-1}(\xi_t) := [\int \xi_t(z) d\lambda(y)]_{x=\int \varsigma dB(\omega)} \quad \text{for } t < T(\int \varsigma dB).$$

Then the process X_t solves the Ito-Skorohod stochastic differential equation

$$(3.15) \qquad dX_t = b^\diamond(X_t)dt + \sigma^\diamond(X_t)\delta B_t \quad ; \quad X_0 = \mathcal{H}^{-1}(\xi_0).$$

Proof. If $z^{(k)} = (z_1, \cdots, z_k, 0, \cdots)$ when $z = (z_1, \cdots, z_k, z_{k+1}, \cdots)$ we have by uniqueness

$$\xi_t^{(k)}(z^{(k)}) = \xi_t^{(m)}(z^{(k)}) \quad \text{for all } m > k.$$

This shows that $\xi_t(z)$ is well-defined. Moreover, note that by antisymmetry of b and σ we have

$$(3.16)\qquad \xi_t(\overline{z}) = \overline{\xi_t(z)} \quad \text{for all } z \in \mathbf{C}_0^{\mathbf{N}}.$$

With X_t defined by (3.14) we have

$$\xi_t(z) = \xi_0 + \int_0^t b(\tilde{X}_s)ds + \int_0^t \sigma(\tilde{X}_s)\sum_j \zeta_j(s)z_j ds \text{ for } t < T(x)$$

for all $z \in \mathbf{C}_0^{\mathbf{N}}$. We integrate this identity with respect to $d\lambda(y)$ and apply the Fubini theorem to obtain

$$(3.17)\quad X_t(\omega) = X_0 + \int_0^t b^\diamond(X_s)ds + \int_0^t \Big(\int \sigma(\tilde{X}_s)\tilde{W}_s d\lambda(y)\Big)_{x=\int \zeta dB} ds \quad \text{for } t < T\Big(\int \zeta dB\Big)$$

By Lemma 3.2 we may replace $\sigma(\tilde{X}_s)$ by its analytic representative and by Theorem 3.3 we obtain (3.15).

If b and σ are analytic then $\xi_t(\cdot)$ becomes analytic and hence ξ_t coincides with its analytic representative $\mathcal{H}(\mathcal{H}^{-1}(\xi_t))$. So in this case Theorem 3.7 simplifies to:

THEOREM 3.8. Let $b : \mathbf{C} \to \mathbf{C}$ and $\sigma : \mathbf{C} \to \mathbf{C}$ be *analytic* functions satisfying $b(\overline{z}) = \overline{b(z)}$ and $\sigma(\overline{z}) = \overline{\sigma(z)}$. Suppose that there exists $T = T(x_1, x_2, \cdots) > 0$ such that for all $z = (z_1, z_2, \cdots) = (x_1 + iy_1, x_2 + iy_2, \cdots)$ there is a unique solution $\xi_t(z) \in L^2(\chi_{[0,T]}dt \times d\lambda \times d\lambda)$ of the equation

$$(3.18)\qquad \frac{d\xi_t}{dt} = b(\xi_t) + \sigma(\xi_t)\cdot\tilde{W}_t(z); \xi_0 \in L^2(\lambda \times \lambda)$$

for $t < T$, where $\xi_0(z_1, z_2, ...)$ is analytic.

Moreover, assume $b(\xi_t)$ and $\sigma(\xi_t)\cdot\tilde{W}_t$ belong to $L^2(\chi_{[0,T]}dt \times d\lambda \times d\lambda)$.

Then

$$(3.19)\qquad X_t := \mathcal{H}^{-1}(\xi_t)$$

solves the Ito-Skorohod stochastic differential equation

$$(3.20)\qquad dX_t = b^\diamond(X_t)dt + \sigma^\diamond(X_t)\delta B_t \quad ; X_0 = \mathcal{H}^{-1}(\xi_0)$$

for $t < T(\omega) := T(\int \zeta_1 dB, \int \zeta_2 dB, \cdots)$. Moreover, this is the unique solution satisfying

$$(3.21)\qquad \int\int\Big(\int_0^T |\tilde{X}_t(z)|^2 dt\Big)d\lambda(x)d\lambda(y) < \infty \quad (z = x + iy)$$

Proof. It only remains to prove uniqueness. If X_t and Y_t both solve (3.20) then $\tilde{X}_t$ and $\tilde{Y}_t$ both solve (3.18). By uniqueness of the solution of (3.18) we have $\tilde{X}_t = \tilde{Y}_t$ and hence

$$X_t = \mathcal{H}^{-1}(\tilde{X}_t) = \mathcal{H}^{-1}(\tilde{Y}_t) = Y_t.$$

Remark. Note that X_0 may be anticipating, so Theorems 3.7 and 3.8 provide a new approach to (this type of) anticipating Skorohod stochastic differential equations. See [P] and the references there for more information about such equations.

§4. Application to population growth in a crowded, stochastic environment.

To illustrate Theorem 3.8 we consider example (3.10) in detail, i.e. we consider the following Ito-Skorohod stochastic differential equation

$$dX_t = rX_t \diamond (1 - X_t)dt + \alpha X_t \diamond (1 - X_t)\delta B_t; X_0 = x \tag{4.1}$$

where x, r, α are constants, r is positive and where we for simplicity have put $N = 1$ and assume $x > 0$.

In view of Theorem 3.8 we are led to consider the following deterministic equation

$$\frac{d\xi_t}{dt} = r\xi_t(1 - \xi_t) + \alpha\xi_t(1 - \xi_t) \cdot \tilde{W}_t(t); \xi_0 = x > 0. \tag{4.2}$$

Put $c = \frac{1-x}{x}$.
First assume $x > \frac{1}{2}$, i.e. $|c| < 1$. Then the (unique) solution of (4.2) is

$$\xi_t(z) = [1 + c\exp(-rt - \alpha F(t, z))]^{-1} \text{ for } t < T \tag{4.3}$$

where

$$F(t, z) = \int_0^t \tilde{W}_s(z)ds = \sum_k Z_k(t)z_k, \tag{4.4}$$

with $Z_k(t) = \int_0^t \zeta_k(s)ds$ and $T = T(x_1, \cdots) = \inf\{t > 0; c\exp(-rt - \alpha\sum_k Z_k(t)x_k) = 1\}$

For $t < T$ we have

$$\int \xi_t(z)d\lambda(y) = \sum_{m=0}^{\infty}(-1)^m c^m \exp(-rmt - \alpha m\sum_k Z_k(t)x_k) \cdot \int \exp(-im\alpha\sum_k Z_k y_k)d\lambda(y) \tag{4.5}$$

Now

$$\int \exp(-im\alpha Z_k y_k - \frac{1}{2}y_k^2)\frac{dy_k}{\sqrt{2\pi}} = \exp(-\frac{1}{2}m^2\alpha^2 Z_k^2) \tag{4.6}$$

So

$$\int \xi_t(z)d\lambda(y) = \sum_{m=0}^{\infty}(-1)^m c^m \exp(-rmt - \alpha m \sum_k Z_k(t)x_k - \frac{1}{2}m^2\alpha^2 \sum_k Z_k^2(t))$$

Substituting $x_k = \int \zeta_k dB$ we conclude that (4.1) has the solution

$$X_t = X_t^{(1)} = \sum_{m=0}^{\infty}(-1)^m c^m \exp(-(rm + \frac{1}{2}\alpha^2 m^2)t - \alpha m B_t) \tag{4.7}$$

This is the solution if $\frac{1}{2} < x < 1$ and $t < T(\omega) = \inf\{t > 0; c\exp(-rt - \alpha B_t) = 1\}$ $(B_0 = 0)$.

Since the series in (4.7) actually converges for all t (for a.a.ω) it is natural to define X_t for all t by this formula. With this definition we see that

$$E^\mu[X_t^{(1)}] = x_t, \tag{4.8}$$

where x_t is the solution of (4.1) in the deterministic case ($\alpha = 0$). Moreover,

$$\lim_{t\to\infty} X_t^{(1)} = 1 \quad \text{a.s.}, \tag{4.9}$$

although for all $t > 0$ we have

$$P^\mu[X_t^{(1)} > 1] > 0 \quad (\text{if } \alpha \neq 0) \tag{4.10}$$

Thus in this stochastic model there is always a positive probability that the population will exceed the limiting value 1.

However, since

$$E^\mu[(X_t^{(1)})^2] = \infty$$

for all t (if $\alpha \neq 0$) $X_t^{(1)}$ is not a global solution of (4.1) in our (L^2) sense. But we shall show below that $X_t^{(1)}$ is a global solution in a weaker sense.

Next assume $0 < x < \frac{1}{2}$, *i.e.* $c > 1$.
Then the unique solution of (4.2) can be written

$$\xi_t(z) = c^{-1}\exp(rt + \alpha F(t,z))[1 + c^{-1}\exp(rt + \alpha F(t,z))]^{-1} \tag{4.11}$$

for $t < T(x)$.

For $t < T(x)$ we have, by a similar calculation as above,

$$\int \xi_t(z)d\lambda(y) = \sum_{m=1}^{\infty}(-1)^{m+1}c^{-m}\exp(rmt + \alpha m \sum_k Z_k(t)x_k - \frac{1}{2}m^2\alpha^2 \sum_k Z_k^2(t))$$

Substituting $x_k = \int \zeta_k dB$ we get the solution

$$X_t = X_t^{(2)} = \sum_{m=1}^{\infty} (-1)^{m+1} c^{-m} \exp((rm - \frac{1}{2}\alpha^2 m^2)t + \alpha m B_t) \tag{4.12}$$

if $0 < x < \frac{1}{2}$ and $t < T(\omega)$.

Again we note that (4.12) actually converges for all t (for a.a. ω) so we define $X_t^{(2)}$ for all t by this formula. In this case ($0 < x < \frac{1}{2}$) we still get

$$E^{\mu}[X_t^{(2)}] = x_t \tag{4.13}$$

and

$$P^{\mu}[X_t^{(2)} > 1] > 0 \quad \text{for all} \quad t > 0 \quad (\alpha \neq 0) \tag{4.14}$$

However, in contrast with (4.9) we now have

$$\lim_{t\to\infty} X_t^{(2)} = 0 \quad \text{a.s. if} \quad r - \frac{1}{2}\alpha^2 < 0 \tag{4.15}$$

Now define X_t by

$$X_t = \begin{cases} X_t^{(1)} & \textit{if } \frac{1}{2} < x \qquad \text{i.e. } |c| < 1 \\ X_t^{(2)} & \textit{if } 0 < x < \frac{1}{2} \quad \text{i.e. } c > 1 \end{cases} \tag{4.16}$$

We claim that X_t actually solves (4.1) for all t, in the sense that X_t is $\mathcal{F}_t$-adapted,

$$P^{\mu}[\int_0^T |X_t \diamond (1 - X_t)|^2 dt < \infty \quad \text{for all } T] = 1 \tag{4.17}$$

and

$$X_T = x + r\int_0^T X_t \diamond (1 - X_t)dt + \alpha \int_0^T X_t \diamond (1 - X_t)dB_t \qquad \text{for all } T \tag{4.18}$$

To verify this we have to compute $X_t \diamond (1 - X_t)$. If $X_t = X_t^{(1)}$ we have by (4.7)

$$\begin{aligned} &X_t \diamond (1 - X_t) = \\ &- \sum_{\substack{m=0\\n=1}}^{\infty} (-c)^{n+m} \exp(-[r(n+m) + \frac{1}{2}\alpha^2(n^2 + m^2)]t) \exp(-\alpha m B_t) \diamond \exp(-\alpha n B_t) \end{aligned} \tag{4.19}$$

To compute the last Wick product we rewrite the last two exponentials as *Wick exponentials*:

Define

$$\text{(4.20)} \qquad Exp(V_t) := \sum_{n=o}^{\infty} \frac{1}{n!} V_t^{\diamond n}$$

for all processes V_t such that the right hand side converges. Then we have (see [LØU, (7.2)])

$$\text{(4.21)} \qquad Exp(\int \varphi dB) = \exp(\int \varphi dB - \frac{1}{2}||\varphi||^2_{L^2(\mathbf{R})}) \quad \forall \varphi \in L^2(\mathbf{R})$$

This gives

(4.22)

$$\begin{aligned} &\exp(-\alpha m B_t) \diamond \exp(-\alpha n B_t) \\ &= \exp(\tfrac{1}{2}\alpha^2(m^2+n^2)t) Exp(-\alpha m B_t) \diamond Exp(-\alpha n B_t) \\ &= \exp(\tfrac{1}{2}\alpha^2(m^2+n^2)t) Exp(-\alpha(m+n)B_t) \\ &= \exp(\tfrac{1}{2}\alpha^2[m^2+n^2-(m+n)^2]t) \exp(-\alpha(m+n)B_t) \\ &= \exp(-\alpha^2 mnt - \alpha(m+n)B_t) \end{aligned}$$

Substituted in (4.18) this gives

$$\begin{aligned} X_t \diamond (1-X_t) &= -\sum_{\substack{m=0 \\ n=1}}^{\infty} (-c)^{n+m} \exp(-[r(n+m) + \frac{1}{2}\alpha^2(n+m)^2]t - \alpha(n+m)B_t) \\ \text{(4.23)} \qquad &= -\sum_{k=1}^{\infty} (-c)^k k \exp(-[rk + \frac{1}{2}\alpha^2 k^2]t - \alpha k B_t) = -\sum_{k=1}^{\infty} (-c)^k k Y_t^{(k)}, \\ &\text{where} \quad Y_t^{(k)} = \exp(-[rk + \frac{1}{2}\alpha^2 k^2]t - \alpha k B_t) \end{aligned}$$

It follows that (4.17) holds.

By Ito's formula we have

$$\text{(4.24)} \qquad \begin{aligned} dY_t^{(k)} &= Y_t^{(k)}(-[rk + \frac{1}{2}\alpha^2 k^2]dt) + Y_t^{(k)}(-\alpha k dB_t) + \frac{1}{2} Y_t^{(k)} \alpha^2 k^2 dt \\ &= Y_t^{(k)}(-rkdt - \alpha k dB_t) \end{aligned}$$

Thus, using (4.24) in (4.23) we get

$$\begin{aligned} &x + \textstyle\int_o^T X_t \diamond (1-X_t)(rdt + \alpha dB_t) \\ &= x - \textstyle\sum\limits_{k=1}^{\infty} (-c)^k \int_0^T Y_t^{(k)} (rkdt + \alpha k dB_t) \end{aligned}$$

$$= x + \sum_{k=1}^{\infty} (-c)^k (Y_T^{(k)} - 1) = x + X_T - 1 + \tfrac{c}{1+c} = X_T,$$

and (4.18) is verified for $X_t = X_t^{(1)}$.

A similar computation verifies (4.17),(4.18) in the case when $X_t = X_t^{(2)}$.

Remark 1 Note that both the computation for $X_t = X_t^{(1)}$ and for $X_t = X_t^{(2)}$ actually still works if we put $c = 1$, as long as $t > 0$ (and $\alpha \neq 0$). But neither of them converges for $t = 0$ with this value of c. It is an interesting question if equation (4.18) has any solution at all with $x = \frac{1}{2}$ (if $\alpha \neq 0$) and if so, whether it is unique or not. The difficulty at this starting point $x = \frac{1}{2}$ reflects the fact that the corresponding (complex) deterministic equation (4.2) does not have a solution for all $z_k \in \mathbf{C}$. In view of (4.9) and (4.15) it is natural to regard $x = \frac{1}{2}$ as a kind of "stochastic bifurcation point".

Remark 2. It is interesting to note that our solution X_t is closely related to the classical Θ-function. The latter is defined by

$$\Theta(w, \tau) = \sum_{n=-\infty}^{\infty} \exp(\pi i n^2 \tau + 2\pi i n w) \quad ; \tag{4.24}$$

where $w \in \mathbf{C}$ and $\tau \in H = \{z \in \mathbf{C};\ Im\ z = 0\}$ (See e.g. [M]). So, for example, if we choose $c = 1$ (and $t > 0$) in (4.7) and (4.12) we have

$$\begin{aligned} X_t^{(1)} - X_t^{(2)} &= \sum_{n=-\infty}^{\infty} (-1)^n \exp(-(rn + \frac{1}{2}\alpha^2 n^2)t - \alpha n B_t) \\ &= \sum_{n=-\infty}^{\infty} \exp(-\frac{1}{2}\alpha^2 n^2 t + n(\pi i - rt - \alpha B_t)) = \Theta(w, \tau) \end{aligned} \tag{4.25}$$

with

$$w = \frac{1}{2} + \frac{i}{2\pi}(rt + \alpha B_t), \quad \tau = \frac{i}{2\pi}\alpha^2 t \tag{4.26}$$

Remark 3. Note that the (unique) solutions $X_t^{(1)}$, $X_t^{(2)}$ in (4.7), (4.12) are *not* Markov. For example, if $0 < x < \frac{1}{2}$ we have

$$E^x[X_{t+h}^{(2)} | \mathcal{F}_t] = \sum_{m=1}^{\infty} (-1)^{m+1} c^{-m} \exp((rm - \frac{1}{2}\alpha^2 m^2)t + \alpha m B_t) \exp(rmh) \tag{4.27}$$

while

$$E^{X_t^{(2)}}[X_h^{(2)}] = \sum_{m=1}^{\infty} (-1)^{m+1} \gamma_t^{-m} \exp(rmh), \tag{4.28}$$

where $c = (1-x)/x$, $\gamma_t = (1 - X_t^{(2)})/X_t^{(2)}$ and $\mathcal{F}_t$ is the σ-algebra generated by $\{B_s(\cdot)\}_{s \leq t}$. The equality of (4.27) and (4.28) would imply that

$$\frac{\gamma_t}{c} = \exp(-rt + \frac{1}{2}\alpha^2 mt - \alpha B_t)$$

for all m, which is impossible unless $\alpha = 0$.

Logistic paths

The same sample path with $r = 1$, $\alpha = 1$. Starting points: 0.75, 0.6

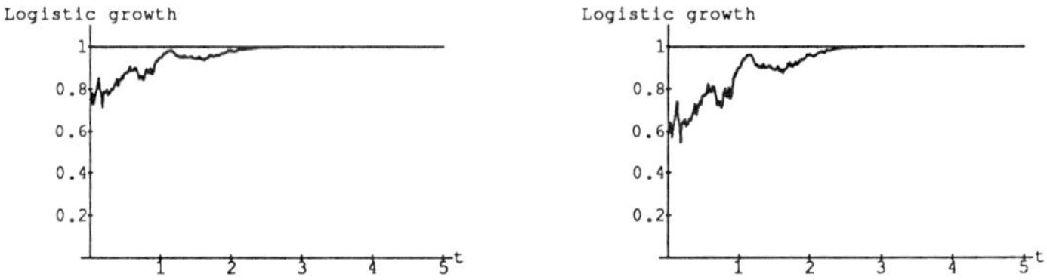

Different sample paths with $r = 1$, $\alpha = 1$. Starting points: 0.55

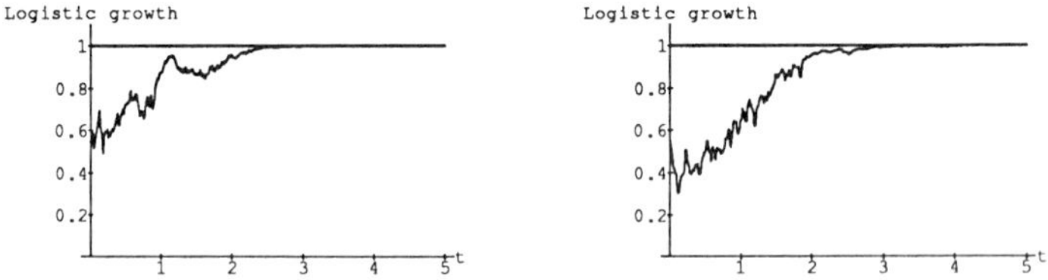

Different sample paths with $r = 1/5$, $\alpha = 1/2$. Starting points: 0.6

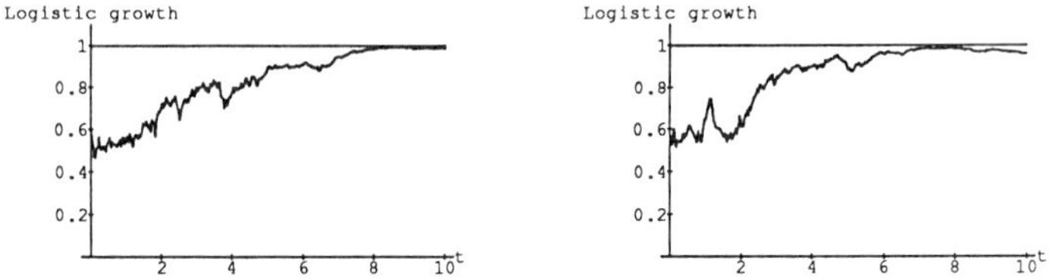

Different sample paths with $r = 1/5$, $\alpha = 1/2$. Starting points: 0.25

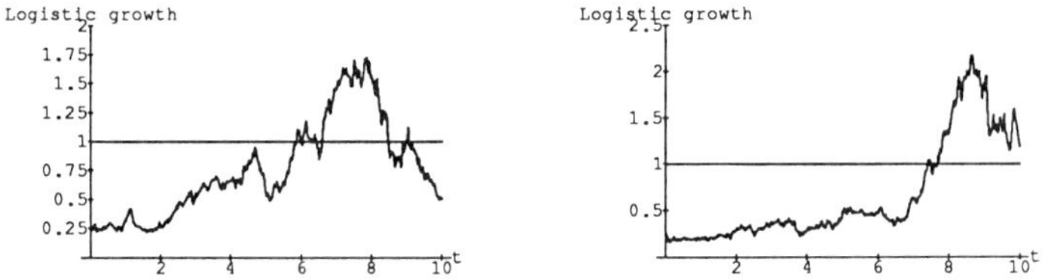

Different sample paths with $r = 1/5$, $\alpha = 1$. Starting points: 0.25

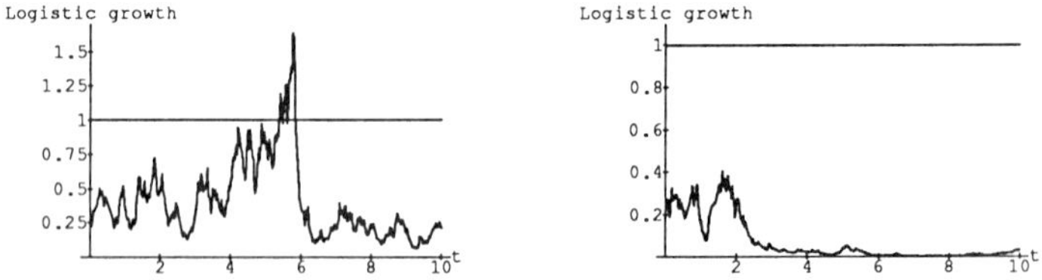

The non-Markovian nature of the solutions reflects the fact that the value of the Wick product $X_t \diamond (1 - X_t)$ at a given $\omega_0 \in \mathcal{S}'$ is not a function of $X_t(\omega_0)$ alone, but depends on the whole set of values $\{X_t(\omega); \omega \in \mathcal{S}'\}$.

The solutions $X_t^{(1)}$, $X_t^{(2)}$ are illustrated on the figure, which shows computer simulations for various choices of r, t and starting point x. In a sense the use of Wick products gives a model of a population with a "memory": If the population reaches the value 1 (the capacity of the environment) from a lower starting point, it has got a momentum which makes it possible to grow even further.

It would be interesting to compare the solutions (4.7), (4.12) to the solution of the "traditional" stochastic model (3.9). Unfortunately this comparison does not seem to be straightforward, because it appears to be difficult to solve (3.9) as explicitly as we have solved (4.1).

Acknowledgements

We are grateful to H. Gjessing, P. Malliavin, P.A. Meyer and J. Potthoff for useful discussions. This work is supported by VISTA, a research cooperation between The Norwegian Academy of Science and Letters and Den Norske Stats Oljeselskap A.S. (Statoil).

REFERENCES

[GV] I.M Gelfand and N.Ya. Vilenkin: Generalized Functions, Vol. 4: Applications of Harmonic Analysis. Academic Press 1964 (English translation).

[H] T. Hida: Brownian Motion. Springer-Verlag 1980.

[HKPS] T. Hida, H.-H. Kuo, J. Potthoff and L. Streit: White Noise Analysis (Forthcoming book).

[I] K. Ito: Multiple Wiener Integral. J. Math. Soc. Japan 3 (1951), 157-169.

[LØU] T. Lindstrøm, B. Øksendal and J. Ubøe: Stochastic differential equations involving positive noise. To appear in M. Barlow and N. Bingham (editors): Stochastic Analysis. Cambridge Univ. Press 1991/92.

[M] D. Mumford: Tata Lectures on Theta I. Birkhäuser 1983.

[MY] P.A. Meyer and J.A. Yan: Les "fonctions characteristiques" des distributions sur l'espace de Wiener. (Séminaire de Probabilités, to appear).

[NZ] D. Nualart and M. Zakai: Generalized stochastic integrals and the Malliavin calculus. Prob. Th. Rel. Fields 73 (1986), 255-280.

[P] E. Pardoux: Applications of anticipating stochastic calculus to stochastic differential equations. In H. Korezliogu and A.S. Ustunel (editors): Stochastic

Analysis and Related Topics II, Springer LNM 1444 (1990), 63-105.

[Si] B. Simon: The $P(\phi)_2$ Euclidean (Quantum) Field Theory. Princeton University Press 1974.

[St] D.W. Stroock: Homogeneous chaos revisited. Seminaire de Prob. XXI, Springer LNM 1247 (1987), 1-7.

Multiscale Analysis In Random Dynamics

Fabio Martinelli

Dipartimento di Matematica, Universita' "La Sapienza", Pz.A.Moro 2,Roma Italy

Abstract

I will report on recent progresses made in the study of the long time behaviour of random dynamics of physical and mathematical interest, like small random perturbations of dynamical systems and Monte Carlo algorithms for ferromagnetic spin systems. The emphasis will be on a general flexible scheme which allows for a multiscale analysis in time or space-time of such models and on concrete examples.

Section 1 Introduction

In this paper we will review some general ideas and techniques originating from percolation theory [1], statistical mechanics of disordered systems [2] and the rigorous theory of Andersol localization [3], that have been succesfully adapted to the study of the long time behaviour of random evolutions like small random perturbations of dynamical systems [4], [5], and Monte Carlo cluster algorithms for the Ising model at low temperature [6], [7]. We will adopt here a rather relax style of exposition and we will not discuss the most general cases; the interested reader is refered to the above mentioned papers for mathematical details and more general hypothesis.

In order to illustrate the problems that we want to discuss, let us start by considering the following ordinary Ito stochastic differential equation in $\mathbf{R^n}$ with small constant diffusion:

$$dx_t^\epsilon = b(x_t^\epsilon)dt + \epsilon w_t \quad x_{t=0}^\epsilon = x \tag{1}$$

Here w_t denotes the ordinary brownian motion in $\mathbf{R^n}$, and the drift b is assumed to be smooth and to satisfy a sequence of assumptions that ensure that the corresponding Markov process has a unique smooth invariant measure which is concentrated as $\epsilon \to 0$ on a finite number of points $x_1 \ldots x_n$ which are asympotically stable equilibria of the unperturbed dynamical system:

$$\frac{dx_t}{dt} = b(x_t) \tag{2}$$

We also assume that the linearization of the above differential equation around each point x_i has eigenvalues with strictly negative real part. The simplest example of such a model would be a gradient system whose potential has finitely many "valleys" arond each point x_i and it grows at plus infinity as $|x| \to \infty$.

The general problem that it was addressed in [4] was the dependence of the process at time t on the initial condition x as t $\to \infty$. This is clearly a fundamental question which is relevant for the study of the rate of convergence to equilibrium, of the properties of the invariant measure, of the asymptotics of the first exit time from an attracted domain [5] and of the behaviour of solutions of elliptic equations with a small parameter in front of the second derivatives in view of their probabilistic representation.

More specifically let us choose a trajectory of the noise $\{w_t\}_{0\leq t<\infty}$ and let us solve the corresponding stochastic differential equation (1) for two different initial conditions x and y; for notations convenience the two corresponding random paths will be denoted by x_t^ϵ and y_t^ϵ respectively. Then we study as a function of time the distance $d(x_t^\epsilon, y_t^\epsilon)$ and ask whether it is going to zero as $t \to \infty$ and how fast.

In order to understand the problems behind such a question, let us suppose that x and y are in the neighborhood of the same equilibrium point x_i. Then, according to the theory of Ventzel and Freidlin [8], the two paths will follow for a very long time the corresponding unperturbed trajectories of (2) and therefore, using the hypotheses on the linearization of (2) around x_i, they will start to join exponentially fast in time. However, after a time of order $\exp(\frac{c}{\epsilon^2})$, $c>0$, a large deviation will force the two processes to jump to another equilibrium point x_j. During such a jump they will necessary cross some region of the space (e.g. a saddle) where the unperturbed dynamical system (2) has the tendency to separate (maybe exponentially fast) nearby trajectories and therefore the determination of the behaviour of $d(x_t^\epsilon, y_t^\epsilon)$ for time scales greater than $\exp(\frac{c}{\epsilon^2})$ becomes a non trivial problem since it might happen that the two paths go into two different equilibria. It should be noted however, and this is the key point for the solution, that typically the time spent by the two paths near the equilibrium point x_i is much larger as $\epsilon \to 0$ than the time spent during the first jump; therefore the distance $d(x_t^\epsilon, y_t^\epsilon)$ will typically be exponentially small in t even after the first jump. By repeating this argument for the next jumps one gets that effectively the distance between the two paths should decay exponentially fast in time t for any t. Since in an infinite time interval any kind of event occurs and in particular jumps which are much longer than their typical time scale, the proof of this fact requires a non trivial analysis of the typical trajectories.

In [4] in collaboration with E.Scoppola, following the ideas and techniques developed for the Anderson localization [9] [10], we carried out this program by means of a complicate multiscale analysis in time . Although this program was rather succesful, the techniques and the geometric constructions involved were rather cumbersome and in some sense not the most natural ones, since in the probabilistic estimates all the scales were present at the same time. Recently however, we have realized in collaboration with L.Sbano [11] that it is possible to apply to the above problems the same kind of simplification operated by Von Dreyfus [2] for the proof of the exponential decay of the Green's function of the Anderson model at high disorder, and perform a multiscale analysis in such a way that the probabilistic estimates of interest propagate recursively in a very natural and simple way from one scale to the next one. As we will show in the sequel, these ideas are quite general and apply in a very flexible way to a variety of very different situations which have some general common denominator.

As anticipated in the title, the second class of models for which it is very interesting to ask the same kind of questions raised above, are Monte Carlo cluster algorithms for e.g. the Ising ferromagnet. Let us consider the d-dimensional nearest neighbor Ising system in a cubic box

$$\Lambda = \{x \in \mathbf{Z}^d,\ x = x_1, ..., x_d\ :\ |x_i| \leq L \quad i = 1, ..., d\}$$

of side $2L + 1$ whose hamiltonian is written as:

$$H_\Lambda^b(\sigma) = -\frac{1}{2} \sum_{x,y \in \Lambda, |x-y|=1} \sigma(x)\sigma(y) - \frac{h}{2} \sum_x \sigma(x) - \frac{1}{2} \sum_{x \in \Lambda, z \notin \Lambda, |x-z|=1} \sigma(x) b(z) \tag{3}$$

where $\sigma \in \{-1, 1\}^\Lambda$, h is the external magnetic field and b is a fixed boundary configuration. On the space $\{-1, +1\}^\Lambda$ we define a Markov process according to the following algorithm : starting from a configuration σ we construct a new configuration σ' in two steps:

(i) First we costruct the "bond configuration" $\{\gamma(b)\}$, b=(x,x'), $|x - x'| = 1$ as follows: a bond (x, x') is defined to be "vacant" if $\sigma(x) \neq \sigma(x')$; if $\sigma(x) = \sigma(x')$ then the bond (x, x') is defined to be "occupied", $\gamma((x, x')) = 1$ with probability 1-exp$(-\beta)$ and "vacant" with probability exp$(-\beta)$, β being the inverse temperature.

(ii) Then, given $\{\gamma(b)\}$, we consider the connected sets of sites C, called "clusters", in the graph whose edges are the occupied bonds b. The second step consists in updating simultaneously all the spins in every cluster C . The updating is such that all the spins in C become either $+1$ with probability given by $\frac{1}{(1 + exp(-\beta h|C|))}$ where $|C|$ is the cardinality of the cluster C, independently for each cluster. Homogenous boundary conditions (b.c.) are taken into account by imposing that the clusters which are connected to the boundary cannot flip and must preserve the same value of the spin at the boundary(e.g. $+1$).

The above algorithm, based on the Fortuin-Kasteleyn representation of the Ising model, is reversible with respect to the Ising Gibbs state and it has been introduced three years ago by Swendsen and Wang [12] in order to reduce or even completely eliminate the critical slowing down that greatly hampered Monte Carlo simulations of critical phenomena in ferromagnetic systems of statistical mechanics like Potts models. Their initial ideas were further developed by a number of people (see e.g. [13] and references therein) and made available for models different from the original ones like plane rotators or completely frustrated systems.

From a theoretical point of view and in connection with numerical simulations, the central point of this subject is to study the rate of approach to equilibrium particularily at the critical point. Unfortunately, with the exception of a rigorous lower bound on the dynamical critical exponent z obtained by Li and Sokal [14], results are available only off the critical point [6]. In section 3 we will discuss this point following the same ideas developed for the " finite dimensional " case of the Ito equation (1). Our approach, as before, is based on the proof of loss of memory of the initial conditions (in the language of Markov chains it is a "coupling argument"). Roughly speaking we show that two different initial spin configurations which evolve under the same noise (i.e. the random numbers involved in the definition of the dynamics are, at each step, the same for the evolution of both configurations) in a finite box of size L, after a time of order of log(L), become identical with a probability bounded below uniformly in L. It is easy to show that this property corresponds to the exponential convergence to the equilibrium measure. Here the loss of memory is studied in terms of the Hamming distance (number of different spins) between two different initial data evolving under the same stochastic noise. The mechanism which is responsible for the loss of memory in the case of homogeneous boundary conditions parallel to the magnetic field h and large β is the following : in a short time scale both configurations will typically consist of a huge cluster of spins parallel to h attached to the boundary of the finite box and of small islands of opposite spins. These islands will rapidly flip in direction of the external field and they will become therefore part of the huge cluster. Of course, due to thermal noise, other new islands will appear in the bulk of the sea of spins attached to the boundary. Since both configurations will have a very large portion of the cluster of the boundary in common, most of the newly formed islands will be identical for both thus producing the loss of memory. Again, in order to prove such a picture, we will make use of a multiscale analysis in space-time. The case of small β is less interesting but it can be discussed by the same techniques. More complicate is the case of boundary conditions antiparallel to the magnetic field. The reason for these difficulties is related to the physical features of the equilibrium state: at low temperature and zero magnetic field the Ising model for $d \geq 2$ exhibits a phase transition and the configurations of the system have a kind of "symmetric double well structure". For small h this feature is preserved even though now one of the two wells becomes deeper. It turns out that when the b.c. are parallel to the magnetic field the structure of the configurations is that of a single well i.e. there is only one "locally stable configuration" (i.e. a spin configuration that can be modified only with a probability going to zero as β tends to infinity) . The reason is that large clusters opposite to the field are immediately flipped. However when b.c. are opposite to h the double well structure is effective; in fact clusters connected to the boundary are pinned in the "wrong phase" and then if h is very small there are two opposite locally stable configurations. Thus final equilibrium can only be reached after the formation in the whole bulk of the "right phase". This phenomenon, which occurs via homogeneous nucleation, requires a much more detailed analysis and it is the subject of [15]. After the appearance of the right phase, the dynamics inside the bulk is practically the same as the one with b.c. parallel to h for time scales much larger than the typical time needed to reach equilibrium and we are back to the previous case. Much more complicate is the case of zero external field h. In this case the Markov process describes the dynamics of many truly interacting clusters and a rigorous, although not completely satisfactory treatment of this situation, has only recently been found again by a multiscale analysis [7].

Let us now sketch the main ideas of our approach. As we have explained above, in both situations the basic mechanism leading to the loss of memory is a local one; by this we mean that if we consider a fixed finite interval of time I (space-time for the S-W dynamics) and we take ϵ or $\frac{1}{\beta}$ very small, then, with large probability, the distance between two configurations at the final time will be much smaller than their distance at time t=0. We are assuming here for simplicity that in the S-W dynamics b.c. are parallel to the external field. We then expect that, if we take a long time interval that we imagine decomposed in many shorter intervals, then typically its contribution to the decrease of the distance between the two random paths will be the "product" of the contribution of each of the short pieces. In order to estimate the probability that this does not happen one proceeds by induction. The main idea is to introduce a sequence of time scales (length and time scales for the S-W dynamics) t_k and a suitably defined probability P_k of "conservation of memory" on scale k, in such a way that P_{k+1} can be estimated in terms of the square of the probability P_k of the same event on scale k.

More precisely the event Ω_{k+1} such that $P(\Omega_{k+1}) = P_{k+1}$ should be defined in a such a way that if Ω_{k+1} occurs during the time interval t_{k+1}, then in the same time interval either some extremely unlikely event whose probability $\tilde{p}_k$ goes to zero very fast as k $\to \infty$ has occured, or there are at least two disjoint time intervals of length t_k inside the given scale t_{k+1} such that in each of them the event Ω_k has occured.

If the above construction is possible and if P_k does not depend on the starting point of the Markov process, then, using the Markov property, we get:

$$P_{k+1} \leq a_k P_k^2 + \tilde{p}_k \tag{4}$$

where in this case a_k counts the number of disjoints pairs of intervals t_k inside t_{k+1}. It is not difficult to see that (4) implies that P_k tends to zero for large k provided that P_{k_0} is sufficiently small for a fixed initial scale k_o. This last condition is in general easily satisfied in the given range of parameters.

The above general scheme has of course to be adapted to each different case by choosing the time scales and the probabilities P_k according to the particular features of the problem. In our opinion it provides a very effective and rather general way to attack problems that are naturally decomposed in a sequence of self similar problems over different scales.

In the next three sections we will discuss concretely the application of the above scheme to the Ito equation and to the S-W dynamics.

Section 2 Small random perturbations of dynamical systems

In this section we apply the ideas illustrated in the previous section to the s.d.e. (1). For the clarity of exposition we restrict ourselves to a very special case of a one dimensional system. Although this might appear as an excessive simplification, there is no serious modification to the arguments given below when one treats more general systems in $\mathbf{R}^n$ with a non gradient drift b(x).

Let us consider the stochastic differential equation (1) with n=1 and the drift b given by :

$$\text{b(x)} = -\frac{d}{dx}V(x) \tag{5}$$

where V(x) is an even polynomial with only quadratic relative minima $\{x_1, ...x_n\}$.

If we denote with $x_t(x)$ the solution of eq. (2) starting from x, then we observe that in this case:

a) max $\{\sup_x \frac{d}{dx}x_t^\epsilon(x), 1\}$ < exp(Bt) for some finite constant B independent of ϵ.

b) There exists an interval C_i around each x_i such that $\sup_{x \in C_i} \frac{d}{dx}x_t(x)$ < exp(-mt) for some positive constant m.

Let now $L(x,t) \equiv L(\epsilon, x, t)$ be the derivative with respect to x of the solution x_t^ϵ of (1) at time t starting from x, and let $C = \cup_i C_i$. Then we have the following:

Theorem 1

For any $m' < m$ there exist positive constants k, t_o, ϵ_o, such that for any $\epsilon < \epsilon_o$ and any $t > t_o$

$$P(\sup_{x \in C} L(x,t) \leq \exp(\text{-mt}) \quad \forall t > t_o) \geq 1 - \exp(\frac{-k}{\epsilon^2})$$

Remark 1 It is clear that the above result proves the loss of memory in the sense explained in the previous section at least for initial conditions x and y starting from the same neighborhoods C_i. A more detailed discussion concerning the loss of memory for arbitrary initial conditions can be found in [1].

Proof

As a starting point let us discretize the time by partitioning the positive half line $[0,\infty)$ into intervals $I_1, ..I_j, ...$ of equal length l_o, with $l_o = \exp(\frac{a}{\epsilon^2})$ for some positive "small" a that will be chosen conveniently later on. We next define the time scales of our multiscale analysis:

$$t_k = l_o d_k$$

where $d_k = d_o^{\alpha^k}$ for some $\alpha \in (1,2)$ and $d_o > 1$. The final step consists in defining the quantities p_k as :

$$p_k = \sup_x P(\Omega(k,x)) \tag{6}$$

where

$$\Omega(k,x) \equiv \{w; \sup_{y;|y-x|<\exp(\text{-At}_{k-1})} L(y,t_k) > \exp(\text{-m}_k t_k)\} \tag{7}$$

Here A is greater than e.g. twice the constant B appearing in a) above and m_k is defined recursively by: $m_{k+1} = m_k - 2B\frac{t_k}{t_{k+1}}$.

Note that because of our definition of the time scales the " masses " m_k are uniformly bounded from below provided that d_o is chosen large enough depending on m.

Given x in the set C and a realization of the brownian motion w_t let x^j denote the solution of (1) computed at times s_j defined recursively by: $s_j = s_{j-1} + t_j \quad s_o = 0$. Then it is not difficult to see that, using the Markov property, we can bound from below the probability appearing in theorem 1 by :

$$1 - \text{const.}\ \Sigma_k p_k \tag{8}$$

It is our goal therefore to prove that p_k goes rapidly to zero as $k \to \infty$ and that for finite k p_k goes exponentially fast to zero as $\epsilon \to 0$. To this end we will prove, following our general scheme, that :

$$p_{k+1} \leq t_{k+1}^2 p_k^2 \tag{9}$$

for any k.

Remark 2 In the simple case we are considering the term $\tilde{p}_k$ is missing. The technical reason is that for the gradient case (5) with V(x) an even polynomial, the time necessary for the unperturbed solution of (2) to enter a finite interval containing all the equilibria of the system is bounded uniformly in the initial starting point , unless V(x) is quadratic. In the more general case this requirement is in general in conflict with **a)** above and the extra term $\tilde{p}_k$ appears (see [11]). The quadratic case V(x) = x^2 is explicitly solved without any need to multiscale analysis.

If we define $f_k \equiv t_k^{\lambda} p_k$ with $\lambda > \frac{2\alpha}{2-\alpha}$, then from (9) we obtain:

$$f_{k+1} \leq f_k^2$$

i.e. $f_k \leq f_{k_o}^{2^{k-k_o}}$ for any k_o. Thus, if for some finite k_o f_{k_o} is smaller than one, then p_k will decrease very fast with k and the series (8) will be finite. In order to verify this " initial scale condition " let us choose $k_o = 0$. Then it is not too difficult to show that $f_o < \exp(-\frac{c}{\epsilon^2})$ for some positive c, provided that the constant a is chosen small enough. The reasons are the following:

i) Under our assumptions and using Ventzel and Freidlin results, the process x_t^{ϵ} will enter one of the intervals C_i in a finite time $t(\epsilon)$ which, for any given a', is smaller than $\exp(\frac{a'}{\epsilon^2})$ as $\epsilon \to 0$.

ii) If the process starts in one of the intervals C_i then it will stay there for a time of the order $\exp(\frac{\delta}{\epsilon^2})$ where δ depends on the diameter of C_i.

iii) If the process starts in C_i and stay there for a time t, then an explicit computation show that L(x,t) < exp(-m't) for some m' < m.

Thus p_o can be roughly estimated by the probability that either i) or ii) did not occur and this, using again the Ventzel and Freidlin estimates, has a probability, for a small enough, much smaller than the factor t_o^{λ} appearing in the definition of f_o .

Thus we are left with the proof of (9). Let us write the interval $[0,t_{k+1}]$ as a union of N+1 intervals $J_1...J_{N+1}$, $N=[\frac{t_{k+1}}{t_k}]$, with the first N each of length t_k. We will show below that in order for the event $\Omega(k+1,x)$ to occur, there must exist at least two disjoint intervals J_i and J_j with $i < j \leq N$ such that the events $\Omega(k, x_i^{\epsilon}(x))$ and $\Omega(k, x_j^{\epsilon}(x))$ occured, where $x_i^{\epsilon}(x)$ is the solution of (1) computed at the begining of the interval J_i and the analogously for $x_j^{\epsilon}(x)$. Once this has been proved, (9) follows from the Markov property and the definition of p_k.

Thus let us assume that it is not possible to find J_i and J_j as above. This means that, with the exception of possibly one interval J_{i_o}, all the remaining intervals J_i $i \neq i_o$ $i \leq N$ are "good" in the sense that:

$$\sup_{|y - x_i^{\epsilon}(x)| \leq \exp(-At_{k-1})} L(y, t_k) < \exp(-m_k t_k)$$

Let us now take y such that $|y - x| \leq \exp(-At_k)$ and let us show that for any i=1....N :

$$|x_i^{\epsilon}(y) - x_i^{\epsilon}(x)| \leq \exp(-At_{k-1}) \tag{10}$$

If (10) holds then, using the rule of composition for the derivative, we can estimate $L(y,t_{k+1})$ by :

$$L(y, t_{k+1}) < \exp(-m_k(N-1)t_k)\exp(2Bt_k) < \exp(-m_{k+1}t_{k+1}) \tag{11}$$

where the first factor in the r.h.s. of (11) takes into account the contribution of the N-1 "good" intervals and the second comes from the apriori bound : $\sup_x L(x,t) < \exp(Bt)$.

Thus we have to prove (10). We have to distinguish between two cases:

i) $i \leq i_o$

ii) i > i_o

The first case is trivial since each one of the "good" intervals gives a distance at its right end smaller than the distance at its lefts end. The second case is also trivial since the interval J_{i_o} can separate nearby points only by a factor exp(Bt_k); thus if the distance at the beginning of J_{i_o} was smaller than exp(-At_k) with A > 2B, then at the end of J_{i_o} it will be smaller than exp((-A+B)t_k)<< exp(-At_{k-1}). We have proved (10) and therefore also (9).

Section 3 Swendsen-Wang dynamics at low temperature

In this section we implement our general multiscale scheme to the analysis of the approach to equilibrium of the Swendsen-Wang dynamics in a finite but arbitrarily large box at low temperature and positive magnetic field. At the end of the section we will comment about the case of zero external field.

We start by constructing the dynamics with + boundary conditions in a finite box Λ of the d-dimensional cubic lattice $\mathbf{Z^d}$. Now, given Λ, let ν_b and $\xi(C)$ be numbers in $\{0,1\}$ and (0, 1) associated to each bond b and to each connected set C in $\mathbf{Z^d}$ respectively, and let $\sigma \in \{-1,1\}^\Lambda$ denotes a generic configuration of plus or minus spins in Λ. Given the numbers ν_b and ξ_C we construct out of the configuration σ a new configuration σ' as follows. From σ we first generate a configuration γ of occupied ($\gamma(b) = 1$) and vacant ($\gamma(b) = 0$) bonds, by setting

$$\gamma(b) = (\frac{1+\sigma_b}{2})\nu_b$$

where $\sigma_b = \sigma_x\sigma_y$ if b=(x,y) and $\sigma(x) = +1$ if $x \in \partial\Lambda$. We will say that two n.n. sites (x,x′) are connected in the bond configuration γ if $\gamma(x,x') = 1$ i.e. the bond (x,x′) is occupied in γ. The maximal connected components C (with respect to the configuration γ) are called clusters. They are of course in particular connected clusters and may reduce to a single site.

Now for any cluster C we set:

$\sigma'(x) = 1 \quad \forall x \in C$ if either $\xi(C) \le \frac{1}{(1+\exp(-h|C|))}$ or $C \cap \partial\Lambda \neq \Phi$

$\sigma'(x) = -1 \quad \forall x \in C$ in all the other cases

Let us now consider two sequences of numbers

$$\omega \equiv (\{\nu_b(t)\}_{t\in\mathrm{N}}\,;\ \{\xi(t,C)\}_{t\in\mathrm{N}})$$

that we think of as the realization of two mutually independent process with values in $\{0,1\}$ and (0,1) respectively, each of which is a collection of independent identically distributed random variables (i.i.d. rv) with distribution:

$$\begin{aligned} \nu_b &= 0 \quad \text{with probability exp(-}\beta\text{)} \\ \nu_b &= 1 \quad \text{with probability 1-exp(-}\beta\text{)} \end{aligned}$$

and uniform distribution for the $\xi(s,C)$.

Given ω we finally construct a random flow on $\{-1,1\}^\Lambda$, $\{\phi_t^{\Lambda,\omega}(\cdot)\}_{t\in N}$ by applying at each time step t the above rule with numbers $\nu_b(t), \xi(t,C)$. Sometimes, for notational convenience, we will write:

$$\sigma_t(x) = \phi_t^{\Lambda,\omega}(\sigma)(x) \tag{12}$$

Remark 1

i) the boundary condition +1 at the boundary of Λ are taken into account by the condition that any cluster C touching $\partial\Lambda$ is set equal to +1. Other boundary conditions may be considered e.g. periodic or open.

ii) Notice that if $\Lambda' \subset \Lambda$ then one can compare the random flows $\phi_t^{\Lambda,\omega}$, $\phi_t^{\Lambda',\omega}$ as follows: given σ in Λ one constructs $\hat{\sigma}$ in Λ' by the rule

$$\hat{\sigma}(x) = \sigma(x) \quad \text{if} \quad x \in \Lambda'$$
$$\hat{\sigma}(x) = +1 \quad \text{if} \quad x \in \partial\Lambda'$$

The evolutions $\phi_t^{\Lambda,\omega}(\sigma)$ and $\phi_t^{\Lambda',\omega}(\hat{\sigma})$ are constructed by means of the same random numbers $(\nu_b(t), \xi(t,C))$ if b and C are in Λ'. However a cluster C intersecting $\partial\Lambda'$ is set equal to +1 for the dynamics $\phi_t^{\Lambda',\omega}$ but may be -1 for the dynamics $\phi_t^{\Lambda,\omega}$.

It is easy to see that the above defined dynamics satisfies the detailed balance condition for the Gibbs state of the Ising model on Λ, with + boundary conditions on $\partial\Lambda$, at inverse temperature β. The proof of this statement can be found for example in [6].

Our point of view for studying the approach to the equilibrium is to analyze the time behaviour of the Hamming distance

$$\rho(\sigma_t, \eta_t) \equiv \frac{1}{4}\sum_x (\sigma_t(x) - \eta_t(x))^2 \tag{13}$$

between two configurations σ_t, η_t evolving under the same random flow $\phi_t^{\Lambda,\omega}$. We will prove that if the time t is taken large enough, depending on the size of the box Λ, then, with very large probability, $\rho(\sigma_t, \eta_t) = 0$. The way we control the Hamming distance is by means of a multiscale analysis very close to the scheme used for the Ito case.

For $k \in N$ let $L_k = 10^k$, $t_k = 2^k$ be a sequence of lengths and time scales. We denote by Λ_k the box centered at the origin of side L_k and we set :

$$P_k \equiv \sup_{\sigma,\eta\in\{-1,1\}^{\Lambda_k}} P(\rho(\phi_{t_k}^{\Lambda_k}(\sigma),\ \phi_{t_k}^{\Lambda_k}(\eta)) > 0) \tag{14}$$

Then we will prove:

Theorem 2

Given h > 0 there exist positive constants C, β_0 depending on d and h, such that if $\beta > \beta_0$, there exists $m \equiv m(\beta,h) > 0$ such that for any

$$P_k \leq \text{Cexp(-m } t_k)$$

An easy consequence of the above result is the exponential approach to equilibrium (see [6]).

Let f be a real function on $\{-1,1\}^{\mathbf{Z}^d}$ of compact support S_f, i.e. $f(\sigma) \equiv f(\{\sigma(x)\}_{x\in S_f})$, with $|S_f| < \infty$, and let $\mu(f)$ denotes its average with respect to the Ising Gibbs state at inverse temeperature and external field h. Then, in the same hypotheses of theorem 1, we have:

Corollary 1 There exist constants $C_f > 0$ and $m = m(\beta) > 0$ such that

$$\sup_{\sigma} |\mu(f) - \mathbf{E}f(\sigma_t)| < C_f \exp\{-mt\}$$

Proof of theorem 2

The theorem is proved inductively. We will in fact show that our basic recursive inequality (5) holds also in this case :

$$P_{k+1} \leq L_{k+1}^d P_k^2 + L_{k+1}^d \exp(-\frac{\beta h}{2}\sqrt{L_k/2}) \tag{15}$$

Once (15) is established then the result of the theorem easily follows provided that for some initial scale k_o one has :

$$P_{k_o} << \frac{1}{L_{k_o+1}^d} \tag{16}$$

This last condition is clearly verified since for β large most of the cluster flip in the direction of the magnetic field h.

Thus we will concentrate on the proof of (15). We write :

$$\begin{aligned} P_{k+1} \leq\ & 2\sup_{\sigma} P(\exists x \in \Lambda_{L_{k+1}};\ \phi_{t_{k+1}}^{\Lambda_k^x}(\sigma)(x) \neq \phi_{t_{k+1}}^{\Lambda_{k+1}}(\sigma)(x)) + \\ & + \sup_{\sigma,\eta} P(\exists x \in \Lambda_{L_{k+1}};\ \phi_{t_{k+1}}^{\Lambda_k^x}(\sigma) \neq \phi_{t_{k+1}}^{\Lambda_k^x}(\eta)) \end{aligned} \tag{17}$$

where Λ_k^x is the box centered at x of side L_k.

The first term in r.h.s. corresponds to the term $\tilde{p}_k$ in (4) and it accounts for the occurence of the event that at the end of the time interval t_{k+1} and for some x in Λ_{k+1}, the two dynamics, one in the large box Λ_{k+1} and the other in the smaller box Λ_k^x have produced a different value of the spin at x. This can be shown to be an extremely unlikely event whose probability can be estimated [6] by:

$$L_{k+1}^d \exp(-\beta h\sqrt{L_k/2}) \tag{18}$$

The intuitive reason behind (18) is the following: since the differences between the two dynamics can propagate only through clusters which flip to -1 i.e. opposite to the external field h, their speed of propagation inside the box Λ_k^x can be bounded by the diameter of the largest among the clusters intersecting the given box at some time $0 \leq t \leq t_{k+1}$ and whose sign is -1. Thus in order to observe a difference at x at time t_{k+1} the speed should have been larger than $\frac{L_k}{t_{k+1}} >> \sqrt{L_k/2}$. Since the probability that a cluster of size N flips against the magnetic field is exponentially small in βhN , (18) follows.

The second term in the r.h.s. of (17), using the Markov property and the fact that $t_{k+1} = 2t_k$, is estimated by:

$$L_{k+1}^d P_k^2$$

Thus (15) is proved.

We conclude this note with a short discussion of the case of zero external field. In this more complicate situation the above simple bound on the speed of propagation of differences between the dynamics in the box Λ_{k+1} and the dynamics in the smaller box Λ_k is no longer valid since clusters take the value -1 with probability $\frac{1}{2}$ irrespectively of their geometry. However it is well known that at very low temeprature the Gibbs state of the Ising model with + b.c. consists of a huge cluster of plus spins and small isolated clusters of minus spins. Therefore one expects that also the dynamics after a certain time will look the same independently of the initial configuration σ. A proof of this fact requires however a good control of the probability of having a big cluster of minus spins at time t. More generally, we will establish a basic estimate on the probability $P_\sigma(L,t,x)$ that the configuration σ_t has a path of vacant bonds of length L containing a fixed site x at time t, with t greater than some time scale t(L) related to L. Let us first discuss the case of zero temperature $\beta = \infty$. In this case no bond is made vacant during the dynamics, and the only possibility to observe a path of vacant dual bonds at time $t + \frac{1}{2}$ is that the same path was already present at any previous time including t=0. The probability of this last event is bounded from above by $(\frac{1}{2})^t$; however if the path in consideration is closed and it separates exactly two different clusters at time t=0, then the above bound becomes exact. This discussion suggests that any bound on $P_\sigma(L,t,x)$ will be at most exponential in the time scale t with rate constant $m(\beta)$ of order ln(2), and in particular that to obtain a rigorous bound on $P_\sigma(L,t,x)$ by means of some kind of Peierls argument should be a very difficult task, unless t>>L, since the number of paths grows exponentially fast in the length scale L. Time scales larger than length scales are on the other hand very difficult to treat rigorously since in this case it is no longer exact to think of the dynamics as being built up by many " local " dynamics on smaller space regions, as we did for example in the proof of theorem 2.

The way out of this " impasse " is again a multiscale analysis suitably tuned to the problem. Let us first fix some notations.

For any integer k we define:

i) $L_k = 4^{k^2}$

ii) $t_k = 3^k$

iii) $\Lambda_k = \Lambda_{L_k}$

iv) Given Λ_L with L> L_k we denote with the name (k,+)-dynamics in Λ_L the algorithm described in the previous section with the following extra condition:

$$\xi(s,C) = 0 \quad \text{if diam}(C) > L_k$$

v) $P_k = \sup_{L \geq L_k; x \in \Lambda_L; t \geq t_k; \sigma \in \{-1,1\}^{\Lambda_L}} P(L,x,t,k,\sigma)$

where $P(L,x,t,k,\sigma)$ is the largest between the probability that there exist at time t + $\frac{1}{2}$ a path of vacant bonds of length n $\geq L_k$ containing x for the dynamics in Λ_L with + b.c. starting from σ and the same quantity computed for the (k,+)-dynamics in Λ_L starting from σ.

Then we have the following result [7]:

Theorem 3

There exists $\beta_o > 0$, c> 0 and a> 0 such that for any $\beta \geq \beta_o$ there exists a positive constant m(β) with m(β) $\geq c$ such that:

$$P_k \leq \frac{1}{L_k^{2a}} \exp(-m(\beta)2^k) \quad \forall k > 5$$

The theorem is again proved by establishing our basic recursive inequality (4) for the above quantity p_k and using the " zero temperature " discussion above to show that on a finite initial scale k_o p_{k_o} is sufficiently small. Once the above result is established it is possible to go back to theorem 2 and prove a fast convergence to equilibrium.

Remark It is important to note that the above bound and therefore also the rate of convergence to equilibrium, is only exponential in t_k^α with $\alpha = \frac{ln(2)}{ln(3)}$. This is due to the fact that p_k appears in the r.h.s. of (4) to the square while $t_{k+1} = 3t_k$.

Theorem 3 provides also the technical estimate to give a "dynamical" proof of the existence of a phase transition at low temperature. In fact, if in two dimensions we define the dynamics directly in the infinite volume, with the rule that any infinite cluster takes the value +1 with probability p and -1 with probability 1-p while finite clusters flip as before, then it is possible to show that, starting from a configuration identically equal to plus one, the time distribution will converge weakly to pμ_+ + (1-p)μ_- while there exists another configuration, which is explicitely constructed, whose distribution converges to $\frac{1}{2}(\mu_+ + \mu_-)$. Here μ_+ and μ_- are the usual two extremal states of the two dimensional Ising model.

Acknowledgments

Most of the material discussed in this note is joint work with Elisabetta Scoppola and Enzo Olivieri. It is a pleasure to thank them for the stimulating adn pleasant collaboration.

References

[1] J.T.Chayes and L.Chayes "Percolation and random media" in Critical Phenomena,Random systems and Gauge theories, Les Houches, Session XLIII 1984, eds K.Osterwalder and R.Stora, 1001-1142, Elsevier Amsterdam

[2] Von Dreyfus, "On the effect of randomness in ferromagnetic models and Schröedinger operators", Ph. D. Thesis New York (1987).

[3] F.Martinelli and E.Scoppola "Introduction to the mathematical theory of Anderson localization" La Rivista del Nuovo Cimento **10** (1987)

[4] F. Martinelli, E. Scoppola, "Small random perturbations of dynamical systems: exponential loss of memory of the initial condition", Comm. Math. Phys. **120** 25–69 1988.

[5] F. Martinelli, E. Oliveri, E. Scoppola, "Small random perturbations of finite ad infinite dimensional dynamical system: unpredicatbility of exit times", J. Stat. Phys. **55** 3/4 1989.

[6] F. Martinelli, E. Olivieri, E. Scoppola "On the Swendsen and Wang dynamics I. Exponential convergence to equilibrium"Journal of Stat. Phys. **62** N. 1/2 117 1991 .

[7] F.Martinelli "Low temperature behaviour of Swedsen-Wang Monte Carlo algorithm" in preparation.

[7] F.Martinelli "Low temperature behaviour of Swedsen-Wang Monte Carlo algorithm" in preparation.

[8] A.D. Ventzel and M.I.Freidlin "On small random perturbations of dynamical systems" Usp.Math.Nauk **25** 3 (1970) (English tranlation:Russ. Math.Surv. **25**, 1 (1970)

[9] J.Froelich and T.Spencer "Absence of diffusion in the Anderson tight binding model for large disorder or low energy" Comm. Math. Phys. **88** 151 (1983)

[10] J.Froelich,F.Martinelli,E.Scoppola and T.Spencer "Constructive proof of localization in the Anderson tight binding model" Comm. Math. Phys. **101** 21 (1985)

[11] L.Sbano Thesis Roma 1990

[12] R.H. Swendsen, J.S. Wang "Nonuniversal critical dynamics in Monte Carlo simulations" Phys. Rev. Lett. **58** (1987).

[13] R.H. Swendsen, J.S. Wang "Cluster Monte Carlo algorithms" Physica A **167** 565-579 (1990)

[14] Xiao-Jian-Li, A.D. Sokal, "Rigorous lower bound on the dynamical critical exponent of the Swendsen–Wang dynamics", Phys. Rev. Lett. **63**, 8, 827 (1989).

[15] F. Martinelli, E. Oliveri, E. Scoppola, "On the Swendsen and Wang dynamics II : Critical droplets and homogeneous nucleation at low temperature for the 2 dimensional Ising model" Journ. Stat. Phy. **62** N. 1/2 135 1991 .

A Limiting Distribution Connected with Fractional Parts of Linear Forms

A.E.Mazel[1] and Ya.G.Sinai[2]

1.International Institute of Earthquakes Prognosis Theory and Mathematical Geophysics, Academy of Sciences of USSR, Moscow, USSR

2.L.D.Landau Institute of Theoretical Physics, Academy of Sciences of USSR, Moscow, USSR

1.Introduction

Consider independent random variables x and y uniformly distributed on the S^1 and the sequence $\{\lambda_m\}\in S^1$, m=0,1,... where $\lambda_m=y+mx \pmod 1$. In this paper a local structure of the sequence $\{\lambda_m\}$ is investigated and the following result is proven.

Theorem. *Consider random variables* $\xi^{(N)}=\#\{0\le m\le N: \lambda_m\in[0,cN^{-1}]\}$, $0<c\le 1$, N=1,2,... *Then* $\Pr\{\xi^{(N)}=k\}$ *tends to the limit* p_k *as* $N\to\infty$ *where* $p_0=1-c+3c^2\pi^{-2}$, $p_1=c-5.25c^2\pi^{-2}$, $p_k=3c^2\pi^{-2}((k-1)^{-2}-2k^{-2}+(k+1)^{-2})$ *with* k=2,3,...

The problem of studying the sequence λ_m arose in the quantum chaos theory in connection with the kicked rotator model introduced in the paper [1]. The monodromy operator of this model has the eigenvalues $\exp\{2\pi i\mu_m\}$ where $\mu_m=y+mx+0.5\alpha m(m-1)$; $\mu_m,x,y,\alpha\in S^1$. Further analysis [2] has shown that the behavior of sequences λ_m and μ_m is closely connected with the behavior of the Weyl's double trigonometric sums well-known in the number theory. In particular the sums:

$$W_{N_1N_2}(x,y,\alpha)=(N_1N_2)^{-0.5}\sum_{n=N_1}^{N_1}\sum_{n=-N_2}^{N_2}\exp\{2\pi in\mu_m\},$$

$$W_{N_1N_2}(x,y)=(N_1N_2)^{-0.5}\sum_{n=-N_1}^{N_1}\sum_{n=-N_2}^{N_2}\exp\{2\pi in\lambda_m\},$$

were investigated in the paper [3] where some results concerning the limiting distribution for the $W_{N_1N_2}(x,y)$ was established when $N_1=N_2=N\to\infty$. More detailed information about this limiting distribution can be obtained by the combination of the methods used in the work [3] with the results of this paper. The above mentioned fact was one of the main motivations for our studies.

2.Geometric Construction, Proof of the Theorem

Consider the two dimensional torus Tor^2 with coordinates $(x\in S^1, y\in S^1)$ and the skew translation $T(x,y)=(x,y+x)$ acting on it. Let $\Pi_0^{(N)}\in Tor^2$ be the strip $\Pi_0^{(N)}=\{(x,y)\in Tor^2 \mid 0\le y\le cN^{-1}\}$ and $\Pi_m^{(N)}=T^{-m}\Pi_0^{(N)}$. It is easy to verify that $T^{-m}(x,y)=(x,y-mx)$, $\Pi_m^{(N)}=\{(x,y)\in Tor^2 \mid -mx\le y\le -mx+cN^{-1}\}$ and $\lambda_m(x,y)\in[0,cN^{-1}]$ iff $(x,y)\in\Pi_m^{(N)}$. Hence $Pr\{\xi^{(N)}=k\}$ is equal to the Lesbeque measure (area) of the set $I_k^{(N)}$ of such points of Tor^2 which belong exactly to k strips of the family $\{\Pi_m^{(N)}, m=0,1,\dots,N\}$. Due to the symmetry the partition of Tor^2 generated by this family is not changed if the definition of the strip $\Pi_m^{(N)}$ is replaced by means of the expression $\Pi_m^{(N)}=T^m\Pi_0^{(N)}=\{(x,y)\in Tor^2 \mid mx\le y\le mx+cN^{-1}\}$. Therefore for the sake of convenience starting with this point we use $\Pi_m^{(N)}=T^m\Pi_0^{(N)}$.

In view of the introduced geometrical construction the problem of calculation of p_k is reduced to finding

$\lim_{N\to\infty} A(I_k^{(N)})$ where here ahd further $A(\cdot)$ denotes the Tor^2 subset's area. Let $I_{k,m}^{(N)}=I_k^{(N)}\cap\Pi_m^{(N)}$. The $I_k^{(N)}$ definition implies that $A(I_k^{(N)})=k^{-1}\sum_{m=0}^{N} A(I_{k,m}^{(N)})$ which leads to the necessity of the detailed investigation of the structure of the intersection of $\Pi_m^{(N)}$ with all other strips. Since from one hand $T^{-m}\Pi_m^{(N)}=\Pi_0^{(N)}$ and from the other one $A(\Pi_m^{(N)}\cap\Pi_{m\pm n}^{(N)})=A(T^{-m}(\Pi_m^{(N)}\cap\Pi_{m\pm n}^{(N)}))$ (the translation T preserves the Lesbeque measure) it is easier to investigate $T^{-m}\Pi_m^{(N)}$ rather than $\Pi_m^{(N)}$.

For the fixed $\Pi_m^{(N)}$ we shall call the strips $\Pi_{m+n}^{(N)}$, $n=1,2,\ldots,N-m$ as the strips from the future and the strips $\Pi_{m-n}^{(N)}$, $n=1,2,\ldots,m$ as the strips from the past.

According to our construction $T^{-m}(\Pi_m^{(N)}\cap\Pi_{m+n}^{(N)})$ consists of n identical parallelograms with coordinates of the vertices: $(ln^{-1},0),(ln^{-1}+c(Nn)^{-1},cN^{-1})$, (ln^{-1},cN^{-1}), $(ln^{-1}-c(Nn)^{-1},0)$, $l=0,1,\ldots,n-1$ (see Fig 1.a) and $T^{-m}(\Pi_m^{(N)}\cap\Pi_{m+n}^{(N)})$ consisting of n identical parallelograms with coordinates of the vertices: $(ln^{-1},0)$, $(ln^{-1}+c(Nn)^{-1},0),(ln^{-1},cN^{-1})$, $(ln^{-1}-c(Nn)^{-1},cN^{-1})$, $l=0,1,\ldots,n-1$ (see Fig 1.b). It is natural to use quantities l/n, $l=0,1,\ldots,n-1$ as coordinates of both types of parallelograms.

Lemma 1. *If $l_1/n_1\neq l_2/n_2$ then area of the intersection of the arbitrary two parallelograms with coordinates l_1/n_1 and l_2/n_2 is equal to zero.*

Proof. The distance between coordinates of parallelograms is obviously greater then $(n_1n_2)^{-1}$, while the length of the horisontal (parallel to the axis x) sides of this parallelograms is equal to the $c(Nn_1)^{-1}$ and $c(Nn_2)^{-1}$.

Hence the intersection of the parallelograms from one and the same time cannot contain more then one point because $c(Nn_1)^{-1} \le (n_1 n_2)^{-1}$ and $c(Nn_2)^{-1} \le (n_1 n_2)^{-1}$ (see Fig 2.a). For the parallelograms from the different times we have the similar result since in this situation $n_1 + n_2 \le N$ and hence $c(Nn_1)^{-1} + c(Nn_2)^{-1} \le (n_1 n_2)^{-1}$ (see Fig 2.b).

Corollary 1. *Parallelograms from one and the same time (i.e. from the future or from the past) have the intersection with non zero area only if their coordinates* l_1/n_1 *and* l_2/n_2 *coincide, i.e.* $l_2 = rl_1$ *and* $n_2 = rn_1$. *In this situation the parallelogram with the coordinate* l_2/n_2 *is embeded into the parallelogram with the coordinate* l_1/n_1.

Corollary 2. *The system of the* $[m/n]$ *embeded parallelograms from the past and the system of the* $[(N-m)/n]$ *embeded parallelograms from the future correspond to each irreducible fraction* l/n, $0 \le l \le n \le N$. *Systems corresponding to different irreducible fractions have intersection. with zero area. (Here and further* $[\cdot]$ *denotes the integer part of the number).*

Lemma 2. *Let* $a=[m/n]$; $b=[(N-m)/n]$; $k=1,2,\dots,a+b+1$;

$$A_{k,a,b} = \begin{cases} 2(k^2-1)^{-1} & \text{with } 1 \le k \le a \\ (2a+k+1)(k^2-1)^{-1}k^{-1} & \text{with } a < k \le b \\ 2(a+b+1)(k^2-1)^{-1}k^{-1} & \text{with } b < k \le a+b \\ (k-1)^{-1} & \text{with } k = a+b+1 \end{cases}.$$

Then for the fixed irreducible fraction l/n *the area of the domain of the k-times intersection of the parallelograms from the future or the past is equal to* $N^{-2}n^{-1}A_{k,a,b}$.

Proof. In the scale $1{:}c(nN)^{-1}$ along x axis and $1{:}cN^{-1}$

along y axis the intersection picture of the systems of parallelograms considered is shown in Fig 3. The calculation of the areas of elements of the obtained partition of the rectangular (2x1) can be easily found and leads to the value of $A_{k,a,b}$ mentioned above.

Let $I^{(N)}_{k,m,n}$ be the part of $I^{(N)}_{k,m}$ belonging to the union of parallelograms with coordinates l/m, where l=1,2,...,n-1. Obviously the number of the non zero irreducible fractions with the fixed denominator n coincides with the value of the Euler function $\varphi(n)$. For odd N we have:

$$A(I^{(N)}_k)=k^{-1}\sum_{m=0}^{N} A(I^{(N)}_{k,m})=2k^{-1}\sum_{m=0}^{(N-1)/2} A(I^{(N)}_{k,m})=$$

$$=2k^{-1}\sum_{m=0}^{(N-1)/2}\sum_{n=1}^{N-m} A(I^{(N)}_{k,m,n})=$$

$$=2k^{-1}\sum_{m=0}^{(N-1)/2}\sum_{n=1}^{N-m} \varphi(n)N^{-2}n^{-1}A_{k,[m/n],[(N-m)/n]}=$$

$$=2k^{-1}\sum_{a=0}^{\infty}\sum_{b=1}^{\infty} A_{k,a,b}c^2N^{-2}\sum_{(m,n)\in D^{(N)}_{a,b}} \varphi(n)/n,$$

where

$$D^{(N)}_{a,b}=\{(m,n)\mid 0\le m\le (N-1)/2,\ 1\le m\le N-m,\ [m/n]=a, [(N-m)/n]=b\}=$$

$$=\{(m,n)\mid 0\le m\le (N-1)/2,\ 1\le m\le N-m,\ an\le m\le an+n,\ N-bn-n\le m\le N-bn\}.$$

(the case of the even N can be investigated in the analogous way).

Lemma 3. *Consider on the plane* $\mathbb{R}^2$ *with the Cartesian coordinates* (u,v) *a polygon defined by means of the system of inequalities:*

$$0\le u\le 0.5,\ 0\le v\le 1-u,\ av\le u\le av+v,\ 1-bv-v\le u\le 1-bv$$

and let $A_{a,b}=((1+\delta_{a,b})(a+b)(a+b+1)(a+b+2))^{-1}$ ($\delta_{a,b}$ *– Kronecker symbol) be the area of this polygon. Then*

$$\lim_{N\to\infty} \sum_{(m,n)\in D^{(N)}_{a,b}} \varphi(n)/n = 6\pi^{-2}A_{a,b} \quad \textit{uniformly in a and b.}$$

Proof. From the theory of multiplicative functions [4] it is known that $N^{-1}\sum_{n=1}^{N} \varphi(n)/n=$

$$=N^{-1}\sum_{n=1}^{N}\sum_{d|n} \mu(d)/d=N^{-1}\sum_{d=1}^{N}\mu(d)d^{-1}[N/d]\xrightarrow[N\to\infty]{}\sum_{d=1}^{\infty}\mu(d)d^{-2}=6\pi^{-2},$$

where $\mu(d)$ is the Möbius function. Hence

$$\sum_{(m,n)\in D^{(N)}} \varphi(n)/n= \sum_{(m,n)\in D^{(N)}} \sum_{d|n} \mu(d)/d= \sum_{d=1}^{N}\mu(d)d^{-1}A_d(D^{(N)}_{a,b}),$$

where $A_d(D^{(N)}_{a,b})$ denotes the number of pairs $(m,n)\in D^{(N)}_{a,b}$ with n multiple to d. Clearly, $A_d(D^{(N)}_{a,b})=d^{-1}A_1(D^{(N)}_{a,b})+\Delta_d(D^{(N)}_{a,b})$, where $|\Delta_d(D^{(N)}_{a,b})|\le 2N$. The latter estimation implies that

$$|N^{-2}\sum_{d=1}^{N}\mu(d)d^{-1}\Delta_d(D^{(N)}_{a,b})|\le N^{-2}\sum_{d=1}^{N} 2Nd^{-1}\le \mathrm{const}N^{-1}\ln N\xrightarrow[N\to\infty]{}0.$$

Obviously $N^{-2}A_1(D^{(N)}_{a,b})\xrightarrow[N\to\infty]{}A_{a,b}$ uniformly in a and b. Hence

$$\lim_{N\to\infty} N^{-2}\sum_{d=1}^{N}\mu(d)d^{-1}A_d(D^{(N)}_{a,b})=\lim_{N\to\infty} N^{-2}A_1(D^{(N)}_{a,b})\sum_{d=1}^{N}\mu(d)d^{-2}=A_{a,b}6\pi^{-2}.$$

Using Lemma 3 we find that for k=2,3,...

$$p_k= \lim_{N\to\infty} 2c^2k^{-1}\sum_{a=0}^{\infty}\sum_{b=1}^{\infty}A_{k,a,b}N^{-2}\sum_{(m,n)\in D^{(N)}_{a,b}}\varphi(n)/n=$$

$$=12c^2\pi^{-2}\sum_{a=0}^{\infty}\sum_{b=1}^{\infty}A_{k,a,b}A_{a,b}$$

Using the expressions for $A_{k,a,b}$ and $A_{a,b}$ given in Lemma 2 and Lemma 3 we can calculate the letter double sum and obtain: $p_k=3c^2\pi^{-2}((k-1)^{-2}-2k^{-2}+(k+1)^{-2})$; $\sum_{k=2}^{\infty}p_k=2.25c^2\pi^{-2}$.

To calculate $A(I_1^{(N)})$ it is necessary to substract the total area of all parallelograms belonging to $\Pi_m^{(N)}$ from the area of the whole $\Pi_m^{(N)}$ and to sum over all m:

$$A(I_1^{(N)})=2\sum_{m=0}^{(N-1)/2} A(I_{1,m}^{(N)})=$$

$$=2\sum_{m=0}^{(N-1)/2}(cN^{-1}-\sum_{n=1}^{m}1.5\varphi(n)c^2N^{-2}n^{-1}-\sum_{n=m+1}^{N-m}\varphi(n)c^2N^{-2}n^{-1})=$$

$$=c-3c^2N^{-2}\sum_{m=0}^{(N-1)/2}\sum_{n=1}^{m}\varphi(n)/n-2c^2N^{-2}\sum_{m=0}^{(N-1)/2}\sum_{n=m+1}^{N-m}\varphi(n)/n;$$

$$p_1=\lim_{N\longrightarrow\infty}A(I_1^{(N)})=c-5.25c^2\pi^{-2}.$$

Finally: $p_0=\lim_{N\longrightarrow\infty}A(I_0^{(N)})=1-\sum_{k=1}^{\infty}p_k=1-c+3c^2\pi^{-2}.$

References

1.G.Casati,B.V.Chiricov,F.M.Izrailev,J.Ford. Stochastic behavior of a quantum pendulum under periodic perturbation, Lecture Notes in Physics, vol.93, Springer Berlin, 1979, p.334-352.

2.Ya.G.Sinai. Mathematical Problems in the Theory of Quantum Chaos, Raymond and Beverly Sacker Distinguished Lectures in Mathematics, Telaviv University, 1989, p.1-19.

3.Ya.G.Sinai. Limit Theorems for Weyl's Double Trigonometric Sums, Proceedings V.A.Steklov Institute, vol.191, Moscow, 1989, p.118-129.

4.M.Kac. Statistical Independence in Probability, Analysis and Number Theory, The Mathematical Association of America, 1959.

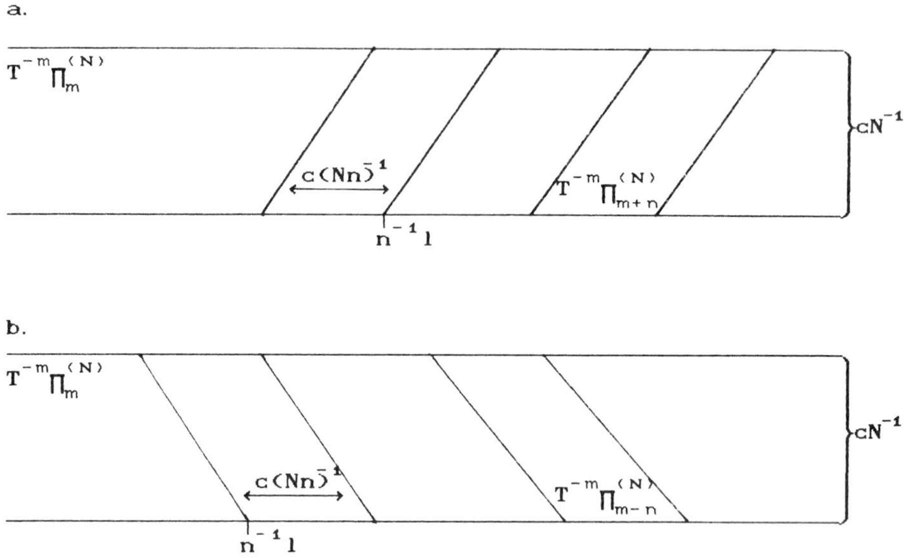
a.
T^{-m}Π_m^{(N)}
cN^{-1}
c(Nn)^{-1}
T^{-m}Π_{m+n}^{(N)}
n^{-1}l
b.
T^{-m}Π_m^{(N)}
cN^{-1}
c(Nn)^{-1}
T^{-m}Π_{m-n}^{(N)}
n^{-1}l

Fig. 1.

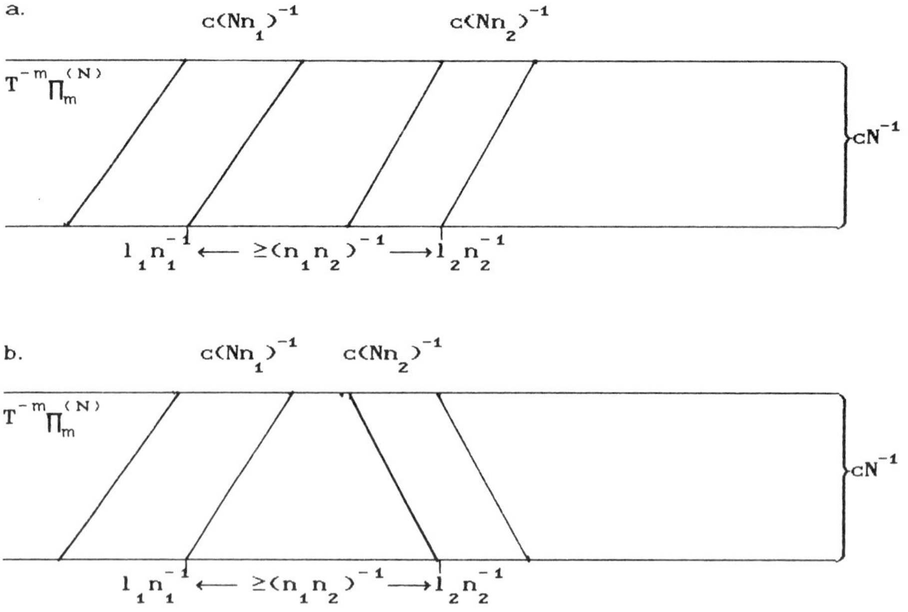
a.
c(Nn_1)^{-1}
c(Nn_2)^{-1}
T^{-m}Π_m^{(N)}
cN^{-1}
l_1n_1^{-1} ← ≥(n_1n_2)^{-1} → l_2n_2^{-1}
b.
c(Nn_1)^{-1}
c(Nn_2)^{-1}
T^{-m}Π_m^{(N)}
cN^{-1}
l_1n_1^{-1} ← ≥(n_1n_2)^{-1} → l_2n_2^{-1}

Fig. 2.

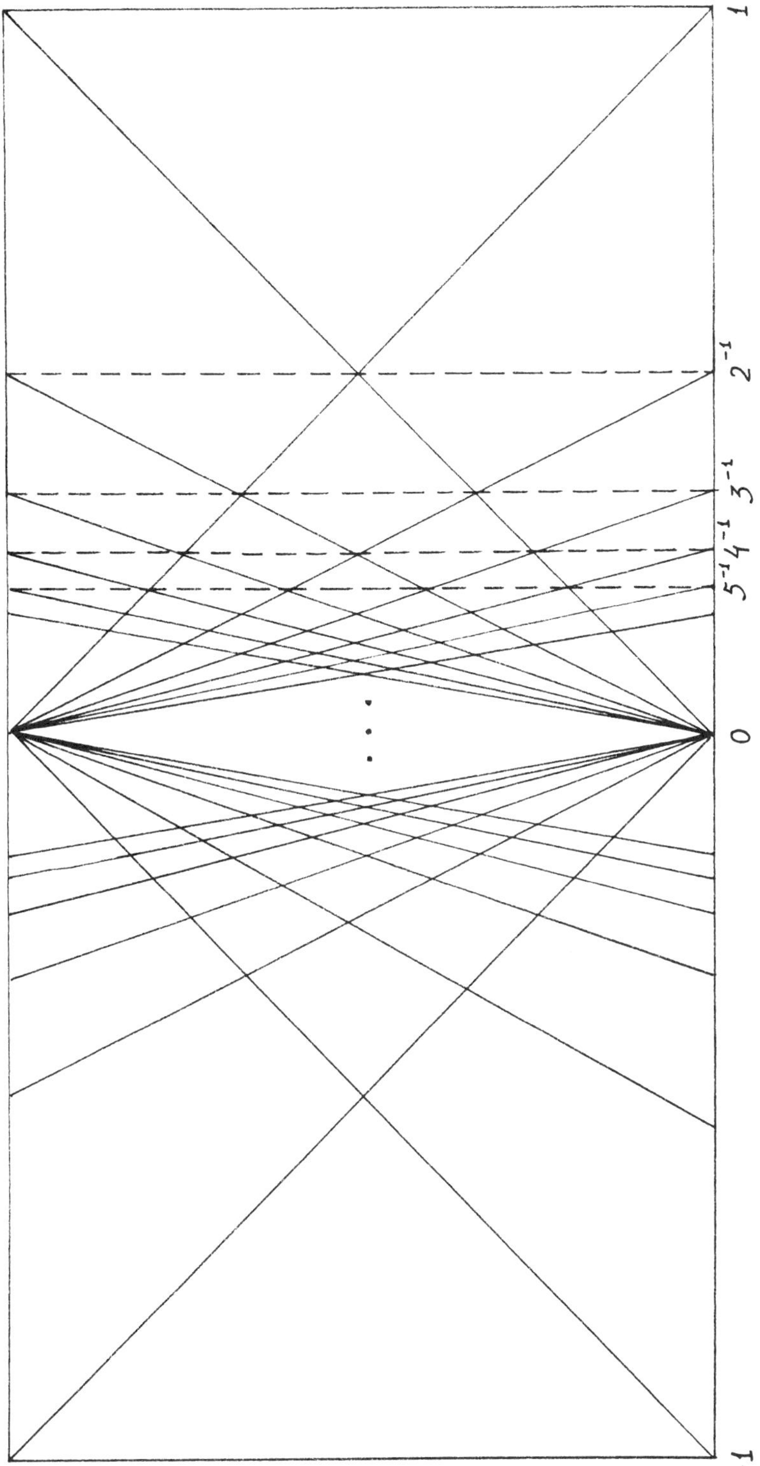

Fig. 3.

Rapport sur les représentations chaotiques

par P.A. Meyer (Strasbourg)

Les célèbres résultats de Wiener [12] sur la possibilité de développer toute fonctionnelle de carré intégrable du mouvement brownien en une série d'intégrales stochastiques multiples datent de 1939, mais ils n'ont été mis sous leur forme définitive que par Kakutani [6], 1950 et Ito [4], 1951. On doit en particulier à Ito la définition correcte des intégrales stochastiques multiples, qui chez Wiener étaient plutôt du type de Stratonovich, de manière à réaliser l'orthogonalité des "chaos de Wiener" d'ordres différents. Par ailleurs, Ito a transporté ce théorème dans le cadre plus général des mesures gaussiennes à accroissements indépendants, en utilisant des intégrales multiples de fonctions symétriques, et sous cette forme le résultat apparaît comme une interprétation probabiliste de l'espace de Fock symétrique en tant qu'espace L^2 d'une mesure gaussienne (Segal [10], 1956). Un second cas de représentation chaotique est celui des processus de Poisson : Wiener semble en avoir eu l'idée depuis le début, mais le résultat définitif semble avoir fait partie du "folklore", et je n'en connais pas de rédaction antérieure au cours de Neveu [8], 1968. Par exemple, dans l'article d'Ito [5] qui utilise les intégrales multiples par rapport à un processus à accroissements indépendants quelconque, la singularité des processus de Wiener et de Poisson n'est pas dégagée. Quoi qu'il en soit, ces deux résultats ont été considérés pendant près de 30 ans comme des théorèmes isolés, propres à certains processus à accroissements indépendants. Les développements de ces toutes dernières années (1988–89) montrent qu'il n'en est pas du tout ainsi, et que les développements chaotiques sont liés à la structure même des processus de Markov.

Le point de départ de ces développements récents est la remarque triviale suivante : pour pouvoir définir les intégrales multiples du type de Wiener

$$I_n(f) = \int_{s_1 > \ldots > s_n} f(s_1, \ldots, s_n)\, dX_{s_1} \ldots dX_{s_n}$$

possédant les deux propriétés cruciales d'isométrie et d'orthogonalité des chaos, il est seulement nécessaire que (X_t) soit une martingale de carré intégrable, nulle en 0 pour fixer les idées, de crochet oblique $< X, X >_t = t$ (autrement dit, $X_t^2 - t$ est une martingale). De même, pour traiter la théorie multidimensionnelle, il suffit de disposer de d martingales X_t^α de crochets obliques $< X^\alpha, X^\beta >_t = \delta^{\alpha\beta} t$. Le problème des représentations chaotiques consiste donc à dégager, sur un espace probabilisé $(\Omega, \mathcal{F}, \mathbb{P}, (\mathcal{F}_t))$, des martingales possédant ces propriétés, puis à examiner si la somme des chaos est dense dans L^2, problème en général beaucoup plus difficile que le précédent.

Par exemple, considérons sur un même espace probabilisé un mouvement brownien Y_t, un temps d'arrêt T de celui-ci, et un processus de Poisson compensé Z_t d'intensité 1 indépendant de Y_t. Le processus

$$X_t = Y_t I_{\{t \leq T\}} + Z_{t-T} I_{\{t>T\}}$$

est une martingale de crochet oblique t, et Stricker [11] a montré que cette martingale possède la *propriété de représentation prévisible* (PRP) : cela signifie que toute martingale est l'intégrale stochastique d'un processus prévisible relativement à la martingale fondamentale (X_t), ou aux martingales fondamentales (X_t^α) dans le cas multidimensionnel. La PRP est une condition nécessaire (probablement non suffisante, bien que l'on ne dispose actuellement d'aucun contre-exemple) pour que la somme des chaos soit tout l'espace L^2 de la tribu engendrée par X *(propriété de représentation chaotique* ou PRC).

Le premier exemple de martingale qui possède la propriété de représentation chaotique, mais n'est pas un processus à accroissements indépendants, est dû à Emery [2]. Azéma avait découvert, en étudiant l'ensemble des zéros du mouvement brownien, deux martingales remarquables de crochet oblique égal à t, et possédant la PRP. Emery a montré que l'une d'entre elles possède la PRC (pour l'autre, la réponse est inconnue). Le chemin suivi par Emery consiste à remarquer d'abord que cette martingale (X_t) satisfait à une *équation de structure*

$$d[X,X]_t = dt - X_{s-}dX_s ,$$

où $[X,X]$ désigne le *crochet droit* de la martingale. Une relation de ce type constitue une condition nécessaire pour la validité de la PRP. En effet, si (X_t) est une martingale de crochet oblique égal à t, le processus $[X,X]_t - t$ est une martingale, donc l'intégrale stochastique $\int_0^t \Phi_s dX_s$ d'un processus prévisible (Φ_t), et cela fournit l'équation de structure cherchée. On ignore si l'existence d'une équation de structure entraîne que la PRP a lieu. Dans le cas de la martingale d'Azéma, et plus généralement des martingales satisfaisant à une équation du meme type avec un coefficient β autre que -1, Emery sait démontrer que l'équation de structure caractérise uniquement la loi de (X_t) (en autorisant une valeur initiale quelconque X_0), et en déduit que ces martingales possèdent effectivement la PRP. La PRC n'est établie que dans l'intervalle $-2 \leq \beta \leq 1$.

Ces remarquables résultats d'Emery datent de 1988, mais il y a des résultats plus récents encore. D'une manière générale, chaque fois que l'on a un processus de Markov (J_t) à valeurs dans un espace d'états E, de semigroupe (P_t) et de générateur A, on sait construire des martingales X_t de la forme

$$X_t^f = f(J_t) - f(J_0) - \int_0^t Af(J_s)\,ds$$

où f appartient au domaine $\mathcal{D}$ du générateur ; le crochet oblique de deux telles martingales est donné par

$$< X^f, X^g >_t = \int_0^t \Gamma(f,g)\circ J_s\,ds$$

où Γ est *l'opérateur carré du champ* $\Gamma(f,g) = A(fg) - f\,Ag - Af\,g$ (cela suppose que le produit des deux éléments f,g du domaine appartient encore à $\mathcal{D}$, mais ce problème ne se posera pas dans le cas traité plus bas). Les martingales X_t^f sont des modèles de *martingales fonctionnelles additives.* Faisons alors un raisonnement heuristique : nous considérons l'espace $\mathcal{H}$ des martingales fonctionnelles additives de carré intégrable $Z = (Z_t)$ du processus comme un module sur l'algèbre des fonctions boréliennes sur E, le produit $f \cdot Z$ étant l'intégrale stochastique $\int_0^t f(J_s)dZ_s$ (sous l'hypothèse de l'existence d'un opérateur carré du champ, le crochet oblique $< Z, Z >$ est absolument continu, ce qui donne un sens à cette intégrale). Nous munissons $\mathcal{H}$ d'une sorte de structure hilbertienne, le "produit scalaire" étant la densité du crochet oblique, *i.e.* un élément de l'algèbre des "scalaires". Et alors la recherche des martingales X_t^α pouvant conduire à une représentation chaotique est exactement celle d'une *base orthonormale* de $\mathcal{H}$ pour cette structure bizarre.

Il y a au moins un cas où l'on sait mener les calculs jusqu'au bout : celui où (J_t) est une chaine de Markov à espace d'états E fini. Dans ce cas, l'algèbre des fonctions sur E est de dimension finie, et le générateur A est un opérateur partout défini. Désignons alors par $a(i,j)$ la matrice du générateur (négative sur la diagonale, positive en dehors), et par $N(i)$ le nombre des j tels que $a(i,j) > 0$ (qui est aussi le nombre des $j \neq i$ tels que le saut de i en j soit autorisé). Il existe alors une base orthonormale au sens décrit ci-dessus si et seulement si $N(i)$ est indépendant de i, et ce nombre est celui des éléments de la base.

Tout cela est très facile, mais il reste à examiner si la représentation chaotique a effectivement lieu relativement à cette base de martingales. Dans le cas des chaines de Markov finies, la réponse positive est due à Biane [1]. La méthode de Biane consiste à remarquer que la représentation chaotique a lieu pour toutes les variables aléatoires dès qu'elle a lieu pour celles de la forme $f(J_t)$. Or les termes du développement chaotique d'une telle fonction sont donnés par une formule du type de la formule d'Isobe–Sato (ou de Krylov–Veretennikov : voir Léandre–Meyer [7]), et il suffit de démontrer que le reste de cette formule (qui est explicitement calculable) tend vers 0. C'est ce que fait Biane, et sa méthode *s'applique, non seulement aux chaînes de Markov finies, mais à toutes les représentations chaotiques actuellement connues.* Il y a donc un grand progrès dans la compréhension des décompositions chaotiques. On notera cependant que les coefficients du développement en chaos d'une v.a. donnée ***dépendent du point initial du processus.*** Autrement dit, on a affaire à des développements en chaos à coefficients dans "l'algèbre des scalaires".

Il y a un dernier point que nous n'avons pas la place d'étudier ici, et qui est la relation de ce qui précède avec la théorie des diffusions quantiques de M.P. Evans et R.L. Hudson [3]. La recherche d'exemples de cette théorie en probabilités classiques a été en fait la motivation qui a amené à l'étude des chaînes de Markov sous l'angle des représentations chaotiques : la décomposition chaotique ci–dessus montre que la structure de l'espace de Hilbert sous–jacent à une chaine de Markov finie est celle d'un espace de Fock, comme pour les solutions d'é.d.s. classiques. Or on peut justifier cela en représentant les chaînes de Markov finies et les diffusions classiques comme solutions commutatives d'équations différentielles stochastiques non commutatives sur l'espace de Fock. Cela redonne un aspect très moderne aux vieilles remarques de Kolmogorov sur les analogies entre chaînes de Markov et diffusions.

REFERENCES

[1] BIANE (Ph.) Chaotic representations for finite Markov chains. *Stochastics*, à paraître, 1989.

[2] EMERY (M.) On the Azéma martingales. *Sém. Prob. XXIII*, LN in M. n° 1372, p. 66–87. Springer 1989.

[3] EVANS (M.P.) et HUDSON (R.L.) Perturbation of quantum diffusions. *Proc. London Math. Soc.*, à paraître, 1990.

[4] ITO (K.) Multiple Wiener Integral. *J. Math. Soc. Japan*, **3**, 1951, p. 157–169.

[5] ITO (K.) Spectral type of the shift transformation of differential processes with stationary increments. *Trans. Amer. Math. Soc.*, **81**, 1956, p. 253–263.

[6] KAKUTANI (S.) Determination of the spectrum of the flow of brownian motion. *Proc. Natl. Acad. Sci. USA*, **36**, 1950, p. 319–323.

[7] LÉANDRE (R.) et MEYER (P.A.) Sur le développement d'une diffusion en chaos de Wiener. *Sém. Prob. XXIII*, LN in M. n° 1372, p. 161–164, Springer 1989.

[8] NEVEU (J.) *Processus aléatoires gaussiens*, Presses de l'Université de Montréal, 1968.

[9] MEYER (P.A.) Diffusions quantiques, d'après Evans et Hudson. A paraître, *Sém. Prob. XXIV*, LN in M., Springer 1990.

[10] SEGAL (I.) Tensor Algebras over Hilbert spaces I. *Trans. Amer. Math. Soc.*, **81**, 1956, p. 106–134.

[11] STRICKER (C.) A propos d'une conjecture de Meyer. *Sém. Prob. XXII*, LN in M. n° 1321, p. 144–146, Springer 1988.

[12] WIENER (N.) The homogeneous chaos. *Amer. J. Math.*, 60, 1939, p. 897–936.

Paul-André Meyer
Institut de Recherche Math. Avancée
Laboratoire Associé au CNRS n° *1*
7 rue René Descartes, F-67084 Strasbourg-Cedex.

MARKOV PROPERTIES OF SOLUTIONS OF STOCHASTIC PARTIAL DIFFERENTIAL EQUATIONS IN A FINITE VOLUME

by
Edward P.Osipov
Departement of Theoretical Physics
Institute for Mathematics
630090, Novosibirsk 90
USSR

Abstract

In the present paper we consider the system of linear stochastic partial differential equations of the form $\partial A = \xi$, where ∂ is the octavic first order differential operator factorizing the 8-dimensional Laplacian, A and ξ are generalized random octave-valued fields, and ξ is a generalized random octave-valued (non-Gaussian) field with independent increments. We show that the solution of this system of equations is a generalized random field with the Markov property with respect to halfspaces. In the present paper we prove the Markov property with respect to the time variable x_0 in the case of a finite volume approximation with periodic boundary conditions in the spatial directions $x_1, ... x_7$.

1. Introduction.

We consider the system of stochastic partial differential equations of the form

$$\partial A = \xi, \tag{1.1}$$

where $\partial = \sum_{\mu=0}^{7} \overline{e}_\mu \partial_\mu$ is the octavic first order differential operator factorizing the 8-dimensional Laplacian [1], ξ is a generalized random octave-valued field with independent increments (a generalized white noise, in general, non-Gaussian), A is a generalized random octave-valued field. The solutions of this equation and of its quaternionic analogue are used in the construction of some models of Euclidean fields [1-6].

We show that the solution of equations (1.1) is the generalized random field having the Markov property with respect to halfspaces $\{\tau \gtrless \text{const}\}$. In the present paper we consider the Markov property with respect to the temporal variable x_0 of the field A_V which is a finite volume approximation of the field A with periodic boundary conditions in the spatial variables.

The quaternionic random fields possess the same Markov property. These fields are the solution of the system $\partial A = \xi$, where $\partial = \sum \overline{e}_\mu \partial_\mu$ is the quaternionic first order differential operator, factorizing the 4-dimensional Laplacian, and ξ is a generalized random quaternionic field with independent increments, cf. also [3-6].

We prove the Markov property of the field A_V in the temporal variable (for the precise statement, see Theorem 2.4). To prove this assertion it is sufficient basically to use the Surgailis' technique [7,8]. The idea of this technique consists in the well known Fourier method of separating the variables. With the help of this separation the system of stochastic partial differential equations can be reduced to an (infinite) decoupled system of ordinary stochastic differential equations for processes which are the Fourier components of the field A_V in the spatial variables.

In Section 2 we introduce notations, introduce our basic objects, describe their properties and formulate the main result - Theorem 2.4.

In Section 3 we prove the Markov property of the coefficient processes (=the components of Fourier transform of the field A_V in the spatial variables). This implies the Markov property of the field A_V.

2. Generalized white noise and stochastic integrals. Octavic first order differential operators and their Green functions.

Let $S^1(V_i)$ be the 1-dimensional circle of length V_i. Let $V = S^1(V_1) \times ... \times S^1(V_7)$ be the 7-dimensional torus.

Let $\mathcal{S}(\mathbb{R}\times V, \mathbb{R}^k) \equiv \mathcal{S}_k(\mathbb{R}\times V)$, $k \geq 1$, be the (Schwartz) space of smooth, rapidly decreasing k-component vector functions $\mathbb{R} \times V \to \mathbb{R}^k$ with topology given by seminorms

$$\|f\|_{M,N} = \sup_{\substack{0\leq m\leq M \\ 0\leq n\leq N}} \sup_{x_k\in\mathbb{R}\times V} |(1+|x|^m)\partial^n f_k(x)|,$$

and let $\mathcal{S}'_k(\mathbb{R}\times V)$ be the dual space of vector generalized functions. $\mathcal{S}_1(\mathbb{R}\times V) \equiv \mathcal{S}(\mathbb{R}\times V)$. $\mathcal{S}'_k(\mathbb{R}\times V)$ can be identified with the subspace of periodic functions from $\mathcal{S}'_k(\mathbb{R}^8)$. This identification can be easily given with the help of partition of unity. The spaces $\mathcal{S}_k(V)$ of smooth functions on the torus can be defined analogously.

Let $\xi_V = (\xi_{V,i},\ i = 1,...k)$ be a homogeneous k-component generalized white noise on $\mathbb{R} \times V$. We have

$$E(\exp(i < \xi_V, f >)) = \exp\Big(\int_{\mathbb{R}\times V} dx \int_{\mathbb{R}^k\setminus\{0\}} \nu(d\alpha)(e^{i<\alpha,f(x)>} - 1 \\ - i < \alpha, f(x) >) - \frac{1}{2}\int_{\mathbb{R}\times V} dx\sigma(f,f)\Big). \tag{2.1}$$

Here $f = (f_1, ..., f_k) \in \mathcal{S}_k(\mathbb{R}\times V), < \xi_V, f >= \sum_{j=1}^k < \xi_{V,j}, f_j >= \sum_{j=1}^k \int dx \xi_{V,j}(x) f_j(x)$, $< \alpha, f(x) >= \sum_{i=1}^k \alpha_i f_i(x), \alpha = (\alpha_1, ..., \alpha_k) \in \mathbb{R}^k, \nu(d_\alpha)$ is a non-negative measure defined on Borel subsets of $\mathbb{R}^k \setminus \{0\}$ such that

$$\int_{\mathbb{R}^k\setminus\{0\}} \nu(d\alpha)|\alpha|^2 < \infty, \qquad |\alpha|^2 = \sum_{i=1}^k \alpha_i^2, \tag{2.2}$$

$\sigma(f,f) = \sum_{i,j=1}^{k} \sigma_{ij} f_i(x) f_j(x)$, $\sigma = (\sigma_{ij})_{i,j=1,\ldots,k}$ is a symmetric non-negative definite matrix.

We call ν the spectral measure or the Lévy measure and σ the diffusion matrix or the Gaussian matrix of the generalized withe noise ξ.

The right hand side of (2.1) gives a continuous positive defined functional on $\mathcal{S}_k(\mathbb{R} \times V)$, cf. [9, ch.III, §4, n.5, Theorem 8]. By the Minlos theorem [9-11] ξ can be realized as a random $\mathbb{R}^k$-valued generalized function, i.e. there exists a probability measure μ , defined on the σ -algebra of subsets of $\mathcal{S}'_k(\mathbb{R} \times V)$ generated by cylinder sets, such that

$\int\limits_{\mathcal{S}'_k(\mathbb{R}\times V)} d\mu(\xi) \exp(i < \xi_V, f >)$ is equal to the right hand side of (2.1). (2.1) implies that $< \xi_{V,i}, f >$ can be extended by continuity to a larger class of functions than $\mathcal{S}_k(\mathbb{R} \times V)$, namely to all $f \in L^2(\mathbb{R} \times V)$. It is true that:

Proposition 2.1 (cf. Proposition 1.1 in [7,8]). For any $f \in L^2(\mathbb{R} \times V)$ and i=1,...,k, there exists a unique (mod P) real random variable $< \xi_{V,i}, f >$, defined as the L^2-limit of $< \xi_{V,i}, f_j >$, $j = 1, 2, \ldots$, as $j \to \infty$, where f_j, $j = 1, 2, \ldots$, is a sequence of functions from $\mathcal{S}(\mathbb{R} \times V)$ convergent to f in $L^2(\mathbb{R} \times V)$ and

(a) $< \xi_{V,i}, f >$ are L^2-continuous with respect to the convergence of f in $L^2(\mathbb{R} \times V)$;

(b) (2.1) is fulfilled;

(c) for any f_{ij}, $i = 1, \ldots, k$, $j = 1, \ldots, j_k$, the joint probability distribution of $< \xi_{V,i}, f_{ij} >$, $i = 1, \ldots, k, j = 1, \ldots, j_k$, is infinitely divisible;

(d) if $2 \leq p < \infty$, $f \in L^p(\mathbb{R} \times V)$ and $\int_{\mathbb{R}^k \setminus \{0\}} \nu(d\alpha)|\alpha|^p < \infty$ then $E(| < \xi_{V,i}, f > |^p) \leq c(p) \int\limits_{\mathbb{R}\times V} dx |f|^p$.

The proof of this statement is analogous to the proof of Proposition 1.1 from [7,8].

Proposition 2.2 (cf. Proposition 1.2 [7,8]. Let the diffusion matrix $\sigma = (\sigma_{ij})_{i,j=1,\ldots,k}$ be strictly positive definite and let $f_{ij,n} \in L^2(\mathbb{R} \times V)$, $n = 1, 2, \ldots$, be sequences of functions, which are convergent in $L^2(\mathbb{R} \times V)$ to f_{ij} as $n \to \infty$ for every $i = 1, \ldots, k$, $j = 1, \ldots, M$. Assume that for every $i = 1, \ldots, k$ the functions f_{ij}, $j = 1, \ldots, M$, are linearly independent. Then for n sufficiently large the distribution of $\{< \xi_{V,i}, f_{ij,n} >, i = 1, \ldots, k, j = 1, \ldots, M\}$ is absolutely continuous with respect to the kM-dimensional Lebesgue measure, the corresponding densities $p_n(x)$ and $p(x)$, $x \in \mathbb{R}^{kM}$, are smooth and positive everywhere in $\mathbb{R}^{kM}$ and

$$sup_{x \in \mathbb{R}^{kM}} |p_n(x) - p(x)| \to 0 \quad \text{as} \quad n \to \infty \tag{2.3}$$

The proof of this statement is also analogous to the proof of Surgailis [7,8, Proposition 1.2].

Now let us consider the octavic first order differential operators factorizing the

8-dimensional Laplacian.

Let $\mathbb{O}$ be the algebra of (real) octaves (=octonions=the algebra of Cayley numbers) [12, ch.13], let $L^2(\mathbb{R}^8, \mathbb{O})$ be the real Hilbert space of octave-valued square integrable functions, $||f||^2 = \int_{\mathbb{R}^8} dx|f(x)|^2 < \infty$. Let e_μ, $\mu = 0, 1, ..., 7$, be octavic units and $e_0 \equiv 1$.

Let $\partial \equiv \sum^7{}_{\mu=0} L(\overline{e}_\mu)\partial_\mu$, $\overline{\partial} = \sum^7{}_{\mu=0} L(e_\mu)\partial_\mu$, where the bar denotes the octave conjugation, $\overline{e}_0 = e_0$ and $\overline{e}_\mu = -e_\mu$, $\mu = 1, ...7$, $L(a)$ is the operator of left multiplication by the octave a and $\partial_\mu \equiv \partial/\partial x_\mu$. ∂ is an operator in $L^2(\mathbb{R}^8, \mathbb{O})$ and $\partial^{tr} = -\overline{\partial}$, where $(.)^{tr}$ denotes the adjoint operator. We have

$$\partial\overline{\partial} = \overline{\partial}\partial = \Delta e_0, \tag{2.4}$$

where $\Delta = \sum_{\mu=0}^7 \partial_\mu^2$, the 8-dimensional Laplacian. The operator ∂ (and $\overline{\partial}$) is invertible in $L^2(\mathbb{R}^8, \mathbb{O})$ and on $\mathcal{S}_8(\mathbb{R}^8) \cong \mathcal{S}(\mathbb{R}^8, \mathbb{O}) \subset L^2(\mathbb{R}^8, \mathbb{O})$

$$\partial^{-1} f = \overline{\partial}\Delta^{-1} f = \int dy S(x-y) f(y) (\equiv \int dy L(S(x-y)) f(y)).$$

Note that here (and in the sequel) $S(x-y)f(y)$ means the octavic product of two octave-valued functions. Furthermore,

$$\begin{aligned} S(x-y) &= (2\pi)^{-8} \int d^8k (i\overline{k})^{-1} \exp(i < k, x-y >) \\ &= -(2\pi)^{-8} i \int d^8k |k|^{-2} k \exp(i < k, x-y >) \\ &= 3\pi^{-4}(x-y)|x-y|^{-8}, \end{aligned}$$

$k = \sum_{\mu=0}^7 e_\mu k_\mu$, $\overline{k} = \sum_{\mu=0}^7 \overline{e}_\mu k_\mu$, $x - y = \sum_{\mu=0}^7 e_\mu(x_\mu - y_\mu)$, $< k, x-y >= \sum_{\mu=0}^7 k_\mu(x_\mu - y_\mu)$, and the Fourier transform is defined in the sense of generalized functions. We remark that

$$\begin{aligned} G(x-y) = (-\Delta)^{-1}(x-y) &= (2\pi)^{-8} \int d^8k |k|^{-2} \exp(i < k, x-y >) \\ &= 2^{-1}\pi^{-4}|x-y|^{-6}, \end{aligned}$$

Now let us go to a periodic approximation in the spatial directions. Now $\partial_V = e_0\partial_0 + \sum_{\mu=1}^7 \overline{e}_\mu \partial_{\mu,V}$, $\overline{\partial}_V = e_0\partial_0 + \sum_{\mu=1}^7 e_\mu \partial_{\mu,V}$, where $\partial_{\mu,V}$, is the partial derivative in x_μ with periodic boundary conditions. As in (2.4)

$$\partial_V \overline{\partial}_V = \overline{\partial}_V \partial_V = (\partial_0^2 + \sum_{\mu=1}^7 \partial_{\mu,V}^2) e_0. \tag{2.5}$$

In the case of periodic boundary conditions a zero-momentum mode appears for the operators $\sum_{\mu=1}^7 \overline{e}_\mu \partial_{\mu,V}$ and $\sum_{\mu=1}^7 \partial_{\mu,V}^2$. Due to this fact we introduce the kernel $S_V(x-y)$

given by the periodic approximation of the kernel $S(x-y)$ with deleted zero-moment mode, namely, in the sense of generalized functions

$$S_V(x-y) = \frac{1}{2\pi|V|i} \sum_{\mathbf{k}\in \mathbb{Z}_\mathbf{V}\setminus\{\mathbf{0}\}} \int dk_0 \frac{e^{i<k,x-y>}}{k_0+\overline{\mathbf{k}}}$$
$$= \frac{1}{2\pi|V|i} \sum_{\mathbf{k}\in \mathbb{Z}_\mathbf{V}\setminus\{\mathbf{0}\}} \int dk_0 \frac{k e^{i<k,x-y>}}{|k|^2},$$

where $\mathbb{Z}_V = \frac{2\pi}{V_1}\mathbb{Z} \times ... \times \frac{2\pi}{V_7}\mathbb{Z}$, $|V| = V_1...V_7$ and we use the same notations for octaves $k = \sum_{\mu=0}^{7} e_\mu k_\mu$, $\mathbf{k} = \sum_{\mu=1}^{7} e_\mu k_\mu$ and vectors $(k_0, k_1, ..., k_7)$, $(k_1, ..., k_7)$. In addition, in the sense of generalized functions

$$(\partial_V)_x S_V(x-y) = \delta(x_0-y_0)\frac{1}{|V|} \sum_{\mathbf{k}\in \mathbb{Z}_\mathbf{V}\setminus\{\mathbf{0}\}} e^{i<\mathbf{k},\mathbf{x}-\mathbf{y}>} \tag{2.6}$$

By Jordans lemma one can easily to show that in the sense of generalized functions

$$S_V(x-y) = \frac{1}{|V|} \sum_{\mathbf{k}\in \mathbb{Z}_V\setminus\{\mathbf{0}\}} \frac{e^{-|x_0-y_0||\mathbf{k}|}e^{i<\mathbf{k},\mathbf{x}-\mathbf{y}>}}{2|\mathbf{k}|}(|\mathbf{k}|\text{sgn}(x_0-y_0) - i\mathbf{k}) = -\overline{\partial}_V G_V(x-y),$$

where $G_V(x-y)$ is the Greenfunction of the Laplacian (2.5) with deleted zero-momentum mode,

$$G_V(x-y) = \frac{1}{2\pi|V|} \sum_{\mathbf{k}\in \mathbb{Z}_\mathbf{V}\setminus\{\mathbf{0}\}} \int dk_0 \frac{e^{i<k,x-y>}}{|\mathbf{k}|^2}$$
$$= \frac{1}{|V|} \sum_{\mathbf{k}\in \mathbb{Z}_V\setminus\{\mathbf{0}\}} \frac{e^{-|x_0-y_0||\mathbf{k}|}e^{i<\mathbf{k},\mathbf{x}-\mathbf{y}>}}{2|\mathbf{k}|}$$

Let us define the random field

$$A_V(x) = \int_{\mathbb{R}\times V} dy L(S_V(x-y))\xi_V(y) \equiv \int_{\mathbb{R}\times V} dy S_V(x-y)\xi_V(y),$$

that is

$$< A_V, f > = < \xi_V, S_V^{tr} f >, \tag{2.8}$$

where S_V^{tr} is the operator adjoint to the operator with kernel $L(S_V(x-y))$,

$(S_V^{tr} f)(y) = \int dx L(\overline{S}_V(x-y)) f(x) \equiv \int dx \overline{S}_V(x-y) f(x)$ and the integral is defined in the sense of generalized functions. The field $A_V(x)$ is the periodic approximation of the solution of eq. (1.1)

$$A(x) = \int_{\mathbb{R}^8} dy S(x-y)\xi(y)$$

and the solution of the periodic approximation of eq. (1.1) with deleted zero-momentum mode, that is,

$$\partial_V A_V(x) = \xi_V(x)$$

for all test functions $f \in \mathcal{S}_8(\mathbb{R} \times V)$ such that $\int d^7 x e^{-i<\mathbf{k},\mathbf{x}>} f(t,\mathbf{x})|_{\mathbf{k}=\mathbf{0}} = 0$.

(2.7) implies that for $f \in \mathcal{S}_8(8)$

$$\| \int d^7 x \overline{S}_V(x_0 - y_0, \mathbf{x} - \mathbf{y}) f(\mathbf{x}) \|_{L^2(V,\mathbb{O})} \leq \|f\|_{L^2(V,\mathbb{O})} e^{-\kappa|x_0 - y_0|} \tag{2.9}$$

where $\kappa = 2\pi(max_{1 \leq i \leq 7} V_i)^{-1}$. We remark that bound (2.9) is very rough, but it is sufficient for us. This bound implies, in particular, that for $f \in L^2(V, \mathbb{O})$

$$\int d^7 x \overline{S}_V(x_0 - y_0, \mathbf{x} - \mathbf{y}) f(\mathbf{x}) \in L^2(\mathbb{R} \times V, \mathbb{O}).$$

Proposition 2.1 and (2.9) imply that (2.8) is defined correctly. In addition bound (2.9) implies

Proposition 2.3. For every $f \in \mathcal{S}_8(V)$ there exists stochastically continuous processes $A_V(t,f)$, $t \in \mathbb{R}$, such that

$$A_V(h \otimes f) = \int_{-\infty}^{+\infty} dt A_V(t,f) h(t) \tag{2.10}$$

holds a.e. for every $h \in C_0^\infty(\mathbb{R})$. For every $t \in \mathbb{R}$

$$A_V(t,f) = < \xi_V, S_V^{tr}(\delta_t \otimes f) > \quad a.e.,$$

where

$$S_V^{tr}(\delta_t \otimes f)(y) = \int d^7 x \overline{S}_V(t - y_0, \mathbf{x} - \mathbf{y}) f(\mathbf{x}).$$

The proof of thes statement may be given analogously to the proof of Proposition 2.1 [7].

Denote by $\sigma(F_k, K \in \mathcal{K})$ the σ-algebra generated by a set $\{F_k | k \in \mathcal{K}\}$ of functions. For $\tau \in \mathbb{R}$ we denote $\mathbb{R}_{\pm,\tau} \times V = \{(x_0, \mathbf{x}) \in \mathbb{R} \times V | x_0 \gtrless \tau\}$ and let $\sigma_\tau^\pm = \sigma(< A_V, f >, f \in$

$\mathcal{S}(\mathbb{R}\times V)$, $\operatorname{supp} f\subset \mathbb{R}_{\pm,\tau}\times V)$ and $\sigma_\tau^0=\sigma(A_V(\tau,f),\ f\in\mathcal{S}(V))$. Note that by stochastic continuity of $A_V(t,f)$, $\sigma_\tau^0\subset\sigma_\tau^+\cap\sigma_\tau^-$.

Given 3 sub-σ-algebras of $\mathcal{F}$(= the σ – algebra generated by cylinder subsets) $\mathcal{F}^+,\mathcal{F}^-$ and $\mathcal{F}^0$ such that $\mathcal{F}^0\subset\mathcal{F}^+\cap\mathcal{F}^-$, we say that $\mathcal{F}^0$ splits $\mathcal{F}^+$ and $\mathcal{F}^-$ if for any bounded random variables F^+, F^- measurable with respect to $\mathcal{F}^+,\mathcal{F}^-$ correspondingly,

$$E(F^+F^-|\mathcal{F}^0)=E(F^+|\mathcal{F}^0)E(F^-|\mathcal{F}^0)\,,$$

where $E(.|\mathcal{F}^0)$ is the conditional expectation with respect to the σ – algebra $\mathcal{F}^0$ ($cf.$ [13, 14, 7, 8]).

Theorem 2.4. (The Markov property of A_V with respect to halfspaces $\mathbb{R}_{\pm,\tau}\times V,\ \tau\in\mathbb{R}$). For every $\tau\in\mathbb{R}$, σ_τ^0 splits σ_τ^+ and σ_τ^-.

The analogous subspaces $\mathbb{R}_{\pm,\tau}\times V$ and σ-algebras σ_τ^0, $\sigma_\tau^\pm$ can be defined in the 2-dimensional ($\mathbb{R}_{\pm,\tau}\times V=\mathbb{R}_{\pm,\tau}\times S^1(V_1)$) and in the 4-dimensional ($\mathbb{R}_{\pm,\tau}\times V=\mathbb{R}_{\pm,\tau}\times S^1(V_1)\times S^1(V_2)\times S^1(V_3)$) cases for the complex-valued, respectively, quaternionic-valued analogue of the random field A_V. In these cases the analogue of Theorem 2.4 is also valid.

The idea of the proof of Theorem 2.4 is to consider, following Surgailis [7,8], the coefficient processes - the Fourier components of $A_V(t,\mathbf{x})$ in spatial variables

$$X_{\mathbf{k}}(t)=A_V(t,e^{-i<\mathbf{k},.>})=\int_V d^7xA_V(t,\mathbf{x})\mathbf{e}^{-\mathbf{i}<\mathbf{k},\mathbf{x}>},\ \mathbf{k}\in\mathbb{Z}_{\mathbf{V}}.$$

The process $X_{\mathbf{k}}(t)$ is a complex-octave-valued random process. Instead of it one can consider the pair

$$\int_V d^7xA_V(t,\mathbf{x})\cos(<\mathbf{k},\mathbf{x}>),\ \int_V d^7xA_V(t,\mathbf{x})\sin(<\mathbf{k},\mathbf{x}>)$$

of real-octave-valued random processes. (2.5) and (2.6) imply that the coefficient processes $X_{\mathbf{k}}(t)$ satisfy the stochastic ordinary differential equations

$$\frac{dX_{\mathbf{k}}(t)}{dt}-i\mathbf{k}X_{\mathbf{k}}(t)=\xi_{V,\mathbf{k}}(t),\ \mathbf{k}\neq\mathbf{0},$$

$$X_{\mathbf{k}}(t)=0,\ \mathbf{k}=\mathbf{0} \tag{2.11}$$

where (in the sense of generalized functions) $\xi_{V,\mathbf{k}}(t)=\int_V d^7x\xi(t,\mathbf{x})e^{-i<\mathbf{k},\mathbf{x}>}$ and $X_{\mathbf{k}}(t)$ are complex-octave-valued random processes. Further, as Surgailis [7,8] we prove that eqs. (2.11) imply that for any finite number of $\mathbf{k},\mathbf{k_1},...,\mathbf{k_N}\in\mathbb{Z}_{\mathbf{V}}(\mathbf{k_n}\neq\mathbf{k_{n'}},\ \mathbf{n}\neq\mathbf{n'})$ the 16N-dimensional process $X_{\mathbf{k_1}}(t),...,X_{\mathbf{k_N}}(t),\ t\in\mathbb{R}$, is Markovian.

Since $X_{\mathbf{k}}(t)$, $\mathbf{k}\in\mathbb{Z}_{\mathbf{V}}$, generate the corresponding σ-algebras σ^0, $\sigma^\pm$ (Proposition 2.5), so if $N\to\infty$ we obtain Theorem 2.4.

The coefficient processes $X_{\mathbf{k}}(t)$ can be expressed as stochastic integrals with respect to $\xi_{V,\mathbf{k}}$

$$X_{\mathbf{k}}(t) = \int ds \xi_{V,\mathbf{k}}(s) e^{-|t-s||\mathbf{k}|}(1/2\mathrm{sgn}(t-s) - \frac{i}{2}\frac{\mathbf{k}}{|\mathbf{k}|}). \tag{2.12}$$

This follows from (2.7)

$$\int_V d^7 x S_V(t-s, \mathbf{x}-\mathbf{y}) e^{-i<\mathbf{k},\mathbf{x}>}$$

$$= \begin{cases} \exp(-|\mathrm{t}-\mathrm{s}||\mathbf{k}| - \mathrm{i} < \mathbf{k}, \mathbf{y} >)(\mathrm{sgn}(\mathrm{t}-\mathrm{s})/2 - \frac{\mathrm{i}}{2}\frac{\mathbf{k}}{|\mathbf{k}|}) & ,\mathbf{k} \neq 0 \\ 0 & ,\mathbf{k} = 0 \end{cases}$$

We note also that as it can be easily seen eq. (2.11) has a unique solution. Indeed, if $X'_{\mathbf{k}}(t)$ and $X''_{\mathbf{k}}(t)$ are two solutions of (2.11) then for the Fourier transformation $Y^{\sim}(k_0)$, where $Y(t) = X'_{\mathbf{k}}(t) - X''_{\mathbf{k}}(t)$, we have $(k_0 e_0 - \mathbf{k})Y^{\sim}(k_0) = 0$ and since the octave $k_0 e_0 - \mathbf{k} \neq 0$ $(|k_0 e_0 - \mathbf{k}| = (k_0^2 + |\mathbf{k}|^2)^{1/2} \neq 0)$ and $(k_0 e_0 - \mathbf{k})^{-1}$ is a smooth function of k_0, so $Y^{\sim}(k_0) = 0$ for all k_0 and $X'_{\mathbf{k}}(t) = X''_{\mathbf{k}}(t)$.

Proposition 2.5 For each $\tau \in \mathbb{R}$ $\mod P$

$$\sigma(X_{\mathbf{k}}(\tau), \mathbf{k} \in \mathbb{Z}_V) = \sigma_\tau^0, \tag{2.13}$$

$$\sigma(X_{\mathbf{k}}(t), t \gtrless \tau, \mathbf{k} \in \mathbb{Z}_V) = \sigma_\tau^\pm. \tag{2.14}$$

Proof. The proof is analogous to that of Surgailis [7]. We note only that for any $\phi \in \mathcal{S}_8(V)$, $\overline{S}_V(\delta_t \otimes \varphi)$ can be approximated in $L^2(\mathbb{R} \times V, \mathbb{O})$ by $\overline{S}_V(\delta_t \otimes \varphi_n)$, where

$$\phi_n(x) = \Sigma_{\mathbf{k}} \varphi^{\sim}(\mathbf{k}) e^{i<\mathbf{k},\mathbf{x}>} c_n(\mathbf{k}),$$

$\varphi^{\sim}(\mathbf{k})$ is the Fourier transform of φ and $c_n(\mathbf{k})$ are some constants. For appropriate $c_n(\mathbf{k})$, $\varphi_{\mathbf{n}} \to \varphi$ as $n \to \infty$ in the L^2-norm. Then, we use bounds of type (2.9). We note also that the zero-momentum Fourier component of φ, that is, $\varphi^{\sim}(\mathbf{0})$ gives no contribution to $\overline{S}_V(\delta_t \otimes \delta)$.

3. Markov property of coefficient processes.

We denote $\mathbb{Z}_{+,V} = \{\mathbf{k} \in \mathbb{Z}_V |, k_1, ..., k_7 \geq 0\}$. Let $\mathbf{k}_1, ..., \mathbf{k}_N \in \mathbb{Z}_{+,V} \setminus \{\mathbf{0}\}$ be fixed, $\mathbf{k}_n \neq \mathbf{k}_{n'}, n \neq n'$, and let $X_n = X_{\mathbf{k_n}}$, $\xi_n = \xi_{V,\mathbf{k_n}}$, $n = 1, ..., N$, be defined as in the previous section. We are going now to discuss the Markov property of the 16N-dimensional process $(X_1(t), ..., X_N(t))$. It is easy to verify that $X_n(t)$ solve ordinary multicomponent stochastic differential equations

$$\dot{X}_n - i\mathbf{k}_n X_n = \xi_n, \quad t \in \mathbb{R},\ \mathbf{k}_n = \Sigma_{\mu=1}^{7} \mathbf{k}_n^{\mu} e_{\mu}, \quad \xi_n = \Sigma_{\mu=0}^{7} \xi_{V,\mathbf{k}_n}^{\mu} e_{\mu}. \tag{3.1}$$

We remark that the coefficient processes X_n defined by (2.12) are not the solutions in K.Ito's sense of the stochastic differential equations (3.1), that is, $X_n(t)$ are not non-anticipating solutions of (3.1). The Markov poperties of $X_n(t)$ seem to be essentially related to the linearity and locality of eq.(3.1).

Our proof of the Markov property of solutions of (3.1) follows Surgailis [7,8]. To consider the Markov property of $X_n(t)$ we approximate $X_n(t)$ by the process $X_{n,T}(t)$ on a finite interval [-T/2, T/2] with appropriate, that is, with periodic boundary conditions and then approximate it by discrete time processes $X_{T,\delta}(t)$. The discrete time approximations of $X_{n,T}(t)$ with temporal periodic boundary conditions are solutions of the corresponding finite-difference stochastic equations. We prove the Markov property of these approximations and then justify the limiting procedure as $\delta \to 0$ and $T \to \infty$. To prove that the Markov property is preserved in this limiting procedure we prove it first for ξ with a non-degenerate Gaussian part (i.e. when $\sigma \neq 0$ in (2.1)) and then we let $\sigma \to 0$ in the resulting formula for conditional expectations with the help of a result on differentiation of measures in $\mathbb{R}^n$ (see Proposition 3.4 in Ref. [7]).

Consider the periodic approximation of eq. (3.1). Let $X_{n,T} = X_{n,T}(t), |t| \leq T/2$, be the solution of the equation

$$\frac{dX_{n,T}}{dt} - i\mathbf{k}_n X_{n,T} = \xi_{n,T}, \ |t| \leq T/2 \tag{3.2}$$

with periodic boundary conditions

$$X_{n,T}(-T/2) = X_{n,T}(T/2). \tag{3.3}$$

Here $\xi_{n,T} = \xi_{V,T,\mathbf{k}_n}(t) = \int_V d^7x \xi_{V,T}(t,\mathbf{x}) \mathbf{e}^{-\mathbf{i}<\mathbf{k}_n,\mathbf{x}>}$ and $\xi_{V,T}$ is a generalized white noise on the torus $S^1(T) \times V$ with the characteristic function

$$E(\exp(i < \xi_{V,T}, f >)) =$$

$$\exp \int_{S^1(T)\times V} dx \Big(\int_{\{\mathbb{R}^8 \setminus \{0\}\}} \nu(d\alpha))(e^{i<\alpha,f(x)>} - 1 - i < \alpha, f(x) >)$$

$$-1/2 \int_{S^1(T)\times V} dx \sigma(f,f)), \quad f \in \mathcal{S}_8(S^1(T) \times V).$$

We note that $\int dt \xi_{n,T}(t) h(t)$, supp $h \subset (-T/2, T/2)$, extends by continuity to all functions $h \in L^2([-T/2,T/2], \mathbb{O})$ and since the characteristic functionals of $< \xi_{n,T}, h >$ and $< \xi_n, h >$ for smooth h with support in (-T/2, T/2) coincide, so $\xi_{n,T}(t)$ can be stochastically identified with $\xi_n(t) (\equiv \xi_{V,\mathbf{k}_n}(t) = \int_V d^7x \xi(t,\mathbf{x}) e^{-i<\mathbf{k}_n,\mathbf{x}>})$ for $|t| \leq T/2$.

It is easy to see that for $\mathbf{k}_n \neq \mathbf{0}$ the solutions X_n and $X_{n,T}$ and the Green functions S_n and $S_{n,T}$ of eqs. (3.1) and (3.2), respectively, are unique (see the one-component variant of eqs. (3.1) in Vladimirov's book [15, ch.11, §7.4b), p.126-127]). Moreover

$$X_n(t) = \int ds S_n(t-s)\xi_n(s)$$

$$X_{n,T}(t) = \int_{-T/2}^{T/2} ds S_{n,T}(t-s)\xi_{n,T}(s) \tag{3.4}$$

$$S_n(t-s) = \frac{1}{2\pi}\int dk_0 \frac{e^{ik_0(t-s)}}{ik_0 - i\mathbf{k}_n} = (\mathrm{sgn(t-s)}/2 - \frac{\mathbf{ik}}{2|\mathbf{k}|})\mathrm{exp}(-|\mathrm{t-s}||\mathbf{k}_\mathrm{n}|)$$

In addition, we rewrite (3.4) in the form convenient to consider the convergence of discrete periodic approximations. Since $\xi_{n,T}(t)$ and $\xi_n(t)$ for $|t| \leq T/2$ are stochastically identified, so $X_{n,T}(t)$ is stochastically identified with $\int_{-T/2}^{T/2} ds S_{n,T}(t-s)\xi_n(s)$. The (unique) solution (3.4) has the form

$$X_{n,T}(t) = \exp(L(i\mathbf{k})(t+T/2))(1-\exp(TL(i\mathbf{k})))^{-1}\int_{-T/2}^{T/2} ds \quad \exp((T/2-s)L(i\mathbf{k}))\xi_n(s)$$

$$+\int_{-T/2}^{t} ds\exp((t-s)L(i\mathbf{k}))\xi_n(s).$$

Here $L(i\mathbf{k})$ is the operator of left multiplication by the complex octave $i\mathbf{k}$ and we have used the stochastic identification $\xi_{n,T}(t)$ and $\xi_n(t)$ for $|t| \leq T/2$. It follows from the relation

$$X_{n,T}(t) = \exp(L(i\mathbf{k})(t+T/2))X_{n,T}(-T/2)$$
$$+\int_{-T/2}^{t} ds \quad \exp((t-s)L(i\mathbf{k}))\xi_n(s)$$

and periodic boundary conditions.

Since by alternativity of the octave multiplication $L(i\mathbf{k})^2 = L(|\mathbf{k}|^2 e_0) = |\mathbf{k}|^2, |\mathbf{k}| = (\Sigma_{\mu=1}^7 k_\mu^2)^{1/2}$, so

$$\exp(tL(i\mathbf{k})) = \mathrm{ch(t}|\mathbf{k}|) + \frac{\mathrm{L}(\mathbf{ik})}{|\mathbf{k}|}\mathrm{sh(t}|\mathbf{k}|) \tag{3.5}$$

and

$$(1-\exp(TL(i\mathbf{k})))^{-1} = [1-\mathrm{ch}(T|\mathbf{k}|) + \frac{L(i\mathbf{k})}{|\mathbf{k}|}\mathrm{sh}(T|\mathbf{k}|)](2-2\mathrm{ch}(T|\mathbf{k}|))^{-1} \tag{3.6}$$

Here we do not write the unity operator explicitly.

This leads to the following expression for $X_{n,T}$

$$X_{n,T}(t) = (2 - 2\mathrm{ch}(T|\mathbf{k}_n|))^{-1} \int_{-T/2}^{T/2} ds[\mathrm{ch}((T+t-s)|\mathbf{k}_n|)$$
$$- \mathrm{ch}((t-s)|\mathbf{k}_n|) + \frac{i\mathbf{k}_n}{|\mathbf{k}_n|}(\mathrm{sh}((T+t-s)|\mathbf{k}_n|) - \mathrm{sh}((t-s)|\mathbf{k}_n|))]\xi_j(s) \qquad (3.7)$$
$$+ \int_{-T/2}^{t} ds[\mathrm{ch}((t-s)|\mathbf{k}_n|) + \frac{i\mathbf{k}_n}{|\mathbf{k}_n|}\mathrm{sh}((t-s)|\mathbf{k}_n|)]\xi_j(s)$$

It is easy to see that for any fixed $t, s \in \mathbb{R}$ and $T \to \infty$

$$\begin{aligned} &(2 - 2\mathrm{ch}(T|\mathbf{k}|))^{-1}[\mathrm{ch}((T+t-s)|\mathbf{k}|) - \mathrm{ch}((t-s)|\mathbf{k})]\chi([-T/2, T/2])(s) + \\ &\mathrm{ch}((t-s)|\mathbf{k}|)\chi([-T/2, t])(s) \to 1/2\exp(-|\mathrm{t}-\mathrm{s}||\mathbf{k}|)\mathrm{sgn}(\mathrm{t}-\mathrm{s}) \end{aligned} \qquad (3.8)$$

and

$$\begin{aligned} &-(2 - 2\mathrm{ch}(\mathrm{T}|\mathbf{k}|))^{-1}[\mathrm{sh}((\mathrm{T}+\mathrm{t}-\mathrm{s})|\mathbf{k}|)]\chi[-\mathrm{T}/2, \mathrm{T}/2])(\mathrm{s}) \\ &- \mathrm{sh}((t-s)|\mathbf{k})\chi([-T/2, t])(s) \to 1/2\exp(-|\mathrm{t}-\mathrm{s}||\mathbf{k}|) \end{aligned} \qquad (3.9)$$

and the convergence is in $L^2(\mathbb{R})$ in s for any fixed $t \in \mathbb{R}$. Here $\chi([a,b])$ is the indicator of $[a,b], \chi([a,b])(s) = 1$ for $s \in [a,b], = 0$ for $s \notin [a,b]$.

Furthermore, let $t_1 < t_2 < ... < t_M$ be given points in $\mathbb{R}$ and $\mathbf{k}_1, ..., \mathbf{k}_N \in \mathbb{Z}_{+,V} \setminus \{0\}$, $\mathbf{k}_n \neq \mathbf{k}_{n'}$, for $n \neq n'$. Let $p_{\infty,T}(x_\infty, x_1, ..., x_M), p(x_1, ..., x_M)$, $p_\infty(x_\infty), x_i \in \mathbb{R}^{16N} = (\mathbb{O}_{\mathbb{C}}^N), i = \infty, 1, ..., M$, denote the distribution densities (if they exist) with respect to the Lebesgue measure of random vectors $(X_T(-T/2), X_T(t_1), ..., X_T(t_M)), (X(t_1), ..., X(t_m)), X(0)$, correspondingly, where $X(t) = (X_{\mathbf{k}_1}(t), ..., X_{\mathbf{k}_N}(t))$ and $X_T(t) = (X_{\mathbf{k}_1,T}(t), ..., X_{\mathbf{k}_N,T}(t))$. The statement analogous to Proposition 4.1 [8] is then valid:

Proposition 3.1 Let σ (the diffusion matrix of the white noise ξ_V) be non-degenerate. Then, for any $-T/2 < t_1 < ... < t_M < T/2$ the densities $p_{\infty,T}$, p and p_∞ exist and are smooth and positive functions of their variables. Moreover

$$\lim\, \sup_{T\to\infty_{x_\infty, x_1, ..., x_M \in \mathbb{R}^{16N}}} |p_{\infty,T}(x_\infty, x_1, ..., x_M) - p_\infty(x_\infty)p(x_1, ..., x_M)| = 0 \qquad (3.10)$$

Proof. We write A_V as the sum of two independent random fields corresponding to the Gaussian and Poissonian white noises, $A_V = A_V^G + A_V^P$. Let us consider the random variable $\Sigma_{\mu,t,\phi_\mathbf{k}} a_\mu(t, \phi_\mathbf{k}) A_{\mu,V}^G(t, \phi_\mathbf{k})$, where $A_{\mu,V}^G(t, \phi_\mathbf{k}) = \int_V dx A_{\mu,V}^G(t, \mathbf{x})\phi_\mathbf{k}(\mathbf{x}), \phi_\mathbf{k}(\mathbf{x}) = \cos(<\mathbf{k}, \mathbf{x}>)$ or $\sin(<\mathbf{k}, \mathbf{x}>)$; $\mu = 0, 1, ..., 7$, $t \in \{t_1, ..., t_M\}, \mathbf{k} \in \{\mathbf{k}_1, .., \mathbf{k}_N\}$. As in the proof of Proposition 2.2 and Proposition 1.2 [8] the smoothness and positivity is implied by the strict positivity of the covariance $b(a,a)$ of this random variable. We have

$$b(a,a) = \int_{\mathbb{R}\times V} dy \sigma(\Sigma_{t,\phi_\mathbf{k}} \overline{S}_V * a(t,.), \Sigma_{t,\phi_\mathbf{k}} \overline{S}_V * a(t,.)),$$

where $a(t,x) = \Sigma_{\mu=0}^{7} a_\mu(t,\phi_{\mathbf{k}})(\delta_t(x_0) \otimes \phi_{\mathbf{k}}(\mathbf{x}))\mathbf{e}_\mu$, and

$$b(a,a) \geq \varepsilon \int_{\mathbb{R}\times V} dy < \Sigma_{t,\phi_{\mathbf{k}}} \overline{S}_V * a(t,.), \Sigma_{t,\phi_{\mathbf{k}}} \overline{S}_V * a(t,.) >$$
$$= \frac{\varepsilon V}{8\pi} \Sigma_{\phi_{\mathbf{k}}} \int dk_0 |k|^{-2} < \Sigma_t \exp(ik_0 t) a(t), \Sigma_t \exp(ik_0 t) a(t) > .$$

Here $a(t) = \Sigma_{\mu=0}^{7} a_\mu(t,\phi_{\mathbf{k}}) e_\mu$ and we have used the relations $< \overline{k}a, a' > = < a, ka' >$ and $\overline{k}(ka) = |k|^2 a$ for $k, a, a' \in \mathbb{O}$. Since $\exp(\mathrm{i}k_0 t)$, $\mathrm{t} \in \{\mathrm{t}_1, ..., \mathrm{t}_M\}$, are linearly independent functions, so for nonzero $a_\mu(t,\phi_{\mathbf{k}})$ the covariance is strictly positive. The case of periodic boundary conditions in the time direction can be considered in the same way.

To deduce (3.10) from (2.3) of Proposition 2.2, we suppose that $\xi'(t)$, $t \in \mathbb{R}$, is an independent copy of the vector generalized white noise $\xi(t) = (\xi_{\mathbf{k}_1}(t), ..., \xi_{\mathbf{k}_N}(t))$ on the same probability space. Write

$$X(t) = \int ds S(t,s)\xi(s),$$

that is

$$X_{k_n}(t) = \int ds S_{k_n}(t,s)\xi_{\mathbf{k}_n}(s), \; n = 1, ..., N,$$

and, analogously,

$$X_T(t) = \int_{-T/2}^{T/2} ds S_T(t,s)\xi(s),$$

where $S(t,s)$ and $S_T(t,s)$ are octave-valued functions on $\mathbb{R}\times\mathbb{R}$ or $[-T/2,T/2]\times[-T/2,T/2]$, which are expressed in terms of Green functions of eqs. (3.1), (3.2)-(3.3).

By translation invariance and independence of the generalized white noise ξ on non-intersecting intervals we have

$$\begin{aligned}
X_T(t) &= \int_{-T/2}^{T/2} ds S_T(t,s)\xi(s) \\
&= \int_{-T/4}^{T/4} ds S_T(t,s)\xi(s) + \int_{T/4}^{T/2} ds S_T(t,s)\xi(s) + \int_{-T/2}^{-T/4} ds S_T(t,s)\xi(s) \\
&\cong \int_{-T/4}^{T/4} ds S_T(t,s)\xi(s) + \int_{T/4}^{T/2} ds S_T(t,s)\xi'(s) + \int_{-T/2}^{-T/4} ds S_T(t,s)\xi'(s) \\
&= \int_{-T/4}^{T/4} ds S_T(t,s)\xi(s) + \int_{-T/4}^{0} ds S_T(t,s+T/2)\xi'(s) \\
&+ \int_{0}^{T/4} ds S_T(t,s-T/2)\xi'(s).
\end{aligned}$$

Here $\cong$ means the equality of distribution laws.

Then, $p_\infty(x_\infty)p(x_1,...,x_M)$ is the probability density of the (joint) distribution of

$$X(0)(\cong X'(0) = \int dsS(0,s)\xi'(s)),\ X(t_1),...,\ X(t_M).$$

Therefore (3.10) follows from (2.3) of Proposition 2.2 and from the convergence

$$\begin{aligned}
&S_T(t_i,.)\chi([-T/4,T/4])(.) \to S(t_i,.),\\
&S_T(-T/2,.)\chi([-T/4,T/4])(.) \to 0,\\
&S_T(t_i,.+T/2)\chi([-T/4,0])(.) + S_T(t_i,.-T/2)\chi([0,T/4])(.) \to 0\\
&S_T(-T/2,.+T/2)\chi([-T/4,0])(.) + S_T(-T/2,-T/2)\chi([0,T/4])(.) \to S(0,.)
\end{aligned} \tag{3.11}$$

as $T \to \infty$, where the convergence is pointwise and in $L^2(\mathbb{R},\mathbb{R}^{16N})$. Equations (3.11) are easy to verify given the explicit expressions (3.4) and (3.7) for $S_n(t,s)$ and $S_{n,T}(t,s)$.

Now let us consider a discrete time approximation with periodic boundary conditions. Let δ be the spacing parameter of the finite periodic lattice, $\delta = T/2J$ and J is integer. Let us consider the following finite difference equation for random variables $X_{T,\delta}(t_\delta)$ with values in complex octaves

$$\partial_{0,\delta}X_{T,\delta}(t_\delta) - i\mathbf{k}_n X_{T,\delta}(t_\delta) = \xi_{T,\delta}(t_\delta),\ t_\delta \in [-T/2,T/2] \cap \delta\mathbb{Z} = \{-J\delta,...,J\delta\} \tag{3.12}$$

with periodic boundary conditions

$$X_{T,\delta}(-T/2) = X_{T,\delta}(T/2),\ i.e. X_{T,\delta}(-J\delta) = X_{T,\delta}(J\delta). \tag{3.13}$$

Here $\xi_{T,\delta}(t_\delta) \equiv \xi_\delta(t_\delta) = \delta^{-1}\xi_{V,n}((t_\delta, t_\delta+\delta])$ for $t_\delta = \{-J\delta,...,(J-1)\delta\}$ and $\xi_{T,\delta}(J\delta) = \xi_{T,\delta}(-J\delta) = \xi_\delta(-J\delta), \partial_{0,\delta}f(t) \equiv \delta^{-1}(f(t+\delta)-f(t))$.

The solution of (3.12)-(3.13) exists, is unique and is given by

$$\begin{aligned}
&X_{T,\delta}(i\delta) = \delta\Sigma_{j=-J+1}^{i}(1+\delta L)^{i-j}\xi_\delta((j-1)\delta)\\
&+(1+\delta L)^{i+J}(1-(1+\delta L)^{2J})^{-1}\delta\Sigma_{j=-J+1}^{J}(1+\delta L)^{J-j}\xi_\delta((j-1)\delta).
\end{aligned} \tag{3.14}$$

Here $L = L(\sqrt{-1}\mathbf{k}_n)$ is the operator of left multiplication by the complex octave $\sqrt{-1}\mathbf{k}_n$ and $\Sigma_{j=-J+1}^{-J}$ means 0.

Introducing for given λ the discrete ch_δ and sh_δ

$$\begin{aligned}
\mathrm{ch}_\delta(j\delta\lambda) &\equiv ((1+\delta\lambda)^j + (1-\delta\lambda)^j)/2\\
\mathrm{sh}_\delta(j\delta\lambda) &\equiv ((1+\delta\lambda)^j - (1-\delta\lambda)^j)/2
\end{aligned}$$

and using the discrete analogous of (3.5), (3.6)

$$(1+\delta L)^j = \mathrm{ch}_\delta(j\delta\lambda) + L'\mathrm{sh}_\delta(j\delta\lambda)$$
$$(1-(1+\delta L)^{2J})^{-1}$$
$$= (1-2\mathrm{ch}_\delta(2J\delta\lambda) + (1-\delta^2\lambda^2)^{2J})^{-1}(1-\mathrm{ch}_\delta(2J\delta\lambda) + L'\mathrm{sh}_\delta(2J\delta\lambda))$$

where $L = L(\sqrt{-1}\mathbf{k})$, $\quad \lambda = |\mathbf{k}|$, $\quad L' = L/|\mathbf{k}| = L/\lambda$, $\quad L^2 = \lambda^2$, we have

$$\begin{aligned} X_{T,\delta}(i\delta) &= \delta\Sigma_{j=-J+1}^{i}(\mathrm{ch}_\delta((i-j)\delta\lambda) + L'\mathrm{sh}_\delta((i-j)\delta\lambda)\xi_\delta((j-1)\delta)) \\ &+ \delta(\mathrm{ch}_\delta((i+J)\delta\lambda) + L'\mathrm{sh}_\delta((i+J)\delta\lambda)) \\ &(1-\mathrm{ch}_\delta(2J\delta\lambda) + L'\mathrm{sh}_\delta(2J\delta\lambda))(1-2\mathrm{ch}_\delta(2J\delta\lambda) + (1-\delta^2\lambda^2)^{2J})^{-1} \\ &\Sigma_{j=-J+1}^{J}(\mathrm{ch}_\delta((J-j)\delta\lambda) + L'\mathrm{sh}_\delta((J-j)\delta\lambda))\xi_\delta((j-1)\delta). \end{aligned} \tag{3.15}$$

It is easy to see that

$$\sup_{|t|\le T/2, |j\delta - t|\le\delta}(|\mathrm{ch}_\delta(j\delta\lambda) - \mathrm{ch}(t\lambda)| + |\mathrm{sh}_\delta(j\delta\lambda) - \mathrm{sh}(t\lambda)|) \to 0 \text{ as } \delta \to 0 \tag{3.16}$$

We set $X_{T,\delta}(j\delta) = (X_{\mathbf{k}_1,T,\delta}(j\delta), ..., X_{\mathbf{k}_N,T,\delta}(j\delta))$,
$\mathbf{k}_n \ne \mathbf{k}_{n'}$, for $n \ne n'$, and let $p_{\infty,T}(x_\infty, ..., x_M)$ be defined as above. (3.7), (3.15), (3.16) and Propositions 2.2 and 3.1 imply.

Proposition 3.2. Assume that σ is non-degenerate. Let $j_m = j_m^\delta, m = 1, ..., M$, $|j_m| \le J$, $j_\infty = -J$ be integers and $|j_m\delta - t_m| \le \delta, m = 1, ..., M$, where $-T/2 < t_1 < ... < t_M < T/2$. Then, for δ sufficiently small the distribution of the random vector $(X_{T,\delta}(j_\infty\delta), X_{T,\delta}(j_1\delta), ..., X_{T,\delta}(j_M\delta))$ has smooth positive density $p_{\infty,T}^\delta = p_{\infty,T}^\delta(x_\infty, x_1, ..., x_M), x_\infty, x_1, ..., x_M \in \mathbb{R}^{16N}$ and

$$\lim \quad \sup_{\delta\to 0_{x_\infty, x_1, ..., x_M \in \mathbb{R}^{16N}}} |p_{\infty,T}(x_\infty, x_1, ..., x_M) - p_{\infty,T}^\delta(x_\infty, x_1, ..., x_M)| = 0.$$

Let $-J < j_- < j_+ < J$ be integers. Set $I = \{j_-, j_- + 1, ..., j_+\}$.
$I^c = \{-J, -J+1, ..., J-1\}\backslash I, \quad \partial I = \{j_- - 1, j_+ + 1\}$.

Proposition 3.3. Assume that σ is non-degenerate, and let F be a bounded measurable function $F : \mathbb{R}^{16N(j_+ - j_- + 1)} \to \mathbb{R}$. Then

$$\begin{aligned} &E(F(X_{T,\delta}(j_-\delta), ..., X_{T,\delta}(j_+\delta))|\sigma(X_{T,\delta}(j\delta), j \in I^c)) \\ &= E(F(X_{T,\delta}(j_-\delta), ..., X_{T,\delta}(j_+\delta))|\sigma(X_{T,\delta}(j\delta), j \in \partial I)). \end{aligned} \tag{3.17}$$

In particular

$$\begin{aligned} &E(F(X_{T,\delta}(j_-\delta), ..., X_{T,\delta}(j_+\delta))|\sigma(X_{T,\delta}(j\delta),\ j = j_+ + 1, ..., J-1)) \\ &= E(F(X_{T,\delta}(j_-\delta), ..., X_{T,\delta}(j_+\delta))|\sigma(X_{T,\delta}((j_+ + 1)\delta), X_{T,\delta}(J\delta))). \end{aligned}$$

In addition, the symmetric Markov property is valid

$$\begin{aligned}&E(F^-(X_{T,\delta}((-J+1)\delta),...,X_{T,\delta}((j_0-1)\delta))\\&F^+(X_{T,\delta}((j_0+1)\delta),...,X_{T,\delta}((J-1)\delta))|\sigma(X_{T,\delta}(j_0\delta),X_{T,\delta}(J\delta)))\\&=E(F^-(X_{T,\delta}((-J+1)\delta),...,X_{T,\delta}((j_0-1)\delta))|\sigma(X_{T,\delta}(j_0\delta),X_{T,\delta}(J\delta)))\\&E(F^+(X_{T,\delta}((j_0+1)\delta),...,X_{T,\delta}((J-1)\delta))|\sigma(X_{T,\delta}(j_0\delta),X_{T,\delta}(J\delta)))\end{aligned} \tag{3.18}$$

Proof. Denote by $p_T^X(x)$ and $p_T^\xi(x)$, $x \in (\mathbb{R}^{16N})^{2J}$, the densities (if they exist) of the probability distributions of the random vectors $(X_{T,\delta}(j\delta), j = -J, ..., J-1)$ and $(\tilde{\xi}_\delta(j\delta), j = -J, ..., J-1)$, where $\tilde{\xi}_\delta(j\delta) = (\delta^{-1}\int dsd^7x\xi(s,x)\chi((t_\delta, t_\delta+\delta])(s)\exp(-i<\mathbf{k}_n,\mathbf{x}>),\quad n = 1,...,N)$. We write $[\partial_{0,\delta}-L]$ for the linear transformation (3.12)-(3.13)

$$[\partial_{0,\delta}-L]x = w, \tag{3.19}$$

where $x = (x_{-J},...,x_{J-1})$ and $x_J = x_{-J}$, $w = (w_{-J},...,w_{J-1})$, $x_j = (x_j^n, n = 1,...,N)$, $w_j = (w_j^n, n = 1,...,N)$, $x_j^n, w_j^n \in \mathbb{O}_\mathbb{C} \cong \mathbb{R}^{16}$, that is

$$\delta^{-1}(x_{j+1}^n - x_j^n) - L_n x_j^n = w_j^n, \qquad j = -J,...,J-1,$$

where $L_n = L(i\mathbf{k}_n)$, $x_J^n = x_{-J}^n$. Since the functions $\chi((j\delta,(j+1)\delta])(.)\cos(<\mathbf{k}_n,.>), \chi((j\delta,(j+1)\delta])(.)\sin(<\mathbf{k}_n,.>)$, $j = -J,...,J-1$, $n = 1,...,N$, $\mathbf{k}_n \neq \mathbf{k}_{n'}$ for $n \neq n'$ and $\mathbf{k}_n \in \mathbb{Z}_{+,V}$, are linearly independent, so Proposition 2.2 in the case of non-degenerate σ implies that the probability density $p_T^\xi(x)$, $x \in (\mathbb{R}^{16N})^{2J}$, exists, is positive, and as the transform (3.19) is one-to-one and linear, the density $p_T^X(x)$ exists too and is positive everywhere in $(\mathbb{R}^{16N})^{2J}$.

Write $(x_{-J},...,x_{J-1}) = x = (x^I, x^{I^c})$, $x^I = (x_j, j \in I)$, $x^{I^c} = (x_j, j \in I^c)$ and consider the distribution density of conditional expectations. We have

$$\begin{aligned}&P((X_{T,\delta}(j\delta),\ j\in I)\in A|X_{T,\delta}(j\delta),\ j\in I^c)\\&=\left.\frac{\int_A dx^I P_T^X(\{x^I,x^{I^c}\})}{\int_{\mathbb{R}^{16N(j_+-j_-+1)}} dx^I p_T^X(\{x^I,x^{I^c}\})}\right|_{x^{I^c}=(X_{T,\delta}(j\delta),j\in I^c)}\end{aligned} \tag{3.20}$$

where $A \in \mathcal{B}(\mathbb{R}^{16N(j_+-j_-+1)})$.

Since

$$p_T^X(x) = p_T^\xi([\partial_{0,\delta}-L]x)|\det[\delta_{0,\delta}-\mathrm{L}]|$$

(which follows from the equality of characteristic functionals of processes $X_{T,\delta}$ and $[\delta_{0,\delta} - L]^{-1}\tilde{\xi}_\delta$), so by linearity of the transform (3.19) the right hand side of (3.20) is equal to

$$\frac{\int_A dx^I p_T^\xi([\partial_{0,\delta}-L]\{x^I,x^{I^c}\})}{\int_{\mathbb{R}^{16N(j_+-j_-+1)}} dx^I p_T^\xi([\partial_{0,\delta}-L]\{x^I,x^{I^c}\})} \tag{3.21}$$

Now since $\widetilde{\xi}_\delta(j\delta)$, $j = -J, ..., J-1$, are independent, it is easy to see that the ratio (3.21) depends only on $(x_j, j \in \partial I)$, that is, on $x_{j_- -1}$ and $x_{j_+ +1}$. Therefore the conditional expectation

$$E(F(X_{T,\delta}(j\delta), j \in I | \sigma(X_{T,\delta}(j\delta), \quad j \in I^c))$$
$$= \int_{\mathbb{R}^{16N(j_+ - j_- +1)}} dx^I F(x^I) p_T^X(x^I | x^{I^c}) \bigg|_{x^{I^c} = (X_{T,\delta}(j\delta), j \in i^c)}$$

is $\sigma(X_{T,\delta}(j\delta), j \in \partial I)$ - measurable, which completes the proof of (3.17).

(3.18) is implied by the symmetric Markov property

$$P((X_{T,\delta}(j\delta), \quad j = \{-J+1, ..., j_0 - 1\}) \in A_-, \quad (X_{T,\delta}(j\delta), j = \{j_0+1,$$
$$..., J-1\}) \in A_+ | X_{T,\delta}(j_0\delta), X_{T,\delta}(J\delta))$$
$$= \quad P((X_{T,\delta}(j\delta), j = \{-J+1, ..., j_0-1\}) \in A_- | X_{T,\delta}(j_0\delta), X_{T,\delta}(J\delta))$$
$$P((X_{T,\delta}(j\delta), \quad j = \{j_0+1, ..., J-1\}) \in A_+ | X_{T,\delta}(j_0\delta), X_{T,\delta}(J\delta))$$

and is considered analogously to the previous.

Proposition 3.4 (Markov property for coefficient process $X(t)$, $t \in \mathbb{R}$). For any $t \in \mathbb{R}$ the σ-algebra $\sigma(X(t))$ splits the σ-algebras $\sigma(X(s), s \geq t)$ and $\sigma(X(s), s \leq t)$.

Proof. We need only to prove that for any $M^+, M^- = 1, 2, ...$ and $t_{M^+} > ... > t_{1^+} > t > t_{1^-} > ... > t_{M^-}$ and any bounded measurable function $F : (\mathbb{R}^{16N})^{M^+} \to \mathbb{R}$,

$$E[F(X(t_{M^+}), ..., X(t_{1^+})) | \sigma(X(t), ..., X(t_{M^-}))]$$
$$= E[F(X(t_{M^+}), ..., X(t_{1^+})) | \sigma(X(t))] \tag{3.22}$$

Assume for a while that σ is non-degenerate. From Propositions 3.2, 3.3 it follows that if T is sufficiently large, the conditional density

$$p_T^X(x_{M^+}, ..., x_{1^+} | x, x_{1^-}, ..., x_{M^-}, x_T)$$
$$= \frac{p_T^X(x_{M^+}, ..., x_{1^+}, x, x_{1^-}, ..., x_{M^-}, x_T)}{p_T^X(x, x_{1^-}, ..., x_{M^-}, x_T)}$$

depends only on x, x_T, that is

$p_T^X(x_{M^+}, ..., x_{1^+} | x, x_{1^-}, ..., x_{M^-}, x_T) = \frac{p_T^X(x_{M^+}, ..., x_{1^+} x, x_T)}{p_T^X(x, x_T)}$ where
$p_T^X(x_{M^+}, ..., x_{1^+}, x, x_{1^-}, ..., x_{M^-}, x_T), p_T^X(x_{M^+}, ..., x_{1^+}, x, x_T), p_T^X(x, x_{1^-}, ..., x_{M^-}, x_T),$

$p_T^X(x, x_T)$, $x_{1\pm}, ..., x_{M\pm}, x, x_T \in \mathbb{R}^{16N}$ are probability densities of distributions of random vectors

$(X_T(t_{M^+}), ..., X_T(t_{1^+}), X_T(t), X_T(t_{1^-}), ..., X_T(t_{M^-}), X_T(T/2)),$

$X_T(t_{M+}), ..., X_T(t_{1+}), X_T(t), X_T(T/2))$, $(X_T(t), X_T(t_{1-}), ..., X_T(t_{M-})$, $X_T(T/2))$ and $(X_T(t), X_T(T/2))$, respectively. Now let $T \to \infty$ then using Proposition 3.1 we obtain (3.22).

The general case of possibly degenerate diffusion matrix σ can be obtained from (3.22) and the result about differentiation of probability measures in $\mathbb{R}^n$, $n \geq 1$, [16, Theorem 1] and [7, Proposition 3.4]. Proposition 3.4 is proved.

Since $\lim_{n\to\infty} E(F|\mathcal{F}_n) = E(F|\lim_{n\to\infty}\mathcal{F}_n)$ for any increasing sequence of σ-algebras $\mathcal{F}_n \to \lim\mathcal{F}_n$ and any bounded random variable F, this fact together with Proposition 2.2 completes the proof of Theorem 2.4.

Acknowledgements.

I would like to thank D.Surgailis for sending me his work[8].

References

1. Osipov, E.P.: Euclidean Markov fields from stochastic partial differential equations in eight-dimensional space. Pre-print TPh-No15(152) of Novosibirsk Institute for Mathematics, 1987

2. Osipov, E.P.: Octavic Markov cosurfaces and relativistic quantum fields in eight-dimensional space. Preprint TPh-No (19)(161) of Novosibirsk Institute for Mathematics, 1988

3. Albeverio, S., Høegh-Krohn, R.: Euclidean Markov fields and relativistic quantum fields from stochastic partial differential equations in four dimensions. Phys. Lett. B177, 175-179 (1986)

4. Albeverio; S., Høegh-Krohn, R.: Quaternionic non-abelian relativistic quantum fields in four space-time dimensions. Phys. Lett. B189, 329-336 (1987)

5. Albeverio, S., Høegh-Krohn; R.: Construction of interacting local relativistic quantum fields in four spacetime dimensions. Phys. Lett. B200, 108-114 (1988); Errata Phys. Lett. B202, 621 (1988)

6. Albeverio S., Høegh-Krohn R., Iwata K.: Covariant Markovian random fields in four space-time dimensions with nonlinear electromagnetic interaction. In "Applications of Self-adjoint Extensions in Quantum Physics", pp. 69-89. Proc. Dubna '87; Eds. P.Exner, P.Seba, Lect. Notes Phys. 324, Springer, Berlin, 1989.

7. Surgailis, D.: On the Markov property of a class of linear infinitely divisible fields. Z. Wahrscheinlichkeitstheorie verw. Gebiete 49, 293-311 (1979)

8. Surgailis, D.: On covariant stochastic differential equations and Markov property of their solutions. Preprint n.144, 23 Marzo 1979, Istituto di Fisica G. Marconi, Universita di Roma, I.N.F.N. - Sezione di Roma

9. Gel'fand, I.M., Vilenkin, N.Ya.: Generalized functions. Vol. 4. Applications of harmonic analysis. New York, London: Academic Press 1968

10. Hida, T.: Stationary stochastic processes. Princeton: Princeton University Press 1970

11. Bourbaki, N.: Integration (Livre VI, chap. 1-9). Paris: Hermann 1952-1969

12. Postnikov, M.M.: Lie groups and algebras (In Russian). Moscow: Nauka 1982

13. Mc Kean, H.P. Jr.: Brownian motion with a several-dimensional time. Teor. Verojat. i Primen. 8, 357-378 (1963)

14. Pitt, L.: A Markov property for Gaussian processes with a multidimensional time parameter. Arch. Rational. Mech. Anal. 43, 367-391 (1971)

15. Vladimirov, V.S.: Generalized functions in mathematical physics (In Russian). Moscow: Nauka 1979

16. Bruckner, A.M., Weiss, M.L.: On approximate identities in abstract measure spaces. Monatsh. Math. 74, 289-301 (1970)

ON STOCHASTIC EVOLUTION EQUATIONS WITH NON-HOMOGENEOUS BOUNDARY CONDITIONS

by

Yu. A. Rozanov
Steklov Mathematical Institute
USSR Academy of Sciences
42 Vavilov Street
Moscow, 117966, GSP-1, USSR

and

Center for Stochastic Processes
Department of Statistics
University of North Carolina
Chapel Hill, NC 27599-3260, USA

We would like to show some applications of a general functional approach to boundary value problems for stochastic partial differential equations in a case of the most well known examples which are associated with *elliptic* operators $\mathcal{A}$ of the second order in a region $G \subseteq \mathbb{R}^d$ and represented by the *Heat Equation*

$$\frac{\partial u}{\partial t} = \mathcal{A}u + f \tag{1}$$

and the *Wave Equation*

$$\frac{\partial^2 u}{\partial t^2} = \mathcal{A}u + f \tag{2}$$

in a cylinder $G\times(t_0,T)$ with initial data at an initial time moment $t=t_0$ in the base-region G and with boundary conditions on its side-boundary.

Looking for a function $u \in W$ from some functional class W one can set the initial/boundary conditions with corresponding data which are provided by a given *sample function* $u^+ \in W$, say. In this way, for example, one can set the initial conditions at $t=t_0$ as

$$u=u^+ \tag{3}$$

for (1) and

$$u=u^+, \quad \frac{\partial u}{\partial t} = \frac{\partial u^+}{\partial t} \tag{4}$$

for (2) in the base-region G, and the boundary conditions as

$$u=u^{+} \tag{5}$$

with respect to the very function $u \in W$ or as

$$\frac{\partial u}{\partial n} = \frac{\partial u^{+}}{\partial n} \tag{6}$$

with respect to its normal derivative on the side-boundary of the cylinder $G\times(t_0,T)$. The initial conditions plus (5)/(6) lead correspondingly to the *first/second boundary* problems which came to be classical in Partial Differential Equations as far as a right side of (1), (2) belongs to $\mathcal{L}_2$ and a proper *Sobolev space* W is concerned. But it is not a case when one considers the stochastic equations (1),(2) with a *generalized random function* $f = \eta$ which represents stochastic (chaotic) disturbances of *white noise* type, say. Dealing with the corresponding *generalized random field* $u=\xi$ in (1), (2) we meet its very irregular realizations which are *non-differential anywhere*, so a new approach to the conditions (4)/(6) has to be developed. And it can be done in the following way.*

One can associate with the corresponding operator

$$\mathcal{L} = \frac{\partial}{\partial t} - \mathcal{A}, \quad \frac{\partial^2}{\partial t^2} - \mathcal{A}$$

in the $\mathcal{L}_2$-space a proper collection of *generalized test functions* (Schwartz distributions) $x = (\varphi,x)$ which are well defined by a property to be continuous over $\varphi \in C_0^\infty$ in some *neighborhood* of the cylinder $G\times(t_0,T) \subseteq \mathbb{R}^{d+1}$ with respect to $\|\mathcal{L}\varphi\|_{\mathcal{L}_2}$. This collection form a Hilbert space with a norm

*For reference see Yu. A. Rozanov, Markov fields and stochastic partial differential equations. Theory Prob. and Appl., 32, 1, 1987, 3-34

$$\|x\| = \sup_{\|\mathcal{L}\varphi\|_{\mathcal{L}_2} \leq 1} |(\varphi, x)|,$$

and it appears to be a closure of all $x \in C_0^\infty$, so one can take the corresponding closure

$$X = [\mathcal{D}]$$

of the Schwartz space

$$\mathcal{D} = C_0^\infty[G \times (t_0, T)]$$

in the cylinder considered and test any generalized function $u = (x,u)$, $x \in \mathcal{D}$, continuous with respect to the norm $\|x\|_X$ in X by all $x \in X$ with a result

$$(x,u) = \lim(x_n, u) \tag{7}$$

which can be obtained by a limit $x_n \longrightarrow x$ in $X = [\mathcal{D}]$ of some $x_n \in \mathcal{D}$, say.

As far as the generalized random functions $u = (x,u)$, $x \in \mathcal{D}$, are concerned one can use a Hilbert $H = \mathcal{L}_2(\Omega)$ of random variables on a probability space Ω and consider the functional class W of all $u = (x,u)$, $x \in \mathcal{D}$ with random values $(x,u) \in H$ continuous in the Hilbert space H with respect to the norm $\|x\|_X$; moreover one can treat these $u \in W$ as the functions

$$u = (x,u) \equiv (u,x), \quad x \in X \tag{8}$$

of the generalized variable $x \in X$ from the space $X = [\mathcal{D}]$. According to this scheme *boundary values* of any generalized function (8) on the boundary Γ of the cylinder $G \times (t_0, T)$ are

$$(x,u) \equiv (u,x), \quad \operatorname{supp} x \subseteq \Gamma, \tag{9}$$

which appear as a result of testing $u \in W$ by the Schwartz distributions $x \in X$ with supports on the boundary Γ. The collection of all boundary x, $\operatorname{supp} x \subseteq \Gamma$, form the certain *subspace* $X(\Gamma) \subseteq X$ and one can call it as a *generalized boundary*. It occurs that there is some part $X^-(\Gamma) \subseteq X(\Gamma)$ of this generalized boundary where the boundary values of any solution $u \in W$ of the corresponding equation (1), (2) are determined by the very equation so one can not set any other conditions on these boundary values

$$(u,x), \quad x \in X^-(\Gamma).$$

In contrast one can set the *boundary conditions*

$$(u,x) = (u^+,x), \quad x \in X^+(\Gamma) \tag{10}$$

on any *direct complement* $X^+(\Gamma)$ to $X^-(\Gamma)$,

$$X(\Gamma) = X(\Gamma)^- + X^+(\Gamma) \tag{11}$$

by means of the arbitrary given *sample function* $u^+ \in W$.

The part $X^+(\Gamma)$ of the generalized boundary in (10) determines a *type of the boundary conditons*, and (10)-(11) with various $X^+(\Gamma) \subseteq X(\Gamma)$ give us a complete characterization of all possible boundary conditions which can be set with the *arbitrary* sample function $u^+ \in W$ for the equation considered with *arbitrary* generalized random function $f = (\varphi, f)$ on its right side continuous in the Hilbert space H over $\varphi \in \mathcal{D}$ with respect to the norm $\|\varphi\|_{\mathcal{L}_2}$, *and any boundary problem of the type* (10)-(11) *has the unique solution* $u \in W$ which can be described by means of the generalized test functions $x \in X$ as

$$(x,u) = (\varphi, f) + (x^+,u^+) \tag{12}$$

according to decomposition

$$x = \mathcal{L}^* \varphi + x^+$$

with $\varphi \in \mathcal{L}_2$ and $x^+ \in X^+$ where $\mathcal{L}_2$ represents a closure $\mathcal{D}$ and f is well defined over all $\varphi \in \mathcal{L}_2$; the formula (12) in particular shows that

$$\sup_{\|x\|\leq 1} \|(x,u)\|_H \leq C[\sup_{\|\varphi\|_{\mathcal{L}_2}\leq 1} |(\varphi,f)\|_H + \sup_{\|x^+\|_X\leq 1} \|(x^+,u^+)\|_H] . \tag{13}$$

One can find out that all *deterministic* functions u ∈ W from the functional class W which serves in the scheme (7)-(13) the parabolic equation (1) represent the well known Sobolev space $W_2^{(2,1)}$ over the cylinder $G\times(t_0^2,T)$ with the bounded regular region G (its boundary to be a differential C^2-manifold, say); moreover the boundary values (9) are determined completely by the *trace* of the very functions $u \in W_2^{(2,1)}$ on the boundary Γ in the whole plus by the trace of the normal derivatives $\partial u/\partial n$ on the side-boundary of the cylinder $G\times(t_0,T)$. Namely a necessary link is given by a fact that the generalized boundary $X(\Gamma)$ contains a *complete system* of the generalized test functions x, supp $x \subseteq \Gamma$, which according to a proper limit process (7) gives us the boundary values

$$(u,x) = \int_\Gamma ux(s)ds, \int \frac{\partial u}{\partial n} x(s)ds \tag{14}$$

by means of the corresponding weight functions x(s), $s \in \Gamma$, on various parts of the boundary Γ, and in this way one can define *the traces* of any generalized *random* function u ∈ W and its normal derivative $\partial u/\partial n$. Thus one can take a proper part of the boundary test functions giving us the boundary values (14) or its linear span $X^+(\Gamma)$, say, and form the corresponding *stochastic boundary problem* (3),(5)/(3),(6) *for the Stochastic Heat Equation* (1) by means of the *traces*
$u^+,\partial u^+/\partial n$ of the random sample function $u^+ \in W$.

In a similar way one can proceed for the *Stochastic Wave Equation* (2). Namely, according to the limit process (7) which leads to (14) in the functional class W which serves the equation (2) one can define the *traces* of any generalized function u ∈ W and its normal derivative $\partial u/\partial n$ on the boundary

Γ of the cylinder $G \times (t_0,T)$, for this boundary does not contain any parts with tangents which form *characteristics* of the hyperbolic equation (2), and the *stochastic boundary problems* (4),(5)/(4),(6) given by means of the *traces* $\overset{+}{u}, \partial\overset{+}{u}/\partial n$ of the *random* sample function $\overset{+}{u} \in W$ become to the well defined.

The question is whether the boundary problems (3),(5)/(3),(6) or (4),(5)/(4),(6) belong to the type of (10)-(11).

This question can be separately analysed for a pure *deterministic case* with respect to the corresponding functional class W which serves in the scheme (7)-(14) the differential equation considered and it is connected with many developments in Partial Differential Equations, for the *deterministic* W is very similar to the corresponding Sobolev space. One can apply the following *criterion* to get a positive answer: the *deterministic* boundary problem with the homogeneous boundary conditions (i.e. with $\overset{+}{u} = 0$) has the unique solution $u \in W$ for any $f \in \mathcal{L}_2$.

There is also another equivalent criterion: the boundary problem with the arbitrary deterministic sample function $\overset{+}{u} \in W$ in the boundary conditions has the unique solution for $u \in W$ for the homogeneous equation with $f = 0$.

These criteria give us a link to various powerful developments in Partial Differential Equations. (See for example J.-L. Lions, E. Mageses, Problems aux limits non homoeénes et applications, I-III, Dunod, Paris, 1968-1970.) The first criterion gives us the following result with respect to the Stochastic Heat Equation (1) with the elliptic operator $\mathcal{A}$ of a farily wide type (in particular with elliptic $\mathcal{A}$ which has constant coefficients).

Theorem. *For arbitrary generalized random* $f = (\varphi, f)$ *continuous in the Hilbert space* H *of random variables over* $\varphi \in \mathcal{D}$ *with respect to* $\|\varphi\|_{\mathcal{L}_2}$ *and for arbitrary generalized random sample function* $\overset{+}{u} \in W$ *in the boundary conditions*

(3),(5)/(3),(6) *there is the unique solution* u ∈ W.

To prove it one can apply a fact*) that for arbitrary $f \in \mathcal{L}_2$ the parabolic equation (1) with the homogeneous boundary conditions (3),(5)/(3),(6) has the unique solution $u \in W_2^{(2,1)}$; namely as was already mentioned the Sobolev space $W_2^{(2,1)}$ over the cylinder $G \times (t_0,T)$ represents all deterministic u ∈ W, so one has only to apply our criterion. (See for example V.h. Mikhaylov, Theory of Partial Differential Equations, 2nd ed., M. Nauka, 1983.)

It is worthy to note that the generalized random field $u = \xi$ which appears in the cylinder $G \times (t_0,T)$ is the unique solution u ∈ W of the stochastic equation (1) with the stochastic boundary conditions (3),(5)/(3),(6) can be extended for all $t \geq t_0$ by means of $T \longrightarrow \infty$. One can easily find out the that the probability model suggested gives us in particular a functional stochastic Itô equation with boundary conditions. Namely, generalized random field $u = \xi$ has the trace ξ_t on all sections $G \times \{t\}$ of the cylinder $G \times (t_0,\infty)$ which can be considered as a generalized random field

$$\xi_t = (\varphi,\xi_t), \quad \varphi \in C_0^\infty(G),$$

in the region G, and the very ξ can be identified by the corresponding random vector process ξ_t, $t \geq t_0$, as

$$(x,\xi) = \int \alpha(t)(\varphi,\xi_t)dt$$

with the complete system of the test functions $x \in \mathcal{D}$ of the form $x \in \varphi \otimes \alpha$ in the direct product $G \times (t_0,\infty)$. Suppose the stochastic source $f = \eta$ in (1) is a derivative of the some random vector-process w_t, $t \geq t_0$ with a phase-state at any time moment t which can be described as a generalized random field

$$w_t = (\varphi,w_t), \quad \varphi \in C_0^\infty(G),$$

in the region G, namely

$$(\varphi \otimes \alpha, \eta) = -\int \alpha'(t)(\varphi, w_t)dt$$

for all test functions of the form $\varphi \otimes \alpha$ in $G \times (t_0, \infty)$. Then the generalized differential equation (1) for $u = \xi$ means

$$-\int \alpha'(t)(\varphi, \xi_t)dt = \int \alpha(r)(\mathcal{A}^*\varphi, \xi_t)dt - \int \alpha'(t)(\varphi, w_t)dt$$

with all test functions $\alpha \in C_0^\infty(t_0, T)$ which in a case $w_{t_0} = 0$ gives us

$$(\varphi, \xi_t) - (\varphi, \xi_{t_0}) = \int_{t_0}^{t} (\mathcal{A}^*\varphi, \xi_t)dt + (\varphi, w_t)$$

and this integral equation determines a well known stochastic Itô equation

$$d\xi_t = \mathcal{A}\xi_t dt + dw_t, \tag{15}$$

so *in our scheme we have the functional Itô equation with the stochastic non-homogeneous boundary conditions* (3),(5)/(3),(6).†

One can find out as well that the *Stochastic Wave Equation* (2) in the *plane* cylinder $(G) \times (t_0, T)$ where $G = (a, b)$ is closely connected with a well known probability model called the *Brownian Sheet*. Namely, for the classical hyperbolic equation (2) with the operator $\mathcal{A} = \partial^2/\partial x^2$ in the interval $a < x < b$,

$$\frac{\partial^2 u}{\partial t^2} - \frac{\partial^2 u}{\partial x^2} = f \tag{16}$$

our scheme (7)-(14) provides a property of $u \in W$ to have *traces* on the boundary Γ of the rectangular $(a, b) \times (t_0, T)$ for the very functions $u \in W$ and their

---

* Note that the homogeneous boundary conditions can be treated in another way by the corresponding semigroup approach

† With respect to connections of (1), (15) see also for example V. A. Buljichev, Markov property of the stochastic evolution. - Theory Prob. and Appl., v. 37, 2, 373-378, 1986.

derivatives $\partial u/\partial n$ along any nontangent direction, and it occurs that the second boundary problem (4),(6) with the boundary conditions (6) on the derivatives

$$\frac{\partial u}{\partial n} = \frac{\partial u^+}{\partial n}$$

along *characteristics*

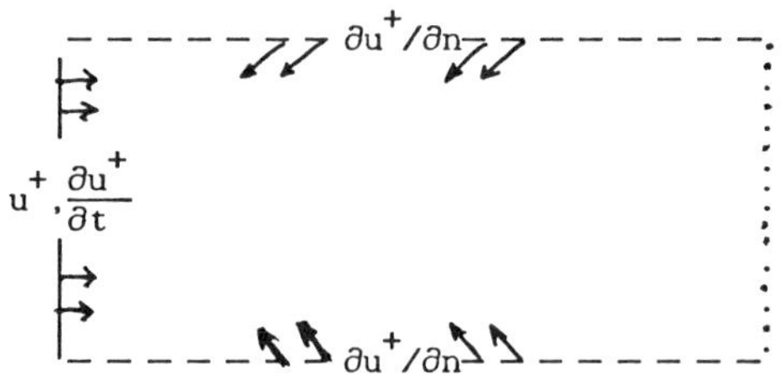

gives us the unique solution $u \in W$ of the equation (16) which represents the very *Brownian Sheet* $u = \xi$ in the rectangular $(a,b) \times (t_0,T)$ when the right side of (16) is the *Gaussian white noise* $f = \eta$ on $\mathcal{L}_2$-space and $u^+ = W$ in (3),(6) is the *sample Brownian Sheet* $u^+ = \xi^+$ *independent* on $f = \eta$ say.

The following result holds true with respect to the *Stochastic Wave Equation* (2) of the form (16).

Theorem. *For the arbitrary generalized random* $f = (\varphi, f)$ *continuous in the Hilbert space* H *of random variables over* $\varphi \in \mathcal{D}$ *with respect to* $\|\varphi\|_{\mathcal{L}_2}$ *and for arbitrary generalized random sample function* $u^+ \in W$ *in the boundary conditions* (4),(5)/(4),(6) *there is the unique solution* $u \in W$.

GREY NOISE

W. R. Schneider

Asea Brown Boveri Corporate Research
CH-5405 Baden, Switzerland

ABSTRACT:

The Mittag-Leffler function E_α is completely monotonic on $\mathbf{R}_-$ for $0 < \alpha \leq 1$. This remarkable fact is exploited to define a probability measure τ_α on a Hilbert triple $K_\alpha \subset H_\alpha \subset K'_\alpha$. This measure is called grey noise. It reduces to white noise for $\alpha = 1$. Mimicking the construction of Brownian motion yields its grey variant. The sample paths of grey Brownian motion are Hölder continuous with index arbitrarily close to $\alpha/2$. The well-known relation between Brownian motion and diffusion carries over to grey Brownian motion and fractional diffusion with time derivative of order α.

1. Introduction

Gaussian white noise is a probability measure τ characterized by [1]

$$\int_{K'} d\tau(T)\, e^{i\langle T,\xi\rangle} = e^{-(\xi,\xi)}, \quad \xi \in K. \tag{1.1}$$

Here $K \subset H \subset K'$ is a triple of real Hilbert spaces obtained as follows: Equip the real Schwartz space $\mathcal{S}(\mathbf{R})$ with the scalar product

$$(\xi,\eta) = \int_{\mathbf{R}} dt\, \xi(t)\, \eta(t)\,. \tag{1.2}$$

The completion of $\mathcal{S}(\mathbf{R})$ with respect to (1.2) is the Hilbert space H. Let $\phi_n \in \mathcal{S}(\mathbf{R})$, $n \in \mathbf{Z}_+$, be an orthonormal basis of H (e.g., the Hermite functions). The subspace $K \subset H$ is defined by

$$K = \Big\{\xi \in H \,\Big|\, \sum_{n=0}^{\infty} (n+1)^2\, (\phi_n,\xi)^2 < \infty\Big\}. \tag{1.3}$$

Equipped with the scalar product

$$(\xi,\eta) = \sum_{n=0}^{\infty} (n+1)^2\, (\phi_n,\xi)\,(\phi_n,\eta)\,, \quad \xi,\eta \in K\,, \tag{1.4}$$

K is a Hilbert space. Finally, K' is the topological dual of K and $\langle T,\xi\rangle$ denotes the value of the continuous linear functional $T \in K'$ at $\xi \in K$. The map Φ_1 defined on H by

$$\Phi_1(\xi) = F_1((\xi,\xi))\,, \quad \xi \in H\,, \tag{1.5}$$

with

$$F_1(z) = e^{-z}\,, \quad z \in \mathbf{C}\,, \tag{1.6}$$

is a particularly simple example of a characteristic functional on a Hilbert space. Some relevant properties of such functionals are discussed in Section 2. The Mittag-Leffler function E_α with $0 < \alpha \le 1$ is used in Section 3 to define a special class of characteristic functionals Φ_α on an arbitrary real Hilbert space X by

$$\Phi_\alpha(\xi) = F_\alpha((\xi,\xi))\,, \quad \xi \in X\,, \tag{1.7}$$

where

$$F_\alpha(z) = E_\alpha(-z) = \sum_{n=0}^{\infty} \frac{(-z)^n}{\Gamma(\alpha n + 1)}, \quad z \in \mathbf{C}. \tag{1.8}$$

In Section 4 grey noise is introduced as the probability measure τ_α characterized by

$$\int_{K'_\alpha} d\tau_\alpha(T)\, e^{i\langle T,\xi\rangle} = F_\alpha((\xi,\xi)_\alpha), \quad \xi \in K_\alpha, \tag{1.9}$$

in analogy to (1.1) with a triple of Hilbert spaces $K_\alpha \subset H_\alpha \subset K'_\alpha$, H_α being the completion of $\mathcal{S}(\mathbf{R})$ with respect to the scalar product

$$(\xi,\eta)_\alpha = \Gamma(1+\alpha)\sin\frac{\pi}{2}\alpha \int_{\mathbf{R}} d\omega\, |\omega|^{1-\alpha}\, \overline{\tilde{\xi}(\omega)}\, \tilde{\eta}(\omega) \tag{1.10}$$

where

$$\tilde{\xi}(\omega) = (2\pi)^{-1/2} \int_{\mathbf{R}} dt\, e^{i\omega t}\, \xi(t). \tag{1.11}$$

Note that (1.10) reduces to (1.2) for $\alpha = 1$. Simultaneously, (1.8) reduces to (1.6), hence (1.9) to (1.1), i.e., grey noise becomes white noise.

The well-known construction of Brownian motion [1] is mimicked in Section 5 to construct a stochastic process B_t^α, $t \in \mathbf{R}_+$, called grey Brownian motion. The investigation of its sample path properties shows that the set of Hölder continuous paths with index smaller than $\alpha/2$ has τ_α-measure one. Furthermore, it is shown that

$$u(x,t) = \int_{K'_\alpha} d\tau_\alpha(T)\, u_0(x + B_t^\alpha(T)) \tag{1.12}$$

is a solution of the fractional diffusion equation [2], [3]

$$u(x,t) = u_0(x) + \frac{1}{\Gamma(\alpha)} \int_0^t ds\,(t-s)^{\alpha-1} \frac{\partial^2}{\partial x^2} u(x,s). \tag{1.13}$$

For $\alpha = 1$ the well-known relation between Brownian motion and (ordinary) diffusion is thus recovered. In Section 6 some possible generalizations are mentioned and open problems are exhibited.

2. Characteristic Functionals

Let X be a real separable Hilbert space with scalar product (ξ, η), $\xi, \eta \in X$. A map $\Phi: X \to \mathbf{C}$ is called positive definite if

$$\sum_{j,k=1}^{n} \bar{z}_j \, \Phi(\xi_j - \xi_k) \, z_k \geq 0 \tag{2.1}$$

for arbitrary $n \in \mathbf{N}$, $z_1, z_2, ..., z_n \in \mathbf{C}$ and $\xi_1, \xi_2, ..., \xi_n \in X$. For $n = 1$ (2.1) implies $\Phi(0) \geq 0$. If $\Phi(0) = 0$ then $\Phi(\xi) = 0$ for all $\xi \in X$ as is easily seen from (2.1) with $n = 2$. Henceforth this trivial case will be excluded. Without restriction of generality we may assume Φ to be normalized, i.e.,

$$\Phi(0) = 1 \,. \tag{2.2}$$

Any probability measure ν on $\mathbf{R}_+$ yields a map Φ satisfying (2.1) and (2.2) via

$$\Phi(\xi) = \int_{\mathbf{R}_+} d\nu(s) \, e^{-s\|\xi\|^2} \,. \tag{2.3}$$

This map depends only on the norm $\|\xi\|$ of ξ, i.e., it is of the form

$$\Phi(\xi) = F(\|\xi\|^2) \,. \tag{2.4}$$

In the case where Φ is given by (2.3) F is given by

$$F(t) = \int_{\mathbf{R}_+} d\nu(s) \, e^{-st} \,. \tag{2.5}$$

Remarkably, the converse is also true if the dimension of X is infinite [1]. One may show [1] that (2.1) and (2.4) imply that F is completely Δ-monotonic, i.e.,

$$\sum_{k=0}^{n} (-1)^k \binom{n}{k} F(t + k\tau) \geq 0 \tag{2.6}$$

for arbitrary $n \in \mathbf{Z}_+$, $t \in \mathbf{R}_+$, $\tau \in \mathbf{R}_+$. According to [4] this is equivalent to complete D-monotonicity, i.e., F continuous on $\mathbf{R}_+$ and

$$(-1)^k \frac{d^k}{dt^k} F(t) \geq 0 \,, \quad t > 0 \,, \quad k \in \mathbf{Z}_+ \,. \tag{2.7}$$

By Bernstein's theorem [4], [5], taking $F(0) = 1$ into account, there exists a probability measure ν on $\mathbf{R}_+$ such that (2.5) holds. If however the dimension of X is finite, (2.1) and (2.4) do not imply (2.5). This is due to the invariance of (2.1), with Φ satisfying (2.4), under the transformation

$$F(t) \;\rightarrow\; F(t)\,\phi_{d/2-1}(t) \tag{2.8}$$

where d is the dimension of X and

$$\phi_\sigma(t) = 2^\sigma\,\Gamma(1+\sigma)\,t^{-\sigma/2}\,J_\sigma(t^{1/2}) \tag{2.9}$$

with J_σ the Bessel function of the first kind. Taking

$$\phi_{d/2-1}(t) = \frac{1}{|S^{d-1}|}\int_{S^{d-1}} d\Omega(\xi)\,e^{it^{1/2}(\xi,\eta)}\,, \quad \eta \in S^{d-1}\,, \tag{2.10}$$

into account it is easily verified that (2.1) holds also after the replacement (2.8) in (2.4). On the other hand complete monotonicity is destroyed by (2.8).

A continuous map $\Phi: X \rightarrow \mathbf{C}$ satisfying (2.1) and (2.2) is called characteristic functional. An example is Φ defined by (2.3). As we have seen in the case of $\dim X = \infty$ the properties (2.1), (2.2) and (2.4) imply (2.3).

Assume now X to be decomposed into the direct sum of finite dimensional subspaces X_k, $k \in \mathbf{Z}_+$,

$$X = \bigoplus_{k=0}^{\infty} X_k\,, \quad \dim X_k = d_k < \infty\,. \tag{2.11}$$

Let c_k, $k \in \mathbf{Z}_+$, be a sequence of positive numbers satisfying

$$1 = c_0 < c_1 < c_2 < \dots < c_k < \dots \tag{2.12}$$

and

$$\sum_{k=0}^{\infty} d_k/c_k < \infty\,, \quad (\sum_{k=0}^{\infty} d_k = \infty)\,. \tag{2.13}$$

We equip each subspace X_k with two new scalar products

$$(\xi,\eta)_\pm = c_k^{\pm 1}\,(\xi,\eta)\,, \quad \xi,\eta \in X_k,\ k \in \mathbf{Z}_+\,. \tag{2.14}$$

Denote the corresponding Hilbert spaces by $X_k^{\pm}$ and define the Hilbert spaces $X_{\pm}$ by

$$X_{\pm} = \bigoplus_{k=0}^{\infty} X_k^{\pm} \tag{2.15}$$

with scalar products $(\cdot,\cdot)_{\pm}$ given by

$$(\xi,\eta)_{\pm} = \sum_{k=0}^{\infty} (P_k^{\pm}\xi, P_k^{\pm}\eta)_{\pm} \ . \tag{2.16}$$

Here $P_k^{\pm}$ are the orthogonal projections of $X^{\pm}$ onto $X_k^{\pm}$, respectively. Obviously we have

$$X_+ \subset X \subset X_- \ . \tag{2.17}$$

The inclusions are proper in view of (2.13). Furthermore, X_+ is dense in X and X is dense in X_-. This is due to the fact that

$$X_0 = \{\xi \in X \mid \exists\ c(\xi) \geq 0 : \ P_k\xi = 0, \ k \geq c(\xi)\} \tag{2.18}$$

is dense in X_+, X, X_- in the respective topology. Here P_k is the orthogonal projection of X onto X_k. Define the bilinear map $\langle\,\cdot\,,\,\cdot\,\rangle : X_- \times X_+ \to \mathbf{R}$ by

$$\langle \eta, \xi \rangle = \sum_{k=0}^{\infty} (P_k^- \eta, P_k^+ \xi) \ , \quad \eta \in X_- \ , \ \xi \in X_+ \ . \tag{2.19}$$

The terms in the sum are well-defined as we have $X_k^- = X_k = X_k^+$ in the sense of vector spaces. Furthermore, the sum is absolutely convergent and satisfies

$$|\langle \eta, \xi \rangle| \leq \|\eta\|_- \|\xi\|_+ \tag{2.20}$$

as is easily verified applying twice Schwarz's inequality. Hence, the linear functional $l: X_+ \to \mathbf{R}$ defined for fixed $\eta \in X_-$ by

$$l(\xi) = \langle \eta, \xi \rangle \ , \quad \xi \in X_+ \tag{2.21}$$

is continuous. Conversely, any continuous linear functional on X_+ is of this form. This is seen as follows. By Riesz's representation theorem there exists $\zeta \in X_+$ such that

$$l(\xi) = (\zeta, \xi)_+ \ . \tag{2.22}$$

This may be rewritten as

$$l(\xi) = \sum_{k=0}^{\infty} (P_k^+ \zeta, P_k^+ \xi)_+ = \sum_{k=0}^{\infty} (c_k P_k^+ \zeta, P_k^+ \xi) \ . \tag{2.23}$$

As

$$\sum_{k=0}^{\infty} c_k^{-1} \| c_k P_k^+ \zeta \|^2 = \sum_{k=0}^{\infty} c_k \| P_k^+ \zeta \|^2 = \| \zeta \|_+^2 < \infty \tag{2.24}$$

there exists $\eta \in X_-$ with $P_k^- \eta = c_k P_k^+ \zeta$, $k \in \mathbf{Z}_+$ and (2.23) takes the form

$$l(\xi) = \sum_{k=0}^{\infty} (P_k^- \eta, P_k^+ \xi) = \langle \eta, \xi \rangle \ . \tag{2.25}$$

Setting $\eta = A^{-1} \zeta$ defines an isometry $A^{-1}: X_+ \to X_-$ as is seen from (2.24). Accordingly, the isometry $A: X_- \to X+$ is given by

$$P_k^+ A \eta = c_k^{-1} P_k^- \eta \ , \quad k \in \mathbf{Z}_+ \ , \ \eta \in X_- \ . \tag{2.26}$$

As $X_+ \subset X_-$ we may consider A as an operator from X_- into itself, which will be denoted by B to distinguish it from A as operator from X_- onto X_+. From

$$P_k^- B \eta = c_k^{-1} P_k^- \eta \ , \quad k \in \mathbf{Z}_+ \ , \ \eta \in X_- \ , \tag{2.27}$$

we obtain

$$(\eta, B\eta)_- = \sum_{k=0}^{\infty} c_k^{-1} \ (P_k^- \eta, P_k^- \eta)_- \geq 0 \tag{2.28}$$

with equality only for $\eta = 0$. In addition, we have

$$\operatorname{tr} B = \sum_{k=0}^{\infty} c_k^{-1} \ \operatorname{tr} P_k^- = \sum_{k=0}^{\infty} \frac{d_k}{c_k} < \infty \tag{2.29}$$

in view of the assumption (2.13): The operator B is nonnegative and has finite trace, i.e., by definition, B is a nuclear operator.

After these preliminaries we may formulate the following theorem (a reformulation of a theorem due to Yu. L. Daletskii [6]).

Theorem. *Let $X_+ \subset X \subset X_-$ be a triple of Hilbert spaces as considered above and let Φ be a characteristic functional on X. Then there is a unique σ-additive probability measure μ on $(X_-, \mathcal{B})$, $\mathcal{B}$ being the σ-algebra of the Borel sets of X_-, such that*

$$\int_{X_-} d\mu(\eta)\, e^{i\langle\eta,\xi\rangle} = \Phi(\xi) \,, \quad \xi \in X_+ \,. \tag{2.30}$$

Proof. Define the map $\Phi_-: X_- \to \mathbf{C}$ by $\Phi_- = \Phi \circ A$ with A being the isometry $A: X_- \to X+$ defined by (2.26). It follows immediately that Φ_- is a characteristic functional on X_-. Furthermore, to $\epsilon > 0$ there exists a nuclear operator $S_\epsilon: X_- \to X_-$ such that

$$\mathrm{Re}\,\{\Phi_-(0) - \Phi_-(\eta)\} < \epsilon \quad \text{for} \quad (\eta, S_\epsilon\eta)_- \le 1 \,, \quad \eta \in X_- \,. \tag{2.31}$$

This is seen as follows: As Φ is continuous there exists $\delta = \delta(\epsilon) > 0$ such that

$$\mathrm{Re}\,\{\Phi(0) - \Phi(\xi)\} < \epsilon \quad \text{for} \quad (\xi, \xi) \le \delta \,, \ \xi \in X \,. \tag{2.32}$$

Hence,

$$\mathrm{Re}\,\{\Phi(0) - \Phi(A\eta)\} < \epsilon \quad \text{for} \quad (A\eta, A\eta) \le \delta \,, \ \eta \in X_- \,. \tag{2.33}$$

Setting $S_\epsilon = B/\delta(\epsilon)$ we obtain (2.31). By Minlos-Sazonov's theorem [6] there exists a unique σ-additive probability measure μ on $(X_-, \mathcal{B})$ such that

$$\int_{X_-} d\mu(\eta)\, e^{i(\eta,\zeta)_-} = \Phi_-(\zeta) \,, \quad \zeta \in X_- \,. \tag{2.34}$$

Setting $\zeta = A^{-1}\xi$, $\xi \in X_+$ and taking

$$(\eta, A^{-1}\xi)_- = \langle\eta, \xi\rangle \tag{2.35}$$

into account yields (2.30).

3. The Mittag-Leffler Function

The Mittag-Leffler function $E_\alpha: \mathbf{C} \to \mathbf{C}$, $\alpha > 0$, is defined by its power series [7]

$$E_\alpha(z) = \sum_{n=0}^{\infty} \frac{z^n}{\Gamma(\alpha n + 1)}, \quad z \in \mathbf{C} . \tag{3.1}$$

It follows that E_α is an entire function of order $1/\alpha$ [7]. For the sake of convenience we introduce

$$F_\alpha(z) = E_\alpha(-z) , \quad z \in \mathbf{C} . \tag{3.2}$$

According to H. Pollard [8] W. Feller obtained the result that F_α is completely monotonic on $\mathbf{R}_+$ for $0 < \alpha \leq 1$. For $\alpha = 1$ this is obvious as

$$F_1(z) = e^{-z} . \tag{3.3}$$

By Bernstein's theorem [5] F_α is the Laplace transform of a probability measure on $\mathbf{R}_+$. Its explicit form was determined in [8] (see also [5] for a different derivation) where it was shown that for $0 < \alpha < 1$

$$F_\alpha(t) = \frac{1}{\alpha} \int_0^\infty dx \, e^{-tx} \, x^{-1-1/\alpha} \, f_\alpha(x^{-1/\alpha}) . \tag{3.4}$$

Here f_α denotes the one-sided stable (or Lévy) probability density [5], [9], [10] characterized by

$$\int_0^\infty dx \; e^{-tx} \; f_\alpha(x) = e^{-t^\alpha} , \quad 0 < \alpha < 1 . \tag{3.5}$$

A "simple" proof runs as follows: In [10] it was shown that stable densities ($\alpha \neq 1$) possess representations in terms of H-functions [10], [11]. For the particular density in (3.4) one obtains

$$\alpha^{-1} \, x^{-1-1/\alpha} \, f_\alpha(x^{-1/\alpha}) = H_{11}^{10}\left(x \,\middle|\, \begin{matrix} (1-\alpha, \alpha) \\ (0,1) \end{matrix} \right) . \tag{3.6}$$

Applying the rule for the Laplace transform of H-functions [11] yields

$$\int_0^\infty dx \, e^{-tx} \, H_{11}^{10}\left(x \,\middle|\, \begin{matrix} (1-\alpha, \alpha) \\ (0,1) \end{matrix} \right) = H_{12}^{11}\left(t \,\middle|\, \begin{matrix} (0,1) & \\ (0,1) & (0,\alpha) \end{matrix} \right) . \tag{3.7}$$

The series expansion of (3.7) is [10], [11]

$$H^{11}_{12}\left(t\,\middle|\,\begin{matrix}(0,1) & \\ (0,1) & (0,\alpha)\end{matrix}\right) = \sum_{k=0}^{\infty} \frac{(-t)^k}{\Gamma(1+\alpha k)} = F_\alpha(t)\ . \tag{3.8}$$

A complementary result is the following theorem announced as a conjecture in [12].

Theorem. *For $\alpha > 1$ the function F_α is not completely monotonic.*

Proof. Assume the contrary. Hence, there exists a probability measure ν_α on $\mathbf{R}_+$ for some $\alpha > 1$ such that F_α is its Laplace transform. From (3.1), (3.2) we obtain that ν_α has moments of any order, explicitly given by

$$m_k(\alpha) = \int_{\mathbf{R}_+} d\nu_\alpha(x)\, x^k = \frac{k!}{\Gamma(1+\alpha k)}\ , \qquad k \in \mathbf{Z}_+\ . \tag{3.9}$$

Applying the inequality

$$m_2 - m_1^2 = \int_{\mathbf{R}} d\nu(x)\,(x - m_1)^2 \geq 0\ , \tag{3.10}$$

holding for any probability measure ν with finite moments m_1 and m_2, to (3.9) yields

$$f(\alpha) := 2\,\Gamma(1+\alpha)^2 - \Gamma(1+2\alpha) \geq 0\ . \tag{3.11}$$

We now show that (3.11) is violated for any $\alpha > 1$. Using $z\Gamma(z) = \Gamma(1+z)$ we obtain

$$f(\alpha+1) < (1+\alpha)^2\, f(\alpha)\ . \tag{3.12}$$

Hence, it is sufficient to prove $f(\alpha) < 0$ for $1 < \alpha \leq 2$. From well-known properties of the Γ-function [13] we have

$$\Gamma(1+\alpha) \leq \alpha\ , \qquad 1 \leq \alpha \leq 2 \tag{3.13}$$

with equality holding at the endpoints $\alpha = 1$ and $\alpha = 2$ only. From (3.13) and (3.11) we obtain for $1 < \alpha \leq 2$

$$f(\alpha) \leq -2\alpha \int_{1+\alpha}^{2\alpha} \Gamma'(\beta)\, d\beta < 0 \tag{3.14}$$

as $\Gamma'(\beta) > \Gamma'(2) = \psi(2) = 1 - \gamma > 0$, $\gamma = 0.577...$ being Euler's constant.

The result (3.4) may be generalized as follows. Together with F_α also $G_{\alpha,\beta}$ is completely monotonic and normalized [5] where

$$G_{\alpha,\beta}(t) = F_\alpha(t^\beta) , \quad 0 < \beta < 1 . \tag{3.15}$$

Again the underlying probability measure may be explicitly determined.

Theorem. *The function $G_{\alpha,\beta}$ defined in (3.15) is the Laplace transform of the probability density $g_{\alpha,\beta}$ on $\mathbf{R}_+$ with*

$$g_{\alpha,\beta}(x) = \frac{\beta}{\alpha}\, x^{-1-\beta/\alpha} \int_0^\infty dy\, y^{\beta/\alpha}\, f_\beta(y)\, f_\alpha((y/x)^{\beta/\alpha}) \tag{3.16}$$

in terms of the stable densities f_α and f_β. Its H-function representation and series expansion is given by

$$\begin{aligned} g_{\alpha,\beta}(x) &= \frac{1}{\beta}\, H^{11}_{22}\left(x \,\middle|\, \begin{matrix} (1-\frac{1}{\beta},\frac{1}{\beta}) & (1-\frac{\alpha}{\beta},\frac{\alpha}{\beta}) \\ (1-\frac{1}{\beta},\frac{1}{\beta}) & (0,1) \end{matrix} \right) \\ &= \sum_{k=0}^{\infty} \frac{(-1)^k}{\Gamma(\beta+\beta k)\Gamma(1-\alpha-\alpha k)}\, x^{\beta(k+1)-1} , \quad \beta \geq \alpha \end{aligned} \tag{3.17}$$

and

$$\begin{aligned} g_{\alpha,\beta}(x) &= \frac{1}{\beta}\, H^{11}_{22}\left(\frac{1}{x} \,\middle|\, \begin{matrix} (\frac{1}{\beta},\frac{1}{\beta}) & (1,1) \\ (\frac{1}{\beta},\frac{1}{\beta}) & (\frac{\alpha}{\beta},\frac{\alpha}{\beta}) \end{matrix} \right) \\ &= \sum_{k=0}^{\infty} \frac{(-1)^k}{\Gamma(-\beta k)\Gamma(1+\alpha k)}\, x^{-\beta k-1} , \quad \beta < \alpha . \end{aligned} \tag{3.18}$$

For $\beta = \alpha$ (3.17) reduces to

$$g_{\alpha,\alpha}(x) = \frac{\sin \pi\alpha}{\pi}\, \frac{x^{\alpha-1}}{1 + 2\cos\pi\alpha\, x^\alpha + x^{2\alpha}} . \tag{3.19}$$

We refrain from reproducing the proof of this theorem [14] and indicate only that the main ingredient is the Mellin transform.

Let $X_+ \subset X \subset X_-$ be a Hilbert space triple as described in Section 2. The probability measures on X_- satisfying (2.30) with

$$\Phi(\xi) = F_\alpha((\xi,\xi)) \quad \text{or} \quad G_{\alpha,\beta}((\xi,\xi)) \tag{3.20}$$

will be denoted by μ_α and $\mu_{\alpha,\beta}$, respectively. In Section 4 we shall consider concrete realizations of the triple $X_+ \subset X \subset X_-$ instead of an abstract setting until now.

We terminate this section by considering the measure μ_α determined by (2.30), (3.20) for the case $\dim X = d < \infty$. Here we may identify the spaces X_+ and X_- with X. Without loss of generality we may set $X = \mathbf{R}^d$. We have the following result [3].

Theorem. *Let μ_α be the measure on $\mathbf{R}^d$ with characteristic functional*

$$\int_{\mathbf{R}^d} d\mu_\alpha(x)\, e^{ip\cdot x} = F_\alpha(p\cdot p) \tag{3.21}$$

where

$$p\cdot x = \sum_{k=1}^{d} p_k\, x_k \tag{3.22}$$

denotes the scalar product on $\mathbf{R}^d$. Then μ_α is absolutely continuous with respect to the Lebesgue measure on $\mathbf{R}^d$ and its density is given by

$$\rho_\alpha^d(x) = (4\pi)^{-d/2}\, H_{12}^{20}\left(\frac{r^2}{4}\,\middle|\,\begin{matrix}(1-\alpha d/2,\alpha) \\ (0,1) \qquad (1-d/2,1)\end{matrix}\right), \qquad r = |x|\,, \tag{3.23}$$

which reduces to

$$\rho_1^d(x) = (4\pi)^{-d/2}\, e^{-r^2/4} \tag{3.24}$$

for $\alpha = 1$. The one-dimensional case $d = 1$ is peculiar as ρ_α^1 defined by (3.23) is a probability density in the extended interval $0 < \alpha \le 2$ instead of the original interval $0 < \alpha \le 1$. For $\alpha > 1$ ($d \ge 2$) and $\alpha > 2$ ($d = 1$) (3.23) becomes indefinite.

4. Grey Noise

Let H^0 be the Schwartz space $\mathcal{S}(\mathbf{R})$ of real-valued infinitely differentiable functions of rapid decrease. Equipped with the scalar product

$$(\xi,\eta)_\gamma = C_\gamma \int_{\mathbf{R}} d\omega\, |\omega|^{1-\gamma}\, \overline{\tilde{\xi}(\omega)}\, \tilde{\eta}(\omega) \tag{4.1}$$

H^0 becomes a prehilbert space. Here C_γ is a positive constant, γ a parameter satisfying $-\infty < \gamma < 2$ and the tilde denotes the Fourier transform

$$\tilde{\xi}(\omega) = (2\pi)^{-1/2} \int_{\mathbf{R}} dt\, e^{i\omega t}\, \xi(t)\ . \tag{4.2}$$

The completion of H^0 with respect to (4.1) will be denoted by H_γ. It will play the role of X in a Hilbert space triple $X_+ \subset X \subset X_-$ as introduced in Section 2. We define the functions $\phi_k \in H^0$, $k \in \mathbf{Z}_+$, in terms of their Fourier transforms as follows

$$\begin{aligned} \tilde{\phi}_{2k}(\omega) &= a_{k,\gamma}\, e^{-\omega^2/2}\, L_k^{-\gamma/2}(\omega^2)\ , \quad k \in \mathbf{Z}_+ \\ \tilde{\phi}_{2k+1}(\omega) &= b_{k,\gamma}\, e^{-\omega^2/2}\, L_k^{1-\gamma/2}(\omega^2)\,\omega\ , \quad k \in \mathbf{Z}_+\ . \end{aligned} \tag{4.3}$$

Here L_k^ϵ, $k \in \mathbf{Z}_+$, $\epsilon > -1$, are the Laguerre polynomials [13] normalized as follows

$$\int_0^\infty dx\, e^{-x}\, L_j^\epsilon(x)\, L_k^\epsilon(x) = \delta_{j,k}\, \frac{\Gamma(k+\epsilon+1)}{\Gamma(k+1)}\ . \tag{4.4}$$

With the choice

$$\begin{aligned} a_{k,\gamma} &= \left(\frac{\Gamma(k+1-\gamma/2)}{C_\gamma \Gamma(k+1)}\right)^{1/2} \\ b_{k,\gamma} &= \left(\frac{\Gamma(k+2-\gamma/2)}{C_\gamma \Gamma(k+1)}\right)^{1/2} \end{aligned} \tag{4.5}$$

the system of functions (4.3) becomes an orthonormal basis of H_γ which we decompose into a direct sum

$$H_\gamma = \bigoplus_{k=0}^{\infty} H_{\gamma,k}\ , \quad \dim H_{\gamma,k} = d_k = 1\ , \tag{4.6}$$

where $H_{\gamma,k}$ is the subspace spanned by ϕ_k. With the choice

$$c_k = (k+1)^2\ , \quad k \in \mathbf{Z}_+ \tag{4.7}$$

satisfying the condition (2.13) in view of

$$\sum_{k=0}^{\infty} \frac{d_k}{c_k} = \sum_{k=0}^{\infty} (k+1)^{-2} = \pi^2/6 \tag{4.8}$$

we construct the Hilbert space triple $K_\gamma \subset H_\gamma \subset K'_\gamma$ corresponding to $X_+ \subset X \subset X_-$ of Section 2.

Definition. Generalized grey noise is the probability measure $\tau_{\alpha,\beta,\gamma}$ on K'_γ defined by

$$\int_{K'_\gamma} d\tau_{\alpha,\beta,\gamma}(T)\, e^{i\langle T,\xi\rangle} = G_{\alpha,\beta}(\|\xi\|_\gamma^2)\ , \quad \xi \in K_\gamma\ , \tag{4.9}$$

where

$$0 < \alpha \leq 1, \quad 0 < \beta \leq 1, \quad -\infty < \gamma < 2 \tag{4.10}$$

and $G_{\alpha,\beta}$ the completely monotonic function on $\mathbf{R}_+$ defined by

$$G_{\alpha,\beta}(t) = \sum_{k=0}^{\infty} \frac{(-t^\beta)^k}{\Gamma(1+\alpha k)}\ , \quad t \in \mathbf{R}_+\ . \tag{4.11}$$

Ordinary grey noise is the special case $\beta = 1$. It is called standard if $\gamma = \alpha$ and

$$C_\alpha = \Gamma(1+\alpha)\, \sin \frac{\pi}{2}\alpha\ . \tag{4.12}$$

For the sake of simplicity $\tau_{\alpha,1,\gamma}$ will be denoted by $\tau_{\alpha,\gamma}$ and $\tau_{\alpha,\alpha}$ by τ_α. Obviously, τ_1 is Gaussian white noise.

Theorem. *Ordinary grey noise has moments of any order. They are given by*

$$\int_{K'_\gamma} d\tau_{\alpha,\gamma}(T) \prod_{k=1}^{2m+1} \langle T,\xi_k\rangle = 0\ , \quad m \in \mathbf{Z}_+ \tag{4.13}$$

and

$$\int_{K'_\gamma} d\tau_{\alpha,\gamma}(T) \prod_{k=1}^{2m} \langle T,\xi_k\rangle = \frac{2^m\, m!}{\Gamma(1+\alpha m)} \sum_{SP} \prod_{k=1}^{m} (\xi_{r_k},\xi_{s_k})_\gamma\ , \quad m \in \mathbf{Z}_+\ , \tag{4.14}$$

where SP is the set of the special permutations

$$(1,2,3,4,...,2m) \quad \rightarrow \quad (r_1,s_1,r_2,s_2,...,r_m,s_m) \tag{4.15}$$

satisfying

$$r_1 < r_2 < ... < r_m\ , \qquad r_k < s_k\ , \quad k = 1,2,...,m\ . \tag{4.16}$$

Proof. Apply the operator D_n defined by

$$D_n f = i^{-n} \frac{\partial^n}{\partial a_1 \partial a_2 \ldots \partial a_n} f \Big|_{a_1=a_2=\ldots=a_n=0} \tag{4.17}$$

to

$$\int_{K'_\gamma} d\tau_{\alpha,\gamma}(T) \exp\left(i\langle T, \sum_{k=1}^{n} a_k \xi_k\rangle\right) = F_\alpha(\|\sum_{k=1}^{n} a_k \xi_k\|_\gamma^2) . \tag{4.18}$$

This yields (4.13) for $n = 2m + 1$ and (4.14) for $n = 2m$.

Remark. From (4.11) it is evident that generalized grey noise with $\beta < 1$ does not possess moments of any order.

5. Grey Brownian Motion

We denote by (L^2_α) the Hilbert space of functions $F: K'_\alpha \to \mathbf{C}$ square integrable with respect to standard grey noise τ_α:

$$(L^2_\alpha) = L^2\left(K'_\alpha, \tau_\alpha\right) . \tag{5.1}$$

Let p be a polynomial in n indeterminates with complex coefficients. For $\xi_1, \ldots, \xi_n \in K_\alpha$ define $P: K'_\alpha \to \mathbf{C}$ by

$$P(T) = p(\langle T, \xi_1\rangle, \ldots, \langle T, \xi_n\rangle) . \tag{5.2}$$

In the case $n = 1$ and $p(x) = x$ we denote P simply by ξ. From the theorem of Section 4 it follows immediately that all polynomials P belong to (L^2_α). In particular, we have from (4.14)

$$\int_{K'_\alpha} d\tau_\alpha(T)\, |\xi(T)|^2 = \frac{2}{\Gamma(1+\alpha)} \|\xi\|^2_\alpha , \quad \xi \in K_\alpha . \tag{5.3}$$

This relation allows to define new (L^2_α)-functions as limits. Take a sequence $\xi_k \in K_\alpha$, $k \in \mathbf{N}$, converging to $\xi \in H_\alpha$ in the norm $\|\cdot\|_\alpha$ of H_α. It follows from (5.3) that this sequence is Cauchy in (L^2_α), hence converges to a limit function conveniently also denoted by ξ. In this way (5.3) extends to all of H_α. In particular, the indicator functions $\chi_{[a,b)}$ of half-open intervals with $0 \le a < b$ belong to H_α. From

$$\chi_{[a,b)}(t) = \begin{cases} 1, & a \le t < b \\ 0, & \text{else} \end{cases} \tag{5.4}$$

we obtain

$$\tilde{\chi}_{[a,b)}(\omega) = (2\pi)^{-1/2}\, e^{i\omega a}\,(i\omega)^{-1}\,(e^{i\omega(b-a)} - 1)\ . \tag{5.5}$$

This leads to

$$\|\chi_{[a,b)}\|_\alpha^2 = (b-a)^\alpha \tag{5.6}$$

taking (4.1), $\gamma = \alpha$ and (4.12) into account. The element in (L^2_α) corresponding to $\chi_{[a,b)}$ will be denoted by $B^\alpha_{[a,b)}$.

Definition. Grey Brownian motion is the stochastic process B^α_t, $t \in \mathbf{R}_+$, with $B^\alpha_t := B^\alpha_{[0,t)}$, $t > 0$, $B^\alpha_0 = 0$. The underlying probability space is $(K'_\alpha, \mathcal{B}, \tau_\alpha)$ with $\mathcal{B}$ the Borel sets of K'_α and τ_α standard grey noise. For $\alpha = 1$ standard Brownian motion [1] is recovered.

From the definition it follows that $B^\alpha_t \in (L^2_\alpha)$. Furthermore, the relations

$$\int_{K'_\alpha} d\tau_\alpha(T)\, e^{i\lambda(B^\alpha_t(T) - B^\alpha_s(T))} = F_\alpha(\lambda^2(t-s)^\alpha) \tag{5.7}$$

and

$$\int_{K'_\alpha} d\tau_\alpha(T)\,(B^\alpha_t(T) - B^\alpha_s(T))^{2m} = \frac{(2m)!}{\Gamma(1+\alpha m)}\,(t-s)^{\alpha m}\ , \quad m \in \mathbf{N}, \tag{5.8}$$

hold for $0 \le s < t$. Actually, (5.8) has a generalization:

Theorem. *For any $p > 0$ and $0 \le s < t$ the equality*

$$\mathrm{E}_\alpha\left(|B^\alpha_t - B^\alpha_s|^p\right) = \frac{\Gamma(1+p)}{\Gamma(1+p\alpha/2)}\,(t-s)^{p\alpha/2} \tag{5.9}$$

holds.

Here E_α denotes expectation with respect to τ_α, i.e.,

$$\mathrm{E}_\alpha(F) = \int_{K'_\alpha} d\tau_\alpha(T)\, F(T)\ . \tag{5.10}$$

Proof. The r.h.s of (5.7) has the representation

$$F_\alpha(\lambda^2(t-s)^\alpha) = \int_{\mathbf{R}} dx\, \rho_\alpha(x)\, e^{i(t-s)^{\alpha/2}\lambda x} \tag{5.11}$$

where $\rho_\alpha = \rho_\alpha^1$ is defined in (3.23). Hence we obtain

$$\mathrm{E}_\alpha\left(|B_t^\alpha - B_s^\alpha|^p\right) = (t-s)^{p\alpha/2} \int_{\mathbf{R}} dx\, \rho_\alpha(x)\, |x|^p\ . \tag{5.12}$$

The explicit form (3.23) with $d = 1$ yields [3]

$$\int_{\mathbf{R}} dx\, \rho_\alpha(x)\, |x|^p = \frac{\Gamma(1+p)}{\Gamma(1+\alpha p/2)}\ . \tag{5.13}$$

Combining (5.12) and (5.13) leads to (5.9).

Corollary. *Let $p > 2/\alpha$. Then*

$$|B_t^\alpha(T) - B_s^\alpha(T)| \le \rho(T)\ |t-s|^\sigma \tag{5.14}$$

where $\rho(T)$ is finite τ_α-almost everywhere and σ can be any number satisfying $0 < \sigma < \alpha/2 - 1/p$. As p can be chosen arbitrarily large, σ can be arbitrarily close to $\alpha/2$.

This generalizes the well-known result for standard Brownian motion [1], [6] to the grey one. The proof is standard [6].

There is a well-known connection between Brownian motion B_t^1 and diffusion [1] which may be formulated as follows. Let $u_0 \in \mathcal{S}(\mathbf{R})$ and consider the expectation

$$u(x,t) = \mathrm{E}_1(u_0(x + B_t^1))\ . \tag{5.15}$$

Then $u\colon \mathbf{R} \times \mathbf{R}_+ \to \mathbf{R}$ is a solution of the diffusion equation

$$\frac{\partial u}{\partial t} = \frac{\partial^2 u}{\partial x^2} \tag{5.16}$$

satisfying the initial condition

$$u(\cdot, 0) = u_0\ . \tag{5.17}$$

Equivalently, u satisfies the integrodifferential equation

$$u(x,t) = u_0(x) + \int_0^t ds\, \frac{\partial^2}{\partial x^2} u(x,s)\ . \tag{5.18}$$

In [3] the fractional diffusion equation was introduced by generalizing (5.18) to

$$u(x,t) = u_0(x) + \frac{1}{\Gamma(\alpha)} \int_0^t ds\,(t-s)^{\alpha-1} \frac{\partial^2}{\partial x^2} u(x,s) \tag{5.19}$$

with $0 < \alpha < 1$. The connection between grey Brownian motion B_t^α and fractional diffusion is given in the following proposition.

Proposition. *Let $u_0 \in \mathcal{S}(\mathbf{R})$ and*

$$u(x,t) = \mathrm{E}_\alpha(u_0(x + B_t^\alpha)) \, . \tag{5.20}$$

Then $u: \mathbf{R} \times \mathbf{R}_+ \to \mathbf{R}$ solves the fractional diffusion equation (5.19).

Proof. Insert the representation

$$u_0(y) = (2\pi)^{-1/2} \int_{\mathbf{R}} d\lambda\, e^{-i\lambda y}\, \tilde{u}_0(\lambda) \tag{5.21}$$

into (5.20) and interchange λ-integration and expectation E_α. Taking (5.7) with $s = 0$ into account leads to

$$u(x,t) = (2\pi)^{-1/2} \int_{\mathbf{R}} d\lambda\, e^{-i\lambda x}\, F_\alpha(\lambda^2 t^\alpha)\, \tilde{u}_0(\lambda) \, . \tag{5.22}$$

Now, $F_\alpha(\lambda^2 t^\alpha)$ satisfies the integral equation

$$F_\alpha(\lambda^2 t^\alpha) = 1 - \lambda^2 \frac{1}{\Gamma(\alpha)} \int_0^t ds\,(t-s)^{\alpha-1}\, F_\alpha(\lambda^2 s^\alpha) \tag{5.23}$$

as is easily verified using the series representation

$$F_\alpha(\lambda^2 t^\alpha) = \sum_{k=0}^{\infty} \frac{(-\lambda^2 t^\alpha)^k}{\Gamma(1+\alpha k)} \, . \tag{5.24}$$

Obviously, the combination of (5.22) and (5.23) yields (5.19) using the inverse relation of (5.21).

Remark. With u_0 also $u(\cdot, t)$ belongs to $\mathcal{S}(\mathbf{R})$ for fixed t. However, the proposition holds for a larger class of initial conditions which we do not specify here. In any case positivity and normalization are conserved [3]: If $u_0(x) \geq 0$ then $u(x,t) \geq 0$ and if

$$\int_{\mathbf{R}} dx\, u(x,t) = U_0 < \infty \tag{5.25}$$

then

$$\int_{\mathbf{R}} dx\, u(x,t) = U_t < \infty \quad \text{and} \quad U_t = U_0 \ . \tag{5.26}$$

6. Remarks and Outlook

In Section 4 the construction of the Hilbert space triple $K_\gamma \subset H_\gamma \subset K'_\gamma$ was based on the Schwartz space $H^0 = \mathcal{S}(\mathbf{R}) = \mathcal{S}(\mathbf{R},\mathbf{R})$ (exhibiting explicitly the range of the functions under consideration) equipped with the scalar product (4.1). A possible generalization is to start from $H^0 = \mathcal{S}(\mathbf{R},\mathbf{R}^d)$, $d > 1$, and replacing (4.1) by

$$(\xi,\eta)_\gamma = \sum_{k=1}^{d} (\xi_k,\eta_k)_\gamma \ , \quad \xi = (\xi_1, \xi_2, \dots, \xi_d) \in \mathcal{S}(\mathbf{R},\mathbf{R}^d) \ . \tag{6.1}$$

In this way we obtain grey noise and grey Brownian motion with values in $\mathbf{R}^d$. Most of the equations of Sections 4 and 5 remain unchanged. They only have to be reinterpreted, e.g., (5.7) with $s = 0$

$$\int_{K'_\alpha} d\tau_\alpha(T) \exp\Big(i \sum_{k=1}^{d} \lambda_k B_t^{\alpha(k)}(T)\Big) = F_\alpha\Big(t^\alpha \sum_{k=1}^{d} \lambda_k^2\Big) \tag{6.2}$$

with $B_t^{\alpha(k)}$ obtained in the same way as in Section 5 by using the vector-valued indicator functions

$$\chi^{(k)}_{[a,b)} = \begin{cases} e^{(k)}, & a \leq t < b \\ 0, & \text{else} \end{cases} \tag{6.3}$$

where $e^{(1)}, e^{(2)}, \dots, e^{(d)}$ is an orthonormal basis of $\mathbf{R}^d$. We note that only in the white case $\alpha = 1$ the components $B_t^{\alpha(k)}$ are independent. Another peculiarity pertaining only to white Brownian motion is the Markov property.

In (5.9) $|\cdot|$ is now the Euclidean norm on $\mathbf{R}^d$ and the prefactor has to be changed according to [3]

$$\int_{\mathbf{R}^d} d^d x \, \rho_\alpha^d(x) \, |x|^p = \frac{2^p \, \Gamma((d+p)/2) \, \Gamma(1+p/2)}{\Gamma(d/2) \, \Gamma(1+\alpha p/2)} \ . \tag{6.4}$$

This reduces to (5.13) for $d = 1$ taking the doubling formula for the Γ- function [13] into account.

In the fractional diffusion equation (5.19) the second order derivative with respect to x has to be replaced by the Laplace operator.

Instead of raising the "space" dimension we may increase the "time" dimension (or both) by considering $H^0 = \mathcal{S}(\mathbf{R}^m, \mathbf{R})$ (or $H^0 = \mathcal{S}(\mathbf{R}^m, \mathbf{R}^d)$) with appropriate changes in the scalar product (4.1) and in the Fourier transform (4.2). In [1] there is an appendix treating $m > 1$ for the case $\alpha = 1$ (I refrain from citing more recent work mainly due to ignorance).

Another problem not treated is the support property of grey noise, i.e., the determination and characterization of the smallest closed subset $S \subset K'_\alpha$ whose complement has τ_α-measure zero.

Finally, one may consider the problem of introducing a kind of external interaction. For $\alpha = 1$ one possibility consists in replacing the Laplace operator Δ by $\Delta - V$ with $V\colon \mathbf{R}^d \to \mathbf{R}$ suitably chosen (e.g., infinitely differentiable and compact support). The modified diffusion equation

$$\frac{\partial u}{\partial t} = (\Delta - V) \, u \tag{6.5}$$

with initial condition $u(\cdot, 0) = u_0$ is then solved by

$$u(x,t) = \mathrm{E}\Big(\exp\big(- \int_0^t ds \, V(x + B_s)\big) u_0(x + B_t) \Big) \ , \tag{6.6}$$

where $\mathrm{E}(\cdot)$ is the expectation with respect to white noise and B_t is the associated Brownian motion. This is one version of the so-called Feynman-Kac formula [15]. The generalization to $\alpha < 1$ is by no means obvious. One obstacle is the lack of the semigroup property with respect to time of the solutions of the fractional diffusion equation.

A different generalization may be obtained as follows. The generalized Mittag-Leffler function $E_{\alpha,\beta}$ is defined by [16]

$$E_{\alpha,\beta}(z) = \sum_{n=0}^{\infty} \frac{z^n}{\Gamma(\alpha n + \beta)}\,, \quad z \in \mathbf{C}\,, \quad \alpha\,, \beta > 0\,. \tag{6.7}$$

For $\beta = 1$ the ordinary Mittag-Leffler function is recovered. We introduce

$$F_{\alpha,\beta}(z) = \Gamma(\beta)\, E_{\alpha,\beta}(-z)\,, \quad z \in \mathbf{C}\,. \tag{6.8}$$

Then it may be shown [14] that the restriction of $F_{\alpha,\beta}$ to $\mathbf{R}_+$ is completely monotonic and normalized if and only if $0 < \alpha \leq 1$ and $\beta \geq \alpha$. Furthermore, the underlying probability measure (Bernstein's theorem) may be explicitly determined [14]. Thus we may replace F_α by $F_{\alpha,\beta}$ wherever F_α is used to define a characteristic functional.

Interesting aspects of grey noise and grey Brownian motion in relation to other stochastic processes have been pointed out in refs. [17] and [18].

REFERENCES

[1] T. Hida, *Brownian Motion* (Springer, New York, Heidelberg, Berlin, 1980).

[2] W. Wyss, "The fractional diffusion equation," J. Math. Phys. **27**, 2782 (1986).

[3] W. R. Schneider and W. Wyss, "Fractional diffusion and wave equations," J. Math. Phys. **30**, 134 (1989).

[4] D. V. Widder, *The Laplace Transform* (University Press, Princeton, 1946).

[5] W. Feller, *An Introduction to Probability Theory and Its Applications*, Vol. II (John Wiley, New York, 1971).

[6] I. I. Gihman and A. V. Skorohod, *The Theory of Stochastic Processes I* (Springer, Berlin, Heidelberg, New York, 1974).

[7] L. Bieberbach, *Lehrbuch der Funktionentheorie*, Band II (Teubner, Leipzig, Berlin, 1927).

[8] H. Pollard, "The completely monotonic character of the Mittag-Leffler function $E_\alpha(-x)$," Bull. Amer. Math. Soc. **54**, 1115 (1948).

[9] V. M. Zolotarev, *One-dimensional Stable Distributions* (Amer. Math. Soc., Providence, 1986).

[10] W. R. Schneider, "Stable distributions: Fox function representation and generalization," in *Stochastic Processes in Classical and Quantum Systems*, edited by S. Albeverio, G. Casati, and D. Merlini, Lecture Notes in Physics, Vol. 262 (Springer, Berlin, 1986).

[11] H. M. Srivastava, K. C. Gupta, and S. P. Goyal, *The H-Functions of One and Two Variables with Applications* (South Asian, New Delhi, Madras, 1982).

[12] W. Wyss, private communication (1988).

[13] *Handbook of Mathematical Functions*, edited by M. Abramowitz and I. A. Stegun (Dover, New York, 1965).

[14] W. R. Schneider, "Complete Monotonicity of the Generalized Mittag-Leffler Function $E_{\alpha,\beta}(-t)$," preprint (1989).

[15] B. Simon, *Functional Integration and Quantum Physics* (Academic Press, New York, San Francisco, London, 1979).

[16] A. Erdély. W. Magnus, F. Oberhettinger, and F. G. Tricomi, *Higher Transcendental Functions*, Vol. III (McGraw-Hill, New York, Toronto, London, 1955).

[17] T. Kolsrud, "On a Class of Probabilistic Integrodifferential Equations," this volume.

[18] M. Yor, "W. Schneider's Grey Noise and Fractional Brownian Motion," Proceedings of the Easter Meeting on Probability: Edinburgh, 10th - 14th April, 1989 (to appear).

Existence of invariant measures for diffusion process with infinite dimensional state space.

Zhang Tu-sheng *

Department of Mathematics, University of Edinburgh,
The King's Buildings, Mayfield Road,
Edinburgh EH9 3JZ, Scotland.

* Supported by British SERC GR/G 00044.

Abstract. In this note, we consider a diffusion process which "informally" solves the SDE $dX_t = dW_t - A(X_t)dt + b(X_t)dt$ on a Banach space E. We characterize its generator and prove the existence of its invariant measure under some conditions.

§1. Introduction and Framework.

Let E be a Banach space with Borel σ-algebra $\mathcal{B}(E)$ and topological dual E'. Let μ be a mean zero Gaussian probability measure on $(E, \mathcal{B}(E))$, that is, each $l \in E'$ has a mean zero Gaussian distribution in R under μ and $\text{supp}\,\mu = E$. Let H_1 be the real Hilbert space obtained by completing E' w.r.t. the norm associated with the inner product

$$(1.1) \qquad \langle k_1, K_2\rangle_{H_1} := \int_E K_1(z)\, K_2(z)\, \mu(dz) \quad \text{for all} \quad K_1, K_2 \in E'.$$

For $h \in H_1$, we define

$$(1.2) \qquad X_h := \lim_{n\to\infty} K_n(z) \quad \text{in} \quad L^2(E;\mu) \qquad (K_n \in E',\ K_n \to h \ \text{in}\ H_1).$$

Now, given a self-adjoint operator A on H_1 such that $A \geqslant c\, Id_{H_1}$ for some $c \in \,]0,\infty[$ and consider the associated semigroup e^{-tA}, $t \geqslant 0$. Then there exist associated operators $T_t = \Gamma(e^{-tA})$, $t \geqslant 0$, on $L^2(E;\mu)$ satisfying

$$(1.3) \qquad T_t I = 1, \quad \Gamma(e^{-tA})(:\prod_{0=1}^{n} X_{h_j}:) =: \prod_{j=1}^{n} X_{e^{-tA}h_j}:$$

Here $:\prod_{0=1}^{n} X_{h_j}:$ is the wick product of X_{h_j}. We refer readers to [12] for details. From [3], we know that T_t, $t \geqslant 0$ is a strongly continuous semigroup of symmetric contractions on $L^2(E;\mu)$. Let L denote the generator of T_t. L is the so-called second quantization of A. To determine the Dirichlet form $(\mathcal{E}, \mathcal{D}(\mathcal{E}))$ associated with $(T_t)_{t\geqslant 0}$, (i.e. $\mathcal{D}(\mathcal{E}) = D(\sqrt{-L})$, $\mathcal{E}(u,v) = (\sqrt{-L}u, \sqrt{-L}v)_\mu$, $u, v \in \mathcal{D}(\mathcal{E})$), we define

$$(1.4) \qquad H = D(\sqrt{A}) \text{ with inner product } \langle h_1, h_2\rangle_H := \langle \sqrt{A}h_1, \sqrt{A}h_2\rangle_{H_1}.$$

Under the following assumption

$$(1.5) \qquad H \subset E \text{ densely and continuously}$$

the authors in [3] proved that $(\mathcal{E}, \mathcal{D}(\mathcal{E}))$ is the closure of the form

$$(1.6) \qquad \mathcal{E}(u,v) = \int \langle \nabla u, \nabla v\rangle_H \, d\mu, \qquad u, v \in \mathcal{F}C_b^\infty.$$

(Here $\mathcal{F}C_b^\infty := \{u : E \to R \mid u(z) := f(l_1(z)\ldots l_m(z)),\ z \in E$ for some $m \in \mathbb{N},\ f \in C_b^\infty(R^m),\ l_1\ldots l_m \in E'\}$.

$\nabla u(z)$ is the unique element in H representing the continous linear map $h \to \frac{\partial u(z)}{\partial h} = \frac{du(z+sh)}{ds}\Big|_{s=0}$), and informally the diffusion associated with $(\mathcal{E}, \mathcal{D}(\mathcal{E}))$ solves the following SDE

$$(1.7) \qquad \begin{aligned} dX_t &= dW_t - A(X_t)dt \\ X &= z . \end{aligned}$$

More precisely, let $\tilde{\mu}_{t,t\geqslant 0}$ be cylinder measures on H such that

$$(1.8) \qquad \int_H \exp(i\langle h', h\rangle_H \,\tilde{\mu}_t(dh') = \exp(\tfrac{1}{2} t \|h\|_H^2) \text{ for all } h \in H.$$

Assume that

$$(1.9) \qquad \text{Each } \tilde{\mu}_t,\ t > 0 \text{ extends a measure } \mu_t^* \text{ on } (E, \mathcal{B}(E)).$$

$$(1.10) \qquad K = E' \cap \mathcal{D}(A) \text{ is dense in } E'.$$

If we denote by $M = \{\Omega, \mathcal{F}, \mathcal{F}_t, X_t, P_z, \ z \in E\}$ the diffusion associated with $(\mathcal{E}, \mathcal{D}(\mathcal{E}))$ (here $\Omega = C([0,\infty) \to E)$, $X_t(w) = W(t)$; $\mathcal{F}_t = \sigma(X_s, s \leq t))$. Then there exists a $\mathcal{F}_t$-Brownian motion W starting at $0 \in E$ with covariance $< >_H$ such that

$$X_t = z + W_t + N_t, \quad t \geq 0$$

(1.11) $\qquad P_z$- a.e. for q.e. $z \in E$ and $k \in E'$

$$K(N_t) = -\int_0^t X_{AK}(X_s)ds \quad t \geq 0 .$$

Let $b(z)$ be a measurable function $E \to H$ with $\sup_{z \in E} |b(z)|_H = M < +\infty$.

In this note, we study the diffusion process which is a perturbation of M by adding a drift term b, and informally solve the following SDE

(1.12) $$dX_t = dW_t - A(X_t)dt + b(X_t)dt.$$

In order to state it precisely, we need the following condition for the stochastic integral:

(1.13) there exists an increasing sequence $\{P_n, \ n = 1, 2, \ldots\}$ of finite dimensional orthogonal projections of H such that

(i) P_n converges strongly to I in H

(ii) each P_n extends to a projection, still denoted by P_n, of E with $\|P_n\|_{E,E} \leq 1$ and $\|P_n z - z\|_E \to 0$ as $n \to \infty$.

Remark 1. All the conditions above are satisfied if we let μ be a Gaussian measure on $S'(R^d)$ with covariance space H_{-1} which is the Sobolev space of order -1, and let $A = (-\Delta + m^2)^{\alpha/2}$ on H_1, because in this case $H = L^2(R^d; dx)$ and E can be chosen to be $S_{-n}(R^d)$ for some $n \in \mathbb{N}$ (we refer the reader to [2] for this example).

Now we introduce a new probability measure Q_z on the path space Ω by(1.14) $\qquad Q_z\big|_{\mathcal{F}_t} = M_t \, P_z\big|_{\mathcal{F}_t}$ for all $t \geq 0$.

Here $M_t = \exp\{\int_0^t b(X_s)\, dW_s - \frac{1}{2}\int_0^t |b(X_s)|_H^2\, ds\}$ is a martingale. Note that the stochastic integral $\int_0^t b(X_s)\, dW_s$ is well defined because of assumption (1.13), see [7]. Define a semigroup on $\beta_b(E)$ through

(1.15) $$P_t f(z) = E_{Q_z}[f(X_t)] \quad \text{for } \mu\text{- a.e. } z.$$

The aim of this paper is to give the explicit expression of the generator of semi-group P_t and prove the existence of an invariant measure for P_t under some conditions.

Similar problems have been studied by I. Shigekawa, in [11] and R. V. Vintschger in [14], when $A = I$ and (H, E, μ) is an abstract Weiner space. In that situation their method depends heavily on Malliavin's calculus, which is well developed. In our case, no Malliavin calculus can be used. The main tool we use here is the theory of Dirichlet form.

§2. Main Results.

First we have

Lemma 2.1. $\{P_t, t \geqslant 0\}$ can be extended continuously to a semigroup on $L^2(E;\mu)$. Moreover, P_t is strongly continuous on $L^2(E;\mu)$.

Proof. For $u \in \mathscr{B}_b(E)$, we have

$$\int_E (P_t u(z))^2\, \mu(dz) = \int_E (P_z[u(X_t)\exp\{\int_0^t b(X_s)dW_s - \frac{1}{2}\int_0^t |b|_H^2(X_s)ds\}])^2\, \mu(dz)$$

$$\leqslant \int_E P_z[u(X_t)^2 \exp\{\int_0^t |b|_H^2\, (X_s)ds\}]\, P_z[\exp\{2\int_0^t b(X_s)dW_s - \frac{1}{2}\int_0^t |2b|_H^2(X_s)ds\}]\, \mu(dz)$$

$$\leqslant e^{tM^2}\int_E P_z[u(X_t)^2]\, \mu(dz) = e^{tM^2}\int_E T_t u^2(z)\, \mu(dz)$$

$$= e^{tM^2}\int_E u^2(z)\, \mu(dz)\ .$$

By this estimate, the first part of the lemma follows.

If $u \in C_b(E)$, then $P_t u(x) \to u(x)$ μ- a.e. as $t \to 0$, which implies that $P_t u \to u$ in $L^2(E;\mu)$ as $t \to 0$. Since $C_b(E)$ is dense in $L^2(E;\mu)$, we complete the proof of this lemma.

Let $\bar{A}$ be the infinitesimal generator of semigroup P_t in $L^2(E;\mu)$, $\mathscr{D}(\bar{A})$ be the domain of $\bar{A}$. In the following, we will give a complete characterization of $(\bar{A}, \mathscr{D}(\bar{A}))$. For this end, we define the following operator on $L^2(E;\mu)$.

$$Au(z) = Lu(z) + \langle b(z), \nabla u(z)\rangle_H \tag{2.1}$$

$$\mathscr{D}(A) = \mathscr{D}(L) .$$

Here $\mathscr{D}(L)$ is the domain of operator L, which is known to us. The lemma 7.1 in [3] says that $\mathscr{A}$:= linear span of $\{\sin X_{h_1}, \cos X_{h_2} \ h_1, h_2 \in \mathscr{D}(A)\}$ is a core of L. Now we state and prove

<u>Lemma 2.2</u>. $\mathscr{D}(L) \subset \mathscr{D}(\bar{A})$ and $\bar{A}|_{\mathscr{D}(L)} = A$.

<u>Proof</u>. Take $u \in \mathscr{D}(L)$. We know that

$$u(X_t) - u(X_o) = M_t^u + \int_0^t Lu(X_s)ds \quad P_z\text{- a.e. for q.e. } z \in E. \tag{2.2}$$

Here M^u is a continuous square-integrable margingale. Furthermore,

$$M_t^u = \int_0^t \nabla u(X_s)dW_s, \quad P_z\text{- a.e. for } \mu\text{- a.e., } z \in E. \tag{2.3}$$

In fact, first (2.3) holds for $u \in \mathscr{F}C_b^\infty$ by the Theorem 5.4.3. in [5] and the definition of stochastic integral, then by limit procedure one gets (2.3) for $u \in \mathscr{D}(L)$. Therefore, for μ- a.e., $z \in E$

$$
\begin{aligned}
P_t u(z) - u(z) &= P_z[(u(X_t) - u(X_o))M_t] \\
&= P_z[M_t^u\ M_t] + P_z[M_t \int_0^t Lu(X_s)ds] \\
&= P_z[\int_0^t \nabla u(X_s)dW_s \cdot \int_0^t M_s b(X_s)dW_s] + \int_0^t P_z[M_s Lu(X_s)]ds \\
&= P_z[\int_0^t \langle b(X_s), \nabla u(X_s)\rangle_H\ M_s ds] + \int_0^t P_z[M_s Lu(X_s)]ds \\
&= \int_0^t P_s\ Au(z)ds\ .
\end{aligned}
$$

By the strong continuity of P_t, we have

$$
\left|\frac{P_t u - u}{t} - Au\right|_{L^2} \leq \frac{1}{t}\int_0^t \left|P_s Au - Au\right|_{L^2} ds \to 0
$$

$$
\text{as} \quad t \to 0\ .
$$

This means $\mathscr{D}(L) \subset \mathscr{D}(\bar{A})$, $\bar{A}\big|_{\mathscr{D}(L)} = A$.

For $u \in \mathscr{D}(L)$, introduce norm $\|u\|^2 = \int_E u^2(z)\ u(dz) + \int_E (Lu)^2(z)\ \mu(dz)$. Since $(L, \mathscr{D}(L))$ is a closed operator on $L^2(E;\mu)$, $(\mathscr{D}(L), \|\cdot\|)$ is a Hilbert space. Then A is a bounded linear operator from $(\mathscr{D}(L), \|\cdot\|)$ to $L^2(E;\mu)$, indeed we see that by the following

(2.4)
$$
\begin{aligned}
\int_E (Au(z))^2\ \mu(dz) &= \int_E (Lu(z) + \langle b(z),\ \nabla u(z)\rangle_H)^2\ \mu(dz) \\
&\leq 2\int_E (Lu)^2(z)\ \mu(dz) + 2\int_E (\langle b(z),\ \nabla u(z)\rangle_H)^2\ \mu(dz) \\
&\leq 2\int_E (Lu)^2(z)\ \mu(dz) + 2M^2\int_E |\nabla u(z)|_H^2\ \mu(dz) \\
&= 2\int_E (Lu)^2(z)\ \mu(dz) + 2M^2\int_E u(-Lu)(z)\ \mu(dz) \\
&\leq 2\int_E (Lu)^2(z)\ \mu(dz) + M^2\int_E (Lu)^2(z)\ \mu(dz) + M^2\int_E u^2(z)\ \mu(dz) \\
&\leq 2(2VM^2)\ \|u\|^2, \quad \text{for} \quad u \in \mathscr{D}(L).
\end{aligned}
$$

The following theorem is one of our main results.

Theorem 2.1. $(A, \mathscr{D}(L)) = (\bar{A}, \mathscr{D}(\bar{A}))$.

Proof. Take $\lambda > M$, we first show that A is a closed operator on $L^2(E;\mu)$ and λ is in the resolvent set of A. To this end, as in [11], we introduce the bilinear form on $\mathscr{D}(\mathscr{E}) \times \mathscr{D}(\mathscr{E})$ by

$$(2.5)\qquad \Phi_\lambda(u,v) = \int_E \langle \nabla u(z), \nabla v(z)\rangle_H \,\mu(dz) - \int_E \langle b(z), \nabla u(z)\rangle_H \,v(z)\,\mu(dz) + \lambda \int_E u(z)\,v(z)\,\mu(dz).$$

(Here $\nabla u(z)$ represents the Gateaux derivative of u, see [3] for detail.)

By the relation of L and $\mathscr{E}$, we know that for $u \in \mathscr{D}(L)$

$$\Phi_\lambda(u,v) = \int_E (\lambda - A)u(z)\,v(z)\,\mu(dz).$$

On the other hand, it is easy to see that there exist positive constants δ_1, δ_2 such that

$$(2.6)\qquad \Phi_\lambda(u,v) \leqslant \delta_1 |u|_{\mathscr{E}_1} |v|_{\mathscr{E}_1}, \quad \Phi_\lambda(u,u) \geqslant \delta_2 |u|_{\mathscr{E}_1}.$$

(Here $\mathscr{E}_1(u,u) = \mathscr{E}(u,u) + (u,u)_{L^2}$.)

Fix any $v \in L^2(E;\mu)$. Using the Lax-Milgram theorem and (2.6) we obtained that there is a $u \in \mathscr{D}(\mathscr{E})$ satisfying

$$(2.7)\qquad \Phi_\lambda(u,w) = \int_E v(z)\,w(z)\,\mu(dz \quad \text{for all } w \in \mathscr{D}(\mathscr{E}).$$

Hence,

$$\int_E \langle \nabla u(z), \nabla w(z)\rangle_H \,\mu(dz) - \int_E \langle b(z), \nabla u(z)\rangle_H \,w(z)\,\mu(dz) + \lambda \int_E u(z)\,w(z)\,\mu(dz)$$
$$= \int_E v(z)\,w(z)\,\mu(dz),$$

that is,

$$\mathscr{E}(u,w) = \int_E [\langle b(z), \nabla u(z)\rangle_H - \lambda u(z) + v(z)]\,w(z)\,\mu(dz).$$

Since $\bar{u}(z) = \langle b(z), \nabla u(z)\rangle_H - \lambda u(z) + v(z) \in L^2(E;\mu)$, it follows by the theory of Dirichlet form that $u \in \mathscr{D}(L)$ and $-Lu = \bar{u}$. Consequently, $(\lambda - A)\,u(z) = v(z)$, which indicates $\lambda - A$ is surjective. But (2.6) shows that $\lambda - A$ is injective. Hence $\lambda - A$ is bijective. Noting that inclusion $(\mathcal{D}(L), \|\cdot\|) \hookrightarrow L^2(E;\mu)$ is continuous, by closed graph theorem we have that $(\lambda-A)^{-1}$ is a bounded linear operator from $L^2(E;\mu)$ to $L^2(E;\mu)$ and A is closed. Moreover, λ is in the resolvent set of A.

On the other hand, we know by Lemma 2.1 that λ is also in the resolvent set of $\bar{A}$. But Lemma 2.2 implies $A \subset \bar{A}$, hence $(\lambda-A)^{-1} \subset (\lambda-\bar{A})^{-1}$. Since both operator $(\lambda-A)^{-1}$ and $(\lambda-\bar{A})^{-1}$ are defined everywhere on $L^2(E;\mu)$ we deduce that $A = \bar{A}$. The proof is finished.

For our second aim, existence of invariant probability measure of P_t, we restrict ourselves to the class of probability measures which are absolutely continuous with respect to μ. In this case, if an invariant measure exists, it must be unique, this is the Proposition 3.1 in [11] which is still true in the present framework because the kernel of our operator L is also the space of all constants. Moreover, the author of [11] proved that $\nu = \rho d\mu$ $(\rho \in L^2(E;\mu))$ is the invariant measure of P_t if and only if $\rho \in \mathrm{Ker}(A^*)$.

In the following, using the theorem about the stability of index, we will obtain the existence of non-trivial ρ belonging to $\mathrm{Ker}(A^*)$ under some conditions. The following theorem brings us some smooth property of elements in the domain $\mathcal{D}(A^*)$. It has been proved in [9] that $\mathcal{D}(\mathcal{E})$ is the Sobolev space of order 1 with respect to μ.

<u>Theorem 2.2.</u> $\mathcal{D}(A^*) \subset \mathcal{D}(\mathcal{E})$.

<u>Proof.</u> Fix $u \in \mathcal{D}(A^*)$, put $v = A^*u$. Then by definition $(Aw, u)_{L^2} = (w, A^*u)_{L^2} = (w,v)$ for $w \in \mathcal{D}(L)$, that is

$$\int_E w(z)\, v(z)\, \mu(dz) = \int_E (Lw)(z)\, u(z)\, \mu(dz) + \int_E \langle b(z), \nabla w(z)\rangle_H\, u(z)\, \mu(dz) .$$

Set $F(w) = \int_E (Lw)(z)\, u(z)\, \mu(dz)$, then $F(w)$ is a linear functional in $\mathcal{D}(L)$ and

$$(2.8)\quad |F(w)| = \left|\int_E wv\,d\mu - \int_E \langle b(z), \nabla w(z)\rangle_H\, u(z)\,\mu(dz)\right|$$
$$\leq \left(\int_E w^2 d\mu\right)^{1/2} \left(\int_E v^2 d\mu\right)^{1/2} + M\left(\int_E |\nabla w|_H^2(z)\,d\mu\right)^{1/2} \left(\int_E u^2 d\mu\right)^{1/2}$$
$$\leq \left(\left(\int_E v^2 d\mu\right)^{1/2} \vee M\left(\int_E u^2 d\mu\right)^{1/2}\right)\left[\left(\int_E w^2 d\mu\right)^{1/2} + \left(\int_E |\nabla w|_H^2(z)\,d\mu\right)^{1/2}\right]$$
$$\leq C\left(\int_E w^2 d\mu + \int_E |\nabla w|_H^2\,d\mu\right)^{1/2} = C\sqrt{\mathcal{E}_1(w,w)}\ .$$

(Here C is a positive constant.)

Since $\mathcal{D}(L)$ is dense in $\mathcal{D}(\mathcal{E})$, (2.8) implies that F can be extended to a bounded linear functional on $(\mathcal{D}(\mathcal{E}), \mathcal{E}_1)$, which is denoted by $\tilde{F}$. Thus there exists a function g such that

$$(2.9)\qquad g \in \mathcal{D}(\mathcal{E}) \ \text{ and } \ \tilde{F}(w) = \mathcal{E}_1(w,g), \quad \text{for any } w \in \mathcal{D}(\mathcal{E}).$$

In particular, for $w \in \mathcal{D}(L)$ we have

$$F(w) = \mathcal{E}_1(w,g) = (-Lw,g) + (w,g).$$

In other expression,

$$\int (Lw)(z)\ u(z)\ \mu(dz) = \int (-Lw)(z)\ g(z)\ \mu(dz) + \int_E w(z)\ g(z)\ \mu(dz)\ ,$$

i.e.

$$\int_E (Lw)(z)(u+g)(z)d\mu = \int_E wg\ d\mu\ , \quad \text{for } w \in \mathcal{D}(L)\ .$$

This indicates that $u + g \in \mathcal{D}(L^*)$ and $L^*(u+g) = g$. But L is self-adjoint, therefore $f = u + g \in \mathcal{D}(L)$. Consequently $u = f - g \in \mathcal{D}(\mathcal{E})$. This ends the proof.

<u>Lemma 2.3</u>. L is a Fredholm operator from $L^2(E;\mu)$ to $L^2(E;\mu)$ and

$$\int_E (Lu)^2(z)\ \mu(dz) \geq c \int_E |\nabla u|_H^2(z)\ \mu(dz) \quad \text{for } u \in \mathcal{D}(L).$$

(Here c is the positive constant such that $A \geq c1_{dH_1}$.)

<u>Proof</u>. Since $A \geq c1_{dH_1}$, we have that

$$\|e^{-tA}\|_{L^\infty(H_1)} \leqslant \frac{(2-1)^{1/2}}{(4-1)^{1/2}} = \frac{1}{\sqrt{3}} \quad \text{for} \quad t \geqslant \frac{\ln\sqrt{3}}{c} .$$

Applying Theorem 1.1 in [12] to $T_t = \Gamma(e^{-tA})$, we get that T_t is a contraction from $L^2(E;\mu)$ to $L^4(E;\mu)$ when $t \geqslant \frac{1}{c}\ln\sqrt{3}$. By this fact and Theorem 2 in [13], it follows that

$$\|T_t|_{\{1\}^\perp}\| \leqslant \sqrt{\frac{1}{3}} , \quad \text{for} \quad t \geqslant \frac{1}{c}\ln\sqrt{3} , \tag{2.10}$$

where $\{1\}^\perp$ is the orthogonal complement of constant in $L^2(E;\mu)$. Noting that the kernel of L only consists of constants and $T_t = e^{tL}$, by general theory of spectrum, (2.10) yields.

$$\int_E (-Lu)u\, d\mu \geqslant c \int_E (u-Eu)^2 d\mu \quad \text{for} \quad u \in \mathcal{D}(L) \tag{2.11}$$

(see the proof of Theorem 2 on page 320, [15]). For $u \in \bigcap_{n=1}^{\infty} \mathcal{D}(L^n)$, replacing u by $\sqrt{-L}u$ in (2.11), we deduce that

$$\int_E (Lu)^2 d\mu \geqslant c \int_E (\sqrt{-L}u)^2 = c \int_E |\nabla u|_H^2 d\mu . \tag{2.12}$$

On the other hand, Corollary 1 on page 203 [8] says that $\bigcap_{n=1}^{\infty} \mathcal{D}(L^n)$ is a core of L. So (2.12) holds for all $u \in \mathcal{D}(L)$.

Note that L is a self-adjoint operator on $L^2(E;\mu)$. In order to show L Is a Fredholm operator, it suffices to prove that the range of L is closed in $L^2(E;\mu)$, but this is implied by (2.11) and (2.12) because L is a closed operator. Hence the proof is finished.

Now we prove the following

Theorem 2.3. If $\frac{M^2}{c} < 1$, then P_t has an invariant probability measure which is absolutely continuous with respect to μ.

Proof. Let B be an operator on $L^2(E;\mu)$ defined by

$$Bu(z) = \langle b(z), \nabla u(z)\rangle_H , \quad \mathcal{D}(B) = \mathcal{D}(\mathcal{E}).$$

Then

$$\int_E (Bu)^2 d\mu \leqslant M^2 \int_E |\nabla u|_H^2(z)\ d\mu$$
$$\leqslant \frac{M^2}{c} \int_E (Lu)^2 d\mu \qquad \text{(by Lemma 2.3)}$$
$$\text{for} \quad u \in \mathcal{D}(L).$$

This shows that B is L-bounded. Since L Is a Fredholm operator with index zero, and $\frac{M^2}{c} < 1$, as an application of Theorem 5.22 in [6] we see that A is also a Fredholm operator and index $(A) = 0$. Therefore $\dim\text{kernel}(A) = \dim\text{kernel}(A^*) \geqslant 1$ because index $A = \dim\text{kernel}(A) - \dim\text{kernel}(A^*)$ and $A1 = 0$. In other words, $A^*\rho = 0$ has a non-trivial solution and we can assume $\rho \geqslant 0$ (see Lemma 3.1 in [11]), $\rho d\mu$ is the measure we want. The proof is completed.

Remark. We believe that the condition $\frac{M^2}{c} < 1$ can be removed by choosing the appropriate method, which we think merits further study of this subject.

Acknowledgment

I am grateful to Professors S. Albeverio and T. J. Lyons for very helpful suggestions and comments. I would like to give my special thanks to Professor M. Röckner for the stimulating discussion during a very pleasant stay in Bonn, especially for pointing out the proof of Lemma 2.3 to me.

References

[1] Albeverio, S., Høegh-Krohn, R.: Dirichlet forms and diffusion processes on rigged Hilbert space. Z. Wahrscheinlichkeitstheorie verw. Gebeite 40, 1-57, (1977).

[2] Albeverio, S., Röckner, M.: Classical Dirichlet forms on topological vector spaces - construction of an associated diffusion process. Prob. Th. Rel. Fields 83, 405-434 (1989).

[3] Albeverio, S., Röckner, M.: Stochastic differential equations in infinite dimensions: solutions via Dirichlet forms. Preprint Edinburgh 1989. Publication in preparation.

[4] Albeverio, S., Röckner, M.: New developments in theory and applications of Dirichlet forms. To appear in Proc. "Ascona July 1988".

[5] Fukushima, M.: Dirichlet forms and Markov processes. Amsterdam-Oxford-New York: NOrth Holland, 1980.

[6] Kato, T.: Perturbation theory for linear operators. Springer-Verlag Berlin-Heidelberg-New York, 1976.

[7] Kuo Hui-Hsiung: Gaussian Measures in Banach Spaces. Springer-Verlag, Berlin-Heidelberg-New York, 1975.

[8] Reed, M., Simon, B.: Methods of modern mathematical physics II, Fourier analysis, self-adjointness. New York-San Francisco-London, Academic Press 1975.

[9] Röckner, M., Zhang Tu-sheng: On uniqueness of generalized Schrödinger operators and applciation. Preprint, Edinburgh 1990.

[10] Röckner, M., Zhang Tu-sheng: Decomposition of Dirichlet processes on Hilbert space. Preprint, Edinburgh 1990.

[11] Shigekawa, I.: Existence of invariant measures of diffusions on abstract Wiener space. Osaka J. Math. 24, 37-59 (1987).

[12] Simon, B.: The $P(\Phi)_2$ Euclidean (Quantum) field theory. Princeton: Princeton University Press 1974.

[13] Simon, B.: A Remark on Nelson's Best Hypercontractive Estimates: Proc. Amer. Math. Soc., 55:2, 376-378 (1976).

[14] Vintschger, R.: Existence of Invariant measures for C[0,1] - valued Diffusions: Prob. Th. Rel. Fields 82, 307-313 (1989).

[15] Yosida, K.: Functional analysis. Springer-Verlag, New York (1971).

On nonlinear equations associated with Lie algebras of diffeomorphism groups of two-dimensional manifolds

R.M. KASHAEV, M.V. SAVELIEV*, S.A. SAVELIEVA*, A.M. VERSHIK***

* Institute for High Energy Physics, Protvino; ** Leningrad State University, Leningrad (U.S.S.R.)

ABSTRACT

We investigate here nonlinear equations associated through a zero curvature type representation with Lie algebras S_0 Diff T^2 and of infinitesimal diffeomorphisms of $(S^1)^2$, and also with a new infinite-dimensional Lie algebra. The latter is some symbiosis of the former two algebras. In particular, the general solution (in the sense of the Goursát problem) of the heavenly equation which describes self-dual Einstein spaces with one rotational Killing symmetry is discussed, as well as the solutions to a generalized equation. The paper is supplied with Appendix containing the definition of the continuum graded Lie algebras and the general construction of the nonlinear equations associated with them.

1. The symmetries generated by infinite-dimensional Lie algebras of diffeomorphism groups of two-dimensional manifolds have been of increasing interest in theoretical physics in the last few years. Suffice it to mention their applications under investigation of the evolution equations for an incompressible fluid on a manifold; of extended objects (strings, membranes, etc.) in gauge field theories and in statistical physics; of extended conformal symmetries and higher spin fields in a continuous limit; of classical and quantum gravity; etc. It is interesting to note here that in many of these problems one and the same algebra figures, though in different aspects.

This is the algebra S_0 Diff T^2, the infinitesimal area-preserving diffeomorphisms of the two-dimensional torus T^2, which is isomorphic to the Poisson brackets algebra on T^2 as well as to the simplest continuous limit of the series A. Naturally, the realizations of these algebras can be completely different, while their identification as $\mathbb{Z}$-graded algebras is ascertained, probably in the most simple way, in the framework of an axiomatics of the continuum Lie algebras ([1], [2,]).

In this paper, considering the Poisson brackets algebra as a continuum Lie algebra we construct a new infinite-dimensional Lie algebra which looks like a symbiosis of the algebra S_0 Diff T^2 and of the algebra of infinitesimal diffeomorphisms of $(S^1)^2$. The Fourier components of the elements of all these algebras generate natural two-indexed generalizations of the Virasoro algebra, some of which were known previously (see e.g. [3], [5]). Let us stipulate from the very beginning that we will not consider here their central extensions.

In what follows we investigate the nonlinear equations associated with Lie algebras in question by means of a zero curvature type representation. For the present, the only known equation among them is, perhaps, the so-called "heavenly" equation (7) which, for the first time, appeared probably in [5] as an equation completely defining the self-dual (real Euclidean) Einstein spaces with one rotational Killing symmetry. This equation has intensely been discussed for almost a decade in the physical literature (see e.g. [7]) in connection with its role in the theory of relativity. However, only the simplest special solutions, like the Eguchi-Hanson gravitational instanton, to equation (7) have been known up to now. In the discussed context this equation is related, in fact, with the algebra S_0 Diff T^2. Note that equation (7) independently appeared also in paper [8] in view of integrable (in the sense of Liouville) Hamiltonian systems associated with the Poisson brackets algebra on T^2, and in paper [9] as a direct continuous limit of the two-dimensional Toda lattice for the series A_n for $n \to \infty$. Finally, in paper [10] a continuous analogue (9) of the two-dimensional Toda lattice was introduced and integrated (in the sense of a formal solution of the Goursát Problem). The simplest special case of this analogue is just equation (7), for which we will obtain here an expression for the general solution simpler than those in [10].

The paper is supplied with Appendix where the definitions of the continuum Lie algebras are given in a quite general form together with the construction of the nonlinear systems associated with these algebras.

2. Consider the Lie algebra of the functions of two variables (t_1 and t_2) with the product [,] defined by their Poisson bracket

$$[U_1, U_2] = -c_0\{U_1, U_2\} \equiv -c_0(U_{1.1}U_{2.2} - U_{1.2}U_{2.1}) ,$$

where $U_{i,j} \equiv \frac{\partial u_i}{\partial t_j}$. Let us describe some heuristic arguments clarifying a transition from this initial bracket to the bracket of form (1) or (2) which we are interested in. For this aim parametrize these elements as

$$U_i = u_i(t) \exp \mathrm{i} \left[\frac{c_1}{c_0} t_2 - \frac{c_2}{c_0} t_1\right] , \; c_0 \neq 0 .$$

Then, define for the functions u_1 and u_2 a new bracket by the formula

$$\begin{aligned}
[u_1, u_2] = & \\
& - c_0 \left[\frac{\partial u_1}{\partial t_1}\frac{\partial u_2}{\partial t_2} - \frac{\partial u_1}{\partial t_2}\frac{\partial u_2}{\partial t_1}\right] \\
& - ic_1 \left[\frac{\partial u_1}{\partial t_1} u_2 - u_1 \frac{\partial u_2}{\partial t_1}\right] \\
& + ic_2 \left[\frac{\partial u_1}{\partial t_2} u_2 - u_1 \frac{\partial u_2}{\partial t_2}\right] \\
& \equiv -c_0\{u_1, u_2\} - ic_1(u_{1.1}u_2 - u_1 u_{2.1}) + ic_2(u_{1.2}u_2 - u_1 u_{2.2}) \ .
\end{aligned}$$

Note that the imaginary unit at c_0 and c_1 in the brackets given above was introduced to avoid its appearance in formula (1). From this formula we assume that $\hat{c} \equiv (c_0, c) \equiv (c_0, c_1, c_2)$ is an arbitrary three-dimensional vector in $\mathbb{R}^3, \mathbb{C}^3$ or $\mathbb{Z}^3$. Such a description of the Lie algebra is equivalent to its formulation in terms of the elements $X(u(t))$, or, which is the same, via their Fourier components $X_m(\phi)$, for which

$$\begin{aligned}
[X_{m_1}(\phi), X_{n_1}(\psi)] = X_{m_1+n_1}(&-ic_0(m_1\phi\psi_{,2} - n_1\phi_{,2}\psi) \\
&- ic_2(\phi_{,2}\psi - \varphi\psi_{,2}) + c_1(m_1 - n_1)\phi \ .\psi) \ .
\end{aligned} \tag{1}$$

Here the functions ϕ and ψ depend on variable t_2. This form of writing down the algebra under discussion is nothing else but its formulation as the continuum $\mathbb{Z}$-graded Lie algebra $g = \bigoplus_{m_1 \in \mathbb{Z}} g_{m_1}$. If one carries out the repeated Fourier expansion, now over the variable t_2 and with the identification $Y_m = X_{m_1}(e^{im_2t_2})$, then the following relations come from (1):

$$[Y_m, Y_n] = (c_0 m \times n + c(m - n))Y_{m+n} \ . \tag{2}$$

Here $m = (m_1, m_2)$ and $n = (n_1, n_2)$ are 2-dimensional integer vectors, $m \times n \equiv m_1 n_2 - m_2 n_1$.

The introduced algebra can be reduced to several known particular cases by choosing some components of the constant 3-vector $\hat{c}$ equal to zero. For example, if $c = 0$ we come to the algebra S_0 Diff T^2, if $c_0 = 0$ and one of the components of c equals zero we obtain the Witt algebra (i.e. centre-free Virasoro algebra), if $c_0 = 0$ we have to do with the Ramos-Shrock algebra [5]. In other words, the mentioned (linear) manner of "mixing" of the algebras S_0 Diff T^2 and Diff $(S^1)^2$ gives again Lie algebra (2) which in what follows will be denoted as $W^{(c_0,c_1,c_2)}$.

It is evident that the case with $c_0 \neq 0$ is reduced to the case with $c_0 = 1$ through the trivial substitution. Finally, there takes place the following proposition.

Proposition 1.
If $c_0 = 0$ and $\frac{c_1}{c_2}$ is an irrational number then the algebra $W^{(0.c_1.c_2)}$ for this case is not reduced to the Witt algebra being at the same time one-dimensional algebra (in the sense that the corresponding vector fields are taken along a one-dimensional foliation). If $c_0 = 0$ and $\frac{c_1}{c_2}$ is a rational number then such an algebra $W^{(0.c_1.c_2)}$ is reduced to the Witt algebra.

Now consider the nonlinear equations which are generated, in accordance with a group-algebraic approach [11], by means of the zero curvature type representation

$$\left[\frac{\partial}{\partial z_+} + A_+ \ , \ \frac{\partial}{\partial z_-} + A_-\right] = 0 \tag{3}$$

with the functions $A_\pm(z_+ \ , z_-)$ taking values in subspaces $\bigoplus\limits_{m_1 \geq 0} g_{\pm m_1}$ of the algebra $g = W^{(\bar{c})}$. Here we will confine ourselves by consideration of the local part $\hat{g} = g_{-1} \oplus g_0 \oplus g_{+1}$ of $W^{(\bar{c})}$, i.e., we choose

$$A_\pm = X_0(\phi_0^\pm) + X_{\pm 1}(\phi_1^\pm) \ . \tag{4}$$

Substituting expansion (4) in representation (3) we obtain the following system of three equations:

$$\phi^-_{0.z_+} - \phi^+_{0.z_-} + c_2\left[\{\phi_0^+, \ \phi_0^-\} + \{\phi_1^+, \ \phi_1^-\}\right] + \left(c_0\frac{\partial}{\partial t} + 2c_1\right)\phi_1^+\phi_1^- = 0 \ , \tag{5 - 1}$$

$$\phi^\pm_{1.z_\pm} \pm \phi_1^\pm \left(c_0\partial/\partial t + c_1\right)\phi_0^\mp + c_2\{\phi_0^\pm, \ \phi_1^\pm\} = 0 \tag{5 - 2}$$

where $\{\phi^+, \ \phi^-\} \equiv \phi^+_{.t}\phi^- - \phi^+\phi^-_{.t}$. Equations (5-2) serve for determination of the functions $\phi_0^\pm$, which, being substituted in (5-1), lead to the unknown nonlinear equation. It will be convenient to consider the cases $c_2 = 0$ and $c_2 \neq 0$ separately.

3a.
Case of algebras $W^{(c_0, c_1, 0)}$. Here equations (5-2) are rewritten as

$$(\ln\phi_1^\pm)_{,z_\mp} = \pm\left(c_0\frac{\partial}{\partial t} + c_1\right)\phi_0^\mp \ .$$

Owing to this formula the equation we are interested in results from (5-1) and can be presented in terms of the function $\rho \equiv \ln\phi_1^+\phi_1^-$ as

$$\Delta\rho + \left(2c_1^2 + 3c_0c_1\frac{\partial}{\partial t} + c_0^2\frac{\partial^2}{\partial t^2}\right)\exp\rho = 0 \ , \tag{6}$$

where $\Delta \equiv \frac{\partial^2}{\partial z_+ \partial z_-}$ is two-dimensional Laplacian. This equation has two important subcases.

i) If $c_1 = 0$, that is for the algebra S_0 Diff $T^2 \simeq W^{(c_0,0,0)}$, then it coincides, after a trivial changing of the variables, with the heavenly equation ([6])

$$\Delta \rho_h = (\exp \rho_h)_{,tt} \; . \tag{7}$$

ii) If $c_0 = 0$, that is for the Witt algebra $W \simeq W^{(0,c_1,0)} \simeq W^{(0,0,c_1)}$, then equation (6) reduces to the Liouville equation

$$\Delta \rho_L = 2 \exp \rho_L \; , \tag{8}$$

whose general solution was constructed about 150 years ago. It can be expressed as

$$\exp\left[-\frac{\rho_L}{2}\right] = \exp\left[-\frac{\rho_0}{2}\right]\left[1 - \int^{z_+} \int^{z_-} dz'_+ dz'_- \exp \rho_0(z'_+,\, z'_-)\right]$$

in terms of the solution $\rho_0(z_+,\, z_-)$ of the Laplace equation $\Delta \rho_0 = 0$.

Note that equation (6) is a particular case of a continuous analogue

$$\Delta \rho - K \exp \rho = 0 \tag{9}$$

of the two-dimensional Toda lattice. This analogue was proposed and integrated in a formal series in Ref. [10] for invertible operator K. It is remarkable that for equation (6) the corresponding K cannot be reduced (for any value of the parameters c_0 and c_1) to the differential operator of the first order. In this connection let us also remind that the case with the Cartan operator K proportional to $\frac{\partial}{\partial t}$ corresponds to the continuum Lie algebra of temperate but not constant (as for $K \sim \frac{\partial^2}{\partial t^2}$) growth ([1]).

Before going to the description of the general solution for equation (7) let us say a few words on it. Firstly, this equation can be rewritten in the following two equivalent forms

$$\Delta \Theta \equiv \exp \Theta_{,tt}, \; \Delta \phi = \phi_{,tt} \exp \phi_{,t} \; . \tag{10}$$

Here Θ and ϕ, as well as $\rho_h = \Theta_{,tt} = \phi_{,tt} = \phi_{,t}$, are functions of three spatial variables $x_1 \equiv z_+ + z_-$, $x_2 \equiv -i(z_+ - z_-)$ and $x_3 \equiv it$. Secondly, the Lagrangian density for equation (7) has the form

$$L = \int dt \left[-\frac{1}{2}\Theta_{,z_+t},\; \Theta_{,z_-t} + \exp \Theta_{,tt}\right] \; .$$

The system under consideration possesses an improved energy momentum tensor

$$W_2^{\pm\pm} = c \int dt \left[\frac{\partial^2 \Theta(t)}{\partial z_\pm^2} - \frac{1}{2}\frac{\partial \Theta(t)}{\partial z_\pm}\frac{\partial^2}{\partial t^2}\frac{\partial \Theta(t)}{\partial z_\pm}\right]$$

with a vanishing trace, $W_2^{+-} = 0$, on shell. This tensor is nothing but the integral of the second order ([10]) of the characteristic equation which corresponds to equation (7).

And, finally, the Riemannian metric corresponding to such a system is given by the formula ([6], [7])

$$ds^2 = \frac{\partial \rho_h}{\partial x_3}\left[(dx_1^2 + dx_2^2)\exp \rho_h + dx_3^2\right]$$
$$+ \left(\frac{\partial \rho_h}{\partial x_3}\right)^{-1}\left[\pm\left(\left(-\frac{\partial \rho_h}{\partial x_2}\right)dx_1 + \left(\frac{\partial \rho_h}{\partial x_1}\right)dx_2\right) + dx_4\right]^2 .$$

Clearly, equation (7) admits a special solution of the form

$$\exp \rho_h(z_+,\ z_-;\ t) = \left(d_0 + d_1 t + \frac{t^2}{2}\right)\exp \rho_L(z_+,\ z_-)\ ,$$

which, in particular, describes the Eguchi-Hanson gravitational instanton

$$\exp \rho_h = \frac{1}{2}\,(x_3^2 - a^2)\,(1 + x_1^2 + x_2^2)^{-1} \ \text{with}\ x_3^2 \geq a^2\ .$$

A formal solution to this eqaution which is obtained from those constructed in [10] if $K = \frac{\partial^2}{\partial t^2}$ contained the corresponding number of integrations over the variables of type t at each term of the infinite series. It is clear that these integrations can be performed explicitly (in fact, as for the case of an arbitrary operator K with a support on the diagonal).

Let us introduce the following notations:

$$\Phi_{m,\omega^{-1}(m)} = \exp\left[\rho_0^+(z_m^+;\ t) + \rho_0^-\left(z_{\omega^{-1}(m)}^-;\ t\right)\right]\ ,$$

where $\rho_0 \equiv \rho_0^+(z_+;\ t) + \rho_0^-(z_-;\ t)$ is the solution of the Laplace equation $\Delta\rho_0 = 0$; $z_0^\pm \equiv z_\pm$, ω is any permutation of the indices from 1 to $n-1$; $\theta(z)$ is the Heaviside function;

$$D_{n,m}^{l_m} = \begin{cases} \varepsilon_m(\omega) - \frac{\partial^2}{\partial t^2}, & l_m = 0 \\ -\theta\left[\omega^{-1}(m) - \omega^{-1}(l_m)\right]\frac{\partial^2}{\partial t^2}, & l_m \neq 0; \end{cases}$$

$$\varepsilon_m(\omega) = \begin{cases} 2 & \text{for all } 1 \leq l \leq m-1 \\ 1 & \text{not for all } 1 \leq l \leq m-1 \end{cases}$$

the inequality $\omega^{-1}(m) < \omega^{-1}(l_m)$ takes place.

Then we come to the

Proposition 2.
The solution of the Goursát (boundary value) problem for equation (7) has the form

$$\rho(z_+,\ z_-;\ t) = \rho_0(z_+,\ z_-;\ t) - \frac{\partial^2}{\partial t^2} \ln\left[1 + \sum_{n\geq 1}(-1)^n Q_n\right], \qquad (11)$$

$$Q_n = \int\cdots\int \prod_{m=1}^{n-1} dz_m^{\pm}\theta\left(z_{m-1}^{\pm} - z_m^{\pm}\right)\Phi_{oo}\sum_{\omega}\sum_{l_1=0}^{0}\sum_{l_2=0}^{1}\cdots$$
$$\cdots\sum_{l_{n-1}=0}^{n-2}\left[D_{n,1}^{l_1}\Phi_{1,\omega^{-1}(1)}\right]\left[D_{n,2}^{l_2}\Phi_{2,\omega^{-1}(2)}\right]\cdots\left[D_{n,n-1}^{l_{n-1}}\Phi_{n-1,\omega^{-1}(n-1)}\right].$$

In this, the operator $D_{n,m}^{l_m}\Phi_{m,\omega^{-1}(m)}$ acts on all operator factors $\left[D_{n,s}^{l_s}\Phi_{s,\omega^{-1}(s)}\right]$ with $s > m$, for which $l_s = m$.

The proof of the given proposition is based on the following formula from [10]:

$$Q_n = \int\cdots\int dz_n^{\pm}\Theta(z_{n-1}^{\pm} - z_n^{\pm})\prod_{m=1}^{n-1} dt_m dz_m^{\pm}\theta(z_{m-1}^{\pm} - z_m^{\pm})\cdot\sum_{\omega}\Phi_{m.\omega^{-1}(m)}(t_m)\Phi_{00}\times$$
$$\times\left\{\varepsilon_m(\omega)\delta(t-t_m) - K(t,\ t_m) - \sum_{l=1}^{m-1}K(t_l,\ t_m)\theta\left[\omega^{-1}(m) - \omega^{-1}(l)\right]\right\}.$$

Then the integration over the variables t_m, $1 \leq m \leq n-1$ can be performed if the kernel $K(t,t')$ of the operator K is of the δ-type, i.e. $K(t,\ t') = \sum\limits_{m\geq 0} c_m\left(\frac{\partial}{\partial t}\right)^m\delta(t-t')$. To clarify the structure of the terms in series (11) let us write the integrands $\hat{Q}_n$, for example, for the first three functions Q_n:

$$\hat{Q}_1 = \Phi_{00}\ .$$
$$\hat{Q}_2 = \Phi_{00}\left(2 - \frac{\partial^2}{\partial t^2}\right)\Phi_{11}\ ,$$
$$\hat{Q}_3 = \Phi_{00}\left\{\left[\left(2 - \frac{\partial^2}{\partial t^2}\right)\Phi_{22}\right]\left[\left(2 - \frac{\partial^2}{\partial t^2}\right)\Phi_{11}\right] - \right.$$
$$\left. - \left(2 - \frac{\partial^2}{\partial t^2}\right)\left(\Phi_{11}\frac{\partial^2}{\partial t^2}\Phi_{22}\right) + \left[\left(2 - \frac{\partial^2}{\partial t^2}\right)\Phi_{21}\right]\left[\left(2 - \frac{\partial^2}{\partial t^2}\right)\Phi_{12}\right]\right\}.$$

Let us make the following remark in connection with the convergence problem for the series in (11). This formula results from the representation [10]

$$\exp\left[-\Theta(z,\ t)\right] = \exp\left[-\Theta_0(z;\ t)\right]\left\{1 + \sum_{n\geq 1}(-1)^n\int\cdots\int\prod_{m=1}^{n} dz_m^{\pm}\times\right. \qquad (12)$$
$$\left.\times\theta(z_{m-1}^{\pm} - z_m^{\pm}) < t\left|X_+^{(1)}\cdots X_+^{(n)}X_-^{(n)}\cdots X_-^{(1)}\right|t>\right\}$$

for the general solution of the Goursát problem to equation (9). Here

$$X_{\pm}^{(m)} \equiv \int dt X_{\pm}(t) \exp \rho_0^{\pm}(z_m^{\pm};\ t),\ \ \rho_0^{\pm} \equiv K\Theta_0^{\pm}\ ;$$

vectors $|t>$ satisfy the relations

$$X_0(\phi)|t> = \phi(t)|t>,\ \ X_+(\phi)|t> = 0\ ,$$

and play the role of the highest weight vectors in the corresponding modules space. At the same time in the case under consideration, i.e. for $K = \frac{\partial^2}{\partial t^2}$, expression (12) is directly related with the analogous formula [11] for the solution of the Toda lattice for the series A_n. Here, in fact, if the functions $\Theta(t)$ and $X_{\pm}(t)$ have a support at the points $t = 1, 2, \ldots, n$, then (12) leads to the solution of the discrete case A_n in the form of a finite polynomial. In other words, the solutions for the series A_n are the partial sums of their continuous limit.

An interesting problem is to construct the solutions like the gravitational instantons with the topological charge N which are special (parametric) solutions. They correspond to a definite choice of the arbitrary functions $\rho_0^{\pm}$.

Note that solution (11) of the boundary value problem for equation (7) is determined via the solution of the Laplace equation which corresponds to the free Lagrangian, in other words, to the asymptotic values of noninteracting (free) fields $\rho_0^{\pm}$. Moreover, the fact that the general solution to this equation (as well as, in accordance with [10], for equation (9) with an arbitrary invertible K) depends on two arbitrary functions $\rho_0^{\pm}(z_{\pm};\ t)$ of two variables is conformed to that for the heavens of type III (see J.D. Finley, J.F. Plebanski. J. Math. Phys. 20 (1979), 1938).

3b. Case of algebras $W^{(c_0,c_1,c_2)}$ with $c_2 \neq 0$.
For this case it will be convenient to pass to the gauge with $\phi_0^+ = 0$, in which ϕ_1^- can be equated, for example, to 1. Here, of course, one sacrifices the symmetry of writing equations (5) over z_+ and z_-. Note, that the way back is always possible due to the form-invariance of representation (3) with respect to the gauge transformations $A_{\pm} \longrightarrow G^{-1}(A_{\pm} + \frac{\partial}{\partial z_{\pm}})G$ which do not violate the gradation spectrum of the functions $A_{\pm}$ iff $G(z_+, z_-)$ are generated by subalgebra g_0. Moreover, it will be also convenient for us to consider the examples with $c_0 = c_2$ and $c_0 \neq c_2$ separately.

i) $W^{(c_0,c_1,c_0)}$ with $c_0 \neq 0$.

It follows from equation (5-2) that $\phi_0^- = c_0^{-1}\frac{f_{,z_-}}{f_{,t}}$ where $f \equiv \phi_1^+ \exp\left[\frac{c_1}{c_0}t\right]$. Substituting this expression into (5-1) we obtain the sought equation

$$\left(\frac{f_{,z_-}}{f_{,t}}\right)_{,z_+} + 2c_0^2 f_{,t} \exp\left[-\frac{c_1}{c_0}t\right] = 0\,. \tag{13}$$

ii) $W^{(c_0,c_1,c_2)}$ **with** $c_0 \neq c_2,\ c_2 \neq 0$.
Let us put $c(c_0 - c_2)^{-1} \equiv d$ and denote

$$\exp\sigma \equiv (\phi_1^+)^{d_2} \exp(d_1 t)\,.$$

Then equation (5-2) in this case takes the form

$$(\exp\sigma)_{,z_-} = c_2(\phi_0^- \exp\sigma)_{,t},$$

that is

$$c_2^{-1}\exp\sigma = \Phi_{,t}\ ,\ \ \phi_0^- \exp\sigma = \Phi_{,z_-}.$$

whereof

$$\phi_0^- = \frac{c_2^{-1}\phi_{,z_-}}{\phi_{,t}},\ \ (\phi_1^+)^{d_2} = c_2\Phi_{,t}\exp(-d_1 t)\,.$$

Substituting these formulas into eq. (5-1) with $\phi_0^+ = 0$ and $\phi_1^- = 1$, we come to the equation

$$\left(\frac{\Phi_{,z_-}}{\Phi_{,t}}\right)_{,z_+} + c_2^{1+d_2^{-1}}\left[(c_0+c_2)\frac{\partial}{\partial t} + 2c_1\right][\Phi_{,t}\exp(-d_1 t)]^{\frac{1}{d_2}} = 0\,. \tag{14}$$

For the case of the Ramos-Shrock algebra, i.e., $W^{(0,c_1,c_2)}$], eq. (14) essentially becomes simpler. Here $d_2 = -1$, and using the variables τ, $\frac{\partial}{\partial\tau} = \exp\left[-\frac{c_1}{c_2}t\right]\frac{\partial}{\partial t}$, we have

$$\left(\frac{\Phi_{,z_-}}{\Phi_{,\tau}}\right)_{,z_+} + c_2\left(\frac{1}{\Phi_{,\tau}}\right)_{,\tau} = 0\,. \tag{15 - 1}$$

Hence, solving this equation as

$$\frac{\Phi_{,z_-}}{\Phi_{,\tau}} = c_2\Psi_{,\tau},\ \ \frac{1}{\Phi_{,\tau}} = -\Psi_{,z_+}\,,$$

it can be rewritten in an equivalent form

$$\Delta\Psi = c_2\Psi_{,\tau}^2\left(\frac{\Psi_{,z_+}}{\Psi_{,\tau}}\right)_{,\tau}\,. \tag{15 - 2}$$

Up to now we managed to find only special solutions to equation (15), namely

$$\begin{aligned} \phi &= (\mu + \lambda\tau)\exp(-\rho_0) \ , \\ \phi &= c_2(\tau + \mu)^{-1}\exp(-\rho_L) \ , \\ \phi &= c_2\lambda \mathrm{th}\ (\mu + \lambda\tau)\exp(-\rho_L) \ ; \end{aligned} \tag{16}$$

and

$$\begin{aligned} \phi &= \rho_0 + \lambda\tau \ , \\ \phi &= \rho_0 + \lambda_1 z_+ z_- + \lambda_2 \exp\left(\frac{\lambda_1}{c_2}\tau\right) \end{aligned} \tag{17}$$

$\mu, \lambda = \mathrm{const.}, \lambda \neq 0$.

APPENDIX.

The notion of the continuum Lie algebras has not yet entered into usual vocabulary of theoreticians. At the same time an inquisitive reader would be naturally interested in realizing Lie algebras of diffeomorphism groups of two-dimensional manifolds and nonlinear equations associated with them (and discussed in this paper) as a part of the general construction of the continuum Lie algebras and equations generated by these algebras. Therefore we will briefly remind the reader in the Appendix of the definition and the main relations of the continuum Lie algebras ([1], [2]), however in a more general formulation.

Let E be an arbitrary associative algebra over the field Φ; $g^{(m_0)} \equiv \bigoplus_{|i|\leq m_0} g_i$ is a direct sum of one dimensional (in a functional sense) subspaces over E whose elements $X_i(\phi)$, $\phi \in E$, satisfy the relations

$$[X_i(\phi),\ X_j(\psi)] = X_{i+j}\left[K_{ij}(\phi,\psi)\right] \tag{A.1}$$

for all $|i|$, $|j|$, $|i+j| \leq m_0$ and for all ϕ, $\psi \in E$. Here K_{ij} are some bilinear mappings $E \times E \to E$. We call $\hat{g} = g^{(m_0)}$ a (modified) local Lie algebra if its elements satisfy the anticommutativity condition and the Jacobi identity, i.e., in terms of the operators K,

$$\begin{aligned} &K_{ij}(\phi,\psi) = -K_{ji}(\psi,\phi) \ ; \\ &K_{k,j+i}(\chi,\ K_{ji}(\psi,\phi)) + K_{j,i+k}(\psi,\ K_{ik}(\phi,\chi)) + \\ &+ K_{i,k+j}(\phi,\ K_{kj}(\chi,\psi)) = 0 \ . \end{aligned} \tag{A.2}$$

Here all indices and their pair sums take values from $-m_0$ to $+m_0$; $\phi, \psi, \chi \in E$.

Now, let $\hat{g}$ be the minimal (in accordance with m_0) local Lie algebra which freely generates a Lie algebra $g'(E;K)$, i.e. $\hat{g}$ is the local part of g'. Denote by J the largest homogeneous ideal which has a trivial intersection with g_0. Then, by analogy with the contragredient case ($m_0 = 1$) [1], it is natural to call an algebra $g'(E;K)/J \equiv g(E;K)$ a continuum Lie algebra, and relations (A.1) with condition (A.2) the defining relations. Clearly, it is a $\mathbb{Z}$-graded Lie algebra, $g = \bigoplus_{m\in\mathbb{Z}} g_m$. Further we will consider only such algebras $g(E;K)$, for which E is an associative commutative algebra over the field $\mathbb{R}$ and $\mathbb{C}$; the mappings K_{ij} will as a rule be realized by the linear operators $E \to E$. Moreover, we will confine ourselves to the equations generated (in the framework of the group-algebraic approach ([11])) by the zero curvature type representation (3) with the functions

$$A_\pm = \sum_{0\le i\le m_0} X_{\pm i}(\phi_i^\pm) , \qquad (A.3)$$

taking values in the subspaces $\bigoplus_{0\le i\le m_0} g_{\pm i}$ of the local part $\hat{g}$ of Lie algebra g. Substituting expansion (A.3) into (3) with account of (A.1) and (A.2) we have

$$\phi^-_{0,z_+} - \phi^+_{0,z_-} + \sum_{0\le i\le m_0} K_{i,-i}(\phi_i^+,\ \phi_i^-) = 0 , \qquad (A.4-1)$$

$$\phi^\pm_{j,z_\pm} \mp \sum_{0\le i\le m_0-j} K_{\pm(i+j),\pm i}(\phi^\pm_{i+j},\ \phi_i^\mp) = 0 \ , 1\le j\le m_0 \qquad (A.4-2)$$

(cf.(5)). Here two equations in (A.4-2) serve for finding the functions $\phi_0^\pm$ (under appropriate conditions for the operators K_{ij}).

The remaining $2(m_0-1)$ equations in (A.4-2) together with (A.4-1), in which $\phi_0^\pm$ are expressed via the functions $\phi_j^\pm$, $1\le j\le m_0$, and their derivatives, represent the sought nonlinear system associated with the algebra $g(E;K)$.

Under the natural assumptions of the contragredient case, namely

$$\begin{aligned} &K_{01}(\phi,\psi) = -K_{0-1}(\phi,\psi) = \psi K_{01}\phi, K_{1-1}(\phi,\psi) = K_{1-1}(\phi\cdot\psi) , \\ &K_{00}(\phi,\psi) = 0 , \end{aligned} \qquad (A.5)$$

this system is reduced to the continuous analogue (9) of the Toda lattice. Here

$$Kf(t) \equiv K_{01}K_{1-1}f(t) \ , \ \exp\rho \equiv \phi_1^+\phi_1^- \ ,$$

and all functions depend on the variables z_+, z_- and t. (In accordance with the terminology adopted in [1], the operator K in this case is called the

Cartan operator). Let us especially note once more that here we do not speak about a continuous limit of the Toda lattice, for which K equals $\frac{\partial^2}{\partial t^2}$ if one considers the series A. We speak about an essentially more general situation when K is in arbitary integro-differential operator. Nevertheless, here it is possible, as we have already mentioned, to obtain a formal solution of the boundary value problem for equation (9) when the operator K is invertible. This solution is represented as an infinite series whose convergence properties are related with the subsequent restrictions on the form of the operator K. Moreover, by the analogy with the discrete case, i.e., for example, with the equations associated with the affine Kac-Moody algebras, it is natural to assume [1,10] that the integrability (or the convergence of the corresponding series) of their continuous analogous is also related with the conditions on the growth of the corresponding algebra. Of course, here the growth of the algebra is understood in the functional sense [1].

ACKNOWLEDGEMENTS.
In conclusion one of the authors (M.S.) would like to thank D.B. Fairlie and A.N. Leznov for the useful discussions. He is also indebted to I. Bakas and R.E. Shrock who acquainted him with their latest results.

REFERENCES

1. M.V. Saveliev, A.M. Vershik. Comm. Math. Phys. 126 (1989) 367; Preprints IHEP, Serpukhov, ICTP 89/55
2. M.V. Saveliev, A.M. Vershik. IHEP Preprint 89-193, Serpukhov 1989. (Phys. Lett. 143A (1990) 121)
3. D.B. Fairlie, P. Fletcher, C.K. Zachos. Preprint DTP-89/37; D.B. Fairlie, C.K. Zachos. Phys. Lett. 224B (1989) 101
4. V. Arnold. Ann. Inst. Fourier XVI – 1 (1966) 319
5. E. Ramos, R.E. Schrock. Preprint ITP-SB-88-67; E. Ramos, C.H. Sah, R.E. Shrock. Preprint ITP-SB-89-16
6. C. Boyer, J. Finley. J. Math. Phys. 23 (1982) 1126
7. J.D. Gedenberg, F.Das. Gener. Relat. Gravit. 16 (1984) 817; I. Bakas. UMDEPP 90-033, 1989
8. M.I. Golenisheva-Kutuzova, A.G. Reyman. Zap. Nauch. Semin. LOMI 169 (1988) 44
9. O.I. Bogoyavlensky. Izv. Acad. Nauk SSR, ser. Mat. 52 (1988) 712
10. M.V. Saveliev. IHEP Preprint 88-39, Serpukhov, 1988; Comm. Math. Phys. 121 (1989) 283

11. A.N. Leznov, M.V. Saveliev. Group Methods for Integration of Nonlinear Dynamical Systems. Moscow, Nauka, 1985; Acta Appl. Math. 16 (1989) 1

FIELDS OF LEFT - INVARIANT STANDARD BROWNIAN MOTION PROCESSES ON A SMOOTH BUNDLE OF COMPACT SEMI SIMPLE LIE GROUPS

J. Marion
D. Testard

Centre de Physique Théorique
CNRS Luminy - Case 907
13288 Marseille Cedex 9 - France

INTRODUCTION

The connection with the theory of brownian motion process on a compact semi simple Lie group G was the main tool in [3] in order to study the reducibility properties and, actually, to perform a canonical reduction of the Energy representation of $C^\infty(I,G)$, at least in the case $I = \mathbb{R}$ or $\mathbb{R}^+$ or S^1 and $G = SU(n)$. More precisely, the associated measure on C (I,G) is quasi-invariant with respect to left or right pointwise translations by smooth elements of $C^\infty(I,G)$; this simple remark allows to use, in this particular context of infinite dimensional Lie groups, classical techniques and methods of harmonic analysis of finite dimensional Lie - groups. This point of view was an important progress and introduced new methods to deal these problems, which were, up to this time, confined in an abstract non-commutative functional analysis point of view [11], [12]. In this subject - among others (see the S. ALBEVERIO's article in this volume) the contribution of R.HØEGH-KROHN was of considerable importance.

In this note, we show that by a straitforward extension of the method, one can treat not only the case of the "trivial" current group $C^\infty(M,G)$ on a compact connected manifold M but also current groups of smooth sections of a bundle $F(G) = \bigcup_{x\in M} G^x$ with basis M and fibers all isomorphic to the same fixed compact semi-simple Lie group G (for instance the bundle $P(G) = (PxG)/_G$ associated to a (non - trivial) smooth G-principal bundle P over M (see eg [10], [15]).

Let us briefly describe the results contained in this note.
As usually we denote by g the Lie - algebra of G : g is a finite dimensional Euclidean space with a bi-invariant scalar product <, > and the associated euclidean norm $|\ \ |$. With F(G) as before, we prove in section I, the existence of global measurable sections $\Lambda = (\Lambda_x)_x$ of Hom (F(G), MxG) *(a measurable trivialization of F(G))* such that $\forall x\in M$, Λ_x is a Lie groups isomorphism from G^x onto G. The infinitesimal version $(\lambda_x)_{x\in M}$, in this context, appears as a global measurable section of Hom ($F(\mathfrak{g})$, $Mx\mathfrak{g}$) satisfying for all $x\in M$.

c) λ_x is a Lie algebra isomorphism from $\mathfrak{g}^x$ (the Lie algebra of G^x) onto $\mathfrak{g}$.

ü) $\lambda_x \, Ad_x (\gamma_x) = Ad(\Lambda_x (\gamma(x)) \, \lambda_x$ for any $\gamma = (\gamma_x)_{x\in M} \in F(G)$.

In section II, using a particular measurable trivialization of F(G), one introduces the field $(\eta^x)_{x\in M}$ of left-invariant standard Brownian motion processes η^x on G^x and, as an application, one constructs new non-located, order 1 unitay representations of the current group $CH^{(1)}$ ([0,1]xM, F(G)) of continuous mappings ψ from [0,1]xM into F(G) such that, for any $x\in M$, the mapping $\psi(.,x)$ is in the Sobolev path group H^1 ([0,1], G^x) , in connection with the approach initiated in [4].

In section III, we construct Borel probability measures on the goup B(F(G)) of measurable sections fo F(G). This constitutes an approach completely different from the classical Hilbert Lie groups constructions [5], [6], [7] and appears as an introductory work to the study of quasi-invariance of these measures by left-pointwise-translation by smooth sections . Following an idea of R.HØEGH-KROHN this question together with an analysis concerning the characterisation amongs representations of B(F(G)) of those which remain irreducible when they are restricted to smooth sections, should be of great interest with respect to the general problem of classification of multiplicative integrals (see [2] [9], [11] and references therein).

1 MEASURABLE TRIVIALIZATIONS OF F(G)

Proposition 1

There exist global measurable sections $\Lambda = (\Lambda_x)_{x \in M}$ of the bundle Hom (F(G), MxG)

such that for each $x \in M$, the mapping Λ_x is a Lie group isomorphism from G^x onto G.

Proof

By local triviality of F(G), for each $x \in M$, the set of Isom (G^x, G) is not empty and one can find a finite sequence $\{(0_1, s_1)....(0_q, s_q)\}$ such that 0_i's over M and s_i (i = 1....q) are local smooth sections defined by 0_i and with values in $\bigcup_{x \in 0_i}$ Isom (G^x, G). One can even choose $(0_i)_{i=1...q}$ such that none of the 0_i's is covered by unions of 0_j $j \neq i$.

Then defining recursively $(A_1...A_q)$ by :
$A_1 = O_1$,
$A_{/R} = O_R\text{-}(A_1 \cup .... \cup A_{/R}$,
one gets a partition of M by non-empty Borel subsets $A_i \subset O_i$ and Λ : M→Hom (F(G), MxG) defined by $\Lambda_x = \Lambda_{x_i}$ if $x \in A_i$ satisfies the properties stated in the proposition.

Proposition 2

Let Λ be as in Proposition 1 and $\lambda_x = (d\Lambda_x)_{e_x}$ be the derivative of Λ_x at the unity e_x of G^x. Then :

i) λ is a global measurable section of the bundle Hom $(F(\mathfrak{g}), Mx\mathfrak{g})$ such that for each x in M λ_x is a Lie algebra isomorphism from $\mathfrak{g}^x$ onto $\mathfrak{g}$;

ii) λ commutes with the adjoint representations Ad_x (resp' Ad) of G^x into g^x (resp of G into g) in the following sense ($\forall$ $x \in M$), $\gamma_x \in G^x$:

$$\lambda_x \, Ad_x(\gamma_x) = Ad(\Lambda_x(\gamma_x))\, \lambda_x$$

Proof

The proof is direct.

In the sequel we will use the following terminology.

Definition 1

A pair (Λ, λ) satisfying the properties described in Prop 1 an 2 will be said to be a measurable trivialization of F(G).

Proposition 3

Let (Λ, λ) a measurable trivialization of F(G). Then :

i) the bilinear form $<, >^x$ on $\mathfrak{g}^x$ defined by

$$< u, v >^x = <\lambda_x(u), \lambda_x(v)>$$

where $<, >$ denotes the biinvariant scalar product on $\mathfrak{g}$, is a biinvariant scalar product on g^x.

ii) For any pair (σ, σ') of continuous sections of $F(\mathfrak{g})$, $x \to <\sigma(x), \sigma'(x)>^x$ is V-integrable with respect to any volume measure V on M.

iii) For any orthonormal basis $b = (\tau_1 ..... \tau_p)$ of $\mathfrak{g}$, $b^x = (\lambda_x^{-1}(\tau_1), ..., \lambda_x^{-1}(\tau_p)$ is an orthonormal basis of $\mathfrak{g}^x$ with respect to $<, >^x$

iv) Let ω the Maurer-Cartan form of G. Then the Maurer-Cartan form ω^x of G^x is such that if γ^x is in G^x, then :

$$\omega^x \gamma^x = \lambda_x^{-1} \cdot \omega_{\Lambda_x(\gamma x)} \cdot (d\Lambda_x)_{\gamma x}$$

Proof

i) Follows from Prop 2 ii) ; ii) and iii) are direct.

In order to prove iv), one recalls that the Maurer-Cartan form Ω of a Hilhert - Lie group L with Lie algebra ℓ and tangent bundle $TL = \bigcup_{\gamma \in L} T_\gamma L$ is such that $\forall \gamma \in L$, Ω_γ is the differential at γ of the left translation $l_{\gamma 1} : \gamma \to \gamma^{-1}\gamma$.

It follows directly that :

$$\Lambda_x \, 1^{x}_{\gamma_x^{-1}} \, \Lambda_x^{-1} = 1_{\Lambda_x(\gamma_x^{-1})}$$

and iv) follows, taking the derivative at the point γ^x.

2 MEASURABLE FIELDS OF LEFT-INVARIANT BROWNIAN MOTIONS ASSOCIATED TO THE BUNDLE F(G).

Let us first recall the construction of the standard Brownian motion on G, following [3] [4].

$\xi(t)$ is the stochastic white noise (generalized) process on $\mathfrak{g}$: it is such that for $f \in L^2(I,\mathfrak{g})$ $I = [0,1]$:

$$< \xi, f > = \int_0^1 < \xi(t), f(t) > dt$$

is a gaussian random variable with mean zero and variance $< f, f >$. Taking:

$$Z(t) = \int_0^t \xi(t) \, dt$$

one gets the brownian motion process Z on $\mathfrak{g}$ starting at 0 at $t = 0$.

Let now $C_0(g)$ be the space of $\mathfrak{g}$-valued continuous mappings such that $\sigma(0) = 0$. The mapping

$$\eta : [0,1] \times g \to G$$

which solves the stochastic differential equation :

$$d\eta \; \eta^{-1} = \xi = dZ$$

with initial value

$$\eta \, (0,\sigma) = e$$

where e denotes the unit of G, is, in the Itô description, the *standard left-invariant brownian motion process on G.*

ξ can also be descrited in the following way : taking an orthonormal basis $(\tau_1....\tau_p)$ in $\mathfrak{g}$, one gets a basis $(\theta_1,....,\theta_p)$ of left invariant vertor fields on G such that for $i = 1,...,p$:

$$\theta_i \, (\gamma) = \omega_\gamma^{-1} \, (\tau_i)$$

where ω is the Maurer-Cartan form on G. Then η is the unique strong Markov diffusion process satisfying $\forall f \in C^\infty(G)$ $\forall \sigma \in C_0(g)$:

$$\partial f \, (\eta(t)) = \sum_{i=1}^{P} \, (\theta_i f) \, (\eta(t)) \, \partial \, B_i$$

$$\eta \, (0,\sigma) = e$$

where ∂ stands for the Stratonovitch differential ([16], V-35) and B_i is for each $i = 1...p$ a copy of the standard brownian motion on $\mathbb{R}$.

η appears also as the left-invariant process on G with infinitesimal generator:

$$\Delta = \frac{1}{2} \sum_{i=1}^{P} \theta_i^2$$

In order to construct fields of brownian motions on F(G), let us consider a fixed measurable trivialization (Λ, λ), an orthonormal basis $(\tau_i)_i$ (i= 1... p) in $\mathfrak{g}$, and introduce, for each $x \in M$, the Brownian motion Z^x on $\mathfrak{g}^x$ by :

$$Z^x = \sum_{i=1}^{P} B_i \ \lambda_x^{-1}(\tau_i)$$

then :

$$Z^x(t) = \lambda_x^{-1} Z(t),$$

and $\xi^x = \lambda_x^{-1} \xi = dZ^x$ is the white noise process on $\mathfrak{g}^x$ such that, as a generalized gaussian process :

$$\mathbb{E}(\xi^x(t)) = 0$$

$$\mathbb{E}(\xi^x(t)\,\xi^x(t')) = \delta(t-t')\ \mathbb{I}_{\mathfrak{g}^x}.$$

as before introducing $C_0(\mathfrak{g}^x)$ to be the set of all continuous mappings $\sigma_x : I \to \mathfrak{g}^x$ with $\sigma_x(0) = 0$, a basis for left-invariant vector fields θ_i^x on G^x such that $(\forall \gamma_x \in G^x)$

$$\theta_i^x(\gamma_x) = \left(\omega_{\gamma_x}^x\right)^{-1}\left(\lambda_x^{-1}(\tau_i)\right)$$

where ω^x denote the Maurer-Cartan form on G^x, then one gets a field of strong Markov diffusion processes η^x satisfying

$$d\eta^x(\eta^x)^{-1} = \xi(x) \qquad\qquad \eta^x(0, \sigma_x) = e_x$$

and with infinitesimal generator :

$$\Delta^x = \frac{1}{2}\sum_{i=1}^{P}(\theta_i^x)^2$$

Notice that all ingredients in the construction depend on the choice of the measurable trivialization. But, in contrast, denoting by μ^x the Brownian measure on the path group $C(I,G^x)$ induced by η^x, one easily sees that μ^x is actually independant of the choice of the measurable trivialization due to the invariance of the Killing form by Lie algebra isomorphisms. This remark allows to speak about the field of Brownian measures $(\mu^x)_{x\in M}$ on F(G).

By its very construction, μ^x appears as quasi-invariant by left and right pointwise translations by elements of the Sobolev path group $H^1(I, G^x)$.

Summarizing the preeceding discussion and using Proposition 3 II) (iv), we can state:

Proposition 4

i) For a given measurable trivialization, (Λ, λ), one gets a measurable field of brownian motion process η^x on G^x such that $\forall t \in [0,1]$, $\forall \sigma_x \in C_0(\mathcal{G}^x)$:

$$\eta^x(t, \sigma_x) = \Lambda_x^{-1}(\eta(t, \lambda_x \sigma_x))$$

ii) the corresponding field of Brownian measures $(\mu^x)_{x\in M}$ is independant of the choice of the trivialization and μ^x is quasi-invariant by right or left pointwise translations by Sobolev paths with values in G^x.

As an application, let us consider, in the spirit of [4], the following construction.
Let $CH^1(I \times M, F(G))$ be the current group consisting of all the continuous mappings ψ : $I \times M \to F(G)$ such that for each x in M the mapping $\psi(., x) : t \to \psi(t, x)$ belongs to $H^1(I, G^x)$, and let us point out that, endowed with the pointwise product it is a Banach-Lie group.

Let v be a volume measure on M ; then $(L^2(C(I, G^x), \mu_x)_x$ is a v-integrable field of Hilbert spaces ; we are then led to consider the Hilbert space :

$$\mathcal{H}^v = \int_M^{\oplus} L^2(C(I, G^x) ; \mu_x)\, dv_x$$

consisting of all the ν-integrable mappings Φ such that for almost all x in M, $\Phi(x)$ belongs to $L^2(C(I, G^x)\,;\,\mu_x)$, with Hilbert norm :

$$||\Phi||_{\nu} = \left\{ \int_M \left(\int_{C(I,G^x)} |\Phi(x)(\gamma_x)|^2 \, d\mu_x(\gamma_x) \right) d\nu(x) \right\}^{1/2}.$$

From the quasi-invariance of the measures μ_x, $x \in M$, and from the fact that for each x in M, and all γ_x in in $H^1(I,G^x)$ the Radom-Nikodym derivative $\dfrac{d\mu_x(\psi^{-1}.\gamma_x)}{d\mu_x(\gamma_x)}$ depends on the Maurer-Cartan cocycle $d\psi.\psi^{-1}$ ([4]) one gets :

Proposition 5

For each ψ in CH^1 (I x M, F(G)) let $\pi^{\nu}(\psi)$ be the operator on $\mathcal{H}^{\nu}$ defined by :

$$((\pi^{\nu}(\psi)\Phi)(x))\,(\gamma_x) = \sqrt{\frac{d\mu_x[\psi^{-1}(.,x).\gamma_x]}{d\mu_x(\gamma_x)}}\;\phi(x)\,(\psi^{-1}(.,x).\gamma_x),\;\; \phi \in \mathcal{H}^{\nu},\; x \in \mathcal{M},$$
$$\gamma_x \in G^x.$$

Then $\pi^{\nu} : \psi \to \pi^{\nu}(\psi)$ is a non located unitary representation of order 1 of the current group CH^1(I x M, F(G)).

Remark : The representation π^{ν} is reducible ; such a representation, closely connected with the energy representation of each $C(I,G^x)$, admits, like these representations (cf : [4]), a stochastic decomposition, which, in this context, may be parametrized by the space CH^1(I x M, F(T)) where T is a maximal torus in G, and $F(T) = \bigcup_{xFM} \Lambda_x^{-1}(T)$ (cf : [12]).

3 PROBABILITY MEASURES ON THE CURRENT GROUP B[F(G)]

Let us consider the covariant functor B which associates to any smooth manifold F the set B(M,F) of all the Borel measurable mappings from M into F ; in the case where $F = \bigcup_{x \in M} F^x$ is a smooth bundle over M let us denote by B[F] the set of global sections of F which are Borel measurable :

$$B[F] = \{\varphi \in B(M,F) \, / \, \varphi(x) \in F^x \; \forall x \in M\}$$

Any smooth bundle of Lie groups or Hilbert spaces has global C^∞ sections ; then B[F(G)] and $B[F(\mathfrak{g})]$ are non empty. One easily sees that B[F(G)] equiped with the pointwise product and the topology of the pointwise convergence is a topological group with unit element the section $\varepsilon : x \rightarrow \varepsilon(x) = e_x$, and that $B[F(\mathfrak{g})]$ equipped with the pointwise defined Lie-bracket and with the topology of the pointwise convergence is a topological Lie algebra.

We point out that, although B[F(G)] has not a Lie group structure, it contains, as subgroups, all infinite dimensional Lie groups of sections of F(G), for instance the Banach - Lie groups $C^k[F(G)]$ of sections of class C^k of F(G), k = 0, 1, ..., and the Sobolev - Hilbert Lie groups $H^n[F(G)]$, $n > \frac{1}{2}$ dim(M), of sections of Sobolev class H^n of F(G).

In the particular case where F(G) = (PxG)/G is the associated bundle of a G - principal bundle P over M $B[(PxG)/G] \approx \{g \in B(P,G) / \forall \gamma \in G, \forall x \in P : g(x.\gamma) = \gamma^{-1}.g(x).\gamma\}$ wich appears then as the largest gauge group associated to P.

Let V be a volume measure on M, and, for each integer $s > \frac{1}{2}$ dim (M), let us consider the space $H^s[F(\mathfrak{g})]$ of global sections of Sobolev class H^s of $F(\mathfrak{g})$. We fix two integers m and n strictly greater than $\frac{1}{2}$ dim (M).

It is well-known that the natural injection i from $H^{m+n}[F(\mathfrak{g})]$ in $H^n[F(\mathfrak{g})]$ is Hilbert - Schmidt (see e.g. [13], chap. XVII, § 2), and then the triple :

$$(i, \; H^{m+n}[F(\mathfrak{g})], \; H^n[F(\mathfrak{g})]),$$

is an abstract Wiener space, so that one gets a Gaussian measure $W^{m,n}$ on the Banach space $C_0\,(H^n[F(\mathfrak{g})])$, of all the continuous mappings $\sigma : [0,1] \rightarrow H^n[F(\mathfrak{g})]$ which satisfy $\sigma(0) = 0$ ([8]).

Now for each τ in $C_0\,(H^n[F(\mathfrak{g})])$ and each x in M we define σ^x as the mapping from [0,1] into $\mathfrak{g}^x$ such that $\sigma^x\,(t) = \sigma\,(t)(x)$, $t \in [0,1]$. Then :

Lemma 6

$$\sigma^x \in C_0(\mathfrak{g}^x) = \{\xi \in C\,([0,1], \mathfrak{g}^x) / \xi(0) = 0\}$$

Proof

One has $\sigma^x\,(0) = \sigma(0)\,(x) = 0$; moreover, as $n > \frac{1}{2}$ dim (M), from the Sobolev embedding theorem it follows that $H^n[F(\mathfrak{g})]$ is contained in the Banach space $C[F(\mathfrak{g})]$ of continuous sections of $F(\mathfrak{g})$, and that there exists a constant $k > 0$ such that for any u in $H^n[F(\mathfrak{g})]$: $\sup |u(x)|^x \leq M\, \|u\|_{(n)}$, where $|\ |^x$ denotes the euclidean norm of $\mathfrak{g}^x$, and $\|\ \|_{(n)}$ the Sobolev norm of $H^n[F(\mathfrak{g})]$. In particular, for any pair (t, t') of elements in [0,1] : $|\sigma^x(t) - \sigma^x(t')| \leq M\, \|\sigma(t) - \sigma(t')\|_{(n)}$, so that σ^x is continuous.

Now let us consider the measurable field $(\eta^x)_{x \in M}$ of left - invariant standard Brownian motion processes associated to F(G) (cf : § II). From the above lemma it follows that for any t in [0,1], any σ in $C_0\,(H^n[F(\mathfrak{g})])$ and any x in M $\eta^x\,(t, \sigma^x)$ is a well-defined element of G^x, and then, the mapping :

$$\hat{\eta}\,(t, \sigma) : x \rightarrow \eta^x\,(t, \sigma^x)$$

is a global measurable section of the bundle F(G).

Let us consider now the family Θ of all the Borel subsets of $C_0\,(H^n[F(\mathfrak{g})])$.

Lemma 7

For each t in [0,1] the mapping $\hat{\eta}_t$ defined by $\hat{\eta}_t\,(\sigma) = \hat{\eta}\,(t, \sigma)$ is a measurable mapping from $C_0\,(H^n[F(\mathfrak{g})])$ into B[F(G)], and then, a random variable on the probability space $(C_0\,(H^n[F(\mathfrak{g})]), \Theta, W^{m,n})$.

Proof

One has, for each x in M : $\hat{\eta}(t, \sigma)(x) = \Lambda_x^{-1}[\eta(t, \lambda_x . \sigma_x)]$, where (Λ, λ) is a measurable trivialisation of F(G). Then, by proposition 4, and the assertion follows then from the measurability of η, Λ and λ.

The mapping $\hat{\eta} : [0,1] \times C_o(H^n[F(\mathcal{g})]) \to B[F(G)]$ such that $\hat{\eta}(t, \sigma) = \hat{\eta}_t(\sigma)$ is then a stochastic process. As a direct consequence one gets :

Proposition 8

Let t in [0,1] and let $p_t^{m,n}$ such that for all Borel subset B of B[F(G)] :

$$p_t^{m,n}(B) = W^{m,n}(\{\sigma \in C_o(H^n[F(\mathcal{g})]) / \hat{\eta}(t, \sigma) \in B\}).$$

then $p_t^{m,n}$ is a probability Borel measure on B[F(G)].

ACKNOWLEDGMENTS

The authors are happy to thank the Research Center Bielefield-Bochum-Stochastics, and the Department of Mathematics of the Faculty of Sciences of Marseille - Luminy, in which several works of the authors together with S.Albeverio and R.Hϕegh-Krohn originated or were developped.

REFERENCES

[1] S. Albeverio, R. Hϕegh-Krohn : The energy representation of Sobolev -Lie groups, Comp. Math. **36** (1978), 37 - 52.

[2] S. Albeverio, R. Hϕegh-Krohn, J. Marion, D. Testard, B. Torresani : Non commutative distributions, Book to appear.

[3] S. Albeverio, R. Hϕegh-Krohn, D. Testard : Irreducibility and reducibility for the energy representation of the group of mappings from a Riemannian manifold into a compact Lie group, J. Funct. Anal. **41** (1981), 378-396.

[4] S. Albeverio, R. Hϕegh-Krohn, D. Testard, A. Vershik : Factorial representations of path groups, J.Funct. Anal. **51** (1983), 115-131.

[5] Y.Daletskii : Stochastic differential geometry, Uspekhi Mat. Naut. **81** (1983), 87-111.

[6] Y. Daletskii, J. Schnaiderman : Diffusion and quasi-invariant measures on infinite dimensional Lie groups, Funkt. Anal. i Pril. **3** (1969), 88-90.

[7] B. Gaveau, Ph. Trauber : Mesures et représentations non locales pour les groupes de Lie d'applications, C.R. Acad. Sc. Paris, **291**, série A, (1980), 575-578.

[8] L. Gross : Potential theory on Hilbert spaces, J. Funct. Anal. **1** (1967), 123-181.

[9] A. Guichardet : Représentation de G^X selon Gelfand et Delorme, Sém. Bourbaki **486**, Paris (1976).

[10] W. Kondracki, J. Rogulski : On the stratification of the orbit space for the action of automorphisms on connections, Dissert. Math., Polska Akad. Nauk., vol. CCL, Warszawa (1986).

[11] J. Marion : G - distributions et G - intégrales multiplicatives sur une variété, Annal. Pol. Math. **43** (1983), 79-93.

[12] J. Marion : Outline of harmonic analysis on groups of paths with values in a Sobolev gauge group, Proc. 2d Intern. Conf. on stochastic processes (Geometry and Physics), Ascona (1988).

[13] K. Maurin : Abbildungen vom Hilbert - Schmidtschen Typus und ihre Anwendungen, Math. Scan. **9** (1961), 359-371.

[14] H. McKean : Stochastic Integrals, Academic Press, New York (1969).

[15] P.Mitter, C. Viallet : On the bundle of connections and the gauge orbit manifold in Yang-Mills theory, Commun. Math. Phys. **79** (1981), 457-472.

[16] L.Rogers, D. Williams : Itô calculus, John Wiley and Son, Singapore (1987).

[17] A. Vershik, I. Gelfand, M. Graev : Representations of the group of smooth mappings of a manifold into a compact Lie group, Comp. Math. **35** (1977), 299-334.

UNITARY HIGHEST WEIGHT REPRESENTATIONS OF GAUGE GROUPS

Bruno Torrésani $

Centre de Physique Théorique*
CNRS-Luminy Case 907
13288 Marseille Cedex 09 FRANCE

Abstract: We describe the Lie groups and Lie algebras of mappings of a topological space into a locally compact group (called here gauge groups and gauge algebras), and their unitary representation theory. We essentially describe the highest weight representations, and the constructions based on continuous tensor products of representations. In particular, the connection between these two approaches can be made in some particular cases, namely for structure groups of the type SU(n,1).

I: INTRODUCTION:

Since the early seventies, mathematicians and mathematical physicists have made considerable efforts in order to understand the structure and to develop the representation theory of infinite-dimensional Lie groups, like for instance diffeomorphisms groups or gauge groups (also called current groups). A gauge group is here defined to be a group of mappings

$$G^X = \mathrm{Map}(X,G)$$

from some topological space X into a locally compact group G, the group structure being provided by the pointwise multiplication law. Different functional frameworks have been used to investigate such groups, for instance groups of Borel measurable mappings (with respect to some Borel measure on X), Schwartz Lie groups $\mathcal{D}(X,G) = C_0^\infty(X,G)$, or Sobolev Lie groups. In this context, a gauge algebra will be defined to be the Lie algebra

$ Partially supported by the Alexander von Humboldt foundation, and the DCAN (GERDSM Toulon)

* Unité propre de recherches du Centre National de la Recherche Scientifique, UPR 7061.

of the corresponding gauge group, that is the Lie algebra of mappings:

$$\mathcal{G}^X = \mathrm{Map}(X,\mathcal{G})$$

with pointwise Lie algebra structure, $\mathcal{G}$ being the Lie algebra of G.

Among gauge groups and gauge algebras, the particular case of one-dimensional X space has specific properties. For instance, $\mathcal{G}^{\mathbb{T}^1}$ leads to the so-called affine Kac-Moody algebras, which can be constructed from a generalised Cartan matrix, as finite-dimensional simple Lie algebras; affine Kac-Moody algebras also have a very rich unitary representation theory. Conversely, $\mathcal{G}^{\mathbb{T}^\nu}$ (which leads to the theory of quasisimple Lie algebras) does not admit any generalised Cartan matrix for $\nu > 1$, and seems to have a much poorer unitary representation theory. It must also be stressed that some representations of G^X, with dim(X)=1, admit an interesting functional realisation in terms of G-valued Brownian motion.

One is interested in continuous unitary representations of gauge groups in Hilbert spaces:

$$\pi:\ G^X \to \mathcal{U}(\mathcal{H})$$

($\mathcal{U}(\mathcal{H})$ denoting the unitary operators on the Hilbert space $\mathcal{H}$) which correspond to representations of the gauge algebra by self-adjoint operators, via the exponential mapping.

From the mathematical point of view, such unitary representations are peculiarly interesting, in the sense that they provide a natural and nontrivial non-commutative generalisation of the theory of distributions on X. Indeed, if one takes for the group G the abelian group $\mathbb{R}$ (or $\mathbb{C}$), then the corresponding gauge group will be an infinite-dimensional abelian group. The corresponding unitary representations are then unitary characters, and can then be viewed as forms on the functional space $\mathbb{R}^X$ (or $\mathbb{C}^X$), otherwise stated real (or complex) distributions on X. This way of constructing non-commutative distributions was the original program of Gelfand and his collaborators, and was pursued by several authors (see e.g. [Ar], [De 77], [Gui], [I], [Pa.Sc], [Ge.Gr], [V.Ge.Gr. 73], [V.Ge.Gr. 74]).

The most natural idea, developped first by Araki, Streater, Parthasarathy and Schmidt, and Gelfand and his collaborators, was to try to build such representations as continuous tensor products of representations of G, labelled by X: more precisely, if ϕ is an element of G^X, one has to make sense to the formal expression:

$$\Pi(\phi) = \bigotimes_{x\in X} \pi_x\big[\phi(x)\big]$$

where π_x are unitary representations of G. However, it turns out that the groups such that this formal expression can make sense are highly constrained. They must have a non-trivial unitary representation-valued first cohomology group. Let us quote for instance that all compact groups have trivial unitary representation-valued first cohomology groups, and that among simple groups, only those of the type SU(n,1) and SO(n,1) can give rise to non trivial continuous tensor products of representations (see [V.Ge.Gr. 74], [De 77]). Such representations are infinitely divisible in the sense of Araki, which makes the connection with the infinitely divisible processes of probability theory (see e.g. [Gui]).

For compact groups, it is nevertheless possible to use the continuous tensor product construction, but one has to start with an extension of the gauge group, namely with a Leibnitz extension. It turns out that such extensions have non trivial representation valued first cohomology groups, and then allow to perform the construction. One is then led to the so-called energy representations, introduced in [I],[Pa.Sc], and studied by many authors (see e.g. [A.HK 78],[Ge.Gr.V 81],[A.HK.Te.V],[Mar 83],[Wa]). Note that in the case dim(X)=1, the standard energy representation of $\mathcal{D}$(X,G) admits a natural interpretation in terms of Brownian motion on G.

Another approach consists in considering the highest weight representation theory of gauge groups. In this approach, it is sufficient to work at the level of the gauge algebra, the Lie algebra of the gauge group. Roughly speaking, highest weight representations are obtained by specifying a polarisation on the gauge algebra; a highest weight representation is then a representation such that there exists a vector in the representation space (highest weight vector) on which the action of the positive component of the gauge algebra is trivial, and that the whole representation space can be obtained as the orbit of the negative component of the gauge algebra through it. Highest weight representations of affine Kac-Moody algebras can be classified, as well as those of elliptic quasisimple Lie algebras; the central result is that for compact $\mathcal{G}$, the unitary highest weight representation theory of affine Kac-Moody algebras is much richer than that of elliptic quasisimple Lie algebras, for $\nu > 1$. However, for noncompact $\mathcal{G}$, they are comparable (actually nontrivial only for $\mathcal{G}$=su(n,1)), and the connection can be made in that case with the continuous tensor product representations.

This paper is organised as follows: after the introduction, we describe some basic constructions and results on gauge groups and gauge algebras in section II; In section III, we discuss continuous tensor products representations of noncompact gauge groups, highest weight representations of noncompact gauge algebras, and clarify the connection between them; Section IV is devoted to conclusions.

Throughout this paper, we will discuss complex and real Lie groups and algebras. To avoid confusions, we will adopt the following convention: we will generically denote by L (resp. $\mathcal{L}$) a complex Lie group (resp. Lie algebra), and by G (resp. $\mathcal{G}$) a real Lie group (resp. Lie algebra).

II: GAUGE GROUPS, GAUGE LIE ALGEBRAS:

We give here a brief account of the construction of gauge groups and Lie algebras. For a more detailed presentation, we refer to the first chapter of [A.HK.M.Te.To], and to the references therein.

1°) Gauge groups; Functional Lie groups:

Many problems arising in mathematical physics, and more precisely in quantum field theory (or string theory) in which infinite-dimensional Lie algebras and groups occur, are quite often solved by purely algebraic arguments; that is to say that one essentially uses the group or Lie algebra structure. This is especially the case when one is interested in highest weight representations of such groups, for which the highest weight assumption is strong enough to allow to work at the infinitesimal level, where functional problems are less essential. However, besides highest weight representations, there exists other types of representations, or other realisations of these representations, where the functional structure one provides the gauge group with is used as an important tool. We will give examples of such constructions later on.

Let X be some Riemannian manifold (for simplicity) of dimension m, and let G be some locally compact finite-dimensional Lie group. Denote by

$$\mathcal{D}(X,G) = C_0^{\infty}(X,G)$$

the group of compactly supported smooth mappings from X into G, with pointwise group structure, and identity element:

$$e \ : \ x \rightarrow e(x) = e_0$$

e_0 being the identity element of G.

Let $$\mathcal{D}(X,\mathcal{G}) = C_0^{\infty}(X,\mathcal{G})$$

be the Lie algebra of compactly supported smooth mappings from X into the Lie algebra $\mathcal{G}$ of G. $\mathcal{D}(X,\mathcal{G})$ is a Lie algebra of $\mathcal{D}(X,G)$, and there exists an exponential map between them, defined by:

$$\exp\{\phi\} = \phi \circ \exp_{\mathcal{G}}, \qquad \phi \in \mathcal{D}(X,\mathcal{G})$$

$\exp_{\mathcal{G}}$ being the exponential mapping of $\mathcal{G}$.

$\mathcal{D}(X,\mathcal{G})$ can be endowed with a Schwartz topology, inherited from that of $\mathcal{D}(X)$. It also inherits a structure of nuclear space. This allows to provide $\mathcal{D}(X,G)$ with a topology (see [A.HK.M.Te.To], chapter I), analogous of a Schwartz topology, such that one has:

_ There exists a neighborhood of e which is homeomorphic to a neighborhood of the origin of $\mathcal{D}(X,\mathcal{G})$.

_ $\mathcal{D}(X,G)$ gets a structure of separated topological group.

Moreover, if $C_0^k(X,G)$ is the group of compactly supported k times differentiable mappings from X into G, it is clear that one has:

$$\mathcal{D}(X,G) = \bigcap_{k=1}^{\infty} C_0^k(X,G)$$

When considering continuous tensor products representations, it is more convenient to work with the group B(X,G) of Borel measurable mappings of X into G (see e.g. [Gui],[Mar 89]). Let $\mathcal{B}_0(X)$ denote the set of bounded Borel sets of X, and for any $A \in \mathcal{B}_0(X)$ and $g \in G$, introduce $\chi_A^g \in B(X,G)$, defined by:

$$\chi_A^g(x) = \begin{cases} g \text{ if } x \in A \\ e_0 \text{ otherwise} \end{cases}$$

Let $\overline{\mathcal{M}_0}(X,G)$ be the group generated by the χ_A^g, $A \in \mathcal{B}_0(X)$, $g \in G$, provided with the topology of the uniform convergence [Mar 89]. Then all the groups $C_0^k(X,G)$, and $\mathcal{D}(X,G)$ are dense subgroups of $\overline{\mathcal{M}_0}(X,G)$.

It is well known that extensions of infinite-dimensional Lie groups and algebras have often a much richer representation theory than the infinite-dimensional Lie groups and algebras themselves. This is clearly the case for Kac-Moody algebras and quasisimple Lie algebras (these algebras will be briefly described in the next section), where it is necessary to consider central extensions to have non-trivial unitary highest weight representations. More precisely, it is useful to extend G^X by the one-dimensional torus $\mathbb{T}^1$, according to the group relation:

$$(\phi, a).(\phi', b) = (\phi . \phi', \psi(\phi, \phi')) \qquad \phi, \phi' \in G^X, \quad a,b \in \mathbb{T}^1$$

ψ being a group one-cocycle, fulfilling:

$$\psi(\phi, \phi'\phi'')\, \psi(\phi', \phi'') = \psi(\phi\phi', \phi'')\, \psi(\phi, \phi') \qquad \phi, \phi', \phi'' \in G^X$$

As it will be shown in the next section, central extensions are crucial ingredients for highest weight representation theory of compact gauge algebras, where the triviality of ψ implies the triviality of the representation[1] .

For non highest weight representations of gauge groups, the jet extensions have also been considered, leading to the so-called energy representations. We briefly discuss here the jet extensions. Let $x_0 \in X$, and introduce the group:

$$G^X_{x_0 k} = \left\{ \phi \in G^X \text{ s.t. } \phi(x_0) = e_0 \text{ and } \partial^\alpha \phi(x_0) = 0 \text{ for any } \alpha \text{ with } |\alpha| \leq k \right\}$$

where one has set, in the coordinate system $(x_1, x_2, \ldots x_m)$:

$$\partial^\alpha \phi(x_0) = \frac{\partial^{|\alpha|}\phi}{\partial x_1^{\alpha_1} \partial x_2^{\alpha_2} .. \partial x_m^{\alpha_m}}(x_0)$$

---

[1] However, if the construction of central extensions is easy at the Lie algebra level [Kac 83], it is far from being trivial at the group level (see e.g. [Mi], [Mu], [Ti]).

$G^X_{x_0 k}$ is a normal subgroup of G^X. Then the quotient group:

$$G^k_m = G^X / G^X_{x_0 k}$$

is easily shown to depend only on the dimension m of X, and not on the particular x_0 one starts with (more precisely, all the quotients $G^X / G^X_{x_0 k}$ are isomorphic). G^k_m is called the Leibnitz group of order k and degree m over G. The principal bundle of k-jets over X is then canonically constructed:

$$j^k(X,G) \to X$$

with base X and type fiber G^k_m. One thus denotes by G^X_k the group of differentiable sections of this fiber bundle. It is a functional Lie group, in the sense of [Mar 89]. In the case k=1, one has the following standard result:

Lemma 1:

i) $G^1_m \cong G \times N^1$ *, where* N^1 *is isomorphic to the abelian group*

$$A(m) = \mathcal{G} \oplus \mathcal{G} \oplus ... \oplus \mathcal{G} \qquad \text{(m times)}$$

The group law is given by:

$$(g,a).(g',a') = (g.g',ad(g).a' + a) \qquad a,a' \in A(m), \quad g,g' \in G$$

ii) $G^X_1 \cong G^X \times \Omega^1(X,\mathcal{G})$, $\Omega^1(X,\mathcal{G})$ *being the space of all* $\mathcal{G}$*-valued one-forms on X.*

iii) One has the canonical embedding of G^X *into* G^X_1 *, given by:*

$$\phi \to (\phi \, , \beta(\phi))$$

where $\beta(\phi) = d\phi . \phi^{-1}$ *is the Maurer-Cartan one-cocycle.*

The proof of the lemma follows from straightforward calculations.
The unitary representations of G^1_m, as well as those of higher order Leibnitz groups can be classified by means of Mackey induction-reduction theory [A.HK.To].

Let us also quote that other types of extensions of gauge groups have also been considered in the literature. More precisely, the so-called Faddeev-Mickelsson term has been studied, in connection with the existence of anomalies in quantum gauge theories. For an elementary introduction to this extension, we refer to [Pr.Se].

2) Gauge Lie algebras:

The theory of gauge Lie algebras represents the infinitesimal version of the theory of gauge groups. Gauge Lie algebras, and more precisely those associated with one-dimensional source space X, have received a considerable attention since the introduction, by Kac and Moody independently, of affine Lie algebras [Kac 67], [Mo]. Affine Lie algebras are infinite-dimensional Lie algebras, constructed from a generalised Cartan matrix by means of Serre's presentation theorem, which can also be realised as central extensions of (untwisted or twisted) loop algebras, i.e. gauge algebras of the form $\mathcal{G}^{T^1}$. More precisely, a generalised Cartan matrix (GCM) is an nxn matrix A, with integral entries a_{ij} such that:

. $a_{ii} = 2$ for any i=1,...n.

. $a_{ij} = 0 \Rightarrow a_{ji} = 0$ i,j=1,...n.

. $a_{ij} \leq 0$ for any i,j=1,...n, $j \neq i$.

Assume that the corank CoRk(A) of A is equal to 0 or 1. Then the Kac-Moody algebra $\mathcal{G}$(A) associated with the GCM A is the complex Lie algebra defined by the generators e_i, f_i, h_i, i=1,...n, and relations:

. $[e_i,f_j] = \delta_{ij}\, h_i$ i,j=1,...n.

. $[h_i,e_j] = a_{ij}\, e_j$ i,j=1,...n.

. $[h_i,f_j] = -\, a_{ij}\, f_j$ i,j=1,...n.

. $[h_i,h_j] = 0$ i,j=1,...n.

. $[\mathrm{Ad}(e_i)]^{1-a_{ji}}.e_j = 0$ i,j=1,...n.

. $[\mathrm{Ad}(f_i)]^{1-a_{ji}}.f_j = 0$ i,j=1,...n.

In the case CoRk(A)=0, according to the famous result of Chevalley, the classification of all possible GCM leads to the classification of complex semisimple Lie algebras. The case CoRk(A)=1 corresponds to the class of affine Kac-Moody algebras. The existence of the GCM allows to perform Cartan's decomposition of $\mathcal{G}$(A) into Cartan subalgebra $\mathfrak{h}$ and rootspaces $\mathcal{G}_\alpha$, $\alpha \in \mathrm{Sp}\big[\mathrm{Ad}(\mathfrak{h})\big]$. For a detailed presentation of Kac-Moody algebras, we refer to [Kac 85], and more precisely to the first chapter for the general construction. A remarkable property of affine Kac-Moody algebras is that they can be realised as central extensions of loop algebras of semisimple Lie algebras. This loop algebra realisation will be presented as a particular case of the gauge algebra realisation of elliptic quasisimple Lie algebras we will describe now.

Since the affine Kac-Moody algebras describe the gauge algebras associated with one-dimensional source space X, it is relevant to try to generalise the results to higher-dimensional source spaces. In particular, the quasisimple Lie algebras, proposed and studied in [HK-To] and [To 86], provide an interesting framework to develop a structure theory for these algebras:

Definition 2:

A Quasisimple Lie algebra is a complex Lie algebra $\mathcal{L}$, such that:

_ *$\mathcal{L}$ possesses a finite-dimensional Cartan (maximal abelian) subalgebra $\mathfrak{h}$, such that* Ad($\mathfrak{h}$) *is diagonalisable on $\mathcal{L}$, and has discrete spectrum (the spectrum of* Ad($\mathfrak{h}$) *is called the root system $\mathcal{R}$ of $\mathcal{L}$).*

_ *$\mathcal{L}$ possesses a non-degenerate symmetric* Ad($\mathcal{L}$)*-invariant* [2] *bilinear form,*

[2] Recall that Ad($\mathcal{G}$)-invariant means that for any triple x,y,z of elements of $\mathcal{G}$, one has

called the Killing form and denoted by <,>; one also assumes for simplicity that the Killing form is real on the real linear span of the roots.

_ *For any root* α *of* $\mathcal{L}$*, such that* $<\alpha,\alpha>\neq 0$*,* $\mathrm{Ad}(\mathcal{L}_\alpha)$ *is a nilpotent* $\mathcal{L}$*-subalgebra*[3].

If the restriction of the Killing form to the real linear span of the roots is positive definite, the quasisimple Lie algebra is a semisimple finite-dimensional Lie algebra. If the restriction of the Killing form to the real linear span of the roots is positive semidefinite, the quasisimple Lie algebra is called elliptic. Otherwise, the quasisimple Lie algebra is called indefinite.

It turns actually out that elliptic quasisimple Lie algebras, or more precisely the corresponding root systems can be classified in a simple way [HK.To],[To 86]. Let $\mathcal{L}$ be an elliptic quasisimple Lie algebra, with root system $\mathcal{R}$. Denote by $\mathcal{R}_f$ the quotient of $\mathcal{R}$ by the kernel of the Killing form restricted to the real linear span of the roots (the dimension of this kernel is called the type of $\mathcal{L}$, and denoted by ν). It then follows from direct verification of the axioms that $\mathcal{R}_f$ is a finite-dimensional root system (as classified for instance in [Bou]), called the gradient root system of $\mathcal{L}$. $\mathcal{R}_f$ is in general the root system of a finite-dimensional simple Lie algebra, except in the case where it is non-reduced[4]. Let us focus to this reduced case for simplicity; $\mathcal{R}_f$ admits two kinds of elements: the short roots (generically denoted by α_s), and the long roots (generically denoted by α_l). One then has the following classification result (one has a similar result when $\mathcal{R}_f$ is non-reduced):

<u>*Theorem 3:*</u>

Let $\mathcal{L}$ *be an elliptic quasisimple Lie algebra with root system* $\mathcal{R}$ *and reduced gradient root system* $\mathcal{R}_f$*; then* $\mathcal{R}$ *is isomorphic to the direct product of* $\mathcal{R}_f$ *by an* ν*-dimensional lattice* Λ *(the isotropic part of* $\mathcal{R}$*), and is characterised by* $\mathcal{R}_f$*,* ν *and an additional non-negative integral number* $\tau\leq\nu$ *(called the twist number). If* $\xi^1,..\xi^\nu$ *is any* $\mathbb{Z}$*-basis of* Λ*, all elements of* $\mathcal{R}$ *can be written in the following way:*

$$\alpha_s+\sum_{k=1}^{\nu} a_k\xi^k \qquad \alpha_s \text{ short element of } \mathcal{R}_f\ ,\ a_1,..a_\nu\in\mathbb{Z}$$

<[x,y],z>=<<x,[y,z]>.

[3] Recall that the non-degeneracy of the killing form on $\mathcal{G}$ implies the nondegeneracy on $\mathcal{h}$, and then allows to transport it on $\mathcal{h}'$. If α is any root of $\mathcal{G}$ (i.e. "eigenvalue" of $\mathrm{Ad}(\mathcal{h})$), $\mathcal{G}_\alpha$ is the corresponding eigenspace: $\mathrm{Ad}(h).\mathcal{G}_\alpha=\alpha(h)\ \mathcal{G}_\alpha$ for any h in $\mathcal{h}$.

[4] A root system $\mathcal{R}$ is said to be reduced if one has the following: $r\in\mathcal{R}$ implies that $2r\notin\mathcal{R}$.

$$\alpha_l + \sum_{k=1}^{\tau} k a_k \xi^k + \sum_{k=\tau+1}^{\nu} a_k \xi^k \qquad \alpha_l \text{ long element of } \mathcal{R}_l \text{ , } a_1,..a_\nu \in \mathbb{Z}$$

$$\sum_{k=1}^{\nu} a_k \xi^k \qquad a_1,..a_\nu \in \mathbb{Z}$$

This justifies the notation:

$$\mathcal{R} \cong (\mathcal{R}_l; \nu, \tau)$$

The proof of the theorem [HK.To], [To 86] is similar to that proposed by MacDonald in the affine case [Mac]; for $\nu>1$, it is however necessary to show in addition that the twist number does not depend on the choice of the $\mathbb{Z}$-basis of Λ. Of course, in the case $\nu=1$, one then recovers the classification of affine Kac-Moody algebras with reduced gradient root system.

Many elliptic quasisimple Lie algebras possess a gauge algebra realisation, that we will describe now. Let $\mathcal{L}_0$ be a finite-dimensional complex semisimple Lie algebra, of root system $\mathcal{R}_0$, and consider the ν-dimensional torus $\mathbb{T}^\nu$, and two copies of $\mathbb{C}^\nu$, with respective orthonormal bases $\{c_1,...c_\nu\}$ and $\{d_1,..d_\nu\}$. Then:

$$\mathcal{L} = P(\mathbb{T}^\nu, \mathcal{L}_0) \oplus \sum_{n=1}^{\nu}{}^{\oplus} \mathbb{C}c_n \oplus \sum_{n=1}^{\nu}{}^{\oplus} \mathbb{C}d_n$$

provided with the Lie product: for any $\{x,c,d\},\{x',c',d'\} \in \mathcal{L}$

$$[\{x,c,d\},\{x',c',d'\}] = \left\{[x,x'] + (d'.\nabla).x - (d.\nabla).x', \Psi(x,x'), 0\right\}$$

where: $$[x,x'](t) = [x(t),x'(t)] \qquad \forall\, t \in \mathbb{T}^\nu,$$

$$\Psi(x,x') = \sum_{i=1}^{\nu} \langle x , \frac{\partial x'}{\partial t_i} \rangle$$

$\mathcal{L}$ can be provided with a Killing form, by defining: for any $\{x,c,d\},\{x',c',d'\} \in \mathcal{G}$

$$\langle \{x,c,d\},\{x',c',d'\} \rangle = \langle x,x' \rangle - c.d' - c'.d$$

where: $$\langle x,x' \rangle = \int_{\mathbb{T}^\nu} dm(t) \langle x(t),x'(t) \rangle$$

(dm being the Lebesgue measure on $\mathbb{T}^\nu$, normalised so that $m(\mathbb{T}^\nu) = 1$), and . stands for the standard component by component scalar product on $\mathbb{C}^\nu$.

It is then straightforward to check that $\mathcal{L}$ is an elliptic quasisimple Lie algebra of type ν, and twist 0. Its root system is of the form:

$$\mathcal{R} \cong (\mathcal{R}_0; \nu, 0)$$

Twist one elliptic quasisimple Lie algebras are also easy to realise, like twisted affine Kac-Moody Lie algebras. The construction is similar to the previous one apart from the fact that instead of starting from the current algebra $P(\mathbb{T}^\nu, \mathcal{L}_0)$, one starts with the algebra of equivariant currents on $\mathbb{T}^\vee$, with respect to an outer automorphism τ of $\mathcal{L}_0$ (i.e. an automorphism given up to an inner automorphism by the Dynkin diagram of $\mathcal{L}_0$). More precisely, if τ is an outer automorphism of $\mathcal{L}_0$, of order k (k=2 or 3), and if ρ is an automorphism of $\mathbb{T}^\vee$ of the same order k, the algebra of $\mathcal{L}_0$-valued τ-equivariant currents on $\mathbb{T}^\vee$ is given by the elements x of $P(\mathbb{T}^\vee, \mathcal{L}_0)$ such that:

$$\tau . [x \circ \rho] = x.$$

The construction is then similar to that of the untwisted case, and leads to root systems of the form:

$$\mathcal{R} \cong (\mathcal{R}_\ell; \nu, 1)$$

If τ is inner, the corresponding $\mathcal{R}$ has twist 0. Twist two quasisimple Lie algebras can be realised in a similar way, starting from the automorphisms of the Dynkin diagram of the Kac-Moody Lie algebras. One can then realise all the twist two elliptic quasisimple Lie algebras, as explained in [HK.To].

III: HIGHEST WEIGHT REPRESENTATIONS OF GAUGE GROUPS AND ALGEBRAS:

We describe here the concept of highest weight representations, and the construction of continuous tensor product representations of gauge groups. Then, we develop the theory of unitary highest weight representations of gauge algebras, and make the connection with the continuous tensor product representations in a particular case.

1°) Unitary highest weight representations:

Let $\mathcal{L}$ be a complex Lie algebra, and assume that there exists a subalgebra $\mathcal{b}$ (Borel subalgebra) and an involutive antilinear antiautomorphism ω of $\mathcal{L}$, such that one has:

$$\mathcal{b} + \omega.\mathcal{b} = \mathcal{L}$$

Let Λ be a complex character of $\mathcal{b}$, and let $\pi: \mathcal{L} \to \mathrm{End}(\mathcal{V})$ be a representation of $\mathcal{L}$, such that there exists a vector v_Λ in $\mathcal{V}$ fulfilling:

. $\pi(b).v_\Lambda = \Lambda(b)\, v_\Lambda$ for any b in $\mathcal{b}$,

. $\pi(\omega.\mathcal{b}).v_\Lambda = \mathcal{V}$.

Then π is called a highest weight representation of $\mathcal{L}$, with highest weight Λ.

Moreover, if there exists a Hermitian form $\mathcal{H}$ on $\mathcal{V}$ such that:

. $\mathcal{H}(v_\Lambda, v_\Lambda) = 1$

. $\mathcal{H}(\pi(\ell).u, v) = \mathcal{H}(u, \pi(\omega.\ell).v)$ for all u,v in $\mathcal{V}$, and all ℓ in $\mathcal{L}$.

then the representation is said to be unitarisable. Indeed, if one defines the real form associated with ω by:

$$\mathcal{L}_\omega = \left\{ \ell \in \mathcal{L} \text{ s.t. } \omega.\ell = -\ell \right\}$$

then a unitarisable representation of $\mathcal{L}$, becomes a unitary representation when restricted to $\mathcal{L}_\omega$.

2) Continuous tensor product representations:

Let now G be a real locally compact Lie group, with Lie algebra $\mathcal{G}$, and assume that X is provided with a finite positive Radon measure μ. Then (following e.g. [Gui], [Mar 89]), one can associate to $\overline{\mathcal{M}_0}(X,G)$, and then by restriction to $\mathcal{D}(X,G)$ a unitary representation as follows: Let $\{O_x\}_{x\in X}$ be a continuous μ-measurable field of unitary representations of the group G:

$$O_x:\ G \rightarrow \mathcal{U}(\mathcal{K}_x)$$

($\mathcal{U}(\mathcal{K}_x)$ meaning the unitary operators on $\mathcal{K}_x$) and $\{b_x\}_{x\in X}$ an associated continuous μ-measurable field of one-cocycles:

$$b_x:\ G \rightarrow \mathcal{K}_x$$

i.e. fulfilling: $\quad b_x(g.g') = b_x(g) + O_x(g).b_x(g') \qquad g,g' \in G$

such that for all x in X, $b_x(G)$ is total in $\mathcal{K}_x$. Set, for all f in G^X:

$$\tilde{O}(f) = \int_X^{\oplus} O_x[f(x)]\, d\mu(x)$$

$$\tilde{b}(f) = \int_X^{\oplus} b_x[f(x)]\, d\mu(x)$$

and:

$$\tilde{\mathcal{K}} = \int_X^{\oplus} \mathcal{K}_x\, d\mu(x)$$

$\tilde{O}$ is a linear representation of G^X, and one has that $\tilde{b} \in Z^1(G^X,\tilde{\mathcal{K}})$. One then associates to these data a representation of G^X as follows: consider the Fock space:

$$\tilde{\mathcal{H}} = \mathrm{Exp}\left[\tilde{\mathcal{K}}\right] = \left[S\ \tilde{\mathcal{K}}\right]_C \cong L^2(\tilde{\mathcal{K}}', \nu)$$

(ν being a Gaussian measure on $\tilde{\mathcal{K}}$, defined by its Fourier transform). We recall that the following exponential vectors (or coherent states):

$$\mathrm{Exp}[\tilde{k}] = \sum_{n=0}^{\infty} \frac{\tilde{k}^{\otimes n}}{\sqrt{n!}}, \qquad \tilde{k} \in \tilde{\mathcal{K}}$$

are total in $\mathrm{Exp}\left[\tilde{\mathcal{K}}\right]$.

The representation $\tilde{\mathcal{U}} = \mathrm{Exp}\left[\tilde{O}\right]$ is then defined by its action:

. On $L^2(\tilde{\mathcal{K}}, \nu)$: if $f \in G^X$ and $\Phi \in L^2(\tilde{\mathcal{K}}', \nu)$

$$\left[\tilde{\mathcal{U}}(f).\Phi\right](F) = e^{i<F,\tilde{b}(f)>}\,\Phi\left[\tilde{O}^{-1}(f).F\right] \quad \text{for any F in } L^2(\tilde{\mathcal{K}}, \nu)$$

. On $\mathrm{Exp}\left[\tilde{\mathcal{K}}\right]$:

$$\tilde{\mathcal{U}}(f).\mathrm{Exp}[h] = e^{-\|\tilde{b}(f)\|^2/2 - <\tilde{O}(f).h,\tilde{b}(f)>}\,\mathrm{Exp}\left[\tilde{O}(f).h + \tilde{b}(f)\right]$$

Remark:

<,> denotes here the Hermitian product on $\tilde{\mathcal{K}}$, inherited from those of the $\mathcal{K}_x$ spaces, according to:

$$<\tilde{k}, \tilde{k}'> = \int_X^{\oplus} <k_x,k'_x>\, d\mu(x)\,, \qquad \forall\ \tilde{k},\tilde{k}' \in \tilde{\mathcal{K}}.$$

Generically, $\tilde{\mathcal{U}}$ is a projective representation of G^X, which means that one has:

$$\tilde{\mathcal{U}}(\phi).\tilde{\mathcal{U}}(\phi') = e^{i\omega(\phi,\phi')}\,\tilde{\mathcal{U}}(\phi.\phi')$$

$$\omega(\phi,\phi') = \mathit{Im} <\tilde{b}(\phi.\phi'),\tilde{O}(\phi).\tilde{b}(\phi')>$$

Otherwise stated, $\omega \in Z^2(G,\mathbb{C})$; if $\omega \notin B^2(G,\mathbb{C})$, it is necessary to go to the universal covering group $\hat{G}$ of G; it is then possible to find a one-cocycle $\gamma \in Z^1(\hat{G},\mathbb{C})$ which trivialises ω, i.e. $\hat{\omega}(\phi,\phi') = \gamma(\phi) + \gamma(\phi') - \gamma(\phi.\phi')$ is a lifting of ω to $\hat{G}$.

Introducing now the new representation:

$$\tilde{\mathcal{V}}(\phi) = e^{-\gamma(\phi)}\,\tilde{\mathcal{U}}(\phi)$$

One has that $\tilde{\mathcal{V}}$ is a unitary linear representation of G^X (the proof uses the unitarity of $\tilde{O}$, and the cocycle property of $\tilde{b}$). Moreover, the following result was obtained by Delorme [De 77] and Vershik, Gelfand and Graev [V.Ge.Gr. 73-74].

Theorem 4:

i) *If* μ *is non-atomic and* $\beta_x(K)$ *is identically equal to 0,* K *being some maximal compact subgroup of* G, *then the representation* $\tilde{\mathcal{V}}$ *is irreducible.*

ii) *A necessary condition for* $\tilde{\mathcal{V}}$ *to be irreducible is that* $\beta_x \notin B^1(G,\mathcal{K}_x)$ *for almost all* x *in* X.

Remarks:

. If all the $\mathcal{K}_x$ are isomorphic: $\mathcal{K}_x \cong \mathcal{K}$, then:

$$\tilde{\mathcal{H}} \cong L^2(X,\mathcal{K},d\mu) \cong L^2(X,d\mu) \otimes \mathcal{K}$$

. If p: $\hat{G} \to G$ is the canonical projection, $\hat{O}_x = O_x \circ p$ defines a unitary representation of $\hat{G}$, and $H^1(\hat{G}, \mathcal{K}_x) \cong H^1(G,\mathcal{K}_x)$.(see [De 77]).

To make the connection with the representation theory developed in the framework of quasisimple Lie algebras, it is useful to study the infinitesimal representation of $\tilde{\mathcal{V}}$. Let $\sigma \in \mathcal{G}^X$, and consider the one parameter subgroup $f_t = \exp(t\sigma)$, $t \in \mathbb{R}$. Define by:

$$\pi(\sigma) = \left\{\frac{d}{dt}\ \tilde{\mathcal{V}}(f_t)\right\}_{t=0}$$

the infinitesimal representation π of $\tilde{\mathcal{V}}$, and set:

$$\tilde{\theta}(\sigma) = \left\{\frac{d}{dt}\ \tilde{O}(f_t)\right\}_{t=0} \in \mathrm{End}(\tilde{\mathcal{K}})$$

$$\tilde{v_\sigma} = \left\{\frac{d}{dt}\ \tilde{b}(f_t)\right\}_{t=0} \in \tilde{\mathcal{K}}$$

$$\tilde{\varphi}(\sigma) = \left\{\frac{d}{dt}\ \tilde{\psi}(f_t)\right\}_{t=0} \in \mathbb{C}.$$

One then has:

Proposition 5:

π acts as follows on the coherent vectors:

$$\pi(\sigma) . \mathrm{Exp}[h] = \left\{ -i\tilde{\varphi}(\sigma) - <h,\tilde{v_\sigma}> \right\} \mathrm{Exp}[h]$$

$$+ \sum_{n=0}^{\infty}{}^{\oplus} \frac{n-1}{\sqrt{(n-1)!}} \left[\tilde{\theta}(\sigma).h + \tilde{v_\sigma} \right] \otimes h^{\otimes n} \quad , \forall\, h \in \tilde{\mathcal{K}}$$

The proof of the proposition follows from direct computation of the infinitesimal representation of $\tilde{\mathcal{V}}$. π is then a linear unitary representation of G^X on the Fock space $\tilde{\mathcal{H}}$. Moreover, if μ is non-atomic, π is irreducible.

It is not very difficult to check that if $\tilde{b}$ and $\tilde{b}'$ are cohomologous representation-valued one-cocycles of G^X, the corresponding exponential representations are equivalent. One then has that an exponential representation associated to a trivial one-cocycle (that is a one-cocycle of the form $b(g) = \pi(g).v - v$) is equivalent to a highly reducible representation (see Theorem 4).

Unfortunately, the one-cohomology of usual Lie groups is in general fairly poor, and this implies that one can construct only a few non-trivial continuous tensor product representations of the corresponding gauge groups. More precisely, one has that:

. If G is semisimple compact, then $H^1(G,\mathcal{V})$ is trivial for all unitary G-modules $\mathcal{V}$.

. If G is simple, different than SU(n,1) and SO(n,1), then $H^1(G,\mathcal{V})$ is trivial for all unitary G-modules $\mathcal{V}$.

. If G=SU(n,1) or SO(n,1) for some integral number n, then there exists one or two unitary representations on a Hilbert space $\mathcal{V}$ such that $H^1(G,\mathcal{V})$ is nontrivial.

To construct new representations using the continuous tensor product technique, in particular in the case of compact groups, it is then necessary to go to extensions of the considered groups. In particular, starting from the first order Leibnitz extensions [A.HK.Te],[Ge.Gr.V 77-81],[Mar 83],[Pa.Sc],[Wa], owing to lemma 1, one ends with the so-called energy representations, which will be briefly described in the conclusion. Up to now, higher order Leibnitz extensions [A.HK.To] do not seem to provide non equivalent unitary representations.

3) Unitary highest weight representations of gauge algebras:

We now adopt the infinitesimal point of view, in the framework of elliptic quasisimple Lie algebras. Let us first focus on the type one elliptic quasisimple Lie algebras, i.e. the affine Kac-Moody algebras. Jakobsen and Kac [Ja.Kac] classified the irreducible unitary highest weight representations of such Lie algebras, and we start the discussion by a brief sketch of their result. The essential part of the argument is the analysis of the $sl(2)^{\mathbb{T}^1}$-subalgebras of the considered affine Lie algebra . Let us first consider the complex Lie algebra $\mathcal{L}_0 = sl(2,\mathbb{C})$, and let $\mathcal{L}$ be the associated untwisted affine Kac-Moody Lie algebra, which can be written as:

$$\mathcal{L} = \sum_{n\in\mathbb{Z}}^{\oplus} \mathbb{C}e_n \oplus \sum_{n\in\mathbb{Z}}^{\oplus} \mathbb{C}h_n \oplus \sum_{n\in\mathbb{Z}}^{\oplus} \mathbb{C}f_n \oplus \mathbb{C}c \oplus \mathbb{C}d$$

where one has set:

$$e_n : t\in\mathbb{T}^1 \to t^n\, e \in sl(2,\mathbb{C})$$
$$f_n : t\in\mathbb{T}^1 \to t^n\, f \in sl(2,\mathbb{C})$$
$$h_n : t\in\mathbb{T}^1 \to t^n\, h \in sl(2,\mathbb{C})$$

e,f,h being a set of Chevalley generators of $sl(2,\mathbb{C})$.

The possible inequivalent Borel subalgebras (see [Ja.Kac] for a detailed analysis) are given by:

$$\mathcal{b}^{st} = \sum_{n=0}^{\infty\,\oplus} \mathbb{C}e_n \oplus \sum_{n=0}^{\infty\,\oplus} \mathbb{C}h_n \oplus \sum_{n=1}^{\infty\,\oplus} \mathbb{C}f_n \oplus \mathbb{C}c \oplus \mathbb{C}d$$

$$\mathcal{b}^{nat} = \sum_{n=-\infty}^{\infty\,\oplus} \mathbb{C}e_n \oplus \sum_{n=0}^{\infty\,\oplus} \mathbb{C}h_n \oplus \mathbb{C}c \oplus \mathbb{C}d$$

corresponding to the following complex structures:

$$\omega^{st}:\ e_0 \to f_0 \quad , \quad h_i \to h_i \quad (i\in\mathbb{Z})$$
$$\omega^{nat}:\ e_0 \to -f_0 \quad , \quad h_i \to h_i \quad (i\in\mathbb{Z})$$

and respectively to the following real forms: $su(2)^{\mathbb{T}^1}$ and $su(1,1)^{\mathbb{T}^1}$.

In the first case ("standard case"), one easily shows that:

$$\Lambda(e_n) = 0 \qquad \forall\, n \geq 0$$
$$\Lambda(f_n) = 0 \qquad \forall\, n > 0$$
$$\Lambda(h_n) = 0 \qquad \forall\, n \neq 0$$

Λ is then completely determined by $\Lambda(h_0)$ and $k=\Lambda(c)$ (k is called the level of the representation). This is the usual irreducible integrable representation theory of compact affine Kac-Moody algebras, in the su(2) case.

In the second case ("natural case"), one easily shows that

$$\Lambda(e_n) = 0 \qquad \forall\, n \in \mathbb{Z}.$$
$$\Lambda(c) = 0$$

and Λ is completely determined by the values of $\Lambda(h_n)$, then by a finite positive Radon measure on the circle $\mathbb{T}^1$.

In both cases, according to the standard irreducible highest weight representation theory, the representation is completely determined by Λ, and can be realised via the Verma module construction, i.e. as the unique irreducible quotient of the Verma module:

$$M(\Lambda) = \frac{U(\mathcal{L})}{U(\mathcal{L})\left\{ b-\Lambda(b), b\in \mathcal{b}\right\}}$$

($U(\mathcal{L})$ denoting the universal enveloping algebra of $\mathcal{L}$). Moreover, in the su(1,1) case, the Verma module is itself irreducible.

Considering now a generic affine Kac-Moody algebra $\mathcal{L}$, one has to "patch together" the results obtained in the sl(2,$\mathbb{C}$)-subalgebras. Jakobsen and Kac obtained the following result:

Theorem 6:

The only non-located irreducible highest weight unitary representations of a real affine Kac-Moody algebra $\mathcal{G}$ are given by:

_ $\mathcal{G}$ *compact: integrable representations, characterised by a set of $Rk(\mathcal{G})$ non-negative integral numbers, among which the level* k, *eigenvalue of the canonical central element* c , *the vanishing of which implies the triviality of the representation.*

_ $\mathcal{G}$ = su(n,1)*: exceptional representations, characterised by a positive non-atomic σ-finite measure on the circle.*

The proof of the theorem is given in [Ja.Kac], and a complete description of the integrable representation theory is given in [Kac 85].

Let us now turn to the case of an untwisted elliptic quasisimple Lie algebra $\mathcal{L}$ of type $\nu > 1$. $\mathcal{L}$ has a ν-dimensional center. Let π be an irreducible unitarisable highest weight representation of $\mathcal{L}$, associated with the compact involution. There then exists a (ν-1)-dimensional subspace of the center of $\mathcal{L}$ that is trivially represented by π. Since to each one-dimensional subspace of the center one can associate an affine Kac-Moody subalgebra of $\mathcal{L}$, one then deduces that the representaton π, restricted to the Kac-Moody $\mathcal{L}$-subalgebras associated with this (ν-1)-dimensional subspace of the center are trivially represented. This in turn implies the non-faithfulness of the representation. This result extends to twisted elliptic quasisimple Lie algebras, and one is led to the result:

Proposition 7:

An elliptic quasisimple Lie algebra of type $\nu > 1$ *does not admit faithful unitarisable representation associated with the compact involution.*

This proposition was proved in [To 87] , to which we refer for a more detailed analysis and the classification of the (non-faithful) unitarisable highest weight representations of elliptic quasisimple Lie algebras. According to theorem 6, one has the following corrolary:

Corollary:
The only possible unitary highest weight representations of real elliptic quasisimple Lie algebras occur in the case $\mathcal{G}$=su(n,1).

It turns out that such exceptional representations do exist. This was proved by Jakobsen and Kac in the affine case, and their proof generalises to the general case; the representation is then completely determined by a positive σ-finite measure on the torus $\mathbb{T}^{\nu}$. Moreover, in that case, the Verma module is irreducible.

4) The su(1,1) case:

It is possible to make the connection between the algebraic approach and the continuous tensor product approach in the case of su(n,1). There is a great similarity in these two approaches, since in both cases the central extension disappears, ant the representation is completely determined by a positive σ-finite measure on the source space X. Actually, the exceptional representations of elliptic quasisimple Lie algebras $su(n,1)^{\mathbb{T}^{\nu}}$ are nothing but the infinitesimal representations of the continuous tensor product representations of $SU(n,1)^{\mathbb{T}^{\nu}}$. This was shown first by P. Delorme [De 88] in the case of the affine su(1,1). We will prove here that the continuous tensor product representations of $su(1,1)^X$ are highest weight representations. The generalisation of this result to the su(n,1) case, together with the corollary of the last section, would achieve the classification of unitarisable representations of elliptic quasisimple Lie algebras, begun in [To 87].

Let now G=SU(1,1) (≅SL(2,ℝ)), and consider the space:

$$\mathcal{K} = L^2_{hol}(\mathcal{D}, r\, dr\, d\theta)$$

$\mathcal{D}$ being the unit disk of the complex plane. Denote by λ_0 the unity function on $\mathcal{D}$, defined by:

$$\lambda_0(z) = 1$$

Consider the representation:

$$O: G \to \mathcal{U}(\mathcal{K})$$

defined by: if $g=\begin{pmatrix} a & b \\ \bar{b} & \bar{a} \end{pmatrix} \in SU(1,1)$:

$$O(g): f \in \mathcal{K} \rightarrow O(g).f$$

$$[O(g).f](z) = \left[-\bar{b}\, z + a\right]^{-2} f\left(\frac{\bar{a}\, z - b}{-\bar{b}\, z + a}\right) \qquad \forall\, z \in \mathcal{D}.$$

O lies in the discrete series of unitary representations of sl(2,$\mathbb{R}$), with highest weight -2. Let $\tilde{\mathcal{V}}$ be the associated exponential representation. We will keep the notations of section (III-2).

Let θ be the infinitesimal representation of O. θ acts on $\mathcal{K}$ according to the following: if $\sigma = \begin{pmatrix} \alpha & \beta \\ \bar{\beta} & \bar{\alpha} \end{pmatrix} \in su(n,1)$ (i.e. $\alpha \in i\,\mathbb{R}$, $\beta \in \mathbb{C}$)

$$\left[\theta(\sigma).f\right](z) = \left\{ \left[\bar{\beta}\, z - \alpha\right] f + \left[\bar{\beta}^2 z^2 + (\bar{\alpha} - \alpha) z - \beta\right] \partial_z f \right\}(z) \qquad \forall\, z \in \mathcal{D}.$$

The $\mathcal{K}$-valued one-cocycle is unique up to equivalence, and is given by:

$$[b(g)](z) = \frac{\bar{b}}{-\bar{b} + \bar{a}\, z} \qquad \forall\, z \in \mathcal{D}.$$

Its infinitesimal counterpart reads:

$$v_\sigma = \left\{\frac{d}{dt}\, b(\exp(t\sigma))\right\}_{t=0} = \bar{\beta}\, \lambda_0 \in \mathcal{K}$$

SL(2,$\mathbb{R}$) is not simply connected, and it is necessary to go to the universal covering group SL(2,$\mathbb{R}$)^.The corresponding element in H^1(SL(2,$\mathbb{R}$)^,$\mathbb{C}$) which trivialises ω is given by:

$$\psi(g) = a \in \mathbb{C}, \qquad \forall\, g = \begin{pmatrix} a & b \\ \bar{b} & \bar{a} \end{pmatrix} \in SU(1,1).$$

The corresponding infinitesimal element reads:

$$\varphi(\sigma) = \alpha \in \mathbb{C}, \quad \forall\, \sigma = \begin{pmatrix} \alpha & \beta \\ \bar{\beta} & \bar{\alpha} \end{pmatrix} \in su(1,1).$$

Now, consider the corresponding cocycles associated to the gauge group:

$$\tilde{\psi}(g) = \int_X a(x)\, d\mu(x) \in \mathbb{C}, \qquad \forall\, g = \begin{pmatrix} a & b \\ \bar{b} & \bar{a} \end{pmatrix} \in SU(1,1)^X.$$

At the infinitesimal level:

$$\tilde{\varphi}(\sigma) = \int_X \alpha(x)\, d\mu(x) \in \mathbb{C}, \quad \forall\, \sigma = \begin{pmatrix} \alpha & \beta \\ \bar{\beta} & \bar{\alpha} \end{pmatrix} \in su(1,1)^X.$$

Remark:

It may be convenient to work at the complex level, i.e. with the complexified group $L = G_{\mathbb{C}}$ = SL(2,$\mathbb{C}$), and Lie algebra $\mathcal{L} = \mathcal{G}_{\mathbb{C}}$= sl(2,$\mathbb{C}$). The corresponding complexified $\theta(\sigma)$, v_σ and $\varphi(\sigma)$ are given by: if $\sigma = \begin{pmatrix} \alpha & \beta \\ \gamma & \delta \end{pmatrix} \in sl(2,\mathbb{C})^X$, $(\delta = -\alpha)$:

$$[\theta(\sigma).f](z) = [(\gamma z-\alpha)\, f + (\gamma^2 z^2 + (\delta-\alpha)\, z - \beta)\, \partial_z f](z)$$

$$v_\sigma(z) = \gamma \lambda_0$$

$$\varphi(\sigma) = \alpha.$$

One is then ready to prove the following result:

Proposition 8:

Let π *be the infinitesimal representation of* $\tilde{\mathcal{V}}$*; then* π *is a unitary highest weight representation of* $sl(2,\mathbb{R})^X$*, whose highest weight is completely determined by a positive* σ*-finite measure on* X.

Proof: One only has to check that π admits a highest weight vector. Actually, one already knows that the "vacuum vector" Ω=Exp[0] is cyclic for $\tilde{\mathcal{V}}$. Consider the following generators of $sl(2,\mathbb{C})^X$:

$$e(\xi) = \begin{pmatrix} 0 & \xi \\ 0 & 0 \end{pmatrix}, \quad h(\xi) = \begin{pmatrix} \xi & 0 \\ 0 & -\xi \end{pmatrix}, \quad f(\xi) = \begin{pmatrix} 0 & 0 \\ \xi & 0 \end{pmatrix}, \quad \xi \in \mathcal{D}(X).$$

One then has that:

$$\pi[e(\xi)]\,.\,\Omega = 0$$

$$\pi[f(\xi)]\,.\,\Omega = \tilde{v}_{f(\xi)} = \int_X^{\oplus} \xi(x)\, \lambda_0\, d\mu(x)$$

$$\pi[h(\xi)]\,.\,\Omega = \left[-\, i \int_X \xi(x)\, d\mu(x)\right] \Omega$$

and Ω is a highest weight vector for π.

The corresponding highest weight is then completely determined by the μ measure on X.

One can get an algebraic description of this representation as follows:

Taking as a Borel subalgebra the subalgebra generated by upper triangular matrices and diagonal matrices, the highest weight is equal to 0 on the upper triangular matrices, and to $\left[-\, i \int_X \xi(x)\, d\mu(x)\right]$ on the diagonal matrices (we recall that with respect to the convention we choosed for the definition of the real form of a complex Lie algebra, $h(\xi) \in su(1,1)^X$ if and only if ξ takes it values in $i\,\mathbb{R}$.

One can then build the $L(\Lambda)$ module from these data. Note that in the case $X=\mathbb{T}^1$, this corresponds to the natural Borel subalgebra. It is interesting to notice that in this case (we recall that the μ measure is assumed to be non-atomic), the Verma module is irreducible.

IV: CONCLUSIONS:

We gave here a description of the theory of irreducible unitary highest weight representations of gauge Lie groups and gauge Lie algebras. One of the main conclusions is that in the compact case, such a representation theory is much richer for dim(X)=1. For higher dimensional space manifolds, one only gets degenerate representations. In the noncompact cases, the representation theory is nontrivial if and only if G=SU(n,1) for some positive n, independently of the dimensionality of X. We described here the su(1,1) case, and the su(n,1) case will be discussed elsewhere. As a corollary, the result for G=SU(n,1) achieves the classification of unitarisable highest weight representations of elliptic quasisimple Lie algebras, started in [To 87].

Nontrivial representations can be obtained by relaxing some of the hypotheses: for instance, unitary representations of compact gauge groups which are not of the highest weight type can be constructed, like the energy representations, or by some constructions inspired by quantum field models (see e.g. [Ha], in connection with the non-compact Wess-Zumino-Witten model).

The continuous tensor product construction can also be applied to the compact case. Indeed, owing to lemma 1, it turns out that the cohomology group $H^1(G_m^1, A(m))$ is non trivial. Assume that X is a Riemannian manifold, and that G is compact, so that its adjoint representation on A(m) is unitary. Taking $\mathcal{K}_x = A(m)$, $\sigma_x = Ad$, $\beta_x(g,a) = a$ for all x in X, ant taking for μ the volume measure of X multiplied by some strictly positive C^∞ density ρ, one ends up with a continuous tensor product representation of $\left[G_m^1\right]^X$, called the energy representation, which is irreducible. The next problem is to know whether it remains irreducible when restricted to G^X. This has been studied by many authors, and the answer is the following [A.HK.T],[Wa]:

. For $m=\dim(X) \geq 3$, the representation remains irreducible.

. For m=2, the reducibility of the representation only depends on the value of a parameter $\alpha(x)=\sqrt{8\pi\rho(x)}$, $x \in X$. The energy representation is irreducible if and only if all the roots λ of the Lie algebra $\mathcal{G}$ of G satisfy: $|\lambda| > \alpha(x)$ for all x in X ($|\lambda|$ being the length of the root vector λ in $\mathcal{G}$.

. For m=1, the energy representation is highly reducible, and contains as a cyclic component the left representation of G^X on the space of brownian trajectories on G.

In the general Kac-Moody case, note that a classification result has been obtained for the representations such that the action of the Cartan subalgebra can be diagonalised and that the weight space is finite-dimensional [Ch.Pr].

Some non-unitary representations can also be of some interest, in particular in connection with quantum field theory, and more specially in the framework of modular invariant quantum field theories. Kac and Wakimoto gave a classification of modular invariant representations of affine Kac-Moody Lie algebras (see [Kac.Wak]).

Acknowledgments:

This paper is a report on the representation theory of current groups, with emphasis on the work done in my PhD Thesis, with Raphael Høegh-Krohn as thesis advisor, and recent extensions. I learnt from Raphael almost everything I know on this field, and he was an extraordinary source of inspiration. His contribution to the representation theory of infinite-dimensional groups is very important.

I am also very indebted to S. Albeverio, P. Delorme, V.G. Kac, J. Marion and D. Testard for many stimulating discussions.

REFERENCES:

[A.HK.78]: S. Albeverio, R. Høegh-Krohn: *'The energy representation of Sobolev-Lie groups'*, Comp. Math. 36 (1978) pp. 37-52.

[A.HK.M.Te.To]: S. Albeverio, R. Høegh-Krohn, J. Marion, D. Testard, B. Torrésani: *'Non-Commutative distributions: Unitary representations of gauge groups and algebras'*, book to appear, Marcel Dekker Publ. Comp.

[A.HK.Te]: S. Albeverio, R. Høegh-Krohn, D. Testard: *'Irreducibility and reducibility for the energy representation of the group of mappings of a Riemmannian manifold into a compact semisimple Lie group'*, J. Funct. An. 41 (1981), pp. 378-396.

[A.HK.Te.V]: S. Albeverio, R. Høegh-Krohn, D. Testard, A.M. Vershik: *'Factorial representations of path groups'*, J. Funct. An. 51 (1983), pp. 115-131.

[A.HK.To]: S. Albeverio, R. Høegh-Krohn, B. Torrésani: *'Some remarks on the representation of jet groups and gauge groups'*, to appear (1991).

[Ar]: H. Araki: *'Factorisable representations of current algebras'*, Publ. RIMS A5 (1970) pp. 361-422.

[Bou]: N. Bourbaki: *'Groupes et algèbres de Lie'*, chapitres 4,5,6, Hermann.

[Ch.Pr]: V. Chari, A. Pressley: *'Integrable representations of Kac-Moody Lie algebras: results and open questions'*, in "Infinite-dimensional Lie algebras and groups", Adv. Series in Math. and Phys. 7, V.G. Kac Ed. (1988), World Scientific.

[De 77]: P. Delorme: "*1-cohomologie des représentations unitaires des groupes de Lie semisimples et résolubles. Produits tensoriels continus de représentations'*, Bull. Soc. Math. France 105 (1977), pp. 281-336.

[De 88]: P. Delorme: Private communication.

[Ge.Gr]: I.M. Gelfand, M.I. Graev: *'Representations of quaternion groups over locally compact and functional fields'*, Funk. Anal. Pril. 2 (1968), pp. 20-35.

[Ge.Gr.V 77]: I.M. Gelfand, M.I. Graev, A.M. Vershik: *'Representations of the group of functions taking values in a compact Lie group'*, Comp. Math. 35 (1977), pp. 299-334.

[Ge.Gr.V 81]: I.M. Gelfand, M.I. Graev, A.M. Vershik: *'Remarks o the representation of the group of functions with values in a compact Lie group'*, Comp. Math. 42 (1977), pp. 217-343.

[Gui]: A. Guichardet: *'Symmetric Hilbert spaces and related topics'*, lect. Notes in Math. 261 (1972), Springer.

[Ha]: Z. Haba: *'Representations of the SL(2,C) Kac-Moody current algebra in the Fock space'*, J. Math. Phys. 32 (1991), p. 19.

[HK.To]: R. Høegh-Krohn, B. Torrésani: *'Classification and construction of quasisimple Lie algebras'*, J. Funct. An. 89 (1990), pp. 106-136.

[I]: R. Ismagilov: *'On unitary representations of $C_0(X,G)$, $G=SU(2)$'*, Math. USSR.Sb. 29 (1976), pp. 105-117.

[Ja.Kac]: H.P. Jakobsen, V.G. Kac: *'A new class of unitarizable highest weight representations of infinite-dimensional Lie algebras'*, Lect. Notes in Phys. 226 (1985), pp. 1-20.

[Kac 67]: V.G. Kac: *'Simple graded Lie algebras of finite growth'*, Funct. An. and Appl. 1 (1967), pp. 328-329.

[Kac 85]: V.G. Kac: *'Infinite-dimensional Lie algebras'*, Cambridge University press, 2nd edition (1985).

[Kac.Wak]: V.G. Kac, M. Wakimoto: *'Modular and conformal invariance constraints in representation theory of affine algebras'*, Adv. Math. 70 (1988), pp. 156-236.

[Mac]: I.G. Mac Donald: *'Affine root systems and Dedekind's η function'*, Invent. Math. 15 (1972), pp. 91-143.

[Mar 83]: J. Marion: *'Generalized energy representations for current groups'*, J. Funct. An. 54 (1983), pp. 1-17.

[Mar 89]: J. Marion: *'Introduction aux groupes de Lie fonctionnels et à leurs représentations unitaires'*, IRMA publication 10 (1989), Cote d'Ivoire.

[Mi]: J. Mickelsson: *'Two cocycle of a Kac-Moody group'*, Phys. Rev. Lett. 55 (1985), pp. 2099-2102.

[Mo]: R.V. Moody: *'A new class of Lie algebras'*, Bull Am. Math. Soc. 73 (1967), pp. 217-221.

[Mu]: M.K. Murray: *'Another construction of the central extension of the* *loop group'*, Comm. Math. Phys. 116 (1988), pp. 73-80.

[Pa.Sc]: K. Parthasarathy, K. Schmidt: *'A new method for constructing factorisable representations for current groups and current algebras'*, Comm. Math. Phys. 50 (1976), pp. 167-175.

[Pr.Se]: A. Pressley, G. Segal: *'Loop groups'*, Oxford University press, Oxford (1986).

[Ti]: J. Tits: *'Groupes associés aux algèbres de Kac-Moody'*, Séminaire Bourbaki n° 700 (1988).

[To 86]: B. Torrésani: *'Représentations projectives des groupes de transformations de jauge locales'*, Thesis, Marseille (1986).

[To 87]: B. Torrésani: *'Unitary positive energy representations of the gauge group'*, Lett. Math. Phys. 13 (1987) pp. 7-15.

[V.Ge.Gr. 73]: A.M. Vershik, I.M. Gelfand, M.I. Graev: *'Representations of the group $SL(2,\mathcal{R})$ where $\mathcal{R}$ is a ring of functions'*, Russ. Math. Surv. 28 (1973), pp. 83-132.

[V.Ge.Gr. 74]: A.M. Vershik, I.M. Gelfand, M.I. Graev: *'Irreducible representations of the group G^X and cohomology'*, Funkt. Anal. ego Pril. 8 (1974) pp. 67-69.

[V. Kar]: A.M. Vershik, S.I. Karpushev: *'Cohomology of groups associated to unitary reprtesentations, neighborhoods of the trivial representation, and conditionnaly positive definite functions'*, Math. USSR Sbornik 47 (1984), pp. 513-526.

[Wa]: N.R. Wallach: *'On the irreducibility and inequivalence of unitary representations of gauge groups'*, Comp. Math. 64 (1987), pp. 3-29.

Some Non-Commutative Orbifolds

David E. Evans

Department of Mathematics and Computer Science,
University College of Swansea, Singleton Park,
Swansea SA2 8PP, Wales.

This is a survey of recent work on some non commutative C*-algebras associated to certain orbifolds which have led to new phenomena in the theory of operator algebras. The starting point of this investigation was the constructions of Blackadar [Bla1] and Kumjian [Kum] of symmetries on some AF algebras whose fixed point algebras are not AF. A C*-algebra is AF (approximately finite dimensional) if it can be expressed as an inductive limit

$$A_1 \to A_2 \to \ldots\ldots$$

of finite dimensional C* -algebras. For example, for any integer $Q > 1$, let $\mathcal{F}_Q$ denote the infinite tensor product $\otimes_{\mathbb{N}} M_Q$ obtained as the C*-inductive limit of the sequence $A_n = \otimes^n M_Q \simeq M_Q{}^n$ with embeddings $x \to x \otimes 1_Q$ where M_Q denotes the $Q \times Q$ complex matrices with unit 1_Q. Blackadar [Bla 1] obtained such an exotic symmetry on the Fermion algebra $\mathcal{F}_2 = \otimes M_2$, by expressing this algebra in a non standard and surprising way as an inductive limit of algebras $C(\mathbb{T}) \otimes M_{4^n}$ using winding operations (both negative and positive) on the circle $\mathbb{T}$. Kumjian [Kum1] considered the Bunce-Deddens algebra, namely the crossed product $C(\mathbb{T}) \times D$ of the circle $\mathbb{T}$ by dyadic rotations D. This algebra $C(\mathbb{T}) \times D$ is not AF, but by throwing in the flip $z \to \bar{z}$ on $\mathbb{T}$, on obtains an AF algebra $(C(\mathbb{T}) \times D) \times \mathbb{Z}_2$ [Kum1]. Taking a dual $\hat{\mathbb{Z}}_2$ action one obtains an AF algebra and a symmetry whose fixed point algebra, essentially the Bunce-Deddens algebra, is not AF, but is of real rank zero [BK]. An AF algebra certainly contains many projections, indeed is generated linearly by them as a Banach space

- since a finite dimensional C*-algebra has this property. In fact a commutative C*-algebra C(X) (the continuous functions on a compact Hausdorff space) is AF if and only if X is totally disconnected. Thus the AF property is one non-commutative generalisation of a space being totally disconnected or zero dimensional, but there are others. One such is the notion of real rank zero [BP]. Real rank zero is a weaker notion than being approximately finite dimensional (but for commutative C*-algebras, they coincide). A (unital) C*-algebra is of real rank zero [BP] if the invertible self adjoint elements are dense in all self adjoint elements; equivalently one of the following conditions hold:

Self adjoint elements of finite spectrum are dense in all self adjoint elements.
Hereditary subalgebras have approximate identities consisting of projections.

The Bunce-Deddens algebra has non trivial K_1, (and so cannot be AF for this reason), but for real rank zero algebras built up, as the Bunce-Deddens algebra, and its crossed product with $\mathbb{Z}_2$, are from certain subhomogeneous algebras on the interval or the circle, the K_1 -obstruction is the only obstruction to the algebra being AF [Ell5]. Thus one is naturally led to consider the orbifold $\mathbb{T}/\mathbb{Z}_2$, represented or described via the non-commutative C*-algebra $C(\mathbb{T}) \times \mathbb{Z}_2$ which can be identified with the algebra

$$\mathbb{C}^2 \overset{M_2}{\overline{\qquad\qquad\qquad}} \mathbb{C}^2 \tag{1}$$

of M_2– valued functions on the unit interval $I = [0,1]$ which are diagonal (or lie in subalgebras isomorphic to $\mathbb{C}^2$) at the endpoints. The spectrum of this algebra can be represented diagrammatically by

$$:\overline{\qquad\qquad\qquad}: \tag{2}$$

Thus the crossed product of $C(\mathbb{T}) \otimes M_{2n}$ by the flip $z \to \bar{z}$ on $\mathbb{T}$, can also be represented by (2); it is the algebra of M_{2n+1} valued functions on the unit interval, whose values at the endpoints lie in some subalgebras isomorphic to $M_{2n} \oplus M_{2n}$. Kumjian's algebra $C(\mathbb{T} \times D) \times Z_2$ can be viewed as an inductive limit of algebras $C(\mathbb{T}) \otimes M_{2n}$. Thus began [Bla 1, Kum 1, BEEK1, BEEK2, BEEK3, EK, Ell 5, Kis, Bra Ell] a study of inductive limits of subhomogeneous algebras on intervals, circles, etc using folding and winding operations on the interval and circle respectively to construct embeddings. We will go into more detail of these embeddings below for folding operations on the interval. The following results [BEEK1] which can be regarded as generalisations of the results of Kumjian and Blackadar respectively were obtained by considering inductive limits of algebras

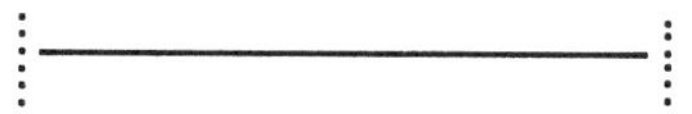

using folding operations on the interval where there are $|G|$ dots at the endpoints for some finite group G:

If G is a finite group with cardinality $|G| > 1$, then there exists an AF algebra $\mathcal{A}$ and an action α of G on $\mathcal{A}$ such that

(3) the dimension group of $\mathcal{A}$ is $\mathbb{Z}[|G|^{-1}] \oplus \mathbb{Z}^{|G|-1}$ and the positive cone is given by strict positivity of the first coordinate,

(4) $K_1(\mathcal{A}^{\alpha}) \cong \mathbb{Z}^{|\hat{G}|-1}$, where $\hat{G}$ is the dual of G, so in particular $\mathcal{A}^{\alpha}$ is not AF

(5) $K_0(\mathcal{A}^{\alpha}) \cong \mathbb{Z}[|G|^{-1}]$.

If G is a finite group with $|G| > 1$, and $\mathcal{A}$ is the UHF algebra $F_{|G|} = \otimes_{\mathbb{N}} M_{|G|}$, then there exists a faithful action α of G on $\mathcal{A}$ such that

(6) $K_1(\mathcal{A}^{\alpha}) \cong \mathbb{Z}[|G|^{-1}]^{|\hat{G}|-1}$, so in particular $\mathcal{A}^{\alpha}$ is not AF.

(7) $K_0(\mathcal{A}^{\alpha}) \cong \mathbb{Z}[|G|^{-1}]$.

[EK] considered another variation of this construction, and could show, for any integer $Q > 1$:

(8) There exists a C*-algegra A which is not AF, but

$$A \otimes (\otimes_{\mathbb{N}} M_Q) \simeq \otimes_{\mathbb{N}} M_Q$$

(9) For any compact group G, $(G \neq \{e\})$, there exists an action of G on the UHF algebra $\otimes_{\mathbb{N}} M_Q$ such that the fixed point algebra $(\otimes_{\mathbb{N}} M_Q)^G$ is not AF.

When G is finite, $Q = |G|$, the existence of an action as in (9) is contained in [BEEK1] (6 - 7) above. In particular any such action of a compact group on an AF algebra with non AF fixed point algebra cannot be locally representible in the sense of [HR]. Moreover (9) answers a question of Effros [Eff1, Eff2]; whether a C*-algebra A whose tensor product with an AF algebra is AF must necessarily be AF itself. Here we sketch the construction of the algebra A in (8) when $Q = 2$. This time the basic building block consists of algebras

$$\text{(10)} \qquad A_n = M_{2^{n-1}} \xrightarrow{\quad M_{2^n} \quad} M_{2^{n-1}}$$

consisting of M_{2^n}-valued functions on the unit interval [0,1], which at the end points lie in unital subalgebras isomorphic to $M_{2^{n-1}}$. Again A_n is embedded in A_{n+1} by a folding operation

$$\text{(11)} \qquad \begin{array}{c} M_{2^{n-1}} \xrightarrow{\quad M_{2^n} \quad} M_{2^{n-1}} \subset C([0,1], M_{2^n}) \\ \downarrow \qquad \qquad \qquad \downarrow \qquad \qquad \qquad \downarrow \\ M_{2^n} \underset{M_{2^{n+1}}}{\supset} M_{2^n} \qquad \begin{pmatrix} f(t/2) & \\ & f(1-t/2) \end{pmatrix} \end{array}$$

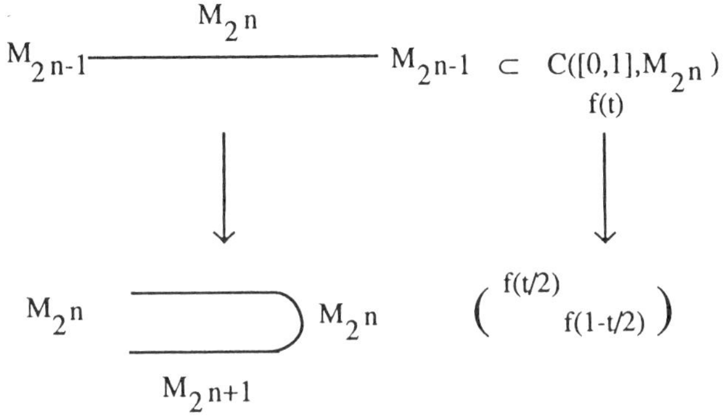

Remark: The position of the subalgebras $M_{2^{n-1}}$ in M_{2^n} is not important in defining A_n, i.e. changing their position will give an isomorphic algebra. To see this consider the following algebras

$$\text{(12)} \qquad A_n^{(i)} = B^{(i)} \xrightarrow{\quad M_{2^n} \quad} C^{(i)}$$

where $B^{(i)} \simeq C^{(i)} \simeq M_{2^{n-1}}$ are unital subalgebras, $i = 0, 1$. The subalgebras at the endpoints will be conjugate by unitaries u_0, u_1 in M_{2^n}, $B^{(1)} = u_0 B^0 u_0^*$,

$C^{(1)} = u_1 C^{(0)} u_1^*$. Choose a continuous path of unitaries $\{u_s : s \in [0,1]\}$ in M_{2n} connecting u_0 to u_1. Then $\mathrm{Ad}\, u_s$ will define an isomorphism of $A_n^{(0)}$ onto $A_n^{(1)}$. In particular, the subalgebras at the end points in (10) could be taken to be in the same position, but if we insisted on that choice, then we could not keep the simple form of the folding embedding in (11). For the precise choice of relative position of the two subalgebras see [BEEK2, EK].

To see that $A \otimes \mathcal{F}_2$ is AF, where $A = \varinjlim A_n$, we need to approximate $f_1, \ldots, f_r$ in A_n by a finite dimensional subalgebra in $A_{n+m} \otimes M_2$ for m large [Bra]. By uniform continuity, given $\varepsilon > 0$,

$$\|f_i(s) - f_i(t)\| < \varepsilon, \text{ for all } i=1,\ldots,r,$$

and $|s-t|$ sufficiently small. Thus after sufficiently many foldings the embedded functions $\tilde{f}_i$ in A_{n+m} satisfy.

$$\|\tilde{f}_i(s) - \tilde{f}_i(t)\| < \varepsilon \text{ for all } i = 1, 2, \ldots, r \text{ and all } s, t.$$

Letting $N = n+m$, consider $F_i(t) = \tilde{f}_i(0)$ in $C([0,1], M_{2N})$. Then $F_i(0) = \tilde{f}_i(0)$ certainly lies in the correct subalgebra (isomorphic to M_{2N-1}) at the left hand endpoint. But $F_i(1) = \tilde{f}_i(0)$ will not necessarily lie in the correct subalgebra (isomorphic to M_{2N-1}) at the right hand endpoint, say $M_{2N-1} \otimes 1 \subset M_{2N}$. But $\tilde{f}_i(0) \in M_{2N-1} \otimes M_2 \otimes 1_2$ a subalgebra of M_{2N}, certainly unitarily equivalent to $M_{2N-1} \otimes 1_2 \otimes M_2$. But $F_i \otimes 1_2$ will lie in

$$\begin{array}{ccc} & M_{2N-1} \otimes M_2 \otimes M_2 & \\ \hline M_{2N-1} \otimes M_2 \otimes 1_2 & & M_{2N-1} \otimes M_2 \otimes 1_2 \end{array}$$

which by the Remark is isomorphic to $A_N \otimes M_2$. Connecting this unitary implementing the equivalence between $M_{2N-1} \otimes M_2 \otimes 1_2$ and $M_{2N-1} \otimes 1_2 \otimes M_2$ to the identity back along the unit interval [0,1], we can adjust $F_i \otimes 1$ in $C([0,1], M_{2N}) \otimes M_2$ to an element G_i in $A_N \otimes M_2$ which approximates $\tilde{f}_i \otimes 1_2$ in $A_N \otimes M_2$, and so that $G_1, \ldots G_r$ generate a finite dimensional subalgebra of $A_N \otimes M_2$.

The K-theory of A_n (and subsequently $K_*(A) = \varinjlim K_*(A_n)$) can be computed from the ideal of functions vanishing at the endpoints, together with the 6-term exact sequence of K-theory. This yields

$$K_0(A_n) = \mathbb{Z}, \quad K_1(A_n) = \mathbb{Z}_2$$

and

$$K_1(A) = \mathbb{Z}[1/2], \quad K_1(A) = \mathbb{Z}_2 \tag{13}$$

The generator of $K_1(A_n) \simeq \mathbb{Z}_2$ is given by the unitary

$$t \to \begin{pmatrix} e^{2\pi i t} & & & & \\ & 1 & & & \\ & & \cdot & & \\ & & & \cdot & \\ & & & & \cdot \\ & & & & & 1 \end{pmatrix} \otimes 1_2 \in M_{2^{n-1}} \otimes M_2$$

The 2-torsion in $K_1(A_n)$ can be seen as follows. If U is any unitary in A_n, then since the interval $[0,1]$ and the unitary group of M_{2^n} are connected, let W be a continuous map from $[0,1]^2$ into $M_{2^{n-1}} \otimes M_2$ such that $W(0,t) = U(t)$, $W(1,t) = 1$. Then $V_s(t) = W(s,t) \otimes 1_2$ defines a path $\{V_s : s \in [0,1]\}$ of unitaries in the algebra

$$\begin{array}{ccc} & M_{2^{n-1}} \otimes M_2 \otimes M_2 & \\ \hline M_{2^{n-1}} \otimes M_2 \otimes 1_2 & & M_{2^{n-1}} \otimes M_2 \otimes 1_2 \end{array} \tag{14}$$

which using the Remark is isomorphic to $A_n \otimes M_2$, (c.f. the proof that $A \otimes \mathcal{F}_2$ is AF) with $V_0 = U \otimes 1_2$, $V_1 = 1_{A_n} \otimes 1_2$. Hence $2[U] = 0$ in $K_1(A_n)$. From the expression for the generator we see that under the isomorphism $K_1(A_n) \simeq \mathbb{Z}_2$, the inclusion $A_n \to A_{n+1}$ induces the identity map on $\mathbb{Z}_2$, and so $K_1(A) \simeq \mathbb{Z}_2$. Thus there is at least a

K_1-obstruction to A being an AF algebra. Tensoring with $\mathcal{F}_2 \simeq \otimes_{\mathbb{N}} M_2$ destroys this K_1 as $K_1(A \otimes \mathcal{F}_2) \cong \varinjlim K_1(A \otimes M_{2^n}) \simeq \varinjlim K_1(A) \simeq 0$ as $K_1(A \otimes M_{2^n}) \simeq K_1(A)$ with $K_1(A \otimes M_{2^n}) \to K_1(A \otimes M_{2^{n+1}})$ corresponding to multiplication by 2. We have noted that $A \otimes \mathcal{F}_2$ is AF. To show that $A \otimes \mathcal{F}_2$ is isomorphic to $\mathcal{F}_2$, it is enough by Elliott's classification of AF algebras by K_0, to show that $K_0(A \otimes \mathcal{F}_2) \simeq K_0(\mathcal{F}_2)$. This can be established from $K_0(A) \simeq \mathbb{Z}[1/2]$ which follows from $K_0(A) \simeq \varinjlim K_0(A_n)$.

In all the examples mentioned above, [Bla1, Kum1, BEEK2, EK, Kis] the fixed point algebra of the AF algebra under the compact group is of real rank zero. The question then naturally arises of whether the fixed point algebra of an AF algebra under a compact group action is necessarily of real rank zero.

We have noted above an apparent similarity between the proof of the AF property of $A \otimes \mathcal{F}_2$ and the 2-torsion of $K_1(A)$. I pointed this out to George Elliott when he was visiting Swansea in the summer of 1989, and he began to investigate whether $K_* = K_0 \oplus K_1$ could be a complete invariant for such algebras. He could show that it is. Assuming we restrict attention to algebras of real rank zero $K_* = K_0 \oplus K_1$ is a complete invariant for C*-algebras which are inductive limits of direct sums of building blocks, each a matrix algebra over the circle or over

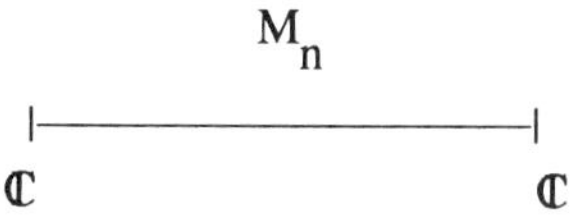

(15)

In particular K_1 is the only obstruction to such an algebra being AF and the classification is a natural extension of the classification [Ell1] of AF algebras by K_0. Progress has been made towards understanding what groups can arise as K_* of such algebras [Ell6] and in classifying algebras arising from building blocks constructed from cylinders [Ell7] using K_*, which can give rise to torsion in K_0. Kishimoto has a classification [Kis] of certain finite group actions on certain inductive limits of algebras which are direct sums of matrix algebras over the circle using $K_* = K_0 \oplus K_1$ which is a natural extension of the classification for locally representable actions of [HR].

It is natural to ask what one can expect in terms of classification or even characterising real rank zero for algebras built up as inductive limits where the basic building blocks have underlying spaces of higher dimension. The method of proof of the classification of Elliott [Ell5] for one-dimensional blocks, and the property of real rank zero for generic non-commutative torii [CE] depend heavily on the underlying spaces being of at most dimension two. For such spaces one can perturb maps into hermitian matrices to avoid those with degenerate spectra - which have some repeated eigenvalues. Such degenerate matrices have co-dimension 3 in the space of hermitian matrices. To understand this better it is convenient to consider the notion of arbitrary real rank of Brown and Pedersen [BP]. A unital C*-algebra A has real rank $\leq n$ if given $\varepsilon > 0$, $(x_1, x_2,..,x_{n+1}) \in A_{sa}^{n+1}$, there exists $(y_1, y_2,...,y_{n+1}) \in A_{sa}^{n+1}$ such that $\|x_i - y_i\| < \varepsilon$ and $\sum_{i=1}^{n+1} y_i^2$ is invertible. Thus, as we have already said, A is of real rank zero if the invertible self adjoint elements are dense in all self adjoint elements. If A is commutative, say C(X), then A being of real rank zero coincides with the property of the algebra being AF. In fact RRC(X) = dim X, the covering dimension of X. Brown and Pedersen based their definition on the notion of topological stable rank of Rieffel [Rie2] : a unital C*-algebra has topological stable rank $\leq$ n if given $\varepsilon > 0$, $(a_1, a_2,...,a_n) \in A^n$, there exists $(b_1,..,b_n) \in A^n$ with $\|a_i - b_i\| < \varepsilon$ and $\sum_{i=1}^{n} b_i^* b_i$ is invertible. Herman and Vaserstein [HV] have shown that (for C*-algebras) topological stable rank is the same as Bass stable rank. Moreover, A is of stable rank 1 if and only if the invertible elements are dense, tsr C(X) = [dim(X)/2] + 1 where [ ] denotes the integer part and $RR(A) \leq 2$ tsr $A - 1$.

It is thus natural to ask when a C*-algebra such as

(16) $$\lim_{\rightarrow} C(X_i) \otimes M_{m_i}$$

is of real rank zero, or even what the real rank of $C(X) \otimes M_m$ is? Beggs and Evans [BE] have shown that if X is a compact Hausdorff space then

(17) $$RR(X) \leq n(2m-1) \text{ if and only if } RR(C(X) \otimes M_m) \leq n$$

This should be contrasted with the situation for topological stable rank [Rie2, Vas] where

(18) $$\mathrm{tsr}(B \otimes M_m) = \left\{\frac{\mathrm{tsr}(B) - 1}{m}\right\} + 1$$

where { } denotes 'least integer greater than.'

In particular, for m large and RR(C(X)) finite

(19) $$RR(C(X) \otimes M_m) = \begin{cases} 0 & \text{if } RR(X) = 0 \\ 1 & \text{otherwise} \end{cases}$$

That $RR(C(X) \otimes M_m) = 0$ if and only if $R(X) = 0$ was shown by Brown and Pedersen [BP]. To show (17) one needs to consider maps from X into H_m^{n+1} (n+1 tuples of $m \times m$ hermitian complex matrices) which avoid the set S:

(20) $$S = \{(a_0, a_1 .., a_n) \in H_m^{n+1} : \bigcap_{j=0}^{n} \mathrm{Ker}\,(a_j) \neq 0\}$$

This is achieved by the following basic lemma, where a closed subset of a manifold is said to be of co-dimension $\geq n$ if it is embedded in a finite union of embedded submanifolds of co-dimension $\geq n$, with the submanifolds imbedded in such a way that the boundary of a given submanifold (its closure minus the submanifold itself) is contained in the union of the submanifolds of strictly lower dimension. The basic lemma then states that if X is a compact Hausdorff space of real rank $\leq m$, and that S is a closed co-dimension $\geq m+1$ subset of the manifold M, with metric d_m, then we can avoid S in maps from X in the following sense. If $f : X \to M$ is a continuous map, and $\varepsilon > 0$, there is a map $g : X \to M$ uniformly within ε of f so that the image of g does not interset S. Computing the co-dimension of S as in (20) then yields the real rank of matrix valued functions. If $M = R^{m+1}$, $S = \{0\}$, then the basic lemma reduces to the definition of real rank. If $M = H_n$, and S the space of degenerate matrices then we see that the basic lemma says we can avoid S when mapping from a space X of real rank ≤ 2 into H_n since the co-dimension of the degenerate matrices is 3. This was needed (at least for spaces which can be covered by a sequence of compact subsets each of which is homeomorphic to a simplex of dimension at most two) by Choi and Elliott [CE] in their proof of the real rank zero property for generic irrational non-commutative two tori, and the basic lemma is a generalisation of their result.

We can now consider again the question of the structure of inductive limits as in (16). Dadarlat, Nagy, Nemethi and Pasnicu [DNNP] have shown that, when the dimension of X_n is uniformly bounded, and the inductive limit C*-algebra is simple then it is automatically of topological stable rank one. (Note that if the algebra is simple, then necessarily $n_i \to \infty$, and so the topological stable rank is at most two by (18)). It is not however necessarily of real rank zero, even when each X_i is the circle. Blackadar, Bratteli, Elliott and Kumjian, [BBEK] have considered inductive limits $\varinjlim C(\mathbb{T}) \otimes M_{n_i}$ using winding embeddings, and have shown that the inductive limit if of real rank zero if and only if the number of standard ± 1 times around embeddings is asymptotically small compared with the total number of embeddings as $i \to \infty$. Thus we can produce simple C*-algebras of the form $\varinjlim C(\mathbb{T})) \otimes M_{n_i}$ with topological stable rank one but non zero real rank - the invertible elements are dense, but the invertible self adjoint elements are not dense in the self adjoint elements. Moreover in [BBEK] a characterisation of real rank zero property of inductive limits $\lim A_n$ which are direct sums of matrix algebras over spaces of at most dimension two in terms of small eigenvalue variation of self adjoint $x \in A_n$ in A_{n+m} for m sufficiently large. This allows one to deduce [Bra Ell] that the tensor product $BD \otimes BD$ is of real rank zero, if BD is the Bunce Deddens algebra $C(\mathbb{T}) \times D$, but it does not allow us to decide whether $BD \otimes BD \otimes BD$ is of real rank zero. This restriction of at most dimension two in [BBEK] again arises because the degenerate matrices have co-dimension 3 in the space of hermitian matrices.

Other exotic symmetries have been analysed in [BEEK2, BEEK3, Kum, Nes] on non-commutative torii. These systems give rise to non-commutative toroidal orbifolds. Albeverio and Høegh-Krohn [AHK] classified (faithful) ergodic actions of a compact abelian group G on unital C*-algebras using skew bicharacters on the dual group. For such an action there are unitaries $\{u_\gamma : \gamma \in \hat{G}\}$ in the C*-algebra, with $g : u_\gamma \to \langle g,\gamma \rangle u_\gamma$, $g \in G, \gamma \in \hat{G}$ and the corresponding bicharacter is given by $(\gamma,\gamma') \to u_\gamma u_{\gamma'} u_\gamma^{-1} u_{\gamma'}^{-1}, \in \mathbb{T}$. Thus if $G = \mathbb{T}^2$, the bicharacters are

$$((m,n), (m', n')) \to q^{mn'-m'n}$$

for some $q \in \mathbb{T}$ $(q = e^{2\pi i\theta})$ with corresponding rotation C*-algebra A_q which for our purposes we will accept to be defined as the universal C*-algebra generated by two unitaries U and V satisfying the commutation relation

(21) $$VU = q\,UV$$

Now $VUV^{-1} = q\,U$; Ad(V) leaves $C^*(U) \simeq C(\mathbb{T})$ invariant and acts by rotation through θ on the circle, and A_q can be identified with the crossed product $C(\mathbb{T}) \times_\theta \mathbb{Z}$. The ergodic $\mathbb{T}^2$ action on A_q is then $(t_1,t_2) : U \to t_1 U, V \to t_2 V$, $(t_1, t_2) \in \mathbb{T}^2$. The K-theory of A_q can be computed from the Pimsner-Voiculescu six-term exact sequence of K-theory [PV] as follows

(22) $$K_0\,(A_q) \simeq \mathbb{Z}^2$$

(23) $$K_1\,(A_q) \simeq \mathbb{Z}^2$$

If $q=1$, $K_0(C(\mathbb{T}^2))$ is generated by the trivial bundle and the Bott element, whilst $K_1(C\,(\mathbb{T}^2))$ is generated by the co-ordinate unitary maps $(z_1\, z_2) \to z_i$, $i = 1, 2$, i.e. U and V. When q is rational (or more precisely when θ is a rational multiple of 2π say P/Q with P,Q relatively prime), A_q can be identified with a homogeneous algebra over $\mathbb{T}^2$, with values in M_Q. This bundle is not trivial unless $q = 1$. Indeed A_q is not isomorphic to $A_{q'}$ unless $q = q'$ or $q = \bar{q}'$, the rational case being due to Høegh-Krohn and Skjelbred [HKS] (see also [Rie3]) and the irrational case to Rieffel, and Pimsner Voiculesu [Rie1, PV]. In fact, in the irrational case, $K_0(A_q)$ (c.f. (22)) is order isomorphic to $\mathbb{Z} + \theta\,\mathbb{Z}$, generated by the identity and a Rieffel projection.

The flip $(z_1, z_2) \to (\bar{z}_1, \bar{z}_2)$ on the classical or commutative 2-torus $\mathbb{T}^2$, gives rise to the orbifold $\mathbb{T}^2/\mathbb{Z}_2$ which is topologically a sphere. The flip corresponds geometrically on $\mathbb{T}^2 = \mathbb{R}^2/\mathbb{Z}^2$ to rotation through 180° in Figure 1.

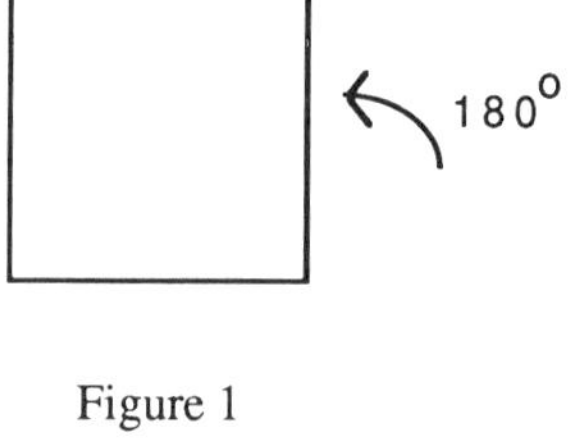

Figure 1

To identify the quotient space $\mathbb{T}^2/\mathbb{Z}_2$ we only need the triangle in Figure 2, with the identifications of the edges as shown. Folding one obtains a tetrahedron, the four verticies corresponding to the fixed points of the flip on $\mathbb{T}^2$.

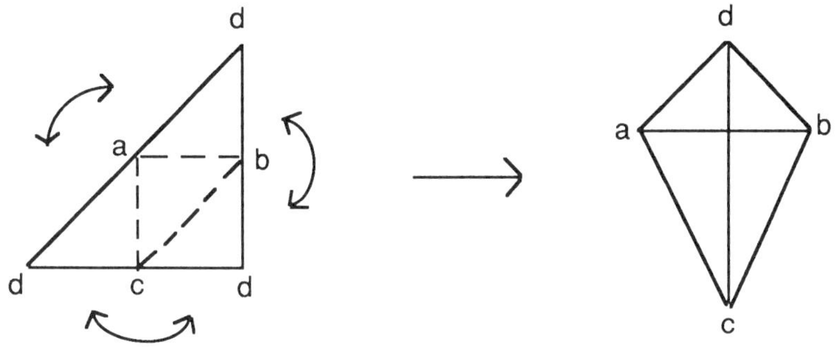

Figure 2 : $\mathbb{T}^2 / \mathbb{Z}_2 = S^2$

A non-commutative toroidal orbifold arises as the fixed point algebra of the non-commutative torus $A_q = C^*(U,V)$ under the symmetry $U, V \rightarrow U^{-1}, V^{-1}$ respectively (which preserves relation (21)). Note that the flip on $\mathbb{T}^2$, induces a transformation on the space of leaves of the Kronecker foliation $\mathcal{F}_\theta$ (i.e. leaves are taken to leaves under the symmetry), and hence a singular foliation of the 2-sphere S^2. The fixed point algebra $B_q = A_q^{\mathbb{Z}_2}$ bears the same relation to this singular foliation, that the non-commutative torus A_q does to the Kronecker foliation $\mathcal{F}_\theta$; $C^*(\mathbb{T}^2, \mathcal{F}_\theta)^{\mathbb{Z}_2}$ is Morita equivalent to B_q. If $q^2 \neq 1$, then the algebra B_q is generated by $U + U^{-1}$ and $V + V^{-1}$. If q is not a root of unity, it is natural to ask, in view of the results described above for exotic symmetries on AF algebras, whether B_q is AF or not. Interest in these algebras also arise because $U + U^{-1} + \lambda (V + V^{-1}), \lambda \in \mathbb{R}$ are the Schrödinger operators with almost periodic potentials considered in [BeS, CEY], which generically have Cantor spectra.

The non-commutative torus A_q is the universal C*-algebra generated by two unitaries U and V satisfying the relation (21). In the irrational case ($q^n \neq 1$, for all integral $n \neq 0$), A_q is simple and is indeed the unique C*-algebra generated by unitaries U and V satisfying the commutation relation (21) [EH]. The non-commutative orbifolds B_q are analogously described as follows [BEEK1]:

There is an action of $SL(2, \mathbb{Z})$ on A_q [Bre, W], where the matrix $\begin{pmatrix} -1 & 0 \\ 0 & -1 \end{pmatrix}$ corresponds to the flip $U \to U^{-1}$, $U \to V^{-1}$. Since $\begin{pmatrix} -1 & 0 \\ 0 & -1 \end{pmatrix}$ generates the centre of $SL(2, \mathbb{Z})$, there is an action of $PSL(2, \mathbb{Z})$, on the fixed point algebra $B_q = A_q^{\mathbb{Z}_2}$. The generators

(24) $$S = \begin{pmatrix} 0 & -1 \\ 1 & 0 \end{pmatrix}, \quad T = \begin{pmatrix} 1 & 1 \\ 0 & 1 \end{pmatrix} \quad \text{of} \quad PSL(2, \mathbb{Z})$$

act on B_q as follows. S merely interchanges A and B, but T acts less trivially by fixing A, but taking B to

(25) $$q^{1/2}(1 - q^2)^{-1} \, [A, B]_q$$

where $[X, Y]_q$ denotes the twisted or q-commutator XY-qYX. Similarly T^{-1} fixes A and takes B to

(26) $$q^{-1/2}(1-q^2)^{-1} \, [A, B]_{q^{-1}}$$

Hence $TT^{-1} B = B$ gives that

(27) $$B = -(q - q^{-1})^2 \, [A,[A, B]_{q^{-1}}]_q$$

or

(28) $$BA^2 + A^2B = (q + q^{-1}) \, ABA - (q - q^{-1})^2 \, B$$

Similarly

(29) $$A = -(q - q^{-1})^2 \, [B,[B,A]_{q-1}]_q$$

or

(30) $$AB^2 + A^2B = (q + q^{-1}) \, BAB - (q - q^{-1})^2 \, A$$

(31) $\quad BABA = (q^2+1-q^{-2})ABAB - (q+q^{-1})A^2B^2 + (q+q^{-1}-(q^3+q^3))(A^2+B^2-2\mathbb{1})$

In the irrational case ($q^n \neq 1$, for all non zero integral n), B_q is simple and is the unique C*-algebra generated by two self adjoint operators A and B (identified with $U+U^{-1}$, $V+V^{-1}$ respectively) satisfying the three independent relations (28), (30) and (31). Note that if $q^4 \neq 1$, then the third relation shows that the unit of B_q is in the polynomial algebra generated by A and B. This actually only occurs when $q^4 \neq 1$. A simpler relation between the unit and a polynomial expression in A and B can be deduced as follows.

If U,V satisfy $UV = q\,VU$ then U^2,V satisfy $VU^2 = q^2\,U^2V$. Replacing A,B,q in (13) with $U^2+U^{-2} = A^2 - 2\mathbb{1}$, B and q^2 we obtain:

$$2(q^2-q^{-2})^2\mathbb{1} = A^2B^2 + B^2A^2 - (q^2+q^{-2})BA^2B$$

$$+2(q-q^{-1})^2B^2 + (q^2-q^{-2})^2A^2 \tag{32}$$

Again this can be solved for $\mathbb{1}$ unless $q^4 = 1$.

In the rational case (more precisely if $q^n = 1$, for some integral n, but $q^2 \neq 1$), B_q is the enveloping C*-algebra of $P(U,V)^{\mathbb{Z}_2}$, where $P(U,V)$ denotes the *-subalgebra of A_q generated by U and V (i.e. the polynomial algebra in U and V). The enveloping C*-algebra of $P(U,V)^{\mathbb{Z}_2}$ does not exist when $q^2 = 1$.

To understand the K-theory of the non-commutative orbifold B_q, it is illuminating to first consider the rational case [BEEK3]. In this case, with $q = \exp(2\pi i P/Q)$, with P, Q relatively prime integers, the algebra B_q can be identified with a subalgebra of the C*-algebra $C(S^2, M_Q)$ of continuous functions from the 2-sphere S^2 into M_Q, which at four distinct points $\omega_0, \omega_1, \omega_2, \omega_3$ in S^2 commute with an associated projection E_i in M_Q of dimension:

(33) $\quad \dim(E_i) = (Q-1)/2$, when Q is odd

(34) $\quad \dim(E_0) = (Q-2)/2$, $\dim(E_i) = Q/2$, for $i = 1,2,3$ when Q is even

Thus when $q = e^{2\pi iP/Q}$, the isomorphism class of B_q depends on Q but not on P, although the non-commutative tori A_q and $A_{q'}$ are isomorphic if and only if $q' = q$ or $q' = q^{-1}$.

It is known [Kum2] that:

(35) $$K_0(B_q) = \mathbb{Z}^6 \text{ if } q^2 \neq 1, (\mathbb{Z}^5 \text{ when } q = -1, \mathbb{Z}^2 \text{ when } q = 1)$$

(36) $$K_1(B_q) = 0$$

with the latter not giving away any information about whether B_q is AF or not, as K_1 always vanishes for an AF algebra. These K-groups can easily be computed [BEEK3] when q is a root of unity using the following simple description of B_q in that case. The algebra $A_{\exp(2\pi iP/Q)}$ is a homogeneous algebra over the 2-torus $\mathbb{T}^2$ with fibre M_Q [HKS Rie3], but the obstruction to isomorphism for the non-commutative torii is eliminated for the non-commutative orbifolds by the behaviour of the four singular points. Consequently, the bundle for $B_{\exp(2\pi iP/Q)}$ is trivial. The K-theory of $B_{\exp(2\pi iP/Q)}$ can then be computed using the ideal of functions which vanish at all the ω_i, together with the 6-term short exact sequence of K-theory. More informally to see $K_0(B_q) = \mathbb{Z}^6$ if $q^2 \neq 1$, a projection in B_q will give an element in $\mathbb{Z}^8$ say $\{(d_1^i, d_2^i) : 1 = 0,1,2,3\}$ with three constraints $d_1^i + d_2^i$ = constant, which gives us $\mathbb{Z}^5$; the additional element of $K_0(B_q)$ being provided by the Bott element on the 2-sphere. (If $p \in C(S^2, M_Q)$ is a projection, then $(d_1^i, d_2^i) = (\dim(p(\omega_i)E_i), \dim(p(\omega_i)(1-E_i)))$ is in $\mathbb{Z}^8$, where the dimension is computed in M_Q, with $d_1^i + d_2^i$ = constant due to the homotopy invariance of $\dim f(\omega)$ for all $\omega \in S^2$). To understand the K-theory and traces of B_q better, it is convenient [BEEK3] to bring the crossed product $C_q = A_q \times \mathbb{Z}_2$ into the game. When $q = e^{2\pi iP/Q}$, and P and Q are relatively prime positive integers C_q is isomorphic a subalgebra of the C*-algebra $C(S^2, M_{2Q})$ of continuous functions from the 2-sphere S^2 into the algebra M_{2Q} determined as follows. There are four distinct points $\omega_0, \omega_1, \omega_2,$

ω_3 in S^2 and an orthogonal projection E in M_{2Q} of dimension Q, such that the subalgebra consists of those functions $f \in C(S^2, M_{2Q})$ such that $f(\omega_i)$ commutes with E for $i = 0,1,2,3$. For any q, the algebraic crossed product $P_q \times \mathbb{Z}_2$ is the *-algebra generated by three unitaries U,V and W satisfying the relations $VU = qUV$, $WU = U^{-1}W$, $WV = V^{-1}W$, and $W^2 = 1$, its completion is the crossed product $C_q = A_q \times \mathbb{Z}_2$. There is a canonical continuous tracial state τ on $A_q \times \mathbb{Z}_2$ given by

(37) $\tau(q^{nm/2}U^nV^m) = 1$ if $n = m = 0$, and 0 otherwise

(38) $\tau(q^{nm/2}U^nV^mW) = 0$

There are four trace functionals $\tau_{p_1p_2}$ where $p_1,p_2 \in \{\text{even, odd}\}$ defined on $P_q \times \mathbb{Z}_2$ by :

(39) $\tau_{p_1p_2}(q^{nm/2}U^nV^m) = 0,$

(40) $\tau_{p_1p_2}(q^{nm/2}U^nV^mW) = 1$, if parity(n) = p_1 and parity(m) = p_2, and 0 otherwise

When q is not a root of unity, the space of trace functionals on $P_q \times \mathbb{Z}_2$ is five dimensional [BEEK1,3], spanned by τ and the $\tau_{p_1p_2}$, with the only continuous trace functionals on C_q (respectively A_q) being scalar multiples of τ (respectively $\tau|A_q$). It is however remarkable, when q is a root of unity, that the other four trace functionals are continuous and extend to C_q. The algebraically defined trace functionals $\tau_{p_1p_2}$ are related to the following geometrically defined trace functionals $\tilde{\tau}_{p_1p_2}$ which are clearly *continuous*. There is a convenient labelling of the four singular points ω_i by $\{\text{even,odd}\}^2$, i.e. $\omega_{p_1p_2}$. Define four continuous trace functionals on C_q by: $\tilde{\tau}_{p_1p_2}(f) = Tr_{2Q}((2E-1)f(\omega_{p_1p_2}))/Q$, where Tr_{2Q} is the un-normalised trace on M_{2Q}, and E is the rank Q projection in M_{2Q} as above. Then, adopting a proper sign convention, the algebraically defined trace functionals are related to the geometrically defined trace functionals by [BEEK3]:

(41) $$\tau_{p_1p_2} = (-1)^{p_1p_2}\tilde{\tau}_{p_1p_2} \quad \text{if } Q \text{ is even and}$$

(42) $$\tau_{nm} = (-1)^{pnm}\{\sum_{p_1p_2} (-1)^{np_1+mp_2}\tilde{\tau}_{p_1p_2}\} \text{ when } Q \text{ is odd}$$

In particular the trace functionals $\tau_{p_1p_2}$ are continuous when q is not a root of unity. The five trace functionals τ and $\tau_{p_1p_2}$ separate the following five projections in C_q written down by Nest [Nes]:

(43) $$1,\ (1+W)/2,\ (1+UW)/2,\ (1+VW)/2,\ (1+q^{1/2}UVW)/2$$

These projections, together with the Rieffel projection show that $K_0(C_q)$ contains $\mathbb{Z}^6$ even when q is not a root of unity.

The four spurious trace functionals on $P_q \times \mathbb{Z}_2$, and in particular the fact that they are unbounded only in the rational case have been exploited to show the following [BEEK3]. The $\mathbb{T}^2$ action on the non-commutative torus A_q gives rise to a family of seminorms $\|\delta_1^m \delta_2^n (.)\|$ from the corresponding derivations δ_1, δ_2 on the smooth subalgebra A_q^∞ and $B_q^\infty = A_q^\infty \cap B_q$. Then an analysis of the spurious traces shows that the linear span of projections in B_q^∞ is *not* dense in B_q^∞ for the natural topology for the seminorms. However Choi and Elliott have shown that (for a dense set of irrational q) that both A_q and B_q have real rank zero - thus the linear span of projections are dense for the norm topology. Note also that Putnam [Put] has shown that all irrational non-commutative 2-torii have topological stable rank one (following Riedel [Rid] who showed this for generic irrational q).

Nest [Nes2] has computed the cyclic cohomology of these non-commutative toroidal orbifolds. In the course of this he has shown that all derivations on the smooth elements of B_q (smooth with respect to the $\mathbb{T}^2$ torus action on the rotation algebra A_q) are approximately inner if q is not a root of unity (c.f. [PS] for the conjectured structure of derivations on AF algebras).

References

[AHK] S. Albererio, R. Høegh-Krohn. Ergodic actions by compact groups on C*-algebras. Math-Zeitschrift. 174 (1980) 1-17.

[AP] J. Anderson and W. Paschke. The rotation algebra. Preprint.

[BCEN] B. Brenken, J. Cuntz, G.A. Elliott, and R. Nest. On the classification of non-commutative tori, III, Contemporary Mathematics 62 (1987) 503-526.

[BEEK1] O. Bratteli, G.A. Elliott, D.E. Evans, A. Kishimoto. Non-commutative spheres. International Journal of Mathematics (to appear).

[BEEK2] O. Bratteli, G.A., Elliott, D.E. Evans, A. Kishimoto. Finite group actions on AF algebras obtained by folding the interval. Preprint Trondheim 1989.

[BEEK3] O. Bratteli, G.A. Elliott, D.E. Evans, A. Kishimoto. Non-commutative spheres II: rational rotations. J. Operator Theory (to appear).

[Be Eva] E. Beggs, D.E. Evans. The real rank of algebras of matrix valued functions. International Journal of Mathematics (to appear).

[BeS] J. Bellisard, B. Simon. Cantor spectrum for the almost Mathieu equation. J. Functional Analysis 48 (1982) 408-419.

[Bla1] B. Blackadar. Symmetries of the CAR algebra. Annals of Math. (to appear).

[Bla2] B. Blackadar. K-theory for operator algebras, Mathematical Sciences Research Institute Publications 5, Springer-Verlag, Berlin-Heidelberg. New York, 1986.

[BP] L. Brown, G.K. Pedersen. C*-algebras of real rank zero. Preprint. Copenhagen 1989.

[BR] O. Bratteli, D.W. Robinson. Operator algebras and quantum statistical mechanics. I and II. Springer-Verlag, 1979 and 1981.

[Bra] O. Bratteli. Inductive limits of finite-dimensional C*-algebras. Trans. Amer. Math. Soc. 171 (1972) 195-234.

[Bra Ell] O Bratteli, G.A. Elliott. An introduction to fractal C*-algebras. Preprint Trondheim 1990.

[Bre] B. Brenken. Representations and automorphisms of the irrational rotation algebra. Pacific J. Math. 111 (1984), 257-282.

[BS] J. Bellissard and B. Simon. Cantor spectrum for the almost Mathieu equation. J. Functional Analysis 48 (1982) 408-419.

[C] A. Connes. Non-commutative differential geometry I-II. Publ. Math. IHES, 62 (1985) 41-144.

[CE] M.-D. Choi, G.A. Elliott. Density of the self adjoint elements with finite spectrum in an irrational rotation C*-algebra. Math. Scand (to appear).

[CEGJ] J. Cuntz, G.A. Elliott, F.M. Goodman, P.E.T. Jorgensen. On the classification of non-commutative tori. II, C.R. Math. Rep. Acad. Sc. Canada, VII, no.3 (1985), 189-194.

[CEY] M.-D. Choi, G.A. Elliott, N. Yui. Gauss polynomials and the rotation algebras. Invent. Math. (to appear).

[Con] A. Connes. C*-algebres et géométrie différentielle. C.R.Acad.Sc. Paris (1980), 559-604.

[CR] A. Connes, M.Rieffel. Yang-Mills for non-commutative tori. Contemporary Mathematics 62 (1985) 237-265.

[D] M. Dadarlat. On homomorphisms of certain C*-algebras. Preprint 1986.

[DN] M. Dadarlat, A. Nemethi. Shape theory and connective K-theory. J. Operator Theory (to appear).

[DNNP] M. Dadarlat, G. Nagy, A. Nemethi, C. Pasnicu. Reduction of topological stable rank in inductive limits of C*-algebras. Preprint INCREST 1990.

[Eff1] E.G. Effros. Dimensions and C*-algebras. Conference Board Math. Sci. 46, American Mathematical Society, Providence, R.I., 1981.

[Eff2] E.G. Effros. On the structure of C*-algebras: some old and new problems. Proceedings of Symposia in Pure Math. 38 Part I (1982) 19-34.

[EH] E.G. Effros, F. Hahn. Locally compact transformation groups and C*-algebras. Memoirs Amer. Math. Soc. 75, 1967.

[EHS] E.G. Effros, D.E. Handelman, C.L. Shen. Dimension groups and their affine representations. Amer. J. Math. 102 (1980) 385-407.

[EK] D.E. Evans, A. Kishimoto. Compact group actions on UHF algebras obtained by folding the interval. J. Funct. Anal. (to appear).

[Ell1] G.A. Elliott. On the classification of inductive limits of sequences of semi-simple finite dimensional algebras. J. Algebra 38 (1976) 29-44.

[Ell2] G.A. Elliott. Gaps in the spectrum of an almost periodic Schrödinger operator. in Geometric methods in operator algebras. Eds. H. Araki, E.G. Effros. Pitman Research Notes in Mathematics 123, Longman, London 1986, 181-191.

[Ell3] G. A. Elliott. On the K-theory of the C*-algebra generated by a projective representation of a torsion free discrete abelian group, in Operator Algebras and Group Representations 1. Pitman, London 1984.

[Ell4] G.A. Elliott. The diffeomorphism group of the irrational rotation C*-algebra, C.R. Math.Rep. Acad. Sci. Canada 8 (1986) 329-334.

[Ell5] G.A. Elliott. On the classification of C*-algebras of real rank zero. Preprint, Copenhagen, January, 1990.

[Ell6] G.A. Elliott. Dimension groups with torsion. Preprint 1990.

[Ell7] G.A. Elliott. On the classification of C*-algebras of real rank II. Work in progress.

[Eva] D.E. Evans. C*-algebraic methods in statistical mechanics and field theory. International Journal of Modern Physics B. 4 (1990) 1069-1118.

[FM1] T. Fack and O. Maréchal. Sur la classification des symétries des C*-algebras UHF. Canad. J. Math 31 (1979) 496-523.

[FM2] T. Fack and O. Maréchal. Sur la classification des automorphismes périodiques des C*-algebras UHF. J. Funct. Anal 40 (1981), 267-301.

[GH] K.R. Goodearl and D.E. Handelman. Classification of ring and C*-algebra direct limits of finite-dimensional semisimple real algebras, Mem. Amer. Math. Soc 372 (1987) vii + 147.

[Gio] T. Giordano. Classification of approximately finite real C*-algebras. J. Reine Angew. Math 385 (1988) 161-194.

[Han1] D.E. Handelman. Positive polynomials and product type actions of compact groups, Mem. Amer. Math. Soc 320 (1985), xi + 79.

[Han2] D.E. Handelman. Extending traces on fixed point C*-algebras under xerox product type actions of compact Lie groups. J. Funct. Anal. 72 (1987) 44-57.

[HKLS] R. Høegh Krohn, M. Landstad, E. Størmer. Compact ergodic groups of automorphisms. Annals of Math. 114 (1981) 75-86.

[HKS] R. Høegh-Krohn, T. Skjelbred. Classification of C*-algebras admitting ergodic actions of the two-dimensional torus. J. Reine Angewandte Math. 328 (1981) 1-8.

[HR1] D.E. Handelman and W. Rossmann. Product type actions of finite and compact groups, Indiana Univ. Math. J. 33 (1984) 479-509.

[HR2] D. Handelman and W. Rossmann. Actions of compact groups on AF C*-algebras. Illinois J. Math. 29 (1985) 51-95.

[HV] R.H. Herman, L.N. Vaserstein. The stable rank of C*-algebras. Inventiones Math. 77 (1984) 553-555.

[Kis] A. Kishimoto. Actions of finite groups on certain inductive limit C*-algebras. Preprint Hokkaido 1990.

[Kum1] A. Kumjian. An involutive automorphism of the Bunce-Deddens algebra, C.R. Math. Rep. Acad. Sci. Canada 10 (1988) 217-218.

[Kum2] A. Kumjian. Non-commutative spherical orbifolds. In preparation.

[Nes1] R. Nest, Cyclic cohomology of non-commutative tori. Can.J. Math. XL (1988) 1046-1057.

[Nes2] R. Nest. Cohomology of a certain non-commutative orbifold. Preprint Copenhagen 1989.

[OPT] D. Olesen, G. Pedersen, M. Takesaki. Ergodic actions of compact abelian groups. J. Operator Thoery 3 (1980) 237-269.

[P] C. Pasnicu. On inductive limits of certain C*-algebras of the form C (X) ⊗ F. Trans. Amer. Math. Soc. 310 (1988) 703-714.

[[PS] R.T. Powers, S. Sakai. Existence of ground states and KMS states for approximately inner dynamics. Commun. math. Phys. 39 (1975) 273-288.

[PSS] I.F. Putnam, K. Schmidt and C. Skau. C*-algebras associated with Denjoy homeomorphisms of the circle. J. Operator Th. 16(1986) 99-126.

[Put2] I.F. Putnam. The C*-algebras associated with minimal homeomorphisms of the Cantor set. Pacific J. Math. 136(1989) 329-353.

[Put3] I.F. Putnam. The invertible elements are dense in the irrational rotation C*-algebras. Preprint Dalhouse 1989.

[Put4] I.F. Putnam. On the topological stable rank of certain transformation group C*-algebras. Ergodic Th. and Dyn. Sys., to appear.

[PV] M. Pimsner, D. Voiculescu. Exact sequences for K-groups and Ext-groups of certain crossed product C*-algebras. J. Operator Theory. 4 (1980) 93-118.

[Rid] N. Riedel, On the topological stable rank of irrational rotation algebras, J. Operator Th. 13 (1985) 143-150.

[Rie1] M. Rieffel. C*-algebras associated with irrational rotations. Pacific J. Math. 93 (1981) 415-429.

[Rie2] M.A. Rieffel. Dimensions and stable rank in the K-theory of C*-algebras, Proc. London Math. Soc. (3), 46 (1983) 301-333.

[Rie3] M.A. Rieffel. The cancellation theorem for projective modules over irrational rotation C*-algebras. Proc. London Math. Soc. (3) 47 (1983) 285-302.

[Rie4] M.A.Rieffel. Projective modules over higher dimensional non-commutative tori. Preprint (1986).

[S] B. Simon. Almost periodic Schrødinger operators: a review, Adv. in Appl. Math. 3 (1982) 463-490.

[Vas] L.N. Vaserstein. Stable rank of rings and dimensionality of topological spaces. Functional Anal. Appl. 5 (1971) 102-110.

[W] Y. Watatani. Toral automorphisms on irrational rotation algebras, Math. Japonica 26 (1981) 479-484.

[Was] A.J. Wassermann. Automorphic actions of compact groups on operator algebras. Ph.D. thesis, University of Pennsylvania 1981.

Ergodic actions of non-abelian compact groups

Magnus B. Landstad

Abstract. We show that an ergodic action of a compact group G on a von Neumann algebra $\mathcal{A}$ is completely characterized by a dual cocycle ω. Each ω defines a multiplication on $A(G)$ and a representation π into $\mathcal{L}(G)$ such that: (1) $\mathcal{A}$ is a factor $\iff$ (2) π is injective $\iff$ (3) π has dense range. The subalgebra $A^\circ(G)$ generated by G's characters is commutative and for $f \in A^\circ(G)$ $\pi(f) = \sum f(x_i(D))e^D_{ii}$ for some $x_i(D) \in G$. (1–3) are then also equivalent to (4) $x_i(D) \neq e$ for $D \not\sim 1$.

1. Introduction

Ergodic actions of compact groups may be thought of as generalizations of ordinary irreducible representations, the correspondence being as follows: If $D \in \widehat{G}$ acts on $\mathbf{C}^d$, $\operatorname{Ad} D_x$ gives an ergodic action of G on $\mathcal{M}(d, \mathbf{C})$. If we also allow D to be a projective representation, this correspondence is 1–1. So the ergodic actions of G on type I factors are the same as the projective representations of of G. The first indication that ergodic actions of compact groups are of a special kind is E. Størmer's result in [**11**]: If a compact abelian group G acts ergodically on a von Neumann algebra $\mathcal{A}$, then $\mathcal{A}$ has a unique invariant trace and is spanned by unitary eigenoperators, i.e. of elements $u(\alpha) \in \mathcal{A}$ where $\alpha \in \widehat{G}$ such that

$$\rho_x(u(\alpha)) = \alpha(x)u(\alpha) \quad \text{for all } x \in G. \tag{1.1}$$

The idea of thinking of ergodic actions of a group G as a dual object to G (following F. Klein's philosophy that an object should be studied by its homomorphisms) and the first classification of all the ergodic actions of a given group G started with the work [**1**] by S. Albeverio and R. Høegh-Krohn, and [**6**] by R. Høegh-Krohn and T. Skjelbred. This was later generalized to arbitrary compact abelian groups by D. Olesen, G.K. Pedersen and M. Takesaki in [**10**].

Abelian groups. The title of this article is chosen to emphasize that our main concern is non-abelian groups, although almost everything is true also for abelian groups. Therefore let us stop shortly to describe the abelian case. So suppose we have a W^*-dynamical system $(\mathcal{A}, \rho, G)$ over an abelian compact group G and let $u(\alpha)$ be the eigenoperators defined in (1.1). Then it is easy to see that there is a function $\omega : \widehat{G} \times \widehat{G} \to \mathbf{T}$ such that

$$u(\alpha)u(\beta) = \omega(\alpha, \beta)u(\alpha + \beta) \quad \text{for } \alpha, \beta \in \widehat{G}. \tag{1.2}$$

ω must be a cocycle, i.e.

$$\omega(\alpha, \beta)\omega(\alpha + \beta, \gamma) = \omega(\alpha, \beta + \gamma)\omega(\beta, \gamma). \tag{1.3}$$

It turns out that ω completely determines $(\mathcal{A}, \rho, G)$, so there is a 1–1 correspondence between ergodic W^*-dynamical systems $(\mathcal{A}, \rho, G)$ and cocycles ω over $\widehat{G}$ such that $(\mathcal{A}, \rho, G) \sim (\mathcal{A}', \rho', G) \iff \omega \sim \omega'$ as cocycles. So the classification of ergodic W^*-dynamical systems is the same as the classification of cocycles on $\widehat{G}$ up to equivalence.

This classification was in fact done by A. Kleppner in [**8**], where the classifying objects are the so called symplectic (or anti-symmetric) bicharacters of $\widehat{G}$. The function π (which is the same as $\omega^{(2)}$ in [**8**]) is defined by

$$\pi(\alpha, \beta) = \omega(\alpha, \beta)\overline{\omega(\beta, \alpha)} \tag{1.4}$$

and satisfies

$$\pi(\alpha, \beta) = \overline{\pi(\beta, \alpha)} \quad \pi(\alpha, \beta + \gamma) = \pi(\alpha, \beta)\pi(\alpha, \gamma) \quad \pi(\alpha, \alpha) = 1. \tag{1.5}$$

So there is a homomorphism (also called π) from $\widehat{G}$ to G such that, $\pi(\alpha, \beta) = \langle \pi(\alpha), \beta \rangle$. Conversely, (and this is the hard part) to every homomorphism $\pi : \widehat{G} \to G$ which is symplectic, that is satisfies

$$\langle \pi(\alpha), \alpha \rangle = 1 \tag{1.6}$$

there is a cocycle ω over $\widehat{G}$ such that $\langle \pi(\alpha), \beta \rangle = \omega(\alpha, \beta)\overline{\omega(\beta, \alpha)}$.

The classification of ergodic actions of a compact abelian group is therefore the same as the classification of all homomorphisms $\pi : \widehat{G} \to G$ which satisfy (1.6). So for instance if $G = \mathbf{T}^n$, then π is completely determined by its values on a set of generators of $\mathbf{Z}^n$.

Since $\pi(\alpha, \beta) = u(\alpha)u(\beta)u(\alpha)^*u(\beta)^*$ we may think of π as a measure on how non-commutative $\mathcal{A}$ is, for instance it should be obvious that $\mathcal{A}$ is commutative if and only if $\pi = 1$. In the opposite direction one has

Theorem 1.1. *The following are equivalent:*

(1) *$\mathcal{A}$ is a factor*
(2) *π is injective*
(3) *$\pi(\widehat{G})$ is dense in G.*

Proof: See [**1**, 3.7],[**10**, 5.8].

So for abelian compact groups one really has a complete description of all ergodic actions.

Non-abelian groups. Let us now turn to the non-abelian compact groups. S. Albeverio and R. Høegh-Krohn made the following observation (although it was not included in the final version of [**1**]):

Theorem 1.2. *If $(\mathcal{A}, \rho, G)$ is an ergodic W^*-dynamical system then there is a closed subgroup H of G and an ergodic W^*-dynamical system $(\mathcal{B}, \beta, H)$ with $\mathcal{B}$ a factor such that*

$$\mathcal{A} \simeq \{f \in L^\infty(G, \mathcal{B}) \mid f(xh) = \beta_{h^{-1}}(f(x)) \quad \text{for } x \in G, h \in H\}$$

and ρ_x corresponds to the map $f \to \lambda(x)f$ where $\lambda(x)f(y) = f(x^{-1}y)$. We say that $(\mathcal{A}, \rho, G)$ is induced from $(\mathcal{B}, \beta, H)$.

So the classification of all ergodic actions of G is the same as classifying all ergodic actions of the closed subgroups of G on factors.

The first step toward a study of ergodic actions of non-abelian compact groups was taken in [**5**] where it was shown that $\mathcal{A}$ must be *finite*: $\mathcal{A}$ has a (unique) G-invariant trace and is an injective von Neumann algebra. So if $\mathcal{A}$ is a factor it is either an I_n-factor with $n < \infty$ or the hyperfinite factor R.

If N is a closed normal subgroup of G there is a natural ergodic action of G/N on the fixed–point algebra $\mathcal{A}^{\rho|N}$. As we noted earlier the ergodic actions of G on I_n-factors are the projective unitary representations of G, so the conjecture made in [**5**] seems quite natural:

Conjecture. *There are no ergodic actions of a simple compact group on the hyperfinite factor R.*

A. Wassermann has in [**14**] defined a *multiplicity map* which he has used together with a clever case by case approach to study the ergodic actions of the groups $SU(2)$, $SO(3)$, $SU(3)$, $SU(2) \times SU(2)$ and $SU(2) \times$any compact abelian group, see [**14**], [**16**] and [**17**]. His work confirms the conjecture so far.

The aim of this article is to show that one to a large extent can develop a theory of cocycles and symplectic bicharacters also in the non-abelian case, at least when we restrict to the case where ρ has what we shall call full multiplicity. The first step, from ergodic actions to so called dual cocycles is very similar to the abelian case when one uses that the non-abelian analogue of $L^\infty(G)$ is $\mathcal{L}(G)$= the von Neumann algebra generated by the left regular representation. A dual cocycle will then be a unitary $\omega \in \mathcal{L}(G) \otimes \mathcal{L}(G)$ satisfying an equation similar to (1.3), and we will also put some normalizing conditions on ω. The main result in part 3 is then that there is a 1–1 correspondence between ergodic W^*-dynamical systems and equivalence classes of normalized dual cocycles over G.

The next step, going from cocycles to bicharacters is less clear. By analogy with (1.4) π should be defined as $\pi = \omega\sigma(\omega^*)$ (σ is the usual flip), but it less clear which properties we should expect from π. In the abelian case, the group homomorhism $\pi : \widehat{G} \to G$ will implement an algebra homomorhism $\pi : l^1(\widehat{G}) \to l^\infty(\widehat{G})$ by

$$\pi(f)(\alpha) = \widehat{f}(\pi(\alpha)).$$

Since the non-abelian analogue of $l^1(\widehat{G})$ is the Fourier algebra $A(G)$, one would expect that π would give a homomorphism of $A(G)$ into $\mathcal{L}(G)$. This is not quite true however, π is only multiplicative when we give $A(G)$ another multiplication which can not be expected to be commutative. $A(G)$ with this multiplication is denoted by $A_\omega(G)$

since it depends on ω. But with these modifications we do get a homomorphism $\pi : A_\omega(G) \to \mathcal{L}(G)$ which satisfies the analogue of Theorem 1.1:

Theorem 4.14. *The following are equivalent:*

(1) *$\mathcal{A}$ is a factor*
(2) *π is injective*
(3) *π has dense image in $\mathcal{L}(G)$*

The classification problem now becomes the following: What are the possible multiplications on $A(G)$ and what are the possible (if any) representations into $\mathcal{L}(G)$ which are symplectic? (Precise definitions are given in part 4.) If G has a faithful irreducible unitary representation D, then π is uniquely determined by the restriction to $\mathcal{B}(D) \otimes \mathcal{B}(D)$. So the problem is reduced to a problem about finite matrices: Determine the matrices $\pi \in \mathcal{B}(D) \otimes \mathcal{B}(D)$ which can occur. The discovery by A. Wassermann that one of the equations π has to satisfy is the so called *Quantum Yang–Baxter equation* [**17**], gives a connection between ergodic actions of compact groups and the extensive theory of what often is called Quantum Groups. We refer to [**3**] for more on that. So far however, none of the many solutions of the Quantum Yang–Baxter equation have given any new ergodic actions.

$A_\omega(G)$ is not abelian, but it turns out that the subalgebra $A^\circ_\omega(G)$ generated by the characters of G is. So if we diagonalize the restriction of π to $A^\circ_\omega(G)$ we get that

$$\pi(f) = \sum_{i,D} f(x_i(D))e^D_{ii} \quad f \in A^\circ_\omega(G),$$

for some elements $x_i(D) \in G$ which are determined only up to conjugacy classes. These are the analogues of the group elements $\pi(\alpha) \in G$ in the abelian case, and we will get in the non-abelian case that (1–3) also are equivalent to

(4) $x_i(D) \neq 1$ for $D \not\sim 1$.

For a group like $\mathbf{T}^n$ it was easy to see what the possible choices for the group elements $\pi(\alpha)$ are. But this is not at all clear in the general case.

We are here presenting results the author has slowly been gathering since the appearance of [**5**]. Many of these have been obtained independently by A. Wassermann and the author would like to acknowledge the various stimulating discussions we have had on this subject during the years. The reader should look in [**14–17**] for more detailed information on material not covered here.

2. Preliminaries

Notation. We shall first list some facts about compact groups and on the way also introduce the notation we are using.

$\lambda(x)$ and $r(x)$ are the left (respectively right) regular representations of G over $L^2(G)$, and $\mathcal{L}(G)$ is the von Neumann algebra generated by the left regular representation of G over $L^2(G)$. $A(G)$ is the predual of $\mathcal{L}(G)$. We have the maps $\delta : \mathcal{L}(G) \to \mathcal{L}(G) \otimes \mathcal{L}(G)$ and $\nu : \mathcal{L}(G) \to \mathcal{L}(G)$ defined by

$$\delta(x) = x \otimes x \quad \text{and} \quad \nu(x) = x^{-1} \quad \text{for } x \in G.$$

$A(G)$ is a commutative Banach $*$-algebra under pointwise multiplication and complex conjugation, and the spectrum of $A(G)$ is equal to G. We shall also need the subalgebra $A^\circ(G)$ of $A(G)$ generated by the characters of G, i.e.

$$f \in A^\circ(G) \iff f(xy - yx) = 0 \quad \text{for all } x, y \in \mathcal{L}(G).$$

The spectrum of $A^\circ(G)$ is the same as that of $C^\circ(G)$ and equals G° = the set of conjugacy classes of G.

We shall also need some facts from the theory of representations of compact groups. If $D \in \widehat{G}$ and $\{\xi_i\}$ is an orthonormal basis of the space $X(D)$ on which D acts, then let ($d = \dim D$)

$$\phi_{ij}^D(x) = \langle \xi_i, D_x \xi_j \rangle \qquad \chi_D = \frac{1}{d}\sum_{i=1}^{d} \phi_{ii}^D$$

$$e_{ij}^D = d\int_G \phi_{ij}^D(x)\lambda(x)\,dx \qquad e^D = \sum_{i=1}^{d} e_{ii}^D.$$

This gives an identification $\mathcal{L}(G) \cong \sum \mathcal{B}(D)$ with e_{ij}^D corresponding to matrix units in $\mathcal{B}(D)$. Here $\mathcal{B}(D) = \mathcal{B}(X(D)) = \mathcal{M}(d, \mathbf{C})$, and we shall often use this identification. One should then note that ν maps $\mathcal{B}(D)$ into $\mathcal{B}(\overline{D})$, and that the character of G as used in group representations is $d^2\chi_D$.

Let e^1 be the projection in $\mathcal{L}(G)$ corresponding to the trivial representation 1 of G, so $e^1 = \int_G \lambda(x)\,dx$. The following are then easily verified for all $x \in \mathcal{L}(G)$:

$$\delta(x) I \otimes e^1 = x \otimes e^1 \tag{2.1}$$

$$(I \otimes x)\delta(e^1) = (\nu(x) \otimes I)\delta(e^1) \tag{2.2}$$

We shall also need Takesaki's operator W defined over $L^2(G \times G)$ by

$$Wf(s,t) = f(s, st).$$

Then $\delta(x) = W^* x \otimes I W$ for all $x \in \mathcal{L}(G)$. Using standard results about representations of compact groups we also have

$$W = \sum_{i,j,D} \phi_{ji}^D \otimes e_{ij}^D,$$

where we have identified a function ϕ with the corresponding multiplication operator $\phi : f \to \phi f$ over $L^2(G)$. Therefore $i \otimes \nu(W)$ is well defined and

$$\nu \otimes \nu(W) = W^* = i \otimes \nu(W). \tag{2.3}$$

In working with non-abelian groups tensor products are an indispensible tool, and the following notation now seems to be standard. The flip σ is defined by $\sigma(a \otimes b) = b \otimes a$, and we then also get maps σ_{12} etc. of $\mathcal{L}(G) \otimes \mathcal{L}(G) \otimes \mathcal{L}(G)$ by

$$\sigma_{12}(a \otimes b \otimes c) = b \otimes a \otimes c \quad \sigma_{13}(a \otimes b \otimes c) = c \otimes b \otimes a.$$

If $U \in \mathcal{A} \otimes \mathcal{L}(G)$ we define $U_1, U_2 \in \mathcal{A} \otimes \mathcal{L}(G) \otimes \mathcal{L}(G) \otimes \mathcal{L}(G)$ by

$$U_1 = U \otimes I \otimes I \qquad U_2 = i \otimes \sigma_{12}(U_1) \quad \text{etc.}$$

If $\omega \in \mathcal{L}(G) \otimes \mathcal{L}(G)$ we shall write similarly

$$\omega_{12} = \omega \otimes I \quad \omega_{13} = \sigma_{23}(\omega \otimes I) \quad \omega_{23} = I \otimes \omega.$$

We shall also need slicemaps $\mathcal{L}(G) \otimes \mathcal{L}(G) \to \mathcal{L}(G)$ defined by

$$i \otimes f(a \otimes b) = f(b)a \quad \text{for } a, b \in \mathcal{L}(G)\ f \in A(G).$$

To increase readability we have tried to cut down on the number of parentheses and therefore certain formulas may at first sight seem ambiguous. For instance

$$i \otimes \nu(U)_2 = (i \otimes \sigma)(i \otimes \nu \otimes i)(U \otimes I).$$

For more about $A(G), \mathcal{L}(G)$ and Hopf–von Neumann algebras we refer to [**12**], and for more on compact groups we refer to [**4**].

Ergodic W^*-dynamical systems. If $(\mathcal{A}, \rho, G)$ is a W^*-dynamical system it is called *ergodic* if the fixed–point algebra

$$\mathcal{A}^\rho = \{a \in \mathcal{A} \mid \rho_x(a) = a \text{ for all } x \in G\} = \mathbf{C}I,$$

and these will be our main objects of study. One can also look at C^*-dynamical systems, but as noted in [**9**, 3B1] there is a 1–1 correspondence between ergodic C^*-dynamical systems and ergodic W^*-dynamical systems preserving all relevant structure. Some proofs are easier in the one case than the other, but we shall concentrate on the von Neumann algebra case.

Actions of compact groups on a locally convex space E are a well developed subject, we refer to [**4**, IX.7] for details. The basic fact we shall need is that one has a decomposition

$$E = \bigoplus_{D \in \widehat{G}} E(D)$$

where the action on each *spectral subspace* $E(D)$ is a multiple of D. In our situation with G acting on $\mathcal{A}$ we shall need not only the spectral subspaces $\mathcal{A}(D)$, but also the following: If D is an irreducible representation of G with $\dim(D)=d$, we call $U \in A \otimes \mathcal{M}(d, \mathbf{C})$ a *D-eigenoperator* for ρ if

$$\rho_x \otimes i(U) = UI \otimes D_x \quad \text{for all } x \in G. \tag{2.4}$$

It was shown in [**5**] that if $(\mathcal{A}, \rho, G)$ is ergodic the unique invariant state τ on $\mathcal{A}$ defined by $\tau(a)I = \int \rho_x(a)\, dx$ is in fact a trace on $\mathcal{A}$. It was also shown there that

$$\dim \mathcal{A}(D) \leq (\dim D)^2 \tag{2.5}$$

and in [9, Theorem 8] that equality holds in (2.5) if and only if there exists a *unitary* D-eigenoperator. If this is true for all $D \in \widehat{G}$ we shall say that ρ has *full* multiplicity, and for the rest of this article we shall always assume this to be the case. If we pick one unitary eigenoperator $U(D)$, any other D-eigenoperator is of the form $(I \otimes \xi)U(D)$ for some $\xi \in \mathcal{M}(d, \mathbf{C})$. If we write

$$U(D) = \sum_{i,j \leq d} u_{ij}^D \otimes e_{ji}^D, \tag{2.6}$$

then $\{u_{ij}^D \mid i,j \leq d\}$ is an orthonormal basis in $\mathcal{A}(D)$ under the inner product defined by τ. Using the theory of representations of G again, one has that

$$\mathcal{A}(D)\mathcal{A}(E) \subset \sum_{F \subset D \otimes E} \mathcal{A}(F) \quad \text{and} \quad \mathcal{A}(D)^* = \mathcal{A}(\overline{D}). \tag{2.7}$$

From this it now follows that $\{u_{ij}^D \mid i,j \leq d, D \in \widehat{G}\}$ spans a dense G-invariant $*$-subalgebra $\mathcal{A}_\circ$ of $\mathcal{A}$. One should note that these last statements are true also if we do not have full multiplicity.

Lemma 2.1. *If we have unitary eigenoperators for both D and E in $\widehat{G}$, then we have unitary eigenoperators for all $F \subset D \otimes E$ and also for $\overline{D}$.*

Proof: $F \subset D \otimes E$ means that there is a partial isometry

$$V : X(F) \to X(D) \otimes X(E) \quad \text{such that} \quad VF_x = D_x \otimes E_x V \quad \text{for } x \in G.$$

Suppose $U(D)$,$U(E)$ are the given unitary eigenoperators, and take

$$U'(F) = U(D)_1 U(E)_2 I \otimes V.$$

Then $\rho_x \otimes i(U'(F)) = U'(F) I \otimes D_x$, so $U'(F)U'(F)^* \in \mathcal{A}^\rho \otimes \mathcal{B}(D) \otimes \mathcal{B}(E)$, i.e. $U'(F)U'(F)^* = I \otimes p$ for some projection $p \in \mathcal{B}(D) \otimes \mathcal{B}(E)$. $U'(F)^*U'(F) = I \otimes e^F$, so there is a partial isometry $Q \in \mathcal{B}(X(D) \otimes X(E), X(F))$ such that $Q^*Q = e^F$ and $QQ^* = p$. Now take

$$U(F) = I \otimes Q^* U(D)_1 U(E)_2 I \otimes V \in \mathcal{A} \otimes \mathcal{B}(F),$$

then $U(F)$ is a unitary eigenoperator. Finally, observe that $i \otimes \nu(U(D)^*)$ is a unitary eigenoperator for $\overline{D}$. ∎

The author thanks E. Gootman and A. Lazar for discovering a mistake in an earlier version of Lemma 2.1.

Corollary 2.2. *Suppose D is a faithful irreducible representation of G. If ρ has a unitary D-eigenoperator, then it has unitary eigenoperators for all $E \in \widehat{G}$, so ρ has full multiplicity. Furthermore $\mathcal{A}(D)$ generates $\mathcal{A}$. In particular if G is a simple compact group then all $D \not\sim 1$ are faithful.*

Proof: This follows from the fact that if $D \in \widehat{G}$ is faithful, then by the Stone–Weierstrass theorem every $E \in \widehat{G}$ is contained in some tensor power of D and $\overline{D}$, (c.f. [**4**, IX,6.1]). So by Lemma 2.1 the $*$-algebra generated by $\mathcal{A}(D)$ equals $\mathcal{A}_\circ$, hence the von Neumann algebra generated by $\mathcal{A}(D)$ equals $\mathcal{A}$.

We are not going to use the theory of crossed products here, so we just mention the following result:

Theorem 2.3. *An ergodic system $(\mathcal{A}, \rho, G)$ has full multiplicity if and only if $\mathcal{A} \times_\rho G \simeq \mathcal{B}(L^2(G))$.*

Proof: See [**9**, Theorem 8] or [**14**, Theorem 15].

3. Ergodic actions and dual cocycles

As earlier G will always be a compact group, and we shall study ergodic W^*-dynamical systems $(\mathcal{A}, \rho, G)$ of full multiplicity. We pick one unitary eigenoperator $U(D)$ for each $D \in \widehat{G}$ and if we now define

$$U = \sum_{D \in \widehat{G}} U(D) \in \mathcal{A} \otimes \mathcal{L}(G), \tag{3.1}$$

then $\rho_x \otimes i(U) = U \; I \otimes \lambda(x)$ for $x \in G$. It is then easy to see that the operator

$$(U \otimes I) \; i \otimes \sigma(U \otimes I) \; i \otimes \delta(U^*)$$

must be invariant under $\rho_x \otimes i \otimes i$, hence it must be equal to $I \otimes \omega$ for some $\omega \in \mathcal{L}(G) \otimes \mathcal{L}(G)$.

Lemma 3.1. *ω satisfies*

$$(\omega \otimes I)\delta \otimes i(\omega) = (I \otimes \omega) i \otimes \delta(\omega) \tag{3.2}$$

Proof:

$$\begin{aligned}
&I \otimes (I \otimes \omega)(i \otimes \delta)(\omega) = (I \otimes \omega_{23}) i \otimes i \otimes \delta[U_1 U_2 i \otimes \delta(U^*)] \\
&= U_1 (I \otimes \omega_{23})(i \otimes i \otimes \delta)(i \otimes \sigma_{12})[U_1 i \otimes \delta(U^*)] \\
&= U_1 (i \otimes \sigma_{12}\sigma_{23})(I \otimes \omega_{12})(i \otimes \sigma_{12}\sigma_{23})(i \otimes \delta \otimes i)[U_1] i \otimes (i \otimes \delta)\delta[U^*] \\
&= U_1 (i \otimes \sigma_{12}\sigma_{23})[(I \otimes \omega) i \otimes \delta(U) \otimes I] i \otimes (\delta \otimes i)\delta[U^*] \\
&= U_1 (i \otimes \sigma_{12}\sigma_{23})[U_1 U_2] i \otimes (\delta \otimes i)\delta[U^*] = U_1 U_2 U_3 i \otimes (\delta \otimes i)\delta[U^*] \\
&= (I \otimes \omega_{12}) i \otimes \delta \otimes i[U_1 U_2 \delta(U^*)] = I \otimes (\omega \otimes I)\delta \otimes i(\omega). \quad \blacksquare
\end{aligned}$$

If G was abelian $\mathcal{L}(G) \otimes \mathcal{L}(G) \simeq L^\infty(\widehat{G} \times \widehat{G})$ and Lemma 3.1 would say that ω is a cocycle over the dual group $\widehat{G}$, so therefore we shall later call ω a dual cocycle. First we want to look a little closer at ω, and for this we are going to use the map $i \otimes \nu$. This should be done with caution: it is only defined on the algebraic tensor product and is neither an automorphism nor an anti-automorphism. But if we are careful we see that $Ui \otimes \nu(U)$ is a unitary operator fixed under $\rho_x \otimes i$, so there is a unitary $u \in \mathcal{L}(G)$ such that

$$i \otimes \nu(U) = U^* I \otimes u. \tag{3.3}$$

Lemma 3.2. *$u = \nu(u)$ and $\nu \otimes \nu(\omega) = \delta(u^*)\sigma(\omega^*)(u \otimes u)$.*

Proof: From (3.3) and $i \otimes \nu(U)^* = i \otimes \nu(U^*)$ it follows that

$$U = I \otimes \nu(u) i \otimes \nu(U^*) = I \otimes \nu(u) i \otimes \nu(U)^*.$$

Thus

$$I \otimes u = Ui \otimes \nu(U) = I \otimes \nu(u) \quad \text{i.e. } u = \nu(u).$$

For the second part we shall use $i \otimes \nu \otimes \nu$ on $U_1 U_2 = (I \otimes \omega) i \otimes \delta(U)$:

$$\begin{aligned}
&i \otimes \nu(U)_1 i \otimes \nu(U)_2 = i \otimes \nu \otimes \nu(i \otimes \delta)(U) I \otimes \nu \otimes \nu(\omega) \\
\iff &(U^* \otimes I) I \otimes u \otimes I (i \otimes \sigma)[(U^* \otimes I) I \otimes u \otimes I] \\
&\qquad = (i \otimes \delta)(i \otimes \nu)(U) I \otimes \nu \otimes \nu(\omega) \\
\iff &i \otimes \sigma[U \otimes I i \otimes \sigma(U \otimes I)]^* I \otimes u \otimes u = i \otimes \delta[U^* I \otimes u] I \otimes \nu \otimes \nu(\omega) \\
\iff &i \otimes \sigma[I \otimes \omega i \otimes \delta(U)]^* I \otimes u \otimes u = i \otimes \delta(U^*) I \otimes \delta(u) \nu \otimes \nu(\omega) \\
\iff &I \otimes \sigma(\omega^*)(u \otimes u) = I \otimes \delta(u) \nu \otimes \nu(\omega)
\end{aligned}$$

so

$$\nu \otimes \nu(\omega) = \delta(u^*)\sigma(\omega^*)(u \otimes u). \quad \blacksquare$$

So far we have seen that to an ergodic W^*-dynamical system with full multiplicity we can define a unitary eigenoperator $U \in \mathcal{A} \otimes \mathcal{L}(G)$, a unitary dual cocycle ω and an operator u. Next we shall see how ω and u vary with our choice of U:

Lemma 3.3. *Let U' be another unitary eigenoperator for ρ (so $U' = I \otimes \xi U$ for some $\xi \in \mathcal{L}(G)$). Then the corresponding elements ω' and u' are given by*

$$\omega' = \xi \otimes \xi \omega \delta(\xi^*) \text{ and } u' = \xi u \nu(\xi)$$

Proof: Straightforward.

The following definitions should therefore be natural and are consistent with the ordinary ones for G abelian:

Definition. *A unitary $\omega \in \mathcal{L}(G) \otimes \mathcal{L}(G)$ is called a dual cocycle if*

$$(\omega \otimes I)\delta \otimes i(\omega) = (I \otimes \omega) i \otimes \delta(\omega) \tag{3.2}$$

Two dual cocycles ω and ω' are said to be equivalent if

$$\omega' = \xi \otimes \xi \, \omega \, \delta(\xi^*) \quad \text{for some unitary } \xi \in \mathcal{L}(G).$$

The next lemma tells us that we can always pick U such that the corresponding $u = I$.

Lemma 3.4. *Let $(\mathcal{A}, \rho, G)$ be as before, then there is an eigenoperator U satisfying*

$$i \otimes \nu(U) = U^*.$$

Proof: If $D \not\sim \overline{D}$ we simply choose $U(D)$ and define $U(\overline{D})$ to be $i \otimes \nu(U(D)^*)$. If $D \sim \overline{D}$, first pick one eigenoperator $U'(D)$. Since the automorphism ν maps $B(D)$ into itself, Lemma 3.3 tells us that

$$U'(D) i \otimes \nu(U'(D)) = I \otimes u \quad \text{and} \quad u = \nu(u)$$

for some unitary $u \in B(D)$. Diagonalizing $u = \sum \lambda_i e_i$ one gets $e_i = \nu(e_i)$, so $v = \sum \lambda_i^{-\frac{1}{2}} e_i$ satisfies $v = \nu(v)$ and $v u \nu(v) = e^D$. Then $U(D) = I \otimes v \, U'(D)$ satisfies $i \otimes \nu(U(D)) = U(D)^*$. Finally, just take U as the sum of all $U(D)$'s as before. ∎

Example. *If $G = SU(2)$ then the only ergodic action of full multiplicity is on $L^\infty(G)$ with $\rho_x(f)(y) = f(x^{-1}y)$.*

Proof: Let D be the 2–dimensional representation of $SU(2)$, then the anti-automorphism ν of $\mathcal{B}(D) = \mathcal{M}(2, \mathbf{C})$ is given by

$$\nu \begin{pmatrix} a & b \\ c & d \end{pmatrix} = \begin{pmatrix} d & -b \\ -c & a \end{pmatrix}.$$

So for some $a, b, c, d \in \mathcal{A}$ we can write

$$U(D) = \begin{pmatrix} a & b \\ c & d \end{pmatrix} \quad \text{and thus } i \otimes \nu(U(D)) = \begin{pmatrix} d & -b \\ -c & a \end{pmatrix}.$$

$$\text{Thus} \quad i \otimes \nu(U(D)) = U(D)^* \iff \begin{pmatrix} d & -b \\ -c & a \end{pmatrix} = \begin{pmatrix} a^* & c^* \\ b^* & d^* \end{pmatrix}.$$

Hence $d = a^*, c = -b^*$ and the argument given in [**16**, Theorem 2] shows that $\{a, b\}$ generates a commutative W^*-algebra which by Corollary 2.2 must be $\mathcal{A}$. Full multiplicity leaves $L^\infty(G)$ as the only possibility. ∎

From now on we shall assume that U is chosen as in Lemma 3.4, so by Lemma 3.3 the corresponding ω will satisfy

$$\nu \otimes \nu(\omega) = \sigma(\omega^*). \tag{3.4}$$

Lemma 3.5.

$$\omega(I \otimes e^1) = I \otimes e^1 \tag{3.5}$$
$$\omega(e^1 \otimes I) = e^1 \otimes I \tag{3.6}$$
$$\omega\delta(e^1) = \delta(e^1) \tag{3.7}$$

Proof: If we use (2.1) we get

$$\begin{aligned} I \otimes \omega(I \otimes e^1) &= U_1 U_2 i \otimes \delta(U^*) I \otimes I \otimes e^1 \\ &= U_1 U_2 (U^* \otimes e^1) = U_1 U_2 (I \otimes I \otimes e^1) U_1^* \\ &= U_1 U_1^* I \otimes I \otimes e^1 = I \otimes I \otimes e^1, \end{aligned}$$

since $UI \otimes e^1 = I \otimes e^1$. So $\omega I \otimes e^1 = I \otimes e^1$ and similarly $\omega e^1 \otimes I = e^1 \otimes I$.

Next we use (2.2) to prove (3.7):

$$\begin{aligned} I \otimes \omega\delta(e^1) &= U_1 U_2 i \otimes \delta(U^*) I \otimes \delta(e^1) = U_1 U_2 I \otimes \delta(e^1) \\ &= U_1 i \otimes \nu(U)_1 I \otimes \delta(e^1) = I \otimes \delta(e^1). \ \blacksquare \end{aligned}$$

We shall need two more properties of ω, and for this one should recall the properties of W given in (2.3).

Lemma 3.6.

$$i \otimes \nu(\omega\sigma(\omega^*)) = \sigma(\omega)\omega^* \tag{3.8}$$
$$i \otimes \nu(\omega W^*) = W\omega^* \tag{3.9}$$

Proof: Take $\pi = \omega\sigma(\omega^*)$, so $I \otimes \pi = U_1 U_2 U_1^* U_2^*$. Then (remember our earlier warnings about $i \otimes \nu$):

$$\begin{aligned} & I \otimes \pi U_2 U_1 = U_1 U_2 \\ \iff & i \otimes i \otimes \nu(U_2) I \otimes i \otimes \nu(\pi) U_1 = U_1 i \otimes i \otimes \nu(U_2) \\ \iff & i \otimes \nu(U)_2 I \otimes i \otimes \nu(\pi) U_1 = U_1 i \otimes \nu(U)_2 \\ \iff & U_2^* I \otimes i \otimes \nu(\pi) U_1 = U_1 U_2^* \\ \iff & I \otimes i \otimes \nu(\pi) = U_2 U_1 U_2^* U_1^* \\ \iff & I \otimes i \otimes \nu(\pi) = I \otimes \sigma(\pi) \end{aligned}$$

To prove (3.9) first note that $U_1 U_2 = (I \otimes \omega) i \otimes \delta(U)$ and the properties of W give

$$U_1 U_2 I \otimes W^* = (I \otimes \omega W^*) U_1,$$

so

$$\begin{aligned} I \otimes i \otimes \nu(\omega W^*)U_1 &= U_1 i \otimes i \otimes \nu(U_2 I \otimes W^*) \\ &= U_1 I \otimes i \otimes \nu(W^*) i \otimes \nu(U)_2 = U_1 I \otimes W\, U_2^* \\ &= I \otimes W\, i \otimes \delta(U) U_2^* = I \otimes W \omega_{23}^* U_1. \end{aligned}$$

Thus $i \otimes \nu(\omega W^*) = W\omega^*$. ∎

Definition. *A normalized dual cocycle is a unitary $\omega \in \mathcal{L}(G) \otimes \mathcal{L}(G)$ which satisfies:*

(3.2) $$(\omega \otimes I)\delta \otimes i(\omega) = (I \otimes \omega) i \otimes \delta(\omega)$$

(3.4) $$\nu \otimes \nu(\omega) = \sigma(\omega^*)$$

(3.5) $$\omega I \otimes e^1 = I \otimes e^1$$

(3.6) $$\omega e^1 \otimes I = e^1 \otimes I$$

(3.7) $$\omega\delta(e^1) = \delta(e^1)$$

(3.8) $$i \otimes \nu(\omega\sigma(\omega^*)) = \sigma(\omega)\omega^*$$

(3.9) $$i \otimes \nu(\omega W^*) = W\omega^*$$

Two normalized dual cocycles ω and ω' are said to be equivalent if

$$\omega' = \xi \otimes \xi\, \omega\, \delta(\xi^*) \quad \text{for some unitary } \xi \in \mathcal{L}(G) \quad \text{with } \xi = \nu(\xi^*).$$

Theorem 3.8. *Suppose $(\mathcal{A}, \rho, G)$ and $(\mathcal{A}', \rho', G)$ are two ergodic W^*-dynamical systems which both have full multiplicity. Let ω and ω' be the corresponding normalized dual cocycles as constructed above. Then*

$$(\mathcal{A}, \rho, G) \sim (\mathcal{A}', \rho', G) \iff \omega \sim \omega' \quad \text{with} \quad \xi = \nu(\xi^*).$$

Proof: $\Longrightarrow$: Suppose $\Phi : \mathcal{A} \to \mathcal{A}'$ is an equivariant isomorphism, $U \in \mathcal{A} \otimes \mathcal{L}(G)$, $U' \in \mathcal{A}' \otimes \mathcal{L}(G)$ such that

$$\rho_x \otimes i(U) = U I \otimes x, \quad \rho'_x \otimes i(U') = U' I \otimes x \quad \text{for } x \in G.$$

Then $V = \Phi \otimes i(U)$ also is a unitary eigenoperator for ρ' so $V = I \otimes \xi U'$ for some $\xi \in \mathcal{L}(G)$.

$$\begin{aligned} I \otimes \omega &= \Phi \otimes i \otimes i[U_1 U_2 i \otimes \delta(U^*)] \\ &= [V_1 V_2 i \otimes \delta(V^*)] = I \otimes \xi \otimes \xi \omega' \delta(\xi^*) \end{aligned}$$

So $\omega \sim \omega'$, and one checks that $\xi = \nu(\xi^*)$.

$\Longleftarrow$: Conversely, suppose $\omega = \xi \otimes \xi\, \omega'\, \delta(\xi^*)$ with $\xi = \nu(\xi^*)$. Let τ be the unique invariant trace on $\mathcal{A}$ and use the GNS-construction to obtain a faithful representation of $\mathcal{A}$ over a Hilbert space H, and define τ' and H' similarly. Now write as before

$$U(D) = \sum u_{ij}^D \otimes e_{ji}^D.$$

So $\{u_{ij}^D\}$ is a linearly independent set which spans a dense subalgebra $\mathcal{A}_\circ$ of $\mathcal{A}$ (c.f.(2.7)). $\mathcal{A}'_\circ$ is constructed the same way, so there is a linear map $\Phi : \mathcal{A}_\circ \to \mathcal{A}'_\circ$ such that

$$\Phi \otimes i(U(D)) = I \otimes \xi\, U'(D).$$

To show that Φ is multiplicative, note that

$$\begin{aligned}
&\Phi \otimes i \otimes i(U(D)_1)\Phi \otimes i \otimes i(U(E)_2)\\
&= (I \otimes \xi \otimes I)U'(D)_1(I \otimes I \otimes \xi)U'(E)_2\\
&= I \otimes (e^D \otimes e^E)(\xi \otimes \xi)\omega'\, i \otimes \delta(U')\\
&= I \otimes (e^D \otimes e^E)\omega\,(i \otimes \delta)[I \otimes \xi\, U']\\
&= [I \otimes (e^D \otimes e^E)\omega](i \otimes \delta)[\Phi \otimes i(U)]\\
&= [I \otimes e^D \otimes e^E]\Phi \otimes i \otimes i[\omega_{23}(i \otimes \delta)(U)]\\
&= \Phi \otimes i \otimes i[U(D)_1 U(E)_2],
\end{aligned}$$

thus $\Phi(u_{ij}^D)\Phi(u_{kl}^E) = \Phi(u_{ij}^D u_{kl}^E)$, hence $\Phi(ab) = \Phi(a)\Phi(b)$ for all $a, b \in \mathcal{A}_\circ$.

$$\begin{aligned}
\Phi \otimes i(U(D))^* &= U'(D)^* I \otimes \xi^* = i \otimes \nu(U'(\overline{D}))I \otimes \nu(\xi)\\
&= i \otimes \nu[I \otimes \xi U'(\overline{D})] = (i \otimes \nu)(\Phi \otimes i)[U(\overline{D})]\\
&= (\Phi \otimes i)(i \otimes \nu)[U(\overline{D})] = \Phi \otimes i(U(D)^*).
\end{aligned}$$

So Φ is a $*$-isomorphism between $\mathcal{A}_\circ$ and $\mathcal{A}'_\circ$. Since $\tau' \circ \Phi = \tau$, Φ will extend to a unitary operator between H and H' which will implement a $*$-isomorphism between $\mathcal{A}$ and $\mathcal{A}'$. Finally $\Phi \circ \rho_x = \rho'_x \circ \Phi$, so Φ is equivariant. ∎

The next result is the opposite of Theorem 3.8: To every normalized dual cocycle there is an ergodic W^*-dynamical system.

Theorem 3.9. *Let ω be a normalized dual cocycle. Then there is an ergodic W^*-dynamical system $(\mathcal{A}, \rho, G)$ with full multiplicity having ω as its normalized dual cocycle.*

Proof: Let $r(x)$ be the right regular representation of G over $L^2(G)$, and let W be as before. Take $U = \omega W^* \in \mathcal{B}(L^2(G)) \otimes \mathcal{L}(G)$. Then

$$r(x) \otimes I\, U = U r(x) \otimes \lambda(x)$$

and

$$\begin{aligned}
U_1 U_2 &= (\omega \otimes I)(W^* \otimes I)\omega_{13} W_{13}^*\\
&= (\omega \otimes I)(\delta \otimes i)(\omega)(W^* \otimes I)W_{13}^*\\
&= (I \otimes \omega)(i \otimes \delta)(\omega)\, i \otimes \delta(W^*)\\
&= (I \otimes \omega)(i \otimes \delta)(U).
\end{aligned}$$

By (3.9) $i \otimes \nu(U) = U^*$. From all this it now follows that the "left coefficients" of U (= the images of U by all slicemaps $\mathcal{B}(L^2(G)) \otimes \mathcal{L}(G) \to \mathcal{B}(L^2(G))$) is a $*$-subalgebra $\mathcal{A}_\circ$ of $\mathcal{B}(L^2(G))$ invariant under $\rho_x = \mathrm{Ad}\nu(x)$. So the weak closure $\mathcal{A}$ of $\mathcal{A}_\circ$ is a W^*-dynamical system under ρ. It remains to show that it is ergodic, and since $\mathcal{A}_\circ^\rho$ is dense in $\mathcal{A}^\rho$, it suffices to show that $\mathcal{A}_\circ^\rho$ is trivial. Every $a \in \mathcal{A}_\circ^\rho$ is of the form $i \otimes \phi(\omega W^*)$ for some $\phi \in A(G)$, so

$$\begin{aligned} a = \int \rho_x(a)\, dx &= i \otimes \phi[\omega \int (r(x) \otimes I) W^* (r(x^{-1}) \otimes I)\, dx] \\ &= i \otimes \phi[\omega(I \otimes e^1)] = \phi(e^1) I. \end{aligned}$$

So $\mathcal{A}_\circ^\rho = \mathbf{C}I$, thus $\mathcal{A}^\rho = \mathbf{C}I$ and therefore ρ is ergodic. Since U is unitary ρ has full multiplicity, and ω is obviously the normalized dual cocycle corresponding to U. ∎

The following corresponds to [**15**,Lemma 21]:

Proposition 3.10. *Let ω be a normalized dual cocycle. Then*

$$\omega \quad \textit{is trivial } (\sim I) \iff \omega \quad \textit{is symmetric } (\omega = \sigma(\omega)).$$

Proof: Only $\Longleftarrow$ needs to be proved. So suppose $\omega = \sigma(\omega)$ and as in Theorem 3.9 take $U = \omega W^*$ and let $\mathcal{B}$ be the C^*-algebra $\subset \mathcal{A}$ generated by the operators $U(\phi) = i \otimes \phi(U)$. Next note that (c.f. the proof of Lemma 3.6)

$$\omega = \sigma(\omega) \iff U_1 U_2 = U_2 U_1.$$

Since $U(\phi)U(\psi) = i \otimes \phi \otimes \psi(U_1 U_2)$ this means that $\mathcal{B}$ is an abelian C^*-algebra. Now let χ be a non-zero multiplicative linear functional on $\mathcal{B}$. Since $\|U(\phi)\| \le \|\phi\|$ we must have

$$\chi(U(\phi)) = \langle \xi, \phi \rangle \quad \text{for some } \xi \neq 0 \quad \text{in } \mathcal{L}(G).$$

One also checks (we omit the details) that $U(\phi)U(\psi) = U(\phi \bigtriangleup \psi)$ where

$$\langle x, \phi \bigtriangleup \psi \rangle = \langle \omega\delta(x), \phi \otimes \psi \rangle \quad \text{for } x \in \mathcal{L}(G). \tag{3.10}$$

So since χ is multiplicative we have

$$\begin{aligned} &\langle \xi, \phi \bigtriangleup \psi \rangle = \langle \xi, \phi \rangle \langle \xi, \psi \rangle \\ \iff &\langle \omega\delta(\xi), \phi \otimes \psi \rangle = \langle \xi \otimes \xi, \phi \otimes \psi \rangle \quad \text{for all } \phi, \psi. \end{aligned}$$

So $\omega\delta(\xi) = \xi \otimes \xi \implies \delta(\xi^*\xi) = \xi^*\xi \otimes \xi^*\xi$. But it follows from [**12**, Lemma 10.4] that then $\xi^*\xi = I$. So ξ is unitary and $\omega = \xi \otimes \xi\delta(\xi)^* \sim I$. ∎

Remark 3.11: This construction of a $*$-algebra from a dual cocycle is the dual version of A. Kleppner's construction in [**7**]. In fact (3.10) defines a product on $A(G)$ which makes it into a Banach algebra which generates Kleppner's left regular ω-representation, even when ω is not symmetric. The product (3.10) is also the one used by A. Wassermann in [**15**,§7].

The following should now be obvious:

Theorem 3.12. *There is a 1–1 correspondence between equivalence classes of ergodic W^*-dynamical systems $(\mathcal{A}, \rho, G)$ of full multiplicity and equivalence classes of normalized dual cocycles ω.*

Remark 3.13: From the observations in [**9**, 3B1] it follows that there also is a 1–1 correspondence between the equivalence classes of ergodic C^*-dynamical systems $(\mathcal{A}, \rho, G)$ of full multiplicity.

Remark 3.14: The problem of when a dual cocycle is equivalent to a normalized one has been treated in [**15**, Theorems 5 and 6], we shall not go into this here.

4. Dual cocycles and symplectic bicharacters

So far we have seen that the study of ergodic W^*-dynamical systems of full multiplicity is the same as that of normalized dual cocycles, exactly as for abelian groups. In [**8**] Kleppner showed that cocycles over $\widehat{G}$ can be classified by so called *symplectic bicharacters*, that is homomorphisms $\pi : \widehat{G} \to G$ which satisfy

$$\langle \pi(\alpha), \beta \rangle = \overline{\langle \pi(\beta), \alpha \rangle} \quad \text{and} \quad \langle \pi(\alpha), \alpha \rangle = 1$$

under the duality between G and $\widehat{G}$. (Observe that the second property implies the first one, but *not* the other way.) This is exactly what is needed to do the classification of ergodic W^*-dynamical systems over abelian compact groups as in [**1**] and [**10**]. We shall here show that we can form a symplectic bicharacter π also in the non-abelian case, and we shall see how properties of π correspond to properties of $(\mathcal{A}, \rho, G)$. The first result shows that dual cocycles (and hence also ergodic W^*-dynamical systems) give interesting examples of Hopf–von Neumann algebras.

Theorem 4.1. *Define $\delta_\omega : \mathcal{L}(G) \to \mathcal{L}(G) \otimes \mathcal{L}(G)$ by $\delta_\omega(x) = \omega\delta(x)\omega^*$, where ω is a normalized dual cocycle. Then $(\mathcal{L}(G), \delta_\omega, \nu, \mathrm{tr})$ is a Hopf–von Neumann algebra.*

Proof: For the definition of a Hopf–von Neumann algebra and for the definition of the trace tr on $\mathcal{L}(G)$, we refer to [**12**]. We omit the details of proof, note for instance that the property $(\delta_\omega \otimes i)\delta_\omega = (i \otimes \delta_\omega)\delta_\omega$ follows from (3.2), and that $\nu \otimes \nu\delta_\omega = \sigma\delta_\omega\nu$ follows from (3.4). ∎

A Hopf–von Neumann algebra may seem like a strange object, here we shall think of δ_ω as a structure giving $A(G)$ another multiplication:

Proposition 4.2. *For $f, g \in A(G)$ define $f \star g \in A(G)$ by*

$$\langle x, f \star g \rangle = \langle \delta_\omega(x), f \otimes g \rangle \quad \text{for } x \in \mathcal{L}(G).$$

Together with the involution $^\#$ given by

$$\langle x, f^\# \rangle = \overline{\langle \nu(x^*), f \rangle}$$

and the usual norm as the predual of $\mathcal{L}(G)$, this makes $A(G)$ a Banach $*$-algebra which we shall denote $A_\omega(G)$.

Proof: This follows readily from [**12**], or directly from the properties of ω.

Remark 4.3: This is the first place we see an important difference between abelian and non-abelian groups: if G is abelian $A(G) = A_\omega(G)$ as Banach $*$-algebras.

Proposition 4.4. *Let* $\pi = \omega\sigma(\omega^*)$, *then*

$$(\pi \otimes I)i \otimes \sigma(\pi \otimes I) = i \otimes \delta_\omega(\pi) \tag{4.1}$$

$$(I \otimes \pi)\sigma \otimes i(\pi \otimes I) = \delta_\omega \otimes i(\pi) \tag{4.2}$$

$$\pi = \sigma(\pi^*) = \nu \otimes \nu(\pi) \tag{4.3}$$

$$\pi(e^1 \otimes I) = e^1 \otimes I \tag{4.4}$$

$$\pi(I \otimes e^1) = I \otimes e^1 \tag{4.5}$$

$$\pi\delta(e^1) = \delta(e^1) \tag{4.6}$$

$$i \otimes \nu(\pi) = \sigma(\pi) = \pi^* \tag{4.7}$$

Proof: We start with (4.1), and we will use only (3.2) and the properties of δ:

$$\begin{aligned} i \otimes \delta_\omega(\pi) &= \omega_{23} i \otimes \delta(\pi)\omega_{23}^* \\ &= \omega_{23} i \otimes \delta(\omega)(i \otimes \delta)\sigma(\omega^*)\omega_{23}^* \\ &= \omega_{12}\delta \otimes i(\omega)\sigma_{12}\sigma_{23}[(\delta \otimes i)(\omega^*)\omega_{12}^*] \\ &= \omega_{12}\delta \otimes i(\omega)\sigma_{12}\sigma_{23}[(i \otimes \delta)(\omega^*)\omega_{23}*] \\ &= \omega_{12}\delta \otimes i(\omega)\sigma_{12}[(i \otimes \delta)(\omega^*)\sigma_{23}(\omega_{23}^*)] \\ &= \omega_{12}\sigma_{12}[\omega_{12}^*\omega_{23}\sigma_{23}(\omega_{23})^*] \\ &= [\omega\sigma(\omega^*) \otimes I]\sigma_{12}[I \otimes \omega\sigma(\omega^*)] = \pi_{12}\pi_{23}. \end{aligned}$$

To prove (4.2), take the adjoint of (4.1) and act on both sides with $\sigma_{23}\sigma_{12}$:

$$\begin{aligned} &\sigma_{23}\sigma_{12}(\pi_{13}^*)\sigma_{23}(\pi_{12}) = \sigma_{23}\sigma_{12}[\omega_{23} i \otimes \delta(\pi^*)\omega_{23}^*] \\ \iff &[I \otimes \sigma(\pi^*)]\pi_{13} = \omega_{12}(\delta \otimes i)\sigma(\pi^*)\omega_{12}^* \\ \iff &\pi_{23}\pi_{13} = (\delta_\omega \otimes i)(\pi). \end{aligned}$$

(4.3–4.7) now follows from (3.4–3.9). ∎

Remark 4.5: A.Wassermann has shown in [**15**, Lemma 26] that π also is a dual cocycle, and satisfies the important Quantum Yang–Baxter equation:

$$\pi_{12}\pi_{13}\pi_{23} = \pi_{23}\pi_{13}\pi_{12} \tag{QYBE}$$

which seems to be of importance in several parts of mathematics. He has also shown that any π satisfying this equation comes from a dual cocycle and hence from an ergodic action. The reader will see that there are many similarities here with ideas described by V.G. Drinfel'd in [**3**], and we refer to that article for references to the large literature on this subject. We shall comment more on this later.

Hopefully the statements of Proposition 4.4 becomes more meaningful when we phrase them with respect to the algebra $A_\omega(G)$. For $f \in A_\omega(G)$ let

$$\pi(f) = i \otimes f(\pi).$$

Then we have:

Theorem 4.6. *$\pi : f \to \pi(f)$ is a bounded $*$-representation of $A_\omega(G)$, i.e.*

(4.8) $$\pi(f \star g) = \pi(f)\pi(g)$$

(4.9) $$\pi(f^\#) = \pi(f)^*$$

(4.10) $$\pi(1) = I$$

(4.11) $$\nu(\pi(f)) = \pi(f \circ \nu)$$

and π is symplectic:

(4.12) $$\langle \pi(f), g\rangle = \langle \pi(g \circ \nu), f\rangle = \overline{\langle \pi(g^\#), f^\#\rangle}$$

Proof: We start with (4.8):

$$\pi(f)\pi(g) = i \otimes f \otimes g(\pi_{12}\pi_{13})$$
$$= i \otimes f \otimes g(i \otimes \delta_\omega(\pi)) = i \otimes f \star g(\pi).$$

(4.11) follows from (4.3):

$$\langle \nu(\pi(f)), g\rangle = \langle \pi, g \circ \nu \otimes f\rangle$$
$$= \langle \pi, g \otimes f \circ \nu\rangle = \langle \pi(f \circ \nu), g\rangle$$

(4.10) follows from (4.5):

$$\langle \pi(1), f\rangle = \langle \pi, f \otimes 1\rangle = \langle \pi(I \otimes e^1), f \otimes 1\rangle$$
$$= \langle I \otimes e^1, f \otimes 1\rangle = \langle I, f\rangle$$

The first half of (4.12) follows from (4.7):

$$\langle \pi(g \circ \nu), f\rangle = \langle i \otimes \nu(\pi), f \otimes g\rangle$$
$$= \langle \sigma(\pi), f \otimes g\rangle = \langle \pi, g \otimes f\rangle = \langle \pi(f), g\rangle,$$

and the last part from (4.3):

$$\langle \pi(f), g\rangle = \langle a, g \otimes f\rangle = \overline{\langle \nu \otimes \nu(\pi^*), g^\# \otimes f^\#\rangle}$$
$$= \overline{\langle \sigma(\pi), g^\# \otimes f^\#\rangle} = \overline{\langle(\pi(g^\#), f^\#\rangle}.$$

We get (4.9) by using (4.12):

$$\langle \pi(f)^*, g\rangle = \overline{\langle \pi(f), g^\# \circ \nu\rangle}$$
$$= \langle \pi(g \circ \nu), f^\#\rangle = \langle \pi(f^\#), g\rangle.$$

Finally, $\|\pi(f)\| \le \|f\|$ should be obvious. ∎

$A_\omega(G)$ is a pre Hopf–von Neumann algebra if we define δ^t and ν^t by

$$\langle x \otimes y, \delta^t\rangle = \langle yx, f\rangle \quad \text{and } \nu^t(f) = f \circ \nu$$

Then

Corollary 4.7.

$$(\pi \otimes \pi)[\delta^t(f)] = \delta_\omega(\pi(f))$$

so $\pi : (A_\omega(G), \delta^t, \nu^t) \to (\mathcal{L}(G), \delta_\omega, \nu)$ *is a Hopf algebra homomorphism.*

We omit the proof (which follows from Theorem 4.1) since we shall not need it.

Corollary 4.8. *The map* π *defined in Theorem 4.6 is injective if and only if its image in* $\mathcal{L}(G)$ *is weakly dense.*

Proof: Follows from (4.12).

Remark: Since the main difference between abelian and non-abelian groups seems to be that in general $\delta_\omega \neq \delta$, it would be natural to investigate what happens if $\delta_\omega = \delta$, or more generally when $A_\omega(G)$ is commutative. By Proposition 4.2 this happens if and only if π commutes with $\delta(\mathcal{L}(G))$. We then have

Proposition 4.9. *If* $A_\omega(G)$ *is abelian then* $(\mathcal{A}, \rho, G)$ *is induced from an ergodic action of a closed abelian subgroup.*

Proof: This is essentially what was done in [**9**, 3B6].

Proposition 4.10. *If* $D \in \widehat{G}$ *is faithful, then* $\pi(e^D \otimes e^D)$ *uniquely determines* π.

Proof: For this first note that from Corollary 2.2 it follows that $\{\phi_{ij}^D\}$ generates $A_\omega(G)$. So from knowing $\pi(e^D \otimes e^D)$ one knows $\pi(\phi_{ij}^D)e^D$ and thus also $\pi(f)e^D$ for all f. From (4.12) we then also know $\pi(\phi_{ij}^D)$, hence $\pi(f)$ for all $f \in A_\omega(G)$ ∎.

When G is abelian, the symplectic bicharacter π is a group homomorphism from $\widehat{G}$ into G. So far our non-abelian analogue is a homomorphism only on the group algebra level. We shall now see that π also can be thought of as a map on the set level from $\widehat{G}$ into G in the general case. We shall see that many of the abelian results still are true, but not so conclusive any longer.

Definition. *Let*

$$A_\omega^\circ(G) = \{f \in A_\omega(G) \mid f(xy - yx) = 0 \quad \text{for all } x, y \in G\}.$$

So $A_\omega^\circ(G)$ is the subalgebra of $A_\omega(G)$ generated by the characters of G. Obviously $f \star g = fg$ for $f, g \in A_\omega^\circ(G)$, so $A_\omega^\circ(G) = A^\circ(G)$ as Banach $*$-algebras. The spectrum of $A^\circ(G)$ is the same as that of $C^\circ(G)$ which equals G° = the set of conjugacy classes of G. So π will be a $*$-representation of $A^\circ(G)$ into $\mathcal{L}(G)$ which can be diagonalized, i.e. there is a basis in the space of each $D \in \widehat{G}$ and elements $x_i(D) \in G^\circ$ such that

$$\pi(f) = \sum_{i,D} f(x_i(D))e_{ii}^D \quad \text{for } f \in A^\circ(G). \tag{4.13}$$

We shall call the set $\{x_i(D) \mid i \leq d, D \in \widehat{G}\}$ the *spectrum* of π, and note that $x_i(D)$ is only determined up to conjugacy. From (4.11) it follows that

$$f(x_i(\overline{D})) = f(x_i(D)^{-1}) \quad \text{for all } f \in A^\circ(G).$$

So $x_i(\overline{D})$ equals $x_i(D)^{-1}$ in G°, i.e. they are conjugate as elements of G. So if $D \not\sim \overline{D}$ we may assume they are equal also as elements of G. However if $D \sim \overline{D}$ we can *not* necessarily assume that we can pick an $x_i(D)$ with $x_i(D) = x_i(D)^{-1}$. For instance in the group $SU(2)$ x and x^{-1} are always conjugate, but usually not equal.

For $D \in \widehat{G}$ let

$$x(D) = \pi(\chi_D) \tag{4.14}$$

so $x(D) = \frac{1}{d}\sum_{i=1}^{d} x_i(D)$ as an element of $\mathcal{L}(G)$. From (4.12) we will now have:

$$\langle x(D), \chi_E\rangle = \overline{\langle x(E), \chi_D\rangle}. \tag{4.15}$$

Since $I \otimes \pi = U_1U_2U_1^*U_2^*$, one should expect that there is a relation between properties of π and the algebraic properties of $\mathcal{A}$. These relations shall be our next theme. First we shall observe that the elements $x_i(D)$ can be described more directly by $\mathcal{A}$. Let $U(D)$ be as before and look at the map $\Phi_D : \mathcal{A} \to \mathcal{A}$ given by

$$\begin{aligned}\Phi_D(a) &= i \otimes \chi_D[U(D)a \otimes IU(D)^*]\\ &= \frac{1}{d}\sum_{i,j} u_{ij}^D a u_{ij}^{D*}.\end{aligned}$$

Φ_D does not depend on which unitary $U(D)$ we have chosen, and

$$\Phi_D \circ \rho_x = \rho_x \circ \Phi_D \quad \Phi_D(a)^* = \Phi_D(a^*) \quad \tau(\Phi_D(a)b) = \tau(a\Phi_{\overline{D}}(b))$$

Furthermore

$$\begin{aligned}\Phi_D \circ \Phi_E(a) &= i \otimes \chi_D \otimes \chi_E[U(D)_1U(E)_2(a \otimes I \otimes I)U(E)_2^*U(D)_1^*]\\ &= i \otimes \chi_D \otimes \chi_E[\omega_{23} i \otimes \delta(Ua \otimes IU^*)\omega_{23}^*]\\ &= i \otimes \chi_D \otimes \chi_E[i \otimes \delta(Ua \otimes IU^*)]\\ &= i \otimes \chi_E \otimes \chi_D[i \otimes \delta(Ua \otimes IU^*)] = \Phi_E \circ \Phi_D(a).\end{aligned}$$

So $\{\Phi_D \mid D \in \widehat{G}\}$ is a commuting selfadjoint (with respect to the inner product defined by τ) family which leaves each spectral subspace $\mathcal{A}(E)$ invariant. Furthermore $\chi_D \to \Phi_{\overline{D}}$ is a representation of $A^\circ(G)$ over this inner product space. Hence the basis $\{u_{ij}^E\}$ of $\mathcal{A}(E)$ can be picked such that

$$\Phi_D(u_{ij}^E) = \overline{\chi_D(x_j(E))}u_{ij}^E \quad \text{for some } x_j(E) \in G^\circ.$$

That these really are the same $x_j(E)$ as above follows from the fact that

$$U_1^* I \otimes \pi = U_2U_1^*U_2^* \implies I \otimes \pi(\chi_D) = U\Phi_D \otimes i(U^*).$$

Lemma 4.11. $\Phi_D(a) = a \iff$ *a commutes with* $\mathcal{A}(D)$.

Proof: $\Longleftarrow$ is obvious. So suppose $\Phi_D(a) = a$ and let $\mathcal{A} \otimes \mathcal{M}(d, \mathbf{C})$ have the norm from the inner product $\langle x, y\rangle = \tau \otimes \chi_D(y^* x)$. Then

$$\begin{aligned}
&\|U(D)(a \otimes I)U(D)^* - a \otimes I\|^2 \\
&= \tau \otimes \chi_D[U(D)(a^*a \otimes I)U(D)^* - U(D)(a^* \otimes I)U(D)^*(a \otimes I) \\
&\quad - (a^* \otimes I)U(D)(a \otimes I)U(D)^* + a^*a \otimes I] \\
&= 2\tau \otimes \chi_D(a^*a \otimes I) - \tau(\Phi_D(a^*)a) - \tau(a^*\Phi_D(a)) = 0
\end{aligned}$$

since $\Phi_D(a) = a \implies \Phi_D(a^*) = a^*$. So

$$U(D)(a \otimes I)U(D)^* = a \otimes I$$

which implies that a commutes with $\{u_{ij}^D\}$, hence with $\mathcal{A}(D)$. ∎

Lemma 4.12. *The center of* $\mathcal{A}$, $\mathcal{Z}(\mathcal{A}) = \mathcal{A} \cap \mathcal{A}'$ *is generated by*

$$\{u_{ij}^E \mid x_j(E) = e\}.$$

Proof: $\mathcal{Z}(\mathcal{A})$ is generated by $\mathcal{Z}(\mathcal{A})_\circ = \mathcal{A}_\circ \cap \mathcal{A}'$, so suppose we have a finite sum

$$a = \sum_{i,j,E} \lambda_{ij} u_{ij}^E \in \mathcal{A}_\circ \cap \mathcal{A}'.$$

Since $\Phi_D(a) = a$ for all D, we must have that

$$\lambda_{ij} \neq 0 \implies \chi_D(x_j(E)) = 1.$$

But if $\chi_D(x_j(E)) = 1$ for all $D \in \widehat{G}$ we must have $x_j(E) = e$. Conversely if $x_j(E) = e$ Lemma 4.11 applies to u_{ij}^E for all $D \in \widehat{G}$, so $u_{ij}^E \in \mathcal{Z}(\mathcal{A})$. ∎

Corollary 4.13. $\mathcal{A}$ *is a factor* $\iff$ $x_j(E) \neq e$ *for all* $E \neq 1$.

Theorem 4.14. *Look at the following statements:*

(1) $\mathcal{A}$ *is a factor,*
(2) $x_i(E) \neq e$ *for* $E \neq 1$,
(3) $\pi(f) = 0 \implies f = 0$,
(4) $\{\pi(f) \mid f \in A_\omega(G)\}$ *is weakly dense in* $\mathcal{L}(G)$,
(5) $\{x_j(D) \mid j \leq d, D \in \widehat{G}\}$ *is dense in* G°.

Then (1) $\iff$ (2) $\iff$ (3) $\iff$ (4) $\implies$ (5)

Remark 4.15: As noted in [**9**, 3B2] (1) is also equivalent with the property that in the corresponding C^*-dynamical system the C^*-algebra is *simple* or *prime*. Theorem 4.14 corresponds to [**15**,Theorem 12].

Proof: (1) $\iff$ (2) is Corollary 4.13, (3) $\iff$ (4) is Corollary 4.8.

(3) $\Longrightarrow$ (2): Suppose (2) is not true, so we have $x_i(E) = 1$ for some $E \neq 1$. Take $\phi = \phi^E_{ii}$. Then $\|\pi(\phi)\| \le \|\phi^E_{ii}\| = 1$. From (4.12) we now have

$$\langle \pi(\phi), \chi_D \rangle = \langle \pi(\chi_{\overline{D}}), \phi^E_{ii} \rangle = \overline{\chi_D(x_i(E))} = 1.$$

So $\pi(\phi)e^D$ has norm ≤ 1 and trace $= 1$, thus $\pi(\phi)e^D = e^D$. This holds for all D, so $\pi(\phi) = I = \pi(1)$. So $\pi(1-\phi) = 0$, but $1 \neq \phi$.

(2) $\Longrightarrow$ (3): For this we will need the following elementary fact:

$$z \in \mathbf{C}, \quad |z| \le 1, \quad z \neq 1 \Longrightarrow \lim_{n\to\infty} \frac{1}{n} \sum_{k=1}^{n} z^k = 0.$$

Now suppose (2) holds and let $\mathcal{M} = \{\pi(\phi) \mid \phi \in A_\omega(G)\}''$. So from (4.12)

$$\langle \mathcal{M}, f \rangle = 0 \iff \pi(f) = 0.$$

Now for $D \in \widehat{G}$, $n \in \mathbf{N}$ look at the following element in $\mathcal{M}$:

$$b(D,n) = \frac{1}{n} \sum_{k=1}^{n} \pi(\chi_D)^k = \frac{1}{n} \sum_{k=1}^{n} \sum_{i,E} \chi_D(x_i(E))^k e^E_{ii}.$$

So $\|b(D,n)\| \le 1$ and

$$\lim_{n\to\infty} \langle b(D,n), f \rangle = \Big\langle \sum_{\chi_D(x_i(E))=1} e^E_{ii}, f \Big\rangle \quad \text{for all } f \in A_\omega(G).$$

So the following weak limit exists in $\mathcal{M}$:

$$b(D) = \lim_{n\to\infty} b(D,n) = \sum_{\chi_D(x_i(E))=1} e^E_{ii}.$$

But since $\chi_D(x_i(E)) = 1 \iff x_i(E) \in \ker D$, it follows from (2) that

$$e^1 = \inf_{D\in\widehat{G}} b(D) \in \mathcal{M}.$$

Thus

$$\begin{aligned} & \pi(f) = 0 \Longrightarrow \pi(f^\# f) = 0 \Longrightarrow f^\# f \in \mathcal{M}^\perp \\ \Longrightarrow \quad & 0 = \langle e^1, f^\# f \rangle = \langle \delta_\omega(e^1), f^\# \otimes f \rangle \\ & = \langle \delta(e^1), f^\# \otimes f \rangle = \int |\langle \lambda(x), f \rangle|^2 \, dx \\ \Longrightarrow \quad & f = 0. \end{aligned}$$

(2) $\Longrightarrow$ (5): Note that $b(D,n)$ and $b(D)$ as defined above in fact define states on $A^{\circ}_{\omega}(G)$:

$$\langle b(D,n), f^{\#}f\rangle \geq 0 \quad \text{and } \langle b(D,n),1\rangle = 1$$

and hence extend to states on $C^{\circ}(G)$. This is true also for all finite products $\prod b(D_i,n_i)$ and a standard approximation argument shows that

$$\langle \prod_{D\in\widehat{G}} b(D), f\rangle = \langle e^1, f\rangle \quad \text{for all } f\in C^{\circ}(G).$$

Thus if $f\in C^{\circ}(G)^+$ vanishes on $\{x_i(D)\}$, then

$$\begin{aligned}\langle\prod_{i=1}^{m} b(D_i,n_i), f\rangle = 0 &\Longrightarrow \langle\prod_{i=1}^{m} b(D_i), f\rangle = 0\\ &\Longrightarrow \langle e^1, f\rangle = 0 \Longrightarrow f=0.\end{aligned}$$

So (2) $\Longrightarrow$ (5). ∎

Example 4.16: For abelian groups we also have (2) $\Longleftarrow$ (5), but that is not true in general:

Let G be a compact connected Lie group and H a maximal abelian subgroup. Let $\theta : \widehat{H} \to H$ be a symplectic bicharacter. Form the corresponding ergodic system $(\mathcal{B},\beta,H)$ as mentioned earlier, and let $(\mathcal{A},\rho,G)$ be the induced system. For each $D\in\widehat{G}$ pick a basis $\{\xi_i^D\}$ such that

$$D_h\xi_i^D = \alpha_i^D(h)\xi_i^D \quad \text{for some } \alpha_i^D\in\widehat{H}.$$

We leave it to the reader to show that the bicharacter corresponding to ρ is given by

$$\pi = \sum_{i,D} e_{ii}^D\otimes\lambda(\theta(\alpha_i^D)).$$

Hence $x_i(D) = \theta(\alpha_i^D)$. Since every element in G is conjugate to an element in H, the map $H\to G^{\circ}$ is onto. Thus if θ is 1–1 then $\{x_i(D)\}$ is dense in G°, but $\mathcal{A}$ is not simple.

Remark 4.17: The following should give an other indication that there must be some difference for non-abelian groups. If G is abelian then ρ_x is inner $\Longleftrightarrow x = \pi(\alpha)$, c.f. [**10**, 5.7]. So if $\mathcal{A}$ is a factor, the elements x for which ρ_x is inner is dense in G. But if G is a simple compact group it is also algebraically simple, so unless $\mathcal{A} = \mathcal{B}(D)$ it is only $x=e$ which will give ρ_x inner. (c.f. [**2**] and [**13**], for which we are indebted to P. de la Harpe.)

Final remarks. We have seen here that the program of classifying the ergodic actions of a non-abelian compact group can be performed along the same line as for abelian groups, we even get the existence of a map from $\widehat{G}$ into G having similar properties. The disappointing thing is that it is not so clear what kind of group elements one will get as solutions.

The encouraging fact is that the seemingly innocent question which started S. Albeverio and R. Høegh-Krohn has led to connections with other parts of mathematics, in particular the quantum groups in [**3**]. One might hope that some of the many solutions to (QYBE) would be helpful also in our situation, but so far none of these seem to correspond to any new ergodic W^*-dynamical systems. For instance, the solutions described in [**3**, 6.2] correspond to systems induced from a dual cocycle on a maximal abelian subgroup, the construction being as in Example 4.16. Proposition 4.4 tells us that the solutions of (QYBE) which we are interested in have to be rather special. If we are aiming at solving the conjecture in part 1, we will in addition be looking for solutions for which the "matrix-coeficients" π_{ij} act irreducibly on $\mathbf{C}^d$. So whether the conjecture is true or not, it should have interesting applications to the theory of (QYBE).

We have here only treated the case with full multiplicity. Some parts of the development also go through in the general case, the main difficulty is that ω and π then just will be partial isometries. One should note that the induction construction in Theorem 1.2 preserves full multiplicity, so also for ergodic actions which are not of full multiplicity the factor case is the generic one.

References

1. S. Albeverio and R. Høegh-Krohn, *Ergodic actions by compact groups on C^*-algebras*, Math. Zeit. **174** (1980), 1–17.
2. E. Cartan, *Sur les représentations linéaires des groupes clos*, Comment. Math. Helv. **2** (1930), 269–283.
3. V. G. Drinfel'd, *Quantum groups*, Proc. Int. Congress of Mathematicians, Berkeley 1986 (1987), 798–820.
4. J. M. G. Fell and R. Doran, "Representations of $*$-algebras, locally compact groups, and Banach $*$-algebraic bundles," Academic Press, San Diego, 1988.
5. R. Høegh-Krohn, M. B. Landstad and E. Størmer, *Compact ergodic groups of automorphisms*, Ann. Math. **114** (1981), 75–86.
6. R. Høegh-Krohn and T. Skjelbred, *Classification of C^*-algebras admitting ergodic actions of the two-dimensional torus*, J.Reine Angew. Math. **328** (1981), 1–8.
7. A. Kleppner, *The structure of some induced representations*, Duke Math. J. **29** (1962), 555–572.
8. ———, *Multipliers on abelian groups*, Math. Ann. **158** (1965), 11–34.
9. M. B. Landstad, *Operator algebras and compact groups*, Proc. of the Int. Conf. in Operator Algebras and Group Representations in Neptun (Romania) 1980, Monographs and Studies in Math. **18, vol.II** (1984), 33–47, Pitman.
10. D. Olesen, G. K. Pedersen and M. Takesaki, *Ergodic actions of compact abelian groups.*, J. Operator Theory **3** (1980), 237–269.
11. E. Størmer, *Spectra of ergodic transformations*, J. Functional Anal. **15** (1974), 202–215.
12. M. Takesaki, *Duality and von Neumann algebras*, Springer Lecture Notes **247** (1972), 665–779.
13. B. L. van der Waerden, *Stetigkeitssätze für halbeinfache Lieschen Gruppen*, Math. Zeit. **36** (1933), 780–786.
14. A. Wassermann, *Ergodic actions of compact groups on operator algebras: I. General theory*, Ann. Math. **130** (1989), 273–319.

15. ———, *Ergodic actions of compact groups on operator algebras: II. Classification of full multiplicity ergodic actions*, Can. J. Math. **XL** (1988), 1482–1527.
16. ———, *Ergodic actions of compact groups on operator algebras: III. Classification for $SU(2)$*, Inv. Math. **93** (1988), 309–354.
17. ———, *Coactions and Yang–Baxter Equations for ergodic actions and subfactors*, Operator Algebras and Applications, vol.II, London Math. Soc. Lecture note series **135** (1988), 203–236, Cambridge Univ. Press.

Department of Mathematics and Statistics, University of Trondheim, AVH, N–7055 Dragvoll, Norway

POSITIVE PROJECTIONS ONTO JORDAN ALGEBRAS AND THEIR ENVELOPING VON NEUMANN ALGEBRAS

by
Erling Størmer

1. Introduction. In their paper [4] Evans and Høegh-Krohn initiated a study of spectral properties of positive linear maps of C^*-algebras into themselves. They restricted attention to finite dimensional algebras and maps with fixed point sets the scalars, called irreducible maps. To make the latter restriction plausible they showed a decomposition theorem [4, Th. 3.1] which to some extent decomposed a map into a direct sum of irreducible ones. In the present paper we shall see how an infinite dimensional extension of this result follows from a general theorem on positive projections of von Neumann algebras. Recall that a positive projection of a von Neumann algebra M into itself is a unital idempotent positive linear map $P : M \to M$. If P is faithful and normal then $P(M_{sa})$ is a JW-algebra, i.e. a weakly closed Jordan subalgebra of the self-adjoint part M_{sa} of M with the Jordan product $a \circ b = \frac{1}{2}(ab + ba)$, see e.g. [3]. We shall investigate the question of whether this implies the existence of a faithful normal conditional expectation of M onto the von Neumann algebra generated by $P(M)$. If we assume P is decomposable, i.e. the sum of a completely positive and a co-positive map, so P has a decomposition $P = V^*\pi V$ with V a bounded linear operator and π a Jordan representation, then we shall solve the problem to the affirmative if there moreover exists an involution α, i.e. an antiautomorphism of order 2 of M, such that $\alpha(P(x)) = P(x)$ for all $x \in M$. As a consequence of this we obtain an extension of Tomiyama's results on the type of the image of a conditional expectation of a von Neumann algebra, [11, 12], to more general projections.

2. Results. Recall that a JW-algebra A is called reversible if it is closed under products $a_1 a_2 \cdots a_n + a_n \cdots a_2 a_1$ with $a_i \in A$. Then A is the self-adjoint part $R(A)_{sa}$ of the weakly closed real $*$-algebra $R(A)$ generated by A.

Theorem. *Let M be a von Neumann algebra, and let P be a faithful normal decomposable projection of M into itself. Suppose there exists an involution α of M such that $\alpha P = P$, and put $A = P(M)_{sa}$. Then A is a reversible JW-algebra, and there exists a faithful normal conditional expectation Q of M onto the von Neumann algebra A'' generated by A.*

Proof. By [9, Cor. 7.3] A is a reversible JW-algebra. By [7, Lem. 2.10] there are three orthogonal central projections e, f, g in A with sum 1 with the following properties: $eA = e(A'')_{sa}, fA''$ has two central projections p and q with sum f such that $pA = p(A'')_{sa}, qA = q(A'')_{sa}$, and there is an involution β of fA'' such that $\beta(p) = q$ and $fA = \{x + \beta(x) : x \in p(A'')_{sa}\}$. The projection g is such that the center of $g(A'')_{sa}$ equals that of gA and $gA'' = R(Ag) + iR(Ag), R(Ag) \cap iR(Ag) = (0)$. We can consider the three cases $x \to P(exe), P(fxf), P(gxg)$ separately because $P(x) = P(exe) + P(fxf) + P(gxg)$. The theorem is trivial for $P(exe)$. In the case of f we let $Q(x) = P_p(pxp) + P_q(qxq)$, where $P_p(pxp) = \lambda^{-1}P(pxp)p$, where $\lambda = P(p)$ is by a trivial extension of [9, Lem. 4.2] a positive operator in the center of pA" with a positive self-adjoint inverse. Similarly P_q is defined.

Then Q is a faithful normal conditional expectation of fMf onto fA''.

Finally we consider gA'', and assume for simplicity $g = 1$. Then $A'' = R(A)+iR(A)$, $R(A)\cap iR(A) = (0)$. Since α is an involution of M, $M = R + iR$ with $R = \{x \in M : \alpha(x) = x^*\}$ a weakly closed real $*$-algebra, $R \cap iR = (0)$ and $\alpha(x + iy) = x^* + iy^*, x, y \in R$ [5, Lem. 7.3.2]. In particular $\alpha(x + iy) = x + iy$ if and only if $x, y \in R_{sa}$. Since $\alpha(x) = x$ for $x \in A$, $R(A) \subset R$, so that the restriction $\alpha|A''$ of α to A'' is the canonical involution $x + iy \to x^* + iy^*$ on $A'' = R(A) + iR(A)$.

Since $\frac{1}{2}(P + P\alpha)$ is a projection of M with the same range as P, we may replace P by $\frac{1}{2}(P + P\alpha)$ and thus assume α is P-invariant, in particular, $P = P\alpha = \alpha P$. Since P is decomposable there are completely positive and copositive maps P_1' and P_2' respectively such that $P = P_1' + P_2'$. Put

$$P_1 = \frac{1}{4}(P_1' + \alpha P_1'\alpha + P_2'\alpha + \alpha P_2')$$
$$P_2 = \frac{1}{4}(P_2' + \alpha P_2'\alpha + P_1'\alpha + \alpha P_1').$$

Then P_1 is completely positive, P_2 is co-positive, and $P_1 + P_2 = P$. Furthermore

$$P_1 = \alpha P_2 = P_2\alpha$$
$$P_2 = \alpha P_1 = P_1\alpha$$

Put

$$Q_0 = P_1 + \alpha P_2 = 2P_1 = 2P_2\alpha.$$

Then Q_0 is completely positive and $Q_0(x) = P(x)$ if $\alpha(x) = x$. Furthermore, by the above equations

$$\alpha Q_0 = Q_0\alpha.$$

Since P is faithful and normal so are P_1 and P_2, and therefore Q_0. Notice also that $P_1P_2 = P_1\alpha P_1 = P_2P_1$, and $P_1^2 = P_2\alpha \cdot \alpha P_2 = P_2^2$, hence we find

$$P = PQ_0.$$

Let $N = \{x \in M : Q_0(x) = x\}$. Then N is a weakly closed self-adjoint subspace of M. We assert that N is a von Neumann subalgebra of M. It suffices to show it is a C^*-algebra. Let by [3, Cor. 1.6] Q be a positive projection of M onto N obtained as a point-ultraweak limit of the sequence $\{n^{-1} \sum_{k=1}^{n} Q_0^k : n \in \mathbb{N}\}$. Since Q_0 is completely positive and $\alpha Q_0 = Q_0\alpha$ the same is true for Q. Furthermore $P = PQ$ by the same reasoning. In particular Q is faithful, hence by [2, §3] and [3, Th. 1.4] its range N is a C^*-algebra, so a von Neumann algebra, as asserted.

Since P is faithful and normal it follows from the equation $P = PQ$ that Q is normal. All that remains is thus to show $N = A''$. But $\alpha Q = Q\alpha$ implies $\alpha(N) = N$. Suppose

$x \in N$ with $\alpha(x) = x$. Then $P(x) = Q(x) = x$, so $x \in A$. Since the spectral subspace $\{x \in N : \alpha(x) = x\}$ generates N as a von Neumann algebra [6], $N = A''$. This completes the proof of the theorem.

Notice that if P is assumed to be α-invariant in the above theorem, then it follows from the proof that $P = PQ$, i.e. P decomposes via A''.

The nontrivial part of the above theorem can be reformulated in terms of the real structure of M.

Corollary 1. *Let R be a weakly closed real $*$-algebra such that $R \cap iR = (0)$. Let $M = R + iR$ be the generated von Neumann algebra (see [6]). Suppose A is a reversible JW-subalgebra of R_{sa} and that P is a faithful normal positive projection of R onto A. Then there exists a faithful normal conditional expectation Q of M onto the von Neumann algebra A'' generated by A such that $P = PQ$.*

Proof. Let α be the involution $\alpha(x+iy) = x^* + iy^*$ of $M, x, y \in R$. Extend P by linearity to all of M. Since $A = P(R), \alpha(P(x)) = P(x)$ for $x \in R$, hence for all $x \in M$. Since A is reversible P is decomposable by [9, Cor. 7.3]. If $x + iy \geq 0$ and $P(x+iy) = 0$ then $P(x) = P(y) = 0$. Since $x \geq 0$ by [6, Cor. 2.2] $x = 0$. But then $y = 0$ by [6, Thm. 2.1]. It follows that P is faithful on M. Thus the corollary is a consequence of the theorem and the above remark since $P(\alpha(x+iy)) = P(x^* + iy^*) = P(x+iy), x, y \in R$.

QED.

In [11, 12] Tomiyama showed that if E is a normal conditional expectation of a von Neumann algebra M onto a von Neumann subalgebra N then N is semifinite if M is, and of type I if M is of type I. The next result is an extension of this to projections of a more general kind.

Corollary 2. *Let M be a von Neumann algebra. Let P be a faithful normal positive projection of M into itself. Suppose there exists an involution α of M such that $\alpha P = P$, and put $A = P(M)_{sa}$. Then A is a JW-subalgebra of M_{sa} such that if M is semifinite (resp. type I) then so is A.*

Proof. By [3, Th. 1.4] A is a JW-subalgebra of M_{sa}. Let e and f be central projections in A with sum 1 such that eA is of type I_2 and fA reversible [5, Th. 5.3.10]. We may thus assume A is reversible. By [9, Cor. 7.3] P is decomposable, so by the theorem there exists a faithful normal expectation Q of M onto A''. By [11, 12] A'' is semifinite if M is, and of type I if M is of type I. But A is semifinite if and only if A'' is semifinite [1], and of type I if and only if A'' is of type I [5, Th. 7.4.3].

QED.

Recall that an abelian von Neumann algebra is called totally atomic if each nonzero projection majorizes a minimal projection.

Corollary 3. *Let M be a von Neumann algebra of type I with totally atomic center on a Hilbert space H. Let P be a faithful normal positive projection of M into itself. Then $A = P(M)_{sa}$ is a JW-algebra of type I with totally atomic center.*

Proof. By assumption on M there exists a faithful normal conditional expectation P' of $B(H)$ onto M. Composing P with P' we may assume $M = B(H)$. From the proof of [8] there exists a faithful family of normal projections of Z' onto Z, where Z is the center of A. Since $A \subset Z'$ we thus obtain by composition with P a faithful family of normal projections of $B(H)$ onto Z. Thus Z is totally atomic, see e.g. [8] or [12]. To show A is of type I we may as in Corollary 2 assume A is reversible, and as in the proof of the theorem consider only the case when Z equals the center of A'', and $A'' = R(A)+iR(A), R(A)\cap iR(A) = (0)$. Let α_0 be the canonical involution on A''. By [7, Th. 3.7] α_0 is spatial, hence extends to an involution α of $B = pB(H)p + (1-p)B(H)(1-p)$ for some central projection p in A" such that $\alpha(P(x)) = \alpha_0(P(x)) = P(x)$ for $x \in B$. Thus A is of type I by Corollary 2.

QED.

We conclude by applying Corollary 3 to obtain an extension of the theorem of Evans and Høegh-Krohn mentioned in the introduction. Instead of considering a single positive linear map we consider a semigroup S of such maps, keeping in mind that the case of a single map ϕ is taken care of by the semigroup $\{\phi^n : n \in \mathbb{N}\}$. We say S is ergodic, or irreducible, if the set $\{x \in M : \phi(x) = x,\ \phi \in S\} = C1$. A family $\mathcal{F}$ of states is called faithful if $x \geq 0$ and $\omega(x) = 0$ for all $\omega \in \mathcal{F}$ implies $x = 0$. If $\omega\phi = \omega$ for all $\omega \in \mathcal{F}, \phi \in S$ we say $\mathcal{F}$ is S-invariant.

Corollary 4. *Let M be a von Neumann algebra of type I with totally atomic center. Let S be a semigroup of normal positive unital maps of M into itself, and suppose there exists a faithful family $\mathcal{F}$ of normal S-invariant states on M. Then there exists an orthogonal family $(p_\nu)_{\nu\in J}$ of projections in M with sum 1 such that $\phi(p_\nu) = p_\nu$ for all $\nu \in J, \phi \in S$, such that for each $\nu \in J$ the set of restrictions $\{\phi|M_{p_\nu} : \phi \in S\}$ is an ergodic semigroup of unital maps of M_{p_ν} into itself.*

Proof. Let $M^S = \{x \in M : \phi(x) = x, \phi \in S\}$. By [10] $A = M^S_{sa}$ is a JW-algebra, and there exists a faithful normal projection P of M onto M^S. By Corollary 3 A is of type I with totally atomic center. Let B be a totally atomic maximal abelian subalgebra of A. Then the set of minimal projections in B is the desired set.

QED.

REFERENCES

1. *S.A. Ajupov*, Extension of traces and type criterions for Jordan algebras of self-adjoint operators. Math. Z. 181 (1982), 253-268.
2. *M.-D. Choi* and *E. Effros*, Injectivity and operator spaces. J. Functional Anal. 24 (1974), 156-209.
3. *E. Effros* and *E. Størmer*, Positive projections and Jordan structure of operator

algebras, Math. Scand. 45 (1979), 127-138.

4. *D.E. Evans* and *R. Høegh-Krohn*, Spectral properties of positive maps on C^*-algebras, J. London Math. Soc. 17 (1978), 345-355.
5. *H. Hanche-Olsen* and *E. Størmer*, Jordan operator algebras, Pitman 1984.
6. *E. Størmer*, Irreducible Jordan algebras of self-adjoint operators. Trans. Amer. Math. Soc. 130 (1968), 153-166.
7. *E. Størmer*, On anti-automorphisms of von Neumann algebras, Pacific J. Math. 21 (1967), 349-370.
8. *E. Størmer*, On projection maps of von Neumann algebras, Math. Scand. 30 (1972), 46-50.
9. *E. Størmer*, Decomposition of positive projections on C^*-algebras. Math. Ann. 247 (1980), 21-41.
10. *K. Thomsen*, Invariant states for positive operator semigroups. Studia Math. 81 (1985), 285-291.
11. *J. Tomiyama*, On the projection of norm one in W^*-algebras, I. Proc. Jap. Acad. 33 (1957), 608-612.
12. *J. Tomiyama*, On the projection of norm one in W^*-algebras. III. Tôhoku Math. J. 11 (1959), 125-129.

Department of Mathematics
University of Oslo

How many singularities can there be in an energy minimizing map from the ball to the sphere?

F. J. ALMGREN, JR. AND E. H. LIEB

Department of Mathematics, Princeton University

ABSTRACT. Energy minimizing harmonic maps from the ball to the sphere arise in the study of liquid crystal geometries and in the classical nonlinear sigma model. We linearly dominate the number of points of discontinuity of such a map by the energy of its boundary value function. Our bound is optimal (modulo the best constant) and is the first bound of its kind. We also show that the locations and numbers of singular points of minimizing maps is often counterintuitive; in particular, boundary symmetries need not be respected.

1 INTRODUCTION

This note is an introduction to and summary of discoveries we have made about the singular behavior of

- A mathematical model of some liquid crystal geometries
- Dirichlet energy minimizing harmonic maps from regions in $\mathbf{R}^3$ to $\mathbf{S}^2$.
- Energy minimizing configurations of a classical nonlinear sigma model ($\mathbf{R}^3 \to \mathbf{S}^2$).

These phenomena are different facets of a common mathematical analysis set forth in detail in our paper [AL]. There we study vector fields ϕ of unit length defined in a reasonable region Ω in $\mathbf{R}^3$. In coordinates we can thus write for each $x = (x^1, x^2, x^3)$ in Ω,

$$\phi(x) = (\phi^1(x), \phi^2(x), \phi^3(x)) \quad \text{with} \quad \sum_{i=1}^{2} \phi^i(x)^2 = 1. \tag{1}$$

Since our target $\mathbf{S}^2$ is 2-dimensional we could, in principle, describe ϕ using two functions instead of our three constrained functions. It is easier, however, to work with three functions and a constraint.

The ϕ's important for us have distribution first derivatives which are square summable. (Caution: the space of such ϕ's satisfying (1) is not the completion of any space of smooth mappings $\Omega \to \mathbf{S}^2$.) The gradients of such ϕ's are defined for almost every x

with norms represented by the formula

$$|\nabla\phi(x)|^2 = \sum_{i=1}^{3}\sum_{\alpha=1}^{3}\left(\frac{\partial\phi^i(x)}{\partial x^\alpha}\right)^2$$

which gives the value of **Dirichlet's integrand** at x. The integral of this integrand is **Dirichlet's energy integral** of ϕ,

$$\mathcal{E}(\phi) = \int_\Omega |\nabla\phi|^2 dV,$$

with $dV = dx^1dx^2dx^3$. Critical points of this energy integral $\mathcal{E}$ are by definition **harmonic functions** and satisfy the associated **Euler-Lagrange partial differential equations**

$$-\Delta\phi^i(x) = \phi^i(x)|\nabla\phi(x)|^2 \quad (i = 1, 2, 3).$$

These equations state that a critical ϕ has vanishing Laplacian in directions in which it is unconstrained. Such an energy functional and associated partial differential equations appear in the physics literature under the rubric of **the nonlinear sigma model.**

Somewhat more generally, reasonable maps $\phi : \mathcal{M} \to \mathcal{N}$ between Riemannian manifolds $\mathcal{M}$ and $\mathcal{N}$ (often submanifolds of Euclidean vector space) have a Dirichlet's energy integral

$$\mathcal{E}_{\mathcal{MN}}(\phi) = \int_\mathcal{M} |D_\phi|^2 dV_\mathcal{M}$$

of which ours is a special case. Alternatively, one can write

$$\mathcal{E}_{\mathcal{MN}}(\phi) = \int_\mathcal{M} g_{ij}(\phi(x))G^{\alpha\beta}(x)\left(\frac{\partial\phi^i}{\partial x^\alpha}(x)\right)\left(\frac{\partial\phi^j}{\partial x^\beta}(x)\right) dV_{\mathcal{M}^x}$$

where g is the metric on N, G is the metric on M and $dV_{\mathcal{M}^x} = (\det G(x))^{1/2}dx$. Extremal mappings for such energies are also called **harmonic mappings**. Such mappings often are not continuous and there is an extensive mathematical theory about them.

The ϕ's mapping Ω to $\mathbf{S}^2$ which are important for us also have well defined **boundary functions** $\psi : \partial\Omega \to \mathbf{S}^2$ having boundary energy

$$\partial\mathcal{E}(\psi) = \int_{\partial\Omega} |\nabla_T\psi|^2 dA$$

which is finite; here $\nabla_T\psi$ is the tangential gradient of ψ and dA is surface area measure. Associated with such a ψ is the number

$$E(\psi) = \inf\{\mathcal{E}(\phi) : \phi \text{ has boundary value function } \psi\}.$$

We call ϕ an **energy minimizing map** for boundary value function ψ if and only if

$$\mathcal{E}(\phi) = E(\psi).$$

If Ω is any reasonable bounded domain and ψ is any boundary value function of finite energy then there will always be at least one minimizer ϕ having ψ as boundary values (a compactness argument). Sometimes, however, there can be more than one minimizer. This is one of the fascinations of this simple nonlinear problem; if the target $\mathbf{S}^2$ were replaced by $\mathbf{R}^3$ (i.e. our constraint were removed) then the Euler-Lagrange partial differential equations are (the unconstrained) linear partial differential equations of Laplace, $\Delta\phi^i = 0$ $(i = 1, 2, 3)$, for which uniqueness is well known.

If our domain Ω is all of $\mathbf{R}^3$ there is no boundary value function ψ, of course. We then say that $\phi : \mathbf{R}^3 \to \mathbf{S}^2$ is a minimizer provided ϕ cannot be modified on a compact set K to decrease energy in some larger bounded open set containing K.

Liquid crystals. The connection of our energy minimizing ϕ's with liquid crystals requires explanation. We imagine that Ω is a container containing a liquid crystal. At points x in Ω the liquid determines a **directrix** $\mathbf{n}(x)$ lying in real projective space $\mathbf{RP}^2$. Since $\mathbf{RP}^2$ is obtained from $\mathbf{S}^2$ by identifying antipodal points, this means intuitively that $\mathbf{n}(x)$ is a unit vector like our $\phi(x)$ except that its head is indistinguishable from its tail. For the liquid crystal with which we are concerned, the energy of $\mathbf{n}$ is defined analogously to our $\mathcal{E}$, e.g. zero energy corresponds to parallel alignment. Like our minimizing ϕ's (as we shall see), any minimizing $\mathbf{n}$ will be continuous except at isolated points. This means, in particular, that any minimizing $\mathbf{n}$ can locally be lifted to become a minimizing ϕ having the same energy; this lifting is global in case Ω is simply connected. (see [BCL], p. 686 for details). Thus, for simply connected Ω's, our original problem is equivalent to the liquid crystal problem. In any case, whether or not Ω is simply connected, our estimates on the number of singular points hold for these liquid crystal minimizers. Line singularities do not occur in our model because they would have infinite Dirichlet energy. They do occur in nature, but to model them one, effectively, has to fatten the line and treat it separately (much as in the liquid helium problem). A further complication for liquid crystals is that there are other, more appropriate, integrands which are quadratic in $\nabla\phi$ and respect rotational symmetry. The general nematic liquid crystal integrand, for example, is of the form

$$K_1(\mathrm{div}\phi)^2 + K_2(\phi \cdot \mathrm{curl}\phi)^2 + K_3(\phi \wedge \mathrm{curl}\phi)^2.$$

Our Dirichlet's energy integrand corresponds (except for a fixed boundary term) to setting $K_1 = K_2 = K_3 = 1$ (see [BCL], p. 653). Our methods give information about such liquid crystal geometries (by a compactness argument) only when K_1, K_2 and K_3 are nearly equal.

2 BASIC FACTS ABOUT MINIMIZERS

2.1 Existence and regularity of minimizers.

As we mentioned above, whenever we have a reasonable domain Ω and boundary function ψ of finite energy, there will always exist a minimizer ϕ having boundary values ψ. Such a result is included in the general analysis of Dirichlet's integral minimizing mappings between manifolds by R. Schoen and K. Uhlenbeck in their basic papers [SU1] [SU2]. They further showed that a minimizing ϕ in our context is a real analytic mapping except at isolated points of discontinuity (which are our singularities). Finally, they concluded that a minimizing ϕ assumes its boundary values smoothly when both $\partial\Omega$ and ψ are comparably smooth.

2.2 Monotonicity of energy and tangential approximations.

One of the basic technical properties of energy minimizing mappings is usually called **monotonicity**. Whenever ϕ is a minimizer in $\Omega, y \in \Omega$, and $0 < r < s < R$ so that the ball $\mathbf{B}_R(y)$ also lies within Ω, then

$$\frac{1}{r}\int_{\mathbf{B}_r(y)} |\nabla\phi|^2 dV \leq \frac{1}{s}\int_{\mathbf{B}_s(y)} |\nabla\phi|^2 dV.$$

For a proof, see [SU1]. (The absence of a corresponding monotonicity estimate is the main reason our analysis of liquid crystals is restricted to the $K_1 = K_2 = K_3$ case.) The monotonicity estimate leads fairly directly to the existence of certain tangential approximations to ϕ at each interior y. A major and deep development occurred in a paper of L. Simon [S] which for our problem guarantees the existence of a **unique tangential approximating mapping**. At regular points this approximating mapping is constant. For a singular point y of ϕ in Ω, Simon's result gives a **unique harmonic mapping** $f : \mathbf{S}^2 \to \mathbf{S}^2$ such that

$$\phi(y + t\omega) \to f(\omega) \quad \text{as} \quad t \to 0_+$$

uniformly for all ω's in $\mathbf{S}^2$ (see [AL]), i.e.

$$\phi(x) \sim f\left(\frac{x - y}{|x - y|}\right)$$

for x's near y. The correspondence here is in several strong senses (see [AL]). In general, if $f : \mathbf{S}^2 \to \mathbf{S}^2$ and $F : \mathbf{R}^3 \to \mathbf{S}^2$ is defined by setting

$$F(x) = f\left(\frac{x}{|x|}\right)$$

for each $x \neq 0$ then f is harmonic if and only if F is.

Example: $f\left(\frac{x}{|x|}\right) = \frac{x}{|x|}$, i.e., f is the identity (see Figure 1).

Figure 1. Here are shown representations of unit vector fields

$$F(x) = \left(\frac{x}{|x|}\right) \qquad \text{and} \qquad G(x) = \mathcal{R}\left(\frac{x}{|x|}\right)$$

in which $\mathcal{R}$ is a counterclockwise rotation through 45^0. Such arrays minimize Dirichlet's integral energy and are also observed as stable liquid crystals geometries [K].

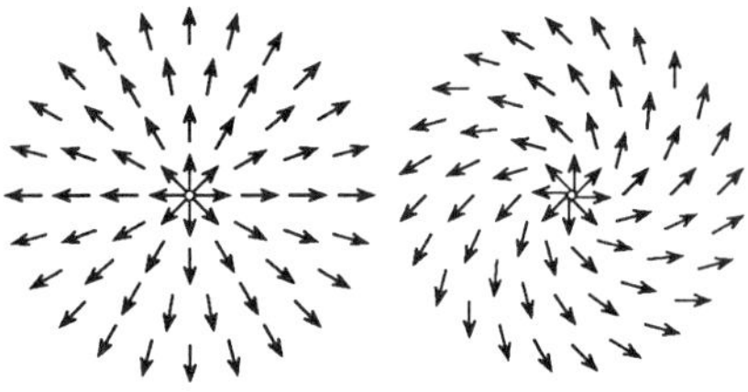

2.3 Harmonic mappings between spheres and mapping degrees.
Any continuous mapping $\mathbf{S}^2 \to \mathbf{S}^2$ has a well defined *topological* degree measuring the number of times the first sphere covers the second, taking into account the orientations. Since the boundary functions ψ under consideration map $\mathbf{S}^2$ to $\mathbf{S}^2$ and have finite energy, they also have a well defined degree given by the *Jacobian integral*

$$\deg(\psi) = \frac{1}{4\pi}\int_{\partial\Omega} J(\psi)dA;$$

here $J(\psi)$ is the Jacobian (determinant) function of ψ whose sign is positive or negative at a point depending on whether $D\psi$ preserves or reverses orientations at that point. For continuous ψ's of finite energy these two notions of degree coincide.

All possible harmonic mappings from $\mathbf{S}^2$ to $\mathbf{S}^2$ have been classified for some time. In complex coordinates (resulting from stereographic projection of the $\mathbf{S}^2$'s onto $\mathbf{C}$) they are all of the form

$$f(z) = \frac{P(z)}{Q(z)} \quad \text{or} \quad f(z) = \frac{P(\bar{z})}{Q(\bar{z})}$$

corresponding to various complex polynomial functions P and Q which are relatively prime. The degree of these f's can be checked to be

$$\deg(f) = \begin{cases} \max(\deg(P), \deg(Q)) & \text{first case;} \\ -\max(\deg(P), \deg(Q)) & \text{second case.} \end{cases}$$

For these harmonic maps $f : \mathbf{S}^2 \to \mathbf{S}^2$ we also set $F(x) = f\left(\frac{x}{|x|}\right)$ as above and compute for each $0 < R < \infty$ that

$$\int_{|x|<R} |\nabla F|^2 dV = 8\pi R |\deg(f)|,$$

i.e. the energy does not depend on P and Q except via the degree.

2.4 Tangential approximations to minimizers.

Suppose $y \in \Omega$ is a singular point of a minimizer ϕ and the tangential approximation is of the form $F(x) = f\left(\frac{x}{|x|}\right)$ corresponding to one of the harmonic f's given in 2.3 above. By the **degree of the singular point** y we mean the mapping degree of the associated f. **Which of the possible f's actually occur in a minimizer?** This question was answered by H. Brezis, J.-M. Coron, and E. Lieb in their paper [BCL]. The *only* f's that occur are rotations $\mathcal{R}$ and reflections of the f in the above example (shown in Figure 1), i.e.

$$f(\omega) = \pm\mathcal{R}(\omega), \quad (\omega \in \mathbf{S}^2) \quad \text{with } \deg(f) = \pm 1. \tag{2}$$

This class does not even include all harmonic maps of degree ± 1, all of which have the same energy—as we have seen. The proof proceeds by a construction of comparison functions. If $|\deg(f)| > 1$ then the energy of F can be decreased by splitting the singularity at the origin into two nearby singularities of lower degree. If $|\deg(f)| = 1$ and $f \neq \pm\mathcal{R}$ then the energy of F can be decreased by moving the singular point slightly.

The paper [BCL] also answered a question that in some sense is complementary to the minimization question we have been studying here. Suppose $y_1, \ldots, y_n$ are *fixed* points in Ω and $d_1, \ldots, d_n$ are *fixed* degrees associated to these points (not necessarily ± 1).

What is the infimum of energy $\mathcal{E}(\phi)$ among all ϕ's which are continuous except at the y_i's and map small spheres around each y_i with degree d_i?

The boundary function ψ is not fixed. This infimum is *not* achieved in general.

The answer is shown in Figure 2.

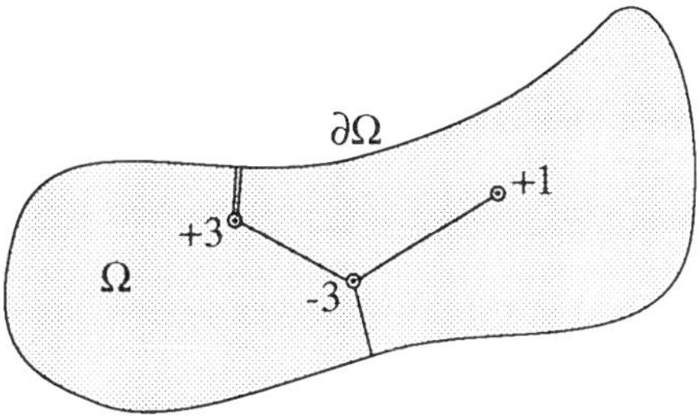

Figure 2. A region Ω is pictured here containing three prescribed singular points whose degrees (+3, -3, +1) are also prescribed. The least energy of unit vector fields having this singular behavior is the least total mass of oriented line segments connecting these singular points (as currents) either to each other or to the boundary. Such a least length array is illustrated.

Think of each singularity as a source or sink of flux and draw lines to carry the flux between singularities, or between a singularity and the boundary. Then

$$\inf \mathcal{E}(\phi) = 8\pi \min \left\{ \sum \text{ lengths of lines} \right\}$$

where the minimum is over all possible ways of constructing the lines. A different proof of this result was later given by F. Almgren, W. Browder, and E. Lieb [ABL] using H. Federer's co-area formula in the context of currents. This result is like quark confinement: a plus and minus quark have an energy proportional to their separation.

From this result with *specified* singularities one is tempted to surmise that, in our original minimization problem, *potential* singularities would tend to annihilate each other (if of opposite degrees) or move to $\partial\Omega$. The number of singularities that will occur will be only that required by topology, i.e.

$$\sum_{\text{singularities}} \deg(\text{singularity}) = \deg(\psi) = \frac{1}{4\pi} \int_{\partial\Omega} J(\psi) dA.$$

This surmise is very wrong, as we shall see later in Example 2, and misled us for a long time. Arbitrarily many singularities (of mixed signs) can occur, even if the Jacobian $J(\psi)$ vanishes identically.

2.5 Boundary regularity and hot spots.

Our main estimates require an extension of the boundary regularity results indicated above in 2.1. These theorems take several pages merely to state precisely, but the essence of the matter is the following. Assume that $\partial\Omega$ is smooth and take a small patch $P \subset \partial\Omega$ which is roughly a 2-dimensional disk of radius R. One consequence of the boundary regularity theory mentioned in 2.1 is the following. There is a fixed $\varepsilon > 0$, independent of R, with the property that whenever the boundary function ψ satisfies

$$\int_P |\nabla_T \psi|^2 dA < \varepsilon$$

then every minimizer ϕ is free of singularities in the region

$$K = \left\{ x : x \in \Omega, \ \operatorname{dist}(x, P) > \frac{1}{2} R\varepsilon, \ \operatorname{dist}(x, P_{\frac{1}{2}}) < 2R\varepsilon \right\},$$

here $P_{\frac{1}{2}}$ is the concentric disk of radius $\frac{1}{2}R$. Note that ε is dimensionless. Our **hot spot boundary regularity** theorem (proved in [AL]) asserts the existence of a fixed number $0 < \delta \ll \varepsilon$ such that whenever $P' \subset P$ is a smaller subpatch of radius δR and

$$\int_{P \sim P'} |\nabla_T \psi|^2 dA < \varepsilon$$

then ϕ is also free of singularities in the region K above. In other words arbitrarily large boundary energy in a very small disk P' cannot by itself induce singularities far away.

3. COUNTING SINGULARITIES

The principal question motivating our work in [AL] is this:

How many singular points $N(\psi)$ is it possible for a minimizing ϕ to have?

The following possibilities seem plausible at the outset:

$$N(\psi) \leq CE(\psi) \qquad \text{FALSE;}$$
$$N(\psi) \leq C\int_{\partial\Omega} |J(\psi)|dA \qquad \text{FALSE;}$$
$$N(\psi) \leq C\partial\mathcal{E}(\psi) \qquad \text{TRUE “\textbf{TheLinearLaw}”.}$$

here C is a constant, possibly depending on Ω.

The first possibility is false by counterexample — see below. The second possibility was suggested by the work in [BCL] and misled us for some time (had it been true it would have led to a beautiful geometric theory). In fact it is quite false as Example 1 below shows; in particular, $N(\psi)$ can be large while $J(\psi)$ vanishes identically.

Our main result, **The Linear Law**, is optimal (modulo the value of $C = C_\Omega$, of which we have no knowledge since our proof is by contradiction based on compactness arguments). It is, to our knowledge, the first global result of its kind.

The following example given by R. Hardt and F.H. Lin in [HL1] shows that $N(\psi)$ can indeed be proportional to $\partial\mathcal{E}(\psi)$. Choose N well separated small disks in $\partial\Omega$. Our ψ is constructed to wrap each disk D around the target sphere once (essentially by the inverse function to stereographic projection while preserving or reversing orientation as one chooses); each ∂D is mapped to the north pole. The complement of these disks in $\partial\Omega$ is mapped by ψ also to the north pole. Then $\partial\mathcal{E}(\psi) \approx CN$; the constant C is independent of the size of the disks since surface energy is scale invariant. Clearly the orientations of ψ on the disks can be arranged so that the total mapping degree of ψ is either zero or one. It is not hard to prove directly that any minimizing ϕ having ψ as boundary value function must have at least one singularity close to each tiny disk — otherwise $\mathcal{E}(\phi)$ would be too large. Thus

$$N(\psi) \geq N \approx C^{-1}\partial\mathcal{E}(\psi).$$

Our first main new result (proved independently by Hardt and Lin in [HL2]) is that singularities cannot be very close if they are well inside Ω.

Theorem 1. *There is a universal constant C (independent of Ω and ψ) such that whenever y and $z \in \Omega$ are singular points of a minimizer ϕ then*

$$dist(y,z) \geq C\ dist(y, \partial\Omega).$$

The idea of the proof is the following. Fix y and suppose the contrary. Then there will be a sequence of minimizers $\phi^{(j)}$ with singular points at $z^{(j)}$ and at y such that $z^{(j)} \to y$ as $j \to \infty$. A compactness argument (contradicting the negation) and monotonicity (2.1) shows that the energy of ϕ in small balls of radius R about y is *uniformly* greater than $8\pi R$. The limit of a subsequence of the minimizers $\phi^{(j)}$ is a minimizer ϕ which thus can have at worst a singularity of degree ± 1 at y (by equation (2) above). The tangential approximation theorem implies that the energy of the limit ϕ must be very close to $8\pi R$ for all small R's. This leads to a contradiction because of the continuity of Dirichlet's integral when minimizers converge.

A consequence of Theorem 1 together with equation (2) above is the following.

Theorem 2. (Complete classification of energy minimizing maps from $\mathbf{R}^3$ to $\mathbf{S}^2$.) *Suppose $\phi : \mathbf{R}^3 \to \mathbf{S}^2$ is a minimizer. Then, either ϕ is a constant mapping or $\phi = \pm\mathcal{R}\left(\frac{x-y}{|x-y|}\right)$ for some y and $\mathcal{R}$.*

Theorem 1 says that if there are many singularities they have to pile up near $\partial\Omega$. This leads to a difficult geometric-combinatorial problem on different scales proportional to δ^k, where δ is given in (2.5) above and $k = 1, 2, 3, \ldots$. We attempt to illustrate this in Figure 3.

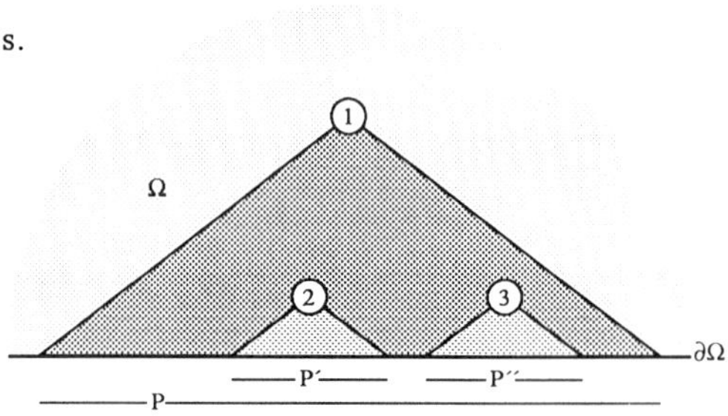

Figure 3. Pictured here are the "cones of influence" in Ω of three singular points. The presence of singular points 1, 2, 3 implies the presence of boundary energy in disks P, P′, P″ in $\partial\Omega$. The problem is that these disks are not disjoint so that the total boundary energy is not a simple sum. Nesting of such cones induces a Cayley tree graph in which a combinatorial analysis overcomes this difficulty.

Referring to the ε and δ of (2.5) consider the points 1, 2 and 3 in Ω at distances $R\varepsilon, R\varepsilon\delta$ and $R\varepsilon\delta$ above a boundary patch P of radius R and two boundary patches P' and P'' of radii $R\delta$ inside P. The hot spot boundary regularity theorem gives us the following lower bounds for the energy of ψ in P if we consider the various possibilities of having singularities at positions 1,2, or 3:

Positions occupied	Local boundary energy
(1 alone) or (2 alone) or (3 alone)	ε
(1 and 2) or (1 and 3) or (2 and 3)	2ε
(1 and 2 and 3)	2ε

The source of all our difficulties is that we cannot infer an energy 3ε if there are singularities at all three points.

If $S^{(k)}$ denotes the strip $\{x : x \in \Omega,\ \mathrm{dist}(x, \partial\Omega) \leq \varepsilon\delta^k\}$, we can effectively decompose each $S^{(k)}$ into cones of height $\varepsilon\delta^k$ and base radius δ^k. We then have a Cayley tree whose vertices represent these cones (i.e. a vertex of order $k+1$ is connected to a vertex of order k in the tree if the smaller cone is inside the larger one). A vertex is occupied if its cone has a singularity near the apex; otherwise it is unoccupied. Each occupied vertex gets an energy ε if and only if no more than one higher order vertex to which it is pathwise connected is occupied.

The actual details of decomposing each $S^{(k)}$ into cones so that due account is taken of overlaps (and all the other problems that will occur to the reader) involves a complicated covering and counting lemma. The final result is The Linear Law for $N(\psi)$ in terms of $\partial\mathcal{E}(\psi)$, as stated at the beginning of this section.

4. THREE EXAMPLES OF COUNTERINTUITIVE BEHAVIOR

Example 1. Zero Mapping Area.
It is easy to prove for any Ω that if ψ takes values only in some closed hemisphere of $\mathbf{S}^2$ then ϕ has no singularities. We, however, are able to construct a single curve Γ in $\mathbf{S}^2$ which is a slight perturbation of the equator and, for each N, a smooth boundary value functions $\psi^N : \partial\Omega \to \mathbf{S}^2$ **having its image equal to** Γ such than any minimizer ϕ^N having boundary values ψ^N must have at least N singular points. In the example of [AL], Ω is taken to be a ball, but the details of Ω are not important. The Jacobian $J(\psi^N)$ of each ψ^N vanishes identically since the image of ψ^N is one dimensional.

The idea behind the construction appears in the following preliminary problem. Consider reasonable mappings $\phi : \mathbf{D}^2 \to \mathbf{S}^2$ from the unit disk $\mathbf{D}^2$ in the plane having two dimensional Dirichlet's integral denoted by $\mathcal{E}_2(\phi)$. Suppose $\Gamma \subset \mathbf{S}^2$ is a smooth embedding of a circle parameterized by a map $P : \partial\mathbf{D}^2 \to \Gamma$. The functions ϕ from $\mathbf{D}^2$ to $\mathbf{S}^2$ having boundary values P can be separated into two homological classes: the + class, in which, heuristically, ϕ "covers the top of $\mathbf{S}^2$ one more time than it covers the bottom" and, the – class in which ϕ "covers the bottom one more time than it covers the top"; see Figure 4.

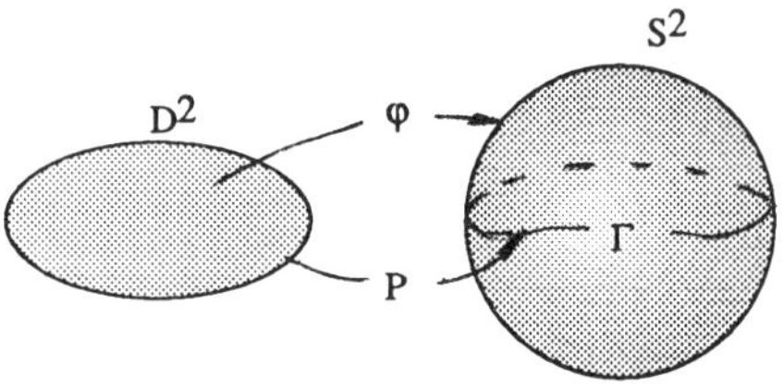

Figure 4. Illustrated here is one of two homologically distinct classes of mappings $\varphi: \mathbf{D}^2 \to \mathbf{S}^2$ corresponding to a given boundary parametrization $\mathrm{P}: \partial\mathbf{D}^2 \to \Gamma$ (the curve Γ is a perturbation of the equator). A "+ function" is one which "covers the northern hemisphere". For some Γ's, the homology type preferred by a least energy mapping can change if the parametrization P is changed. This phenomenon leads ultimately to construction of least energy mappings from the ball to the sphere having many interior singularities but for which the boundary mapping of the sphere to the sphere has zero mapping area (its entire image lies within the curve Γ).

Consider the two numbers

$$E^{\pm}(P) = \inf\{\mathcal{E}_2(\phi) : \phi = P \text{ on } \partial\mathrm{D}^2 \text{ and } \phi \in \pm \text{ class}\}.$$

In general $E^+(P)$ will not be the same as $E^-(P)$.

We construct a single Γ having two different (homotopic) parameterizations P^+ and P^- such that

$$E^+(P^+) < E^-(P^+) - \varepsilon \quad \text{and} \quad E^-(P^-) < E^+(P^-) - \varepsilon$$

for some $\varepsilon > 0$. In other words if the parametrization of Γ changes from P^+ to P^- any *absolute* minimizer ϕ changes from lying in the + class to lying in the – class.

The next step is to let Ω be a very long solid tube, T, of radius 1 and length $N(L+1)$. (Actually, T is bent into a torus so that we can ignore the two ends.) As boundary function ψ we alternately paste P^- and P^+ on sections of length L (i.e. each cross-sectional disk has P^- or P^+ on its boundary). In the transitional regions of length 1 we smoothly interpolate between P^- and P^+ (which can be done since they are homotopic). In the transition region ψ continues to take values only in Γ. See Figure 5.

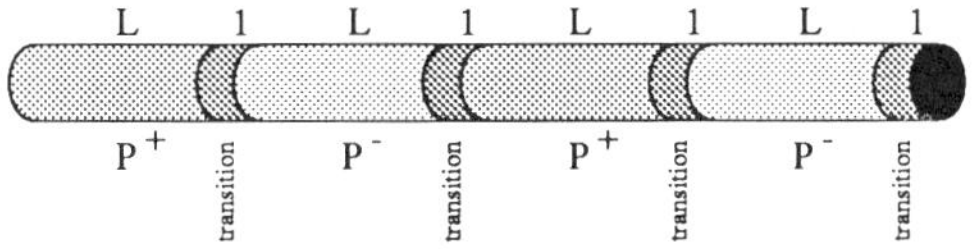

Figure 5. Illustrated here is a boundary value function $\psi: \partial\Omega \to \mathbf{S}^2$ for a long tube domain Ω. The image of ψ is a smooth curve Γ in $\mathbf{S}^2$. On crossectional circles of $\partial\Omega$ the boundary values alternate between intervals of P^+ mappings and intervals of P^- separated by transition intervals. Least energy maps $\varphi: \Omega \to \mathbf{S}^2$ with such boundary values map most crossections in P^+ regions to cover the northern hemisphere and map most crossections in P^- regions to cover the southern hemisphere. The minimizer φ therefore has at least one singular point near each transition region.

If L is large enough (depending only on ε), it is believable (and we prove it) that ϕ must be mostly a – function on the P^- disks and it must be mostly a + function on the P^+ disks, for otherwise $\mathcal{E}(\phi)$ would be unnecessarily large. But when ϕ switches from being a – function to being a + function ϕ must have a singularity for topological reasons. Thus, ϕ will have at least N singularities altogether.

The drawback to this tube example is that the domain T depends on N. To achieve the same result for a fixed domain Ω = unit ball, we first cut the surface ∂T longitudinally (i.e. perpendicular to the disks) and flatten it (key estimates here come from the conformal equivalence of the disk and the upper half plane and the fact that Dirichlet's integral in two dimensions is invariant under conformal reparameterizations of domains). This yields a strip of width 2π and length $N(L+1)$. We also rotate P^+ if necessary so that P^+ and P^- have the same value $\gamma \in \Gamma$ along the cut. Next we shrink the strip to width $(2\pi)^2/N(L+1)$ and length 2π. Finally we past this strip (which is very narrow since N is large) along the equator of Ω and let $\psi : \partial\Omega \to \mathbf{S}^2$ be the old ψ in the strip and let $\psi(x) = \gamma$ for $x \in \partial\Omega$ but $x \notin$ the strip. A somewhat nerve wracking argument shows, as expected, that any minimizer $\phi : \Omega \to \mathbf{S}^2$ must have at least N singularities close to the equator of Ω.

Example 2. Symmetry Breaking.
When ϕ takes values in $\mathbf{R}^3$ instead of $\mathbf{S}^2$, any geometric symmetry of Ω and ψ is inherited by the minimizing ϕ. The reason is simply that minimizers are unique in the linear case ($\Delta\phi = 0$). When, as in our case, ϕ takes values in $\mathbf{S}^2$, the symmetry of Ω and ψ can be broken by ϕ; obviously there must then be several minimizers.

Let Ω be the unit ball in $\mathbf{R}^3$ and let $\psi : \partial\Omega \to \mathbf{S}^2$ be the distortion of the identity map illustrated in Figure 6. In small caps N (resp. S) on $\partial\Omega$, ψ covers the northern (resp.

southern) hemisphere of $\mathbf{S}^2$. The two maps are mirror images of each other. On the rest of $\partial\Omega$ between N and S, ψ takes values in the equator of $\mathbf{S}^2$ in the obvious way, i.e. $\psi(x,y,z) = (x^2+y^2)^{-1/2}(x,y,0)$.

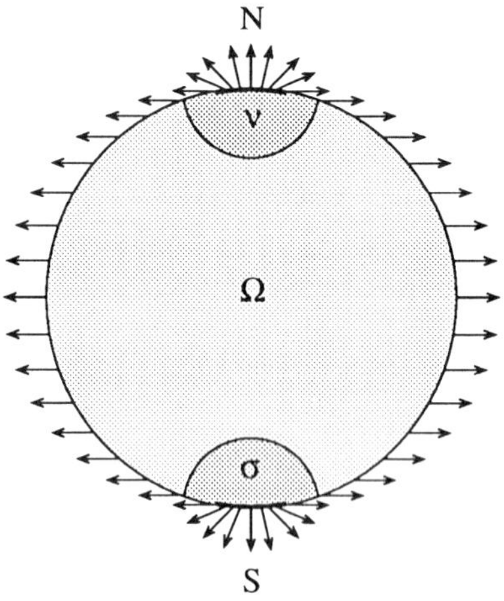

Figure 6. Here our domain Ω is the unit ball so that $\partial\Omega$ is the unit sphere. Pictured schematically is a special boundary value function $\psi\colon \partial\Omega \to \mathbf{S}^2$ having a mirror image symmetry through the equatorial plane. A small cap N around the north pole maps to cover the entire northern hemisphere of $\mathbf{S}^2$ while a small cap S around the south pole covers the entire southern hemisphere. The sphere less these two caps maps entirely to the equator. Longitude is preserved in each of these regions. No minimizing $\varphi\colon \Omega \to \mathbf{S}^2$ having boundary values ψ can possess such a symmetry since the (necessarily odd) number of singular points must be contained within one of the regions ν and σ near the poles.

Theorem 3. *Any minimizer ϕ can have singularities only in small shaded regions in Ω, labelled ν and σ, near the caps N and S.*

Since $\deg(\psi) = 1$, this result implies that ϕ does not inherit the mirror image symmetry through the equatorial disk possessed by ψ. (Our function ϕ necessarily has an odd number of singularities, and, if ϕ were symmetric, it would necessarily have one on the equatorial disk in Ω.)

The proof of Theorem 3 has two parts. First, we show that when N and S are small ϕ has no singularities in a concentric ball Ω' of radius $1 - \varepsilon$ for some small ε. This is done by a variational (or comparison) argument. Second, we show that there are no singularities in $\{x : 1 \geq |x| > 1-\varepsilon \text{ and } \operatorname{dist}(x, \sigma \cup \nu) > \varepsilon\}$ by using the boundary regularity 2.1.

Example 3. Boiling Water.

The [BCL] result mentioned in 2.4 above suggests that + and − singularities tend to annihilate each other. On the other hand, the hot spot boundary regularity mentioned in 2.5 above suggests that different length scales (as measured by the distance to

$\partial\Omega$) are independent, and therefore implying that + and − singularities *can* coexist provided their distances to $\partial\Omega$ are very different. There would appear to be a conflict here and one of our results is that of the two points of view just mentioned the second one is correct. We have proved the following.

Theorem 4. *Let Ω be the unit ball and let $p_1, \ldots, p_M$ be any distinct points in $\partial\Omega$. Also let $N_1, \ldots, N_M$ be any positive integers and for each $i = 1, \ldots, M$ let A_i be any sequence of length N_i consisting of $+1$'s and -1's. Finally, let $\varepsilon > 0$. Then there is a smooth $\psi : \partial\Omega \to \mathbf{S}^2$ such that*

(i) $\partial\mathcal{E}(\psi) \leq \varepsilon + 8\pi\sum_{i=1}^{M} N_i$.

(ii) The minimizer ϕ is unique.

(iii) For each $i = 1, \ldots, M$ there are at least N_i singularities stacked nearly vertically above p_i (like bubbles in a pan of water that is about to boil), and these have the specified sequence of degrees given by A_i.

REFERENCES

[ABL] F. Almgren, W. Browder and E. Lieb, *Co-area, liquid crystals and minimal surfaces*, in *Partial Differential Equations*, ed. S.S. Chern, Springer Lecture Notes in Math. **1306**, 1-12 (1988).

[AL] F. Almgren and E. Lieb, *Singularities of energy minimizing maps from the ball to the sphere: examples, counterexamples and bounds*, Ann. of Math. **128**, 483-530 (1988). See also *Singularities of energy minimizing maps from the ball to the sphere*, Bull. Amer. Math. Soc. **17**, 304-306 (1987).

[BCL] H. Brezis, J.-M. Coron and E. Lieb, *Harmonic maps with defects*, Commun. Math. Phys. **107**, 649-705 (1986).

[HL1] R. Hardt and F.H. Lin, *A remark on H^1 mappings*, Manuscripta Math. **56**, 1-10 (1986).

[HL2] R. Hardt and F.H. Lin, *Stability of singularities of minimizing harmonic maps*, J. Diff. Geom. **29**, 113-123 (1989).

[K] M. Kléman, *Points, lignes, parois dans les fluides anisotropes et les solides cristalline*, Les Éditiones de Physique (Orsay), **I**, 36-37.

[S] L. Simon, *Asymptotics for a class of nonlinear evolution equations with applications to geometric problems*, Ann. of Math. **118**, 525-571 (1983).

[SU1] R. Schoen and K. Uhlenbeck, *A regularity theory for harmonic maps*, J. Diff. Geom. **17**, 307-335 (1982).

[SU2] R. Schoen and K. Uhlenbeck, *Boundary regularity and the Dirichlet problem of harmonic maps*, J. Diff. Geom. **18**, 253-268 (1983).

FRONT TRACKING FOR PETROLEUM RESERVOIRS

FRODE BRATVEDT[1,2], KYRRE BRATVEDT[1], CHRISTIAN F. BUCHHOLZ[1], TORE GIMSE[2], HELGE HOLDEN[3], LARS HOLDEN[4], NILS HENRIK RISEBRO[2]

[1]Technical Software Consultants as
Gaustadalléen 21
N–0371 Oslo 3, Norway
[2]Institute of Mathematics
University of Oslo
N–0316 Oslo 3, Norway
[3]Division of Mathematical Sciences
The Norwegian Institute of Technology
The University of Trondheim
N–7034 Trondheim, Norway
[4]Norwegian Computing Center
P.O. Box 114, Blindern
N–0314 Oslo 3, Norway

ABSTRACT. We describe a newly developed computer simulator for two-dimensional petroleum reservoirs. The simulator is based on the front tracking concept where the interface between water and oil, called a front, is considered as an independent computational object. The mathematical background and the computer implementation are discussed as well as applications to test cases.

This work has been supported over the years by the Norwegian Research Council for Science and the Humanities (NAVF), VISTA, the Royal Norwegian Council for Technical and Industrial Research (NTNF), and Norsk Data.

0. How it all started.

In 1982 Raphael Høegh-Krohn and Helge Holden initiated a program with the ultimate goal of producing an efficient and accurate computer simulator for analyzing the very complicated behavior of the various fluids in a petroleum reservoir. Most of the commercially available simulators are based on traditional numerical techniques with little insight into the mathematical properties of the underlying equations. The idea was to combine recent mathematical understanding of the properties of the equations with modern, specially adapted numerical techniques and state-of-the-art computer technology to develop a better and faster simulator than the existing ones.

With a background from mathematical physics, it was then very reasonable for us to follow the by then fairly recent approach to reservoir simulation initiated by James Glimm, Oliver McBryan and co-workers in 1979. Fundamental in their front tracking method is a combination of deep mathematical and physical insight into the flow in porous media.

Vastly underestimating the complexity of the problem we had set out to solve, we were soon joined by an increasing group of enthusiastic students – Lars Holden, Kyrre Bratvedt, Frode Bratvedt, Christian F. Buchholz, Nils Henrik Risebro, Daniel Gandolfo, Leif Alm, and Tore Gimse – and we immediately started to study the mathematical, physical, numerical and computer technical aspects of the problem and to develop a computer code with the goal of having a three-dimensional simulator available in 'a few years'. The group was supported by Norsk Data with a ND570 computer, and VISTA (a joint research project between Den norske stats oljeselskap, Statoil, and The Norwegian Academy of Science and Letters), The Norwegian Research Council for Science and the Humanities (NAVF), the Royal Norwegian Council for Technical and Industrial Research (NTNF), Saga, and Norsk Hydro.

The original concept and scope of the simulator has been considerably changed since Høegh-Krohn's tragic death. The project is now organized in *Technical Software Consultants as*. As of today we have a computer code available on a commercial basis which is able to simulate realistic two-phase, two-dimensional reservoirs with compressibility. In this paper we are going to survey certain aspects of 'front tracking à la Oslo'.

1. Introduction.

To develop a fast and efficient computer simulator for petroleum reservoirs has been and still is, a problem of major interest. The increased use of enhanced oil recovery techniques and the development of supercomputers have both made it possible and more important to have accurate simulators available. However, in spite of the extensive effort invested in the problem, there are still many challenging open problems.

The fundamental physical problem in the description of petroleum reservoirs is the problem of *scales*. Contrary to stable fluid flow, there is no natural length scale for flow in porous media. The reservoir has many different length scales and on each scale the reservoir description is expressed in terms of different quantities. See e.g. Haldorsen *et al.*(1988). On the microscopic scale one studies e.g. the size and distribution of pores, and the detailed analysis of the rock-fluid interactions. A typical object on the macroscopic scale is the core plug. Here one determines e.g. the permeabilities and compressibility. The next scale can be called the megascopic scale, the size of which is the typical grid block used in the numerical computations. This is the scale that a full field simulator will have to work on. Due to the enormous size of the reservoir, the grid blocks will have to so big that for instance the permeability can vary considerably over a grid block, and as we will see, even the fluid flow may have abrupt changes within a grid block. Finally the gigascopic scale addresses whole regions in which the reservoir itself is only a part. A major theoretical problem is to understand how the quantities on one scale can be understood, and in particular, be computed from the relevant quantities on a finer scale. A transition from a fine scale to a coarser scale involves some averaging where information is lost. The data available for the computations are partly coming from a small number of core samples taken from the reservoir, and partly more global geological information. Even if it had been available, it would not be possible to handle all the detailed information on a microscopic scale when one simulates the whole reservoir.

The fundamental mathematical problem is the lack of mathematical theory for the equations that govern the flow. There are no existence or uniqueness theorems for solutions of the differential equations. However it is reasonable to assume that one has existence and uniqueness. More important from the point of view of simulations is the question of stability. There is considerable uncertainty in the measured quantities that enter the equations. This makes a stability analysis very important. Another approach to this problem is to make a stochastic model of the reservoir. In this way one can specify the qualitative properties of the reservoir with a certain probability distribution. See Haldorsen *et al.*(1988), Holden and Risebro(1991a),(1991b), Glimm and Sharp(1990), Aasen *et al.*(1990).

A fundamental numerical problem is the decoupling of the pressure equation and the saturation equation (see sec. 5). These equations have very different behavior, and we solve them alternately by using the computed solution of one equation as input in the other equation. Furthermore the problem of scales create numerical problem as the variables have considerable subgrid variance. Some parameters may vary by three orders of magnitude from one block to the next. In addition, the enormous amount of data is very time consuming in the simulations.

The goal of the computer simulator from the point of view of management is to give accurately and efficiently the amount of hydrocarbons produced based on the data available and expressed in terms of production conditions. On the engineering level one is also interested in understanding as much as possible of the behavior of the reservoir. This understanding is based on a complicated interaction between experiment, experience (e.g. production history and geological knowledge), and computer simulation. Since the system of equations is highly underdetermined due to lack of data, and hence has many unknown parameters, this interaction is the most important means of validating the simulator.

The reservoir is traditionally modeled by a set of nonlinear partial differential equations (see next section). One equation expresses the conservation of mass in the system. Another central equation, called Darcy's law, expresses the dynamics of the flow in the porous medium. The flow on this scale is characterized by the following important features: *There is a significant change in the flow values over length scales that are small compared to the scale of the flow.* For instance, the saturations, i.e. relative volume fractions of the various phases, change considerably over a distance of the order of centimeters. This abrupt change is due to the non-mixing of oil and water — the transition is not exactly sharp due to *capillary effects*. The sharp transition, called a *front*, will be a subgrid phenomenon in the numerical calculations, and hence the grid approximation will smear out the front. The main idea of front tracking is to add as independent computational objects, or, if one prefers, additional degrees of freedom, these lower dimensional interfaces. Away from this region the flow is fairly smooth. These physical properties are of course reflected in the mathematical properties of the equations. Neglecting capillary effects, the nonlinear conservation law develops singularities, in the sense of steepening and eventually non-existence of derivatives in finite time even for infinitely smooth initial data. This implies that one has to consider weak solutions of the equation, which again makes the uniqueness questions highly non-trivial. Traditional numerical techniques have problems at exactly this point since most error estimates are expressed in terms of the derivatives of the solution. Many adaptive computational methods have been develped to overcome these problems. The general feature of these methods is a local refinement of the underlying grid structure close to the region where it is required. We are here solely going to discuss front tracking. Front tracking is characterized, as we have mentioned, by an additional lower dimensional grid in addition to the underlying grid which has dimension equal to that of the space. For a two-dimensional reservoir the front is a one-dimensional grid, i.e. a curve. The front then corresponds to the sharp transition between say an oil and water phase, and hence the front will develop dynamically with the fluids.

Front tracking was originally suggested by Richtmyer(1957) and Moretti(1975), but made into a major scientific enterprise by James Glimm and Oliver McBryan with co-workers Dan Marchesin, Brent Lindquist, Eli Isaacson, Brad Plohr, John Grove, and others. By now the front tracking concept has been applied to a varity of phenomena. In the reference list we give only some of the papers relevant for petroleum reservoirs.

2. The model.

In this section we briefly state the standard equations that are used to model flow in porous media. A general reference is Scheidegger(1974).

The reservoir is a rock with minute pores filled with fluids. For a petroleum reservoir the fluids will be various hydrocarbons and water. The fluids can be separated into *phases*.

Assume that we have N phases, indexed by $\alpha = 1, \ldots, N$. Conservation of mass for phase α reads

$$\frac{\partial(\phi \rho_\alpha S_\alpha)}{\partial t} + \nabla \cdot (\rho_\alpha v_\alpha) = 0 \tag{1}$$

where ϕ is the porosity, S_α is the saturation of phase α, i.e. the ratio between the volume occupied by phase α and the total pore volume, ρ_α denotes the density of the phase, and finally v_α denotes the velocity of phase α. The phase velocity v_α is governed by Darcy's law

$$v_\alpha = -K\lambda_\alpha(\nabla p_\alpha - \rho_\alpha g \nabla D) \tag{2}$$

where K is the permeability matrix, λ_α is the phase mobility, i.e. the ratio of the relative permeability of the phase α and the viscosity of the phase α, viz. $\lambda_\alpha = k_{r,\alpha}/\mu_\alpha$, p_α is the partial pressure of phase α, g is the constant of gravitation and D measures the vertical distance in the reservoir. Ignoring the term due to gravity, Darcy's law says that the velocity of the fluid is proportional to the pressure gradient. Experimentally this relation has been verified for slowly moving one-phase flow in porous media.

For the other functions we assume the following known functional relations:

$$\begin{aligned} &\phi = \phi(x,p), \quad K = K(x), \quad k_{r,\alpha} = k_{r,\alpha}(S_\beta, \beta = 1, \ldots, N), \\ &\rho_\alpha = \rho_\alpha(p_\alpha), \quad p_\alpha = p_\alpha(p, S_\beta, \beta = 1, \ldots, N), \quad \mu_\alpha = \mu_\alpha(p_\alpha) \end{aligned} \tag{3}$$

where p is the pressure in the rock. Hence we have as unknowns the functions S_α, $\alpha = 1, \ldots, N$ and the pressure of the rock, p. Since we have

$$\sum_\alpha S_\alpha = 1, \tag{4}$$

i.e. the fluids fill the available pore volume, one S_α is determined from the others, and thus the number of unknowns equals the number of equations.

Assume now that both the fluids and the rock are incompressible so that $\phi_t = 0$ and $\rho_\alpha =$ const. Then (1) simplifies to

$$\phi S_{\alpha,t} + \nabla \cdot v_\alpha = 0. \tag{5}$$

We may now rewrite Darcy's law as follows

$$v_\alpha = \phi(F_\alpha - D_\alpha) \tag{6}$$

with

$$F_\alpha = -\frac{K\lambda_\alpha}{\phi\lambda_T}[-\frac{1}{K}v_T + \sum_\beta \lambda_\beta(\rho_\beta - \rho_\alpha)g\nabla D] \tag{7}$$

and

$$D_\alpha = \frac{K\lambda_\alpha}{\phi\lambda_T}\sum_\beta \lambda_\beta \nabla(p_\alpha - p_\beta) \tag{8}$$

where

$$\lambda_T = \sum_\alpha \lambda_\alpha, \quad v_T = \sum_\alpha v_\alpha. \tag{9}$$

Finally we can write (5) as

$$S_{\alpha,t} + \nabla \cdot F_\alpha = \nabla \cdot D_\alpha. \tag{10}$$

Observe that D_α is a sum of derivatives.

Capillary pressure is associated with the difference $p_\alpha - p_\beta$, and hence when we neglect capillary pressure, i.e. set $p_\alpha = p_\beta$ for all α and β, we find that $D_\alpha = 0$. If we specialize to non-capillary, incompressible two-phase flow which we will study in the subsequent sections, we find, when we use subscript o for the oleic phase and w for the aquous phase,

$$S_{o,t} + \nabla \cdot F_o(S_o) = 0 \tag{11}$$

where

$$F_o(S_o) = \frac{K\lambda_o}{\phi}(\nabla p - \rho_o g \nabla D) \tag{12}$$

and

$$S_w = 1 - S_o. \tag{13}$$

Equation (11) will be refered to as the saturation equation, we also obtain a similar equation for the aquous phase. When we add the saturation equations for each phase we obtain

$$\nabla \cdot \left\{ \frac{K}{\phi} [(\lambda_o + \lambda_w)\nabla p - (\lambda_o \rho_o + \lambda_w \rho_w) g \nabla D] \right\} = 0 \tag{14}$$

This will be refered to as the pressure equation. Reservoir simulation consists in the simultaneous solution of the saturation and pressure equations.

3. The conservation law.

We will here give some of the basic mathematical properties of the conservation laws. A general reference is Smoller(1983). The flow may be decomposed in the normal and tangential direction of a surface of discontinuity, and the one-dimensional equation can be considered a model for the flow in the normal direction. For simplicity we start by looking at the scalar conservation law in one space dimension, viz.

$$s_t + f(s)_x = 0 \tag{15}$$

with Riemann initial data

$$s(0,x) = s_0(x) = \begin{cases} s_L, & \text{for } x < 0 \\ s_R, & \text{for } x > 0. \end{cases} \tag{16}$$

The Riemann problem is the simplest case where one can study the dynamics of a simple front, i.e. a discontinuity. From the fundamental work of Glimm(1965) we know that the solution of the Riemann problem is a building block in the solution of the general Cauchy problem also for systems of equations.

The solution to this equation with initial data (16) consists of a succession of *elementary waves*, either *shocks* or *rarefaction waves*. In addition one has to impose *entropy conditions* to ensure a unique, physical solution. Assume for definiteness that $s_L < s_R$ and let f_c denote the lower convex envelope of f relative to the interval $[s_L, s_R]$. Let

$$s_L = s_0 < s_1 < \cdots < s_N = s_R \tag{17}$$

be so that either $f_c(s) = f(s)$ for all $s \in (s_i, s_{i+1})$ or $f_c(s) < f(s)$ for all $s \in (s_i, s_{i+1})$. Consider now an interval (s_i, s_{i+1}). If $f_c(s) = f(s)$, then we have a rarefaction wave which reads $s_{\rho,i}(\xi) = (f_c')^{-1}(\xi)$ for $\xi \in (f_c(s_i), f_c(s_{i+1}))$ where $\xi = x/t$. If however $f_c(s) < f(s)$, we have a shock solution given by

$$s_{\sigma,i}(\xi) = \begin{cases} s_i, & \text{for } f_c(s_i) < \xi < \frac{f_c(s_{i+1})-f_c(s_i)}{s_{i+1}-s_i} \\ s_{i+1}, & \text{for } \frac{f_c(s_{i+1})-f_c(s_i)}{s_{i+1}-s_i} < \xi < f_c(s_{i+1}). \end{cases} \tag{18}$$

The general entropy solution equals

$$s(\xi) = \begin{cases} s_L, & \text{for } \xi < f_c'(s_L) \\ s_{\rho,i}(\xi), & \text{if } f_c = f \text{ on } (s_i, s_{i+1}) \\ s_{\sigma,i}(\xi), & \text{if } f_c < f \text{ on } (s_i, s_{i+1}) \\ s_R, & \text{for } \xi > f_c'(s_R). \end{cases} \tag{19}$$

As we see, the computation of the rarefaction wave amounts to inverting a function, which is cumbersome numerically. Using an idea going back to Dafermos(1972), see also Kale(1982), we approximate f by a piecewise linear function. Then also f_c will be piecewise linear, and we will be left with only shocks. By making the piecewise linear approximation of the flux function finer, we may approximate the solution to any degree of accuracy. This numerical method was presented in Holden *et al.*(1988). Crucial for this to work as a numerical method is the fact that there is only a finite number of shocks in infinite time. Detailed proofs of these results appear in another contribution in this volume, see Holden and Holden(1991).

4. Heterogeneous media.

A major problem in reservoir simulation is heterogeneities in the rock properties. Petroleum reservoirs are often layered, and parameters such as porosity and permeability may vary discontinuously. This means that the flux function $F(S)$ in (16) depends discontinuously on the space variable and one cannot use standard theory, see Kružkov(1970), to infer the existence of a weak solution to the initial value problem. Therefore, the group has studied both theoretically and numerically the problem of flow in porous media with 'discontinuous geology'.

The simplest initial value problem where the flux function depends discontinuously on the space variable is the Riemann problem:

$$\begin{aligned} u_t + f_l(u)_x &= 0 \qquad & u_0(x) &= u_l \text{ for } x < 0 \\ u_t + f_r(u)_x &= 0 \qquad & u_0(x) &= u_r \text{ for } x > 0. \end{aligned} \tag{20}$$

Here f_l and f_r are functions of a single variable u, and u_l and u_r are constants.

In Gimse and Risebro(1991) we show existence of a weak solution to this Riemann problem. This solution is constructed so that it satisfies a Hugoniot relation at $x = 0$. This Hugoniot relation is in general not enough to ensure uniqueness, and we postulate a principle by which we can always select a unique solution: We require the discontinuity in the solution $u(x,t)$ at $x = 0$ to be the smallest possible discontinuity that satisfies the Hugoniot relation here. We can always choose a solution with his property. If one expands (20) by adding a second conservation law describing the conservation of the parameter governing the discontinuity in the flux function at $x = 0$, one obtains a so-called triangular system, see L. Holden and Høegh-Krohn(1990). We have shown that the viscous profile entropy condition for this enlarged system is equivalent to the minimal jump condition postulated here. The proofs of this are constructive and in a natural fashion define a numerical method which is used in the simulator for two-dimensional full scale simulations.

The solution of the Riemann problem can now be used as a building block in a proof of existence of a weak solution to the Cauchy problem for the one-dimensional case.

If (11) is rewritten as

$$s_t + \Big\{f_0(s)\big(v_T - g(x)k_r(s)\big)\Big\}_x = 0 \tag{21}$$

where $s = s(x,t)$ denotes the saturation of one of the phases, f_0 is the fractional flow function, v_T is the total Darcy velocity and $k_r(s)$ is the relative permeability of the phase not denoted by s. The gravitational term $g(x)$ includes the density differences between the phases as well as the absolute permeability of the rock and the angle of dip of the reservoir. This term is therefore not necessarily a continuous function of x. We may consider $g = g(x)$ as an independent parameter with $g_t = 0$. Hence we can write (21) as a triangular system

$$\begin{aligned} u_t + f(u)_x &= 0, \\ u(x,0) &= u_0(x). \end{aligned} \tag{22}$$

when we for simplicity have scaled $v_T = 1$. Here $f(u) = \big(h(s,g),0\big)$ with

$$h(s,g) = f_0(s)\big(1 - gk_r(s)\big), \tag{23}$$

$f_0 = f_0(s)$ being a Lipschitz-continuous, increasing function with one point of inflection (s-shaped) with $f_0(0) = 0$ and $f_0(1) = 1$. The relative permeability $k_r(s)$ is usually assumed to be a decreasing, convex function of the saturation such that $k_r(0) = 1$ and $k_r(1) = 0$. Note that this implies that $h(1,g) = 1$ for all g. Also each $h(\cdot,g)$ is Lipschitz continuous and has (possibly) one minimum and two points of inflection within the interval of definition, and finally $\frac{\partial h}{\partial g} \leq 0$. The scheme we use to analyze the system (22) is a generalization of Dafermos' scheme (Dafermos(1972), Holden *et al.*(1988)). The basic idea of these schemes is to generate a series of exact solutions to equations which approximate (22). The approximate equations are formed by taking a piecewise linear (in s) approximation to $h(s,g)$ for each g. The approximation is denoted by f_δ, and $\lim_{\delta\to 0} ||f - f_\delta||_{L_1} = 0$. Furthermore we approximate the initial data by a piecewise constant function $u_{0,\delta}$ such that $\lim_{\delta\to 0} ||u_0 - u_{0,\delta}||_{L_1} = 0$.

We then generate a weak solution $u_\delta(x,t)$ to the initial value problem

$$u_{\delta t} + f_\delta(u_\delta)_x = 0 \qquad u_\delta(x,0) = u_{0,\delta}(x). \tag{24}$$

The initial function $u_{0,\delta}$ defines a series of finitely many Riemann problems, and by construction the solution to these problems are constant states separated by discontinuities.

We can track these discontinuites and thereby propagate the solution forward in time, until two of them collide. At this point we have a situation similar to what we had initially, namely a sequence of Riemann problems. Therefore we can solve these and propagate the solution until the next collision.

In Gimse and Risebro(1990) we show that the sequence $\{u_\delta\}$ possesses a subsequence which converges to a weak solution of the initial value problem, more precisely we prove the following theorem:

Theorem. *If $g(x)$ has bounded total variation, (22) possesses a weak solution $s(x,t)$ for arbitrary initial data $s_0(x)$ of bounded total variation.*

We close this section with a simple numerical example which illustrate both how the sequence u_δ is constructed and the numerical method used in the full scale simulator. The simulation was done in a vertical, one-dimensional reservoir of length 100 meters. The cross sectioal area was $200m^2$, and the reservoir porosity was constant (0.4). The permeability, however, was piecewise constant, dividing the reservoir into three regions. The bottom region (below $50m$) and the top region (above $60m$) were high permeability zones ($K = 2000mD$), whereas the middle zone had low permeability ($K = 2mD$). We included two phases of water–like and gasous–like properties. For relative permeabilities we use $k_w = s^2$, where s is the saturation of water, and similarily $k_g = (1-s)^2$ for gasous relative permeability. The phase densities were set to $1000kg/m^3$ for water and $100kg/m^3$ for gas (reservoir conditions). We assumed no compressibility. The initial pressure was set to $260bar$ at the top of the reservoir, and the production and injection rates were both $200m^3$/day after water injection was started at $x = 0$. The initial water saturation was 1.0 at the bottom of the reservoir and 0.3 at the top. Initially these states were separated by a front at height 40 meters. Note that this initial situation is unstable in the sense that the saturation 0.3 is not stable across the geological discontinuites. In Figure 1 the two different flow-functions are shown. In Figure 2 the solution is displayed in the $x-t$ plane. We allowed 41 uniformly distributed saturations, and used 100 grid blocks for the numerical solution of the pressure equation. The discontinuities are displayed as lines. In the lower part of figure 2 we have indicated the different regions. In Figure 3 the solution $s(x,t)$ is displayed for different times.

5. The implementation.

In numerical calucuations of flow in porous media, it is common to split the equations describing the flow into a *saturation equation* (11) and a *pressure equation* (14). This is justified by the large differences in characteristic times between these two equations. These equations are then solved sequentially, the result from one equation is used as input in the other. In this way one can advance in time using discrete time-steps ΔT. We will now describe how each of these equations are treated in the reservoir simulator at present. We start with the pressure equation.

The pressure equation

Since the saturation and pressure equations are solved sequentially, and since the pressure equation (14) depends parametrically on the saturation (through the mobilities), it is of great importance to represent the saturation solution accurately. We do not want to smear the sharp saturation profiles, and therefore we use a finite element method to solve the pressure equation. We use quadrangular elements that may be divided into triangular elements. We can approximate the saturation fronts by placing element boundaries along the fronts. In a similar way the geometry of the reservoir is captured by this adaptive gridding. In addition, convergence and stability theory for finite element formulations is well-known and readily applicable to obtain good error estimates.

The coefficients of the element equation are supposed to be constant within each block of calculation. For those of them depending on the pressure, we use explicit values calculated at the previous timestep. The system of linear equations obtained is solved by a standard preconditioned, conjugate gradient method, and finally the velocity field obtained from the pressure gradients is used in the saturation equation.

The saturation equation

The original plan of the project was to develop a full three-dimensional front tracking system. But the idea of tracking general two-dimensional interfaces in three dimensions was soon abandoned due to the enormous difficulties both in representing these interfaces and in determining their interactions. A reduced three-dimensional front tracker was however developed as a thesis by one of us, see Buchholz(1987), and this worked well for three-dimensional reservoirs with mainly vertical interface movement.

As for two-dimensional front tracking, it was found that the solution to the general scalar two-dimensional Riemann problem was much more complicated than originally thought, see Risebro(1987), and the two-dimensional Riemann problem with a varying geology remains unsolved. These difficulties forced us to limit our ambitions somewhat. As of today, we have a working commercial front tracking system for the case of incompressible two-phase flow in two dimensions without gravitation, see Bratvedt *et al.*(1991). For more general problems we do not use full front tracking in several dimensions, but solve the one-dimensional equations in each dimension separately and use *dimensional splitting* to advance the solution in time.

Dimensional splitting or *fractional steps* as the method is also called, is a popular method for the numerical computation of solutions to conservation laws in several dimensions. For a discussion of general dimensional splitting, see Crandall and Majda(1980). In our simulator we represent the saturation as being constant on rectangular blocks, i.e.

$$s(x,y) = s_{ij} \quad \text{for } x_{i-1} < x < x_i \text{ and } y_{j-1} < y < y_j \tag{25}$$

for some discretization $\{x_i\}_0^m$, $\{y_j\}_0^n$. In two-dimensions the saturation equation (11) may be written as

$$s_t + g(s)_x + h(s)_y = 0 \tag{26}$$

where g and h denotes the components of the flux function in the x and y direction respectively. These will now be assumed to be piecewise linear continuous, such flux functions were discussed in sec. 3. Let then $S^x(t)$ be the solution operator which takes an initial function $s_0(x)$ to the unique entropy solution of

$$s_t + g(s)_x = 0 \qquad s(x,0) = s_0(x), \tag{27}$$

i.e. $s(x,t) = S^x(t)s_0(x)$. Similarly let S^y denote the solution operator in the y direction. Note that $S^x(t)s_0(x)$ can be calculated by the method described in Holden and

Holden(1991) or in sec. 4 for any piecewise constant $g(x)$. Let $\pi(s(x,y))$ be a projection onto functions which are constant on blocks of the grid system $\{x_i\}_0^m$, $\{y_j\}_0^n$ such that

$$\iint \pi(u(x,y))\,dxdy = \iint u(x,y)\,dxdy, \tag{28}$$

i.e. the projection is conservative. Now we can formulate the dimensional splitting algorithm

$$s(x,y,t+\Delta t) = \big(\pi \circ S^y(\Delta t) \circ \pi \circ S^x(\Delta t)\big)s(x,y,t). \tag{29}$$

In 'computer code' this looks like:

$t := 0$
do while $t < \Delta T$
  do $i := 1$ **step** 1 **to** m
    $s_{ij} := \pi \circ S^x(\Delta t)\, s_{ij}$
  enddo
  do $j := 1$ **step** 1 **to** n
    $s_{ij} := \pi \circ S^y(\Delta t)\, s_{ij}$
  enddo
  $t := t + \Delta t$
enddo

In this case ΔT would be the time interval between each solution of the pressure equation. It is obvious that the properties of the computed solution $s_{ij}(t)$ depends heavily on the choice of the mapping π. An immediate choice is to take $\pi(s)$ to be the average value of s in each grid block. This choice was found to work well in many examples, and if the time step Δt obeys the CFL–condition, this method is the same as Godunov's first order method. Taking the average does however imply that discontinuities are smeared, and since this is undesirable we have tried to construct π such that the 'correct discontinuities' are preserved. At the moment the construction of π is the subject of intense research within the front tracking group.

6. Some testcases.

A set of simulation test cases has been set up to demonstrate some of the capabilities of the front-tracking simulator. Here we present three of them.

CASE 1: CHANNEL PROBLEM, OIL-WATER. This first case was defined to investigate the behavior of the front tracking concept on flow in channels. The grid is 25×25 with blocks of uniform size ($20m \times 20m$) and reservoir thickness uniform at $10m$. A channel system in the form of a cross is defined by setting the permeability to either 0 (no flow, outside the cross) or $100mD$ (inside the cross), see Figures 4a-c. Water is injected at a uniform rate in the wells at the bottom and on the left, and oil is produced at the two other wells at the top and on the right side. Figures 4a-c show the saturation fronts at times 200 days, 600 days and 900 days. In addition we compared the results of the front tracker with those of a simulator used in the oil industry based on a standard finite difference method. These comparisons showed that the finite difference simulator needed a refined grid (99×99 blocks) to be able to show a comparable resolution of the fronts as the front-tracker at the coarse grid. However, the finite difference simulator used more than 7 hours of computation time on the fine grid compared to a few minutes for the front-tracker on the coarse grid.

CASE 2: CROSS SECTION, OIL-WATER. To investigate the abilities of the front tracker on a set of more realistic data, Statoil supplied us with data from a reservoir in the North-Sea. This was a two-dimensional cross-section model with oil-water and a base case with 10×6 gridblocks defining 6 geological layers. The geological data are presented in the table below. The 6 layers also represent 4 fluid regions wtih differsent fluid properties. The reservoir dimensions are $1000m \times 100m$ in x and z direction. The thickness is $500.0m$. The production is from the 3 lowermost geological layers. There is one water injection well two thirds up along the left boundary of the reservoir and one production well in the right part of the reservoir. The wells are shown as white lines in the figures. The simulations were performed on a 90×54 grid. Figures 5a-c show the saturation distribution of the simulation at 1095 days, 1460 days, and 3285 days.

In this simulation case the front tracking simulator proved that it can handle the complex geological data that are required in real data simulation cases. In addition, comparisons with a reservoir simulator widely used in the oil industry showed that the front tracking simulator produces accurate results from much coarser grids than do finite difference simulators.

CASE 3: CROSS-SECTIONAL CHANNEL SAND SYSTEM. The last simulation case to be presented was a special study done to utilize the ability of the front-tracking simulator to model complex geological structures and simulate the front behavior accurately. A geological cross-section resulting from outcrop studies was established by geologists. The model represents a section through a complex channel system. Eight sand types were identified, with permeabilities in the range of $0mD$ to $2000mD$. A set of contour lines were generated for each sand type by interactive digitalisation of the cross–section drawn by the geologists. The contours formed the basis for the irregular grid, which was automatically generated by means of the grid preprocessor. Figure 6 shows the sand contours digitalised from the geological cross section model. To separate the sand types vertically, a minimum of 40 layers were defined. Horizontally, 100 cells were used. Water flooding was imposed directly in the oil zone by an injector perforated partially on the left boundary of the model. A production well was perforated partially on the right boundary. The model was given a dip of ten degrees in the direction of flow. The pressure was maintained through voidage replacement. The water front behavior at three specified time steps is shown in Figures 7a-c. The figures demonstrate that the heterogeneous geology creates complex front geometries in time. In the high permeable sands, fingers of water are seen to advance rapidly towards the producer.

REFERENCES

Aasen, J.O., Silseth, J.K., Holden, L., Omre, H., Halvorsen, K.B. and Høiberg, J.(1990), *A stochastic reservoir model and its use in evaluations of uncertainties in the results of recovery processes*, North

Sea Oil & Gas Reservoirs II (edited by A.T. Buller, E. Berg, O. Hjelmeland, J. Kleppe, O. Torsæter, J.O. Aasen), Graham & Trotman, London–Dordrecht–Boston, pp. 425–436.

Bratvedt, F., Bratvedt, K., Buchholz, C.F., Holden, H., Holden, L. and Risebro, N.H. (1991), *A new front tracking method for reservoir simulation*, SPE Reservoir Simulation (to appear).

Buchholz, C.F.(1987), *Et fronttrackingskonsept for 3D reservoarsimulering (In Norwegian)*, Cand. scient. thesis, University of Oslo.

Chern, I.-L., Glimm, J.,McBryan, O.,Plohr, B. and Yaniv, S.(1986), *Front tracking for gas dynamics*, J. Comp. Phys. **62**, 83–110.

Crandall, M., Majda, A.(1980), *The method of fractional steps for conservation laws*, Numer. Math. **34**, 285–314.

Dafermos, C. M.(1972), *Polygonal Approximation of solutions of the initial value problem for a conservation law*, J. Math. Anal. Appl. **38**, 33–41.

Gimse, T., Risebro N.H.(1990), *Solution of the Cauchy problem for a conservation law with a discontinuous flux function*, Preprint, University of Oslo.

Gimse, T., Risebro N.H.(1991), *Riemann problems with discontinuous flux functions*, Proceedings of the Third International Conference on Hyperbolic Problems, Uppsala (to appear).

Glimm, J., (1965), *Solutions in the large for nonlinear hyperbolic systems of equations*, Comm. Pure Appl. Math. **18**, 697–715.

Glimm, J., Isaacson, E., Marchesin, D. and McBryan, O. (1981), *Front tracking for hyperbolic systems*, Adv. Appl. Math. **2**, 91–119.

Glimm, J., Marchesin, D. and McBryan, O. (1981), *A numerical method for two phase flow with an unstable interface*, J. Comp. Phys. **39**, 179–200.

Glimm, J., McBryan, O.(1985), *A computational model for interfaces*, Adv. Appl. Math. **6**, 422–435.

Glimm, J., Lindquist, B., McBryan, O.A., Plohr, B and Yaniv, S.(1981), *Front tracking for petroleum reservoir simulation*, Society of Petroleum Engineers, preprint # 12238, 41–49.

Glimm, J., Sharp, D.H.(1987), *Numerical analysis and the scientific method*, IBM J. Research and Development **31**, 169–177.

Glimm, J., Sharp, D.H.(1990), *A random field model for anomalous diffusion in heterogeneous porous media*, Preprint, Los Alamos.

Haldorsen, H.H., Brand, P.J. and Macdonald, C.J.(1988), *Review of the stochastic nature of reservoirs*, Mathematics in Oil Production (Edited S. Edwards, P.R. King), Clarendon Press, Oxford, pp. 109–209.

Holden, L., Høegh-Krohn, R.(1990), *A class of N nonlinear hyperbolic conservation laws*, J. Diff. Eqn. **84**, 73–99.

Holden, H., Holden, L. and Høegh-Krohn, R.(1988), *A numerical method for first order nonlinear scalar conservation laws in one-dimension*, Comput. Math. Applic. **15**, 595–602.

Holden, H., Holden, L.(1991), *On scalar conservation laws in one-dimension*, In these proceedings.

Holden, H., Risebro, N. H.(1991a), *Stochastic properties of the Buckley–Leverett equation*, SIAM J. Appl. Math.(in print).

Holden, H., Risebro, N. H.(1991b), *A stochastic approach to conservation laws*, Proceedings of the Third International Conference on Hyperbolic Problems, Uppsala (to appear).

Kale, D.M.(1982), *Dynamics of moving interfaces in porous media: III. Extension of Buckley–Leverett theory*, Society of Petroleum Engineers, preprint # 11249, 2–11.

Kružkov, S.N.(1970), *Quasi-linear equations of the first order*, Matem. Sbornik **2**, 217–243.

Moretti, G.(1975), *On the matter of shock fitting*, Proceedings of the Fourth International Conference on Numerical Methods in Fluid Dynamics (edited by R. D. Richtmyer), Springer–Verlag, Berlin–New York–Heidelberg, pp. 287–292.

Richtmyer, R. D.(1957), *Difference Methods for Initial Value Problems*, Interscience Publ., New York.

Risebro, N. H.(1987), *The partial differential equation $u_t + \sum_{i=1}^{n} f_i(u)_x = 0$; a numerical method*, Cand. scient. thesis, University of Oslo.

Scheidegger, A. E.(1974), *The Physics of Flow through Porous Media*, University of Toronto Press, Toronto.

Smoller, J.(1983), *Shock Waves and Reaction-Diffusion Equations*, Springer-Verlag, New York–Heidelberg–Berlin.

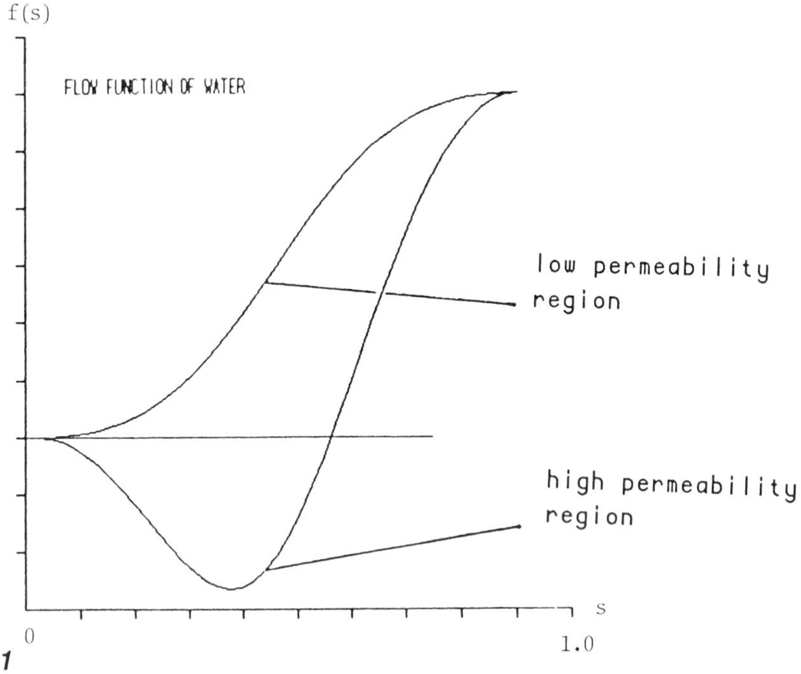
f(s)
FLOW FUNCTION OF WATER
low permeability region
high permeability region
0
s
1.0

Figure 1

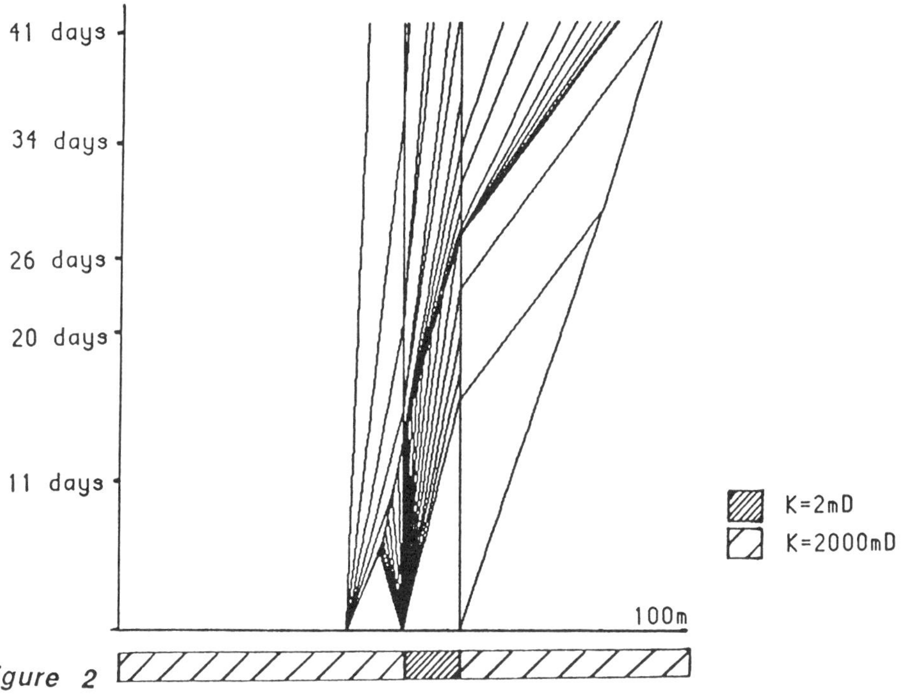

Figure 2

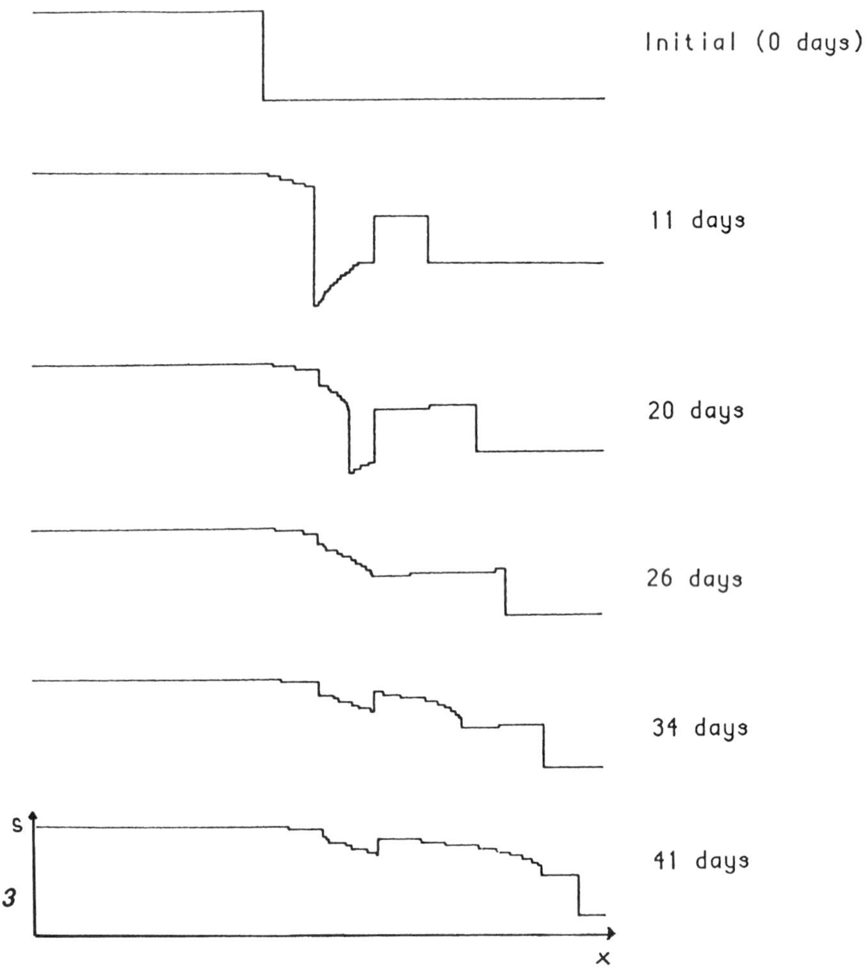

Figure 3

Figure 4a

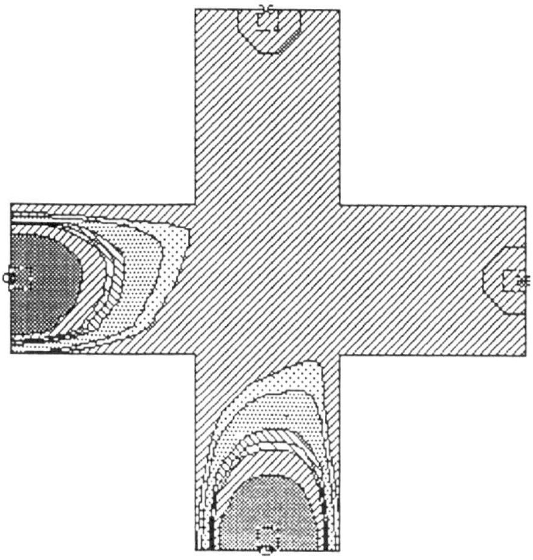

Figure 4b

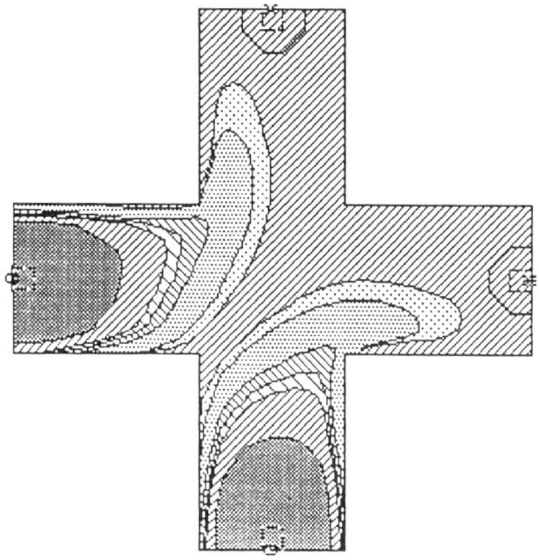

Figure 4c

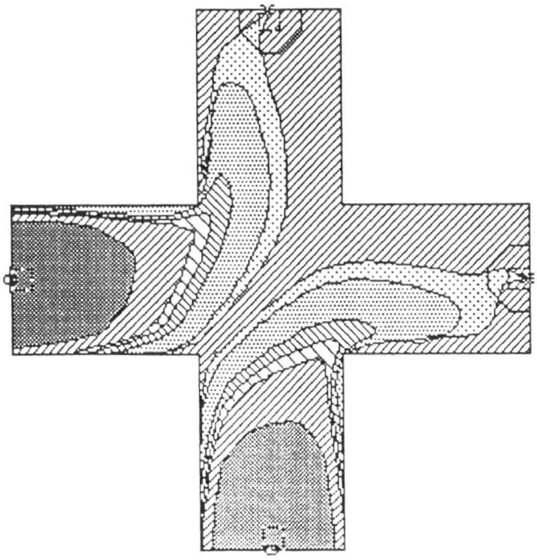

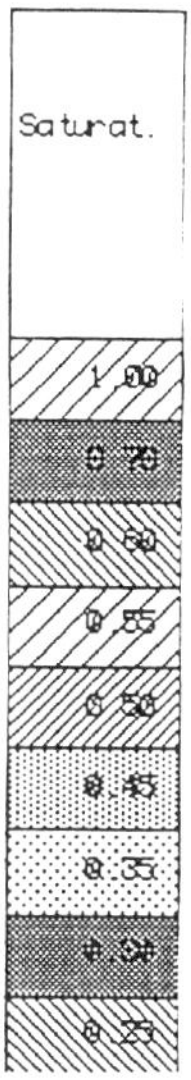

Figure 5a

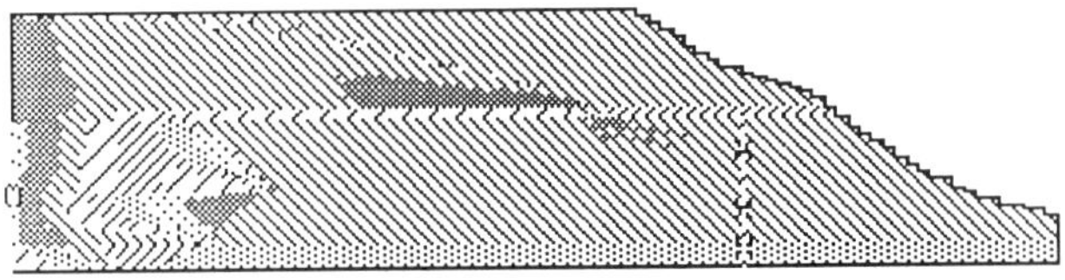

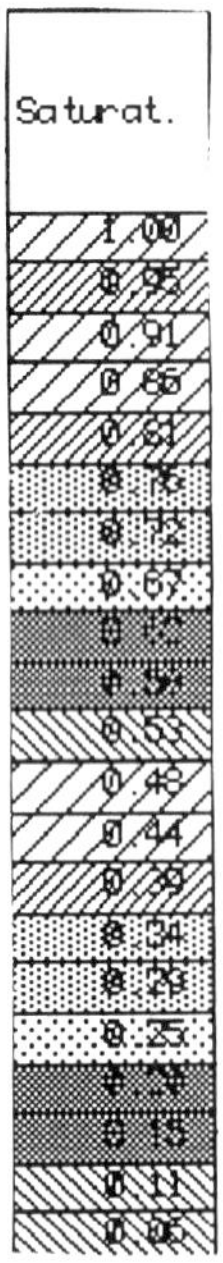

Figure 5b

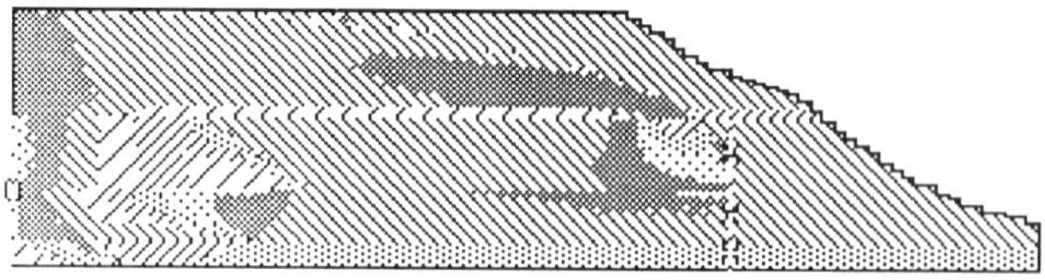

Figure 5c

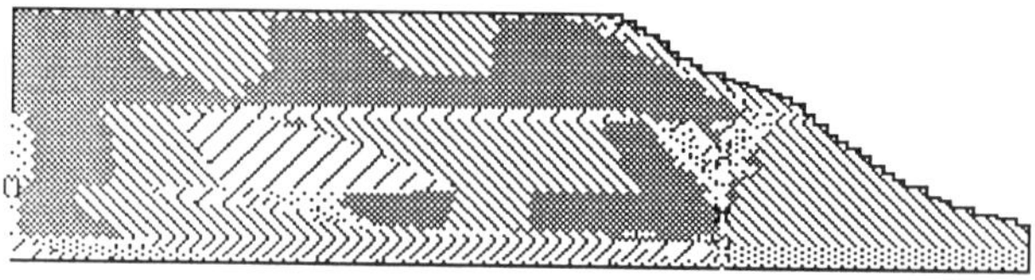

Figure 6

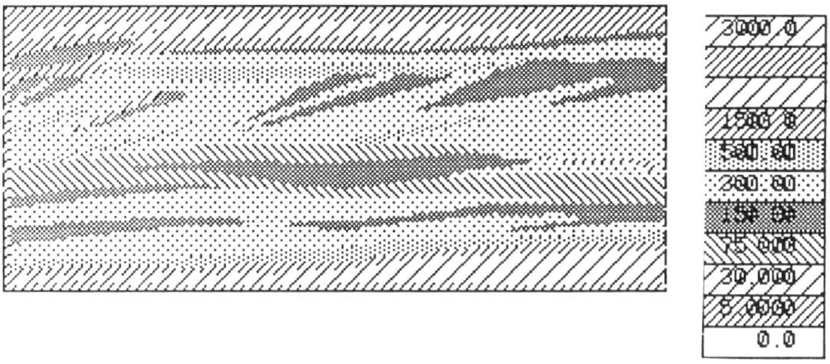

Figure 7a

1.5 days

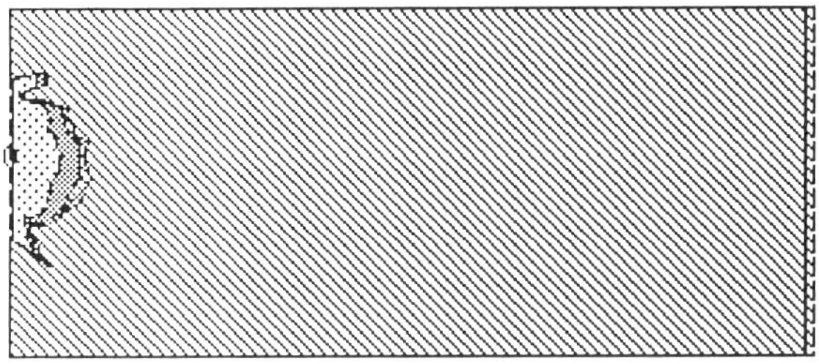

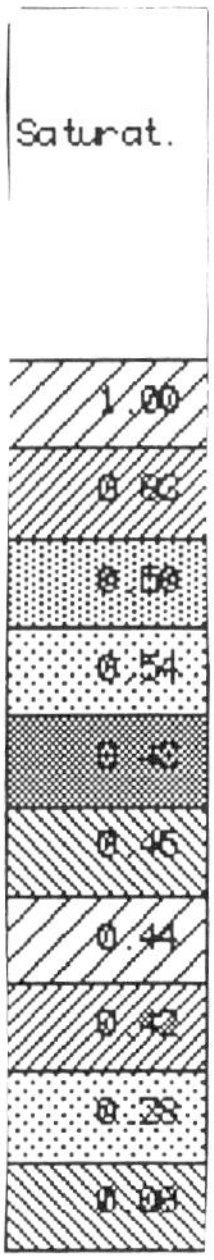

Figure 7b

15 days

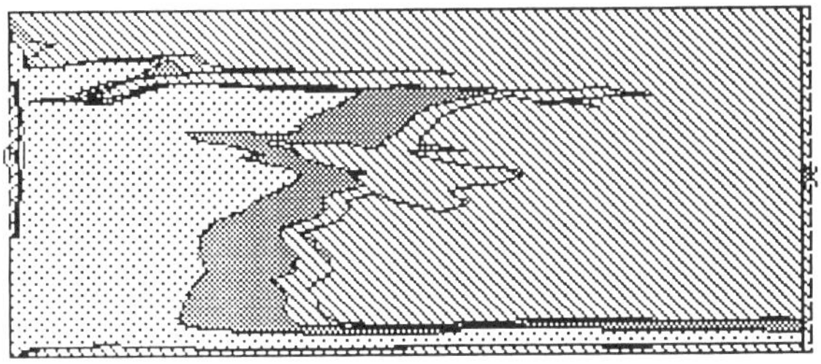

Figure 7c

19.5 days

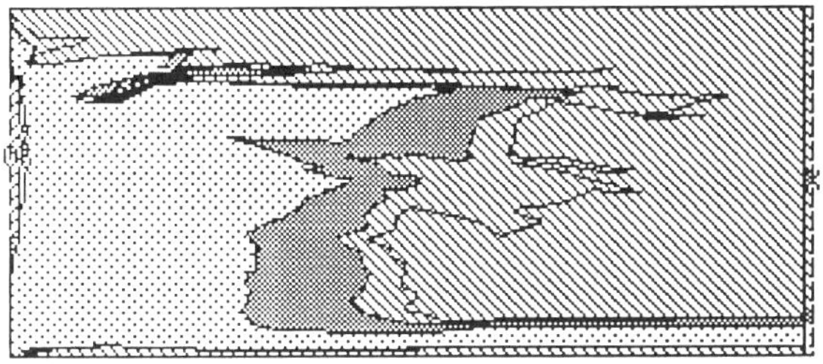

Quasi-periodic, finite-gap solutions of the modified Korteweg-deVries Equation

F. GESZTESY*

BiBoS-Research Center
Faculty of Physics
University of Bielefeld
D-4800 Bielefeld 1
FRG

Division of Mathematical Sciences
Norwegian Institute of Technology
University of Trondheim
N-7034 Trondheim
NORWAY

Abstract. Commutation and τ-function methods together with algebro-geometric techniques are used to derive all real-valued, quasi-periodic, finite-gap solutions of the modified Korteweg-deVries equation.

* Permanent address: Department of Mathematics, University of Missouri, Columbia, MO 65211, USA
E-Mail: MATHFG@UMCVMB.bitnet

1. INTRODUCTION

The main goal of this paper is to derive all real-valued, quasi-periodic, finite-gap solutions of the modified Korteweg-deVries (mKdV) equation. While these solutions, in principle, can be derived by purely algebro-geometric methods as developed e.g. in [Kri 1–4], [Dub 2,4], we shall use a different strategy. More precisely, we shall combine commutation methods and Hirota's τ-function approach together with the explicit theta-function representation for the real-valued, quasi-periodic, finite-gap solutions of the Korteweg-deVries (KdV) equation to achieve our goal. It turns out that a combination of commutation methods (i.e., the comparison of self-adjoint operators of the type A^*A and AA^*) [Dei] and Hirota's τ-function approach yields auto-Bäcklund transformations for the KdV-equation whose space part is the well known Miura-transformation and whose time part is precisely the mKdV-equation.

We also remark that methods applicable to the mKdV-equation (and more generally to AKNS-systems) have been discussed e.g. in [Kri 1–4], [Dub 2,4], [Che], [Co-Jo], [Gi-Jo], [Mar 2], [Gr-Gu]. However, to the best of our knowledge, the representation of quasi-periodic, finite-gap mKdV-solutions in terms of theta-functions has not appeared in print.

Finally we describe the content of this paper. We start with the appendices that also define the notation we use in the main body of this paper. Appendix A reviews classical material on Abelian differentials and their period relations on hyperelliptic Riemann surfaces. Appendix B recalls the notion of divisors, the Riemann-Roch theorem, the Jacobi variety, and Abel's theorem. Appendix C introduces the Riemann theta function associated with hyperelliptic curves, discusses Riemann's vanishing theorem, the solution of Jacobi's inversion problem, and the representation of elementary symmetric functions of the 2nd kind for values of meromorphic functions on the hyperelliptic curve in terms of Riemann theta functions. (We emphasize that almost all results of Appendices A–C directly extend to general, not necessarily hyperelliptic, compact Riemann surfaces. Due to our applications of this material in the context of the (m)KdV-equations we decided to restrict the treatment to hyperelliptic curves.) Appendix D finally recalls the construction of auto-Bäcklund transformations for the KdV-equations employing commutation and τ-function methods.

Based on Appendices A–C we introduce Baker-Akhiezer functions for the KdV-equation in Section 2 and review the construction of all real-valued, quasi-periodic, finite-gap KdV-solutions. Our treatment follows the reviews of [DMN], [Dub 3], [McK 1] though we occasionally deviate from these presentations and provide further details. In our main Section 3 we finally derive all real-valued, quasi-periodic, finite-gap mKdV-solutions on the basis of Section 2 and Appendix D.

2. BAKER-AKHIEZER FUNCTIONS, FINITE-GAP KdV-SOLUTIONS

In this section we review the construction of finite-gap KdV-solutions in terms of Riemann theta functions. While we chose to closely follow the treatments in [DMN], [Dub 3], [McK 1], some of the original work and some variations on this material can be found e.g. in [Ahi], [Da-Ta], [Dub 1–4], [Du-No], [Er-Mc], [It-Ma 1,2], [Kri 1–4], [Lax], [Lev 4], [Mar 1], [McK 2–6], [Mc-Mo], [Nov], [NMPZ], [Rod 2].

We freely employ the notation of Appendices A–D in the following. In order to avoid the trivial case $g = 0$ we shall assume $g \in \mathbb{N}$ for the rest of this section.

Definition 2.1. Let $\mathfrak{D}_{P_1+\cdots+P_g}$ be a positive divisor of degree g on the hyperelliptic curve $K_g, g \in \mathbb{N}$ (defined in (A.9)-(A.14)) with $\zeta_\alpha = z_\alpha^{-1/2}$ the local coordinate near $P_\infty = (\infty, \infty) \in K_g$ and $q : \mathbb{C} \to \mathbb{C}$ a polynomial. Then the Baker-Akhiezer (B–A) function $\psi_y : K_g \backslash \{P_\infty\} \to \mathbb{C}_\infty$, $y \in \mathbb{C}$ associated with $K_g, P_\infty, q, \zeta_\alpha, \mathfrak{D}_{P_1+\cdots+P_g}$ is defined as follows:

(i). ψ_y is meromorphic on $K_g \backslash \{P_\infty\}$ with poles at most at the points $P_1, \ldots, P_g$, i.e., the divisor $\mathfrak{D}$ of poles of ψ_y on $K_g \backslash \{P_\infty\}$ satisfies

$$\mathfrak{D} \geq -\mathfrak{D}_{P_1+\cdots+P_g}. \tag{2.1}$$

(ii). The function

$$(\psi_y \circ z_\alpha^{-1})(\zeta_\alpha) e^{-iyq(\zeta_\alpha^{-1})} = c + O(\zeta_\alpha),\ c \in \mathbb{C} \tag{2.2}$$

is holomorphic near $\zeta_\alpha = 0$ (i.e., $\psi_y \circ z_\alpha^{-1}$ has an essential singularity of the type $e^{iyq(\zeta_\alpha^{-1})}$ near $\zeta_\alpha = 0$).

Theorem 2.2. For a given divisor $\mathfrak{D}_{P_1+\ldots+P_g}$ and given q, ζ_α, y, the set of B-A-functions forms a linear space. Moreover, if $\mathfrak{D}_{P_1+\ldots+P_g}$ is nonspecial and q is such that the divisor $\mathfrak{D}_0$ of the zeros of ψ_y on $K_g \backslash \{P_\infty\}$ is nonspecial too, this linear space is one-dimensional.

Sketch of proof. Existence: Define

$$\psi_y(P) := e^{-iy \int_{P_0}^{P} \omega^{(2)}} \frac{\theta(\underline{\zeta}_{P_0} - \underline{A}_{P_0}(P) + \underline{\alpha}_{P_0}(\mathfrak{D}_{P_1+\cdots+P_g}) + (y/2\pi)\underline{U})}{\theta(\underline{\zeta}_{P_0} - \underline{A}_{P_0}(P) + \underline{\alpha}_{P_0}(\mathfrak{D}_{P_1+\cdots+P_g}))}, P \in K_g \backslash \{P_\infty\}. \tag{2.3}$$

Here $P_0 \in K_g \backslash \{P_\infty\}$ is a fixed base point and $\omega^{(2)}$ is a normalized DSK with a single pole at P_∞ and principal part $-dq(\zeta_\alpha^{-1}) = q'(\zeta_\alpha^{-1})\zeta_\alpha^{-2} d\zeta_\alpha$ at P_∞ whose b-periods are given by $\underline{U} = (U_1, \ldots, U_g)$

$$\int_{a_j} \omega^{(2)} = 0,\ U_j := \int_{b_j} \omega^{(2)},\ 1 \leq j \leq g. \tag{2.4}$$

By (C.3) one easily checks that ψ_y is single-valued on $K_g \backslash \{P_\infty\}$. (2.1) then follows from (C.23) and (C.24), (2.2) is obviously fulfilled.
Uniqueness: Let $\tilde{\psi}_y$ be another B-A-function. Then $\tilde{\psi}_y/\psi_y$ is meromorphic on K_g with poles in $\mathfrak{D}_0$. Since $\mathfrak{D}_0$ is nonspecial by hypothesis, $\tilde{\psi}_y/\psi_y$ must equal a constant. ∎

In the following we shall normalize the B-A-function ψ_y such that $c = 1$ in (2.2). In order to connect the B-A-function with Floquet theory and the KdV-equation we introduce the systems of differential equations

$$\begin{aligned} \mu_{j,x}(t,x) &= 2R_0(\mu_j(t,x))^{1/2}\Big\{\prod_{\substack{\ell=1\\ \ell\neq j}}^{g}[\mu_\ell(t,x)-\mu_j(t,x)]\Big\}^{-1}, \\ \mu_{j,t}(t,x) &= 2\Big[\Lambda - 2\sum_{\substack{\ell=1\\ \ell\neq j}}^{g}\mu_\ell(t,x)\Big]\mu_{j,x}(t,x),\ (t,x)\in\mathbb{R}^2,\ 1\le j\le g, \end{aligned} \tag{2.5}$$

where

$$\Lambda := \sum_{n=0}^{2g} E_n. \tag{2.6}$$

Given the initial condition

$$\hat{\mu}_j(t_0,x_0) := \Big(\mu_j(t_0,x_0),\ R_0(\mu_j(t_0,x_0))^{1/2}\Big) \in K_g,\ 1\le j\le g, \tag{2.7}$$

(2.5) yields a unique solution denoted by

$$\underline{\hat{\mu}}(t,x) := (\hat{\mu}_1(t,x),\dots,\hat{\mu}_g(t,x)) \in K_g^g,\ (t,x)\in\mathbb{R}^2 \tag{2.8}$$

with

$$\mu_j(t,x)\in[E_{2j-1},E_{2j}],\ (t,x)\in\mathbb{R}^2,\ 1\le j\le g. \tag{2.9}$$

In fact, (2.5) is integrated with the help of the Abel map as shown below. For notational brevity we write

$$\underline{\alpha}_{P_0}(\underline{P}) := \underline{\alpha}_{P_0}(\mathfrak{D}_{P_1+\dots+P_g}) \tag{2.10}$$

for $\underline{P} = (P_1,\dots,P_g)\in K_g^g$ with $P_j \neq P_\ell$ for $j\neq \ell$, $1\le j,\ell\le g$ and

$$\underline{c}_j := (c_{1,j},\dots,c_{g,j}),\ 1\le j\le g, \tag{2.11}$$

where $c_{j,\ell}$ are defined in (A.22).

Theorem 2.3.

$$\partial_x\underline{\alpha}_{P_0}\left(\underline{\hat{\mu}}(t,x)\right) = 2(-1)^{g+1}\underline{c}_g, \tag{2.12}$$

$$\partial_t \underline{\alpha}_{P_0}\left(\underline{\hat{\mu}}(t,x)\right) = 4(-1)^{g+1}\left[\Lambda \underline{c}_g + 2\underline{c}_{g-1}\right], \; (t,x) \in \mathbb{R}^2. \tag{2.13}$$

Proof.

$$\begin{aligned}\partial_x \underline{\alpha}_{P_0}\left(\underline{\hat{\mu}}(t,x)\right) &= \left\{\sum_{j=1}^{g} \mu_{j,x}(t,x) \sum_{k=1}^{g} c_{\ell,k}\mu_j(t,x)^{k-1} R_0\left(\mu_j(t,x)\right)^{-1/2}\right\}_{\ell=1}^{g} \\ &= \left\{2\sum_{j,k=1}^{g} c_{\ell,k}\mu_j(t,x)^{k-1} \prod_{\substack{m=1\\ m\neq j}}^{g} \left[\mu_m(t,x)-\mu_j(t,x)\right]^{-1}\right\} = 2(-1)^{g+1}\underline{c}_g\end{aligned} \tag{2.14}$$

by

$$\sum_{j=1}^{g} \mu_j^{k-1} \prod_{\substack{m=1\\ m\neq j}}^{g} (\mu_m - \mu_j)^{-1} = (-1)^{g+1}\delta_{k,g}, \; \mu_j \in \mathbb{C}, \; 1 \leq j, \; k \leq g \tag{2.15}$$

(a special case of Lagrange's interpolation formula). Similarly one proves (2.13) invoking (2.15) and

$$\sum_{j=1}^{g} \Big(\sum_{\substack{n=1\\ n\neq j}}^{g} \mu_n\Big)\mu_j^{k-1} \prod_{\substack{m=1\\ m\neq j}}^{g} (\mu_m - \mu_j)^{-1} = (-1)^{g}\delta_{k,g-1}, \; \mu_j \in \mathbb{C}, \; 1 \leq j,k \leq g. \tag{2.16}$$

∎

In particular, Theorem 2.3 implies

$$\begin{aligned}&\underline{\alpha}_{P_0}(\underline{\hat{\mu}}(t,x)) - \underline{\alpha}_{P_0}\left(\underline{\hat{\mu}}(t_0,x_0)\right) \\ &= \left\{2(-1)^{g+1}(x-x_0)\underline{c}_g + 4(-1)^{g+1}(t-t_0)\left[\Lambda \underline{c}_g + 2\underline{c}_{g-1}\right]\right\} (\operatorname{mod} L_g), \; (t,x) \in \mathbb{R}^2.\end{aligned} \tag{2.17}$$

As discussed in detail e.g. in [Lev 1–4], the $\mu_j(t,x_0)$, $\mu_j(t_0,x)$ are quasi-periodic functions of t and x with the same periods for $j = 1, \ldots, g$ whose number is less or equal to g (the periods and their number differ in general for t and x). As can be inferred from (2.5), the $\mu_j(t,x)$ sweep the closure $\overline{\rho}_j$ of the gap $\rho_j = (E_{2j-1}, E_{2j})$ as t or x varies in $\mathbb{R}$.

Next, using the abbreviations (2.10), we define the normalized B–A-function

$$\begin{aligned}\psi(P,x,x_0,t,t_0) &:= \exp\left[-i(x-x_0)\int_{P_0}^{P}\omega_0^{(2)} - 12i(t-t_0)\int_{P_0}^{P}\omega_2^{(2)}\right]\bullet \\ &\bullet \frac{\theta\left(\underline{\zeta}_{P_0} - \underline{A}_{P_0}(P) + \underline{\alpha}_{P_0}\left(\underline{\hat{\mu}}(t_0,x_0)\right) + ((x-x_0)/2\pi)\,\underline{U}_0 + (6(t-t_0)/\pi)\,\underline{U}_2\right)}{\theta\left(\underline{\zeta}_{P_0} - \underline{A}_{P_0}(P_\infty) + \underline{\alpha}_{P_0}\left(\underline{\hat{\mu}}(t_0,x_0)\right) + ((x-x_0)/2\pi)\,\underline{U}_0 + (6(t-t_0)/\pi)\,\underline{U}_2\right)}\bullet \\ &\bullet \frac{\theta\left(\underline{\zeta}_{P_0} - \underline{A}_{P_0}(P_\infty) + \underline{\alpha}_{P_0}\left(\underline{\hat{\mu}}(t_0,x_0)\right)\right)}{\theta\left(\underline{\zeta}_{P_0} - \underline{A}_{P_0}(P) + \underline{\alpha}_{P_0}\left(\underline{\hat{\mu}}(t_0,x_0)\right)\right)}, \\ &\qquad P_0 = (E_0,0), \; P \in K_g \backslash \{P_\infty\}.\end{aligned} \tag{2.18}$$

Here $\omega_{2k}^{(2)}$, $k = 0, 1$ are normalized DSK's with a single pole at P_∞ and principal part

$$\omega_{2k}^{(2)} = \left[\zeta_\alpha^{-2-2k} + 0(1)\right] d\zeta_\alpha,\ k = 0, 1 \tag{2.19}$$

at P_∞ whose b-periods are $\underline{U}_k$

$$\int_{a_j} \omega_0^{(2)} = \int_{a_j} \omega_2^{(2)} = 0,\ U_{0,j} := \int_{b_j} \omega_0^{(2)},\ U_{2,j} := \int_{b_j} \omega_2^{(2)},\ 1 \le j \le g. \tag{2.20}$$

By (C.33) one infers

$$U_{0,j} = \int_{b_j} \omega_0^{(2)} = 4\pi(-1)^{g+1} c_{j,g},\ 1 \le j \le g. \tag{2.21}$$

Similarly, (A.29) and (C.41) imply

$$U_{2,j} = \int_{b_j} \omega_2^{(2)} = (2\pi/3)(-1)^{g+1}\left[\Lambda c_{j,g} + 2c_{j,g-1}\right],\ 1 \le j \le g \tag{2.22}$$

displaying the connection between (2.17) and the arguments of the θ-functions in (2.18).

Next we consider $\omega_0^{(2)}$, $\omega_2^{(2)}$ in more detail.

Lemma 2.4. For $P = \left(\zeta_\alpha^{-2},\ R_0(\zeta_\alpha^{-2})^{1/2}\right)$ near $P_\infty = (\infty, \infty)$ we have

$$\int_{P_0}^{P} \omega_k^{(2)} = -(1+2k)^{-1}\zeta_\alpha^{-1-2k} + 0(\zeta_\alpha),\ k = 0, 1,\ P_0 = (E_0, 0) \tag{2.23}$$

as $\zeta_\alpha \to 0$.

Proof. Let γ be a simple, closed and smooth contour on the upper sheet Π_+ encircling P_∞ clockwise and returning to P_0. Since the a-periods of $\omega_k^{(2)}$, $k = 0, 1$ vanish by definition, we infer

$$\int_\gamma \omega_k^{(2)} = -\sum_{j=1}^{g} \int_{a_j} \omega_k^{(2)} = 0,\ k = 0, 1. \tag{2.24}$$

Deforming γ in such a way that γ joins P_0 and P_- (near P_∞) on the lower rim of the cut $\overline{\rho}_0 = [\infty, E_0]$ (see (A.8)), encircles P_∞ clockwise with a small circle $C_{R^{-2}}$ of radius R^{-2} to P_+ on the upper rim of $\overline{\rho}_0$ opposite to P_-, returning back to P_0 along the upper rim yields

$$0 = \int_\gamma \omega_k^{(2)} = 2\int_{P_0}^{P_-} \omega_{k,-}^{(2)} + \int_{C_{R^{-2}}} \omega_k^{(2)} \tag{2.25}$$

and hence (2.23) since

$$\int_{C_{R-2}} \omega^{(2)} = \int_0^\pi R^{-1}(-i)e^{-i\theta}\left[R^{2+2k}e^{i(2+2k)\theta} + 0(1)\right] d\theta$$
$$= 2(1+2k)^{-1}R^{1+2k} + 0(R^{-1}). \tag{2.26}$$

(Here $\omega_{k,-}^{(2)}$ denotes $\omega_k^{(2)}$ along the lower rim of ρ_0.) ■

Because of (2.19) and (A.30) we may write

$$\omega_0^{(2)} = (-i/2)R_0(z)^{-1/2}\prod_{j=1}^{g}(\lambda_j - z)\,dz, \tag{2.27}$$

which serves as a definition of the numbers $\{\lambda_j\}_{k=1}^{g} \subset \mathbb{C}$.

Lemma 2.5. The numbers λ_j in (2.27) are real-valued and precisely one of them lies in each nonempty, open gap ρ_j, $1 \le j \le g$, i.e.,

$$\lambda_j \in \rho_j = (E_{2j-1}, E_{2j}),\ 1 \le j \le g \tag{2.28}$$

upon ordering the λ_j in magnitude. (In particular, $\lambda_j \ne \lambda_\ell$ for $j \ne \ell$.)

Proof. Because of the vanishing of the a-periods of $\omega_0^{(2)}$ and the fact that $R_0(z)^{1/2}$ is real-valued and of constant (opposite) sign on the two rims of each nonempty gap ρ_j, we infer from

$$0 = \int_{a_j}\omega_0^{(2)} = (-i)\int_{E_{2j-1}}^{E_{2j}} R_0(z)^{-1/2}\prod_{\ell=1}^{g}(\lambda_\ell - z)dz,\ 1 \le j \le g \tag{2.29}$$

that $\prod_{\ell=1}^{g}(\lambda_\ell - z)$ has precisely one zero in each gap $\rho_j = (E_{2j-1}, E_{2j})$, $1 \le j \le g$. ■

Lemma 2.6.

$$\psi(P,x,x_0,t,t_0)\psi(P^*,x,x_0,t,t_0) = \prod_{j=1}^{g}\left(\frac{\mu_j(t,x)-z}{\mu_j(t_0,x_0)-z}\right),$$
$$x,x_0,t,t_0 \in \mathbb{R},\ P=(z,R_0(z)^{1/2}),\ P^* = (z,-R_0(z)^{1/2}),\ z \in \mathbb{C}. \tag{2.30}$$

Proof. Multiplying $\psi(P)$ and $\psi(P^*)$ cancels the exponentials in (2.18) and leaves the ratios of theta functions symmetric in P and P^* which thus only depend on $z = \Pi(P) = \Pi(P^*)$.

By the properties of the θ-functions involved, the left hand side of (2.30) extends to a meromorphic function of $z \in \mathbb{C}_\infty$. Due to the theta-divisors in (2.18) and the normalization near $z = \infty$ one obtains (2.30). ∎

Theorem 2.7.

$$\sum_{j=1}^{g} \mu_j(t,x) = \sum_{j=1}^{g} \lambda_j + (4\pi)^{-2} \sum_{j,\ell=1}^{g} U_{0,j} U_{0,\ell} \partial^2_{z_j z_\ell} \log \theta(\underline{z})\Big|_{\underline{z}=\underline{\zeta}_{P_0} - \underline{A}_{P_0}(P_\infty) + \underline{\alpha}_{P_0}(\hat{\mu}(t,x))}$$

$$= \sum_{j=1}^{g} \lambda_j + (4\pi)^{-2} \sum_{j,\ell=1}^{g} U_{0,j} U_{0,\ell} \partial^2_{z_j z_\ell} \log \theta(\underline{z})\Big|_{\underline{z}=\underline{\xi}_{P_\infty} + \underline{\alpha}_{P_\infty}(\underline{\hat{\mu}}(t,x))}, \quad (t,x) \in \mathbb{R}^2. \tag{2.31}$$

Proof. Identifying $P_j = \hat{\mu}_j(t,x)$, $1 \le j \le g$ in (C.47), differentiating the result with respect to x yields by (C.48), (2.5), (2.27)

$$2\pi \sum_{j=1}^{g} \prod_{n=1}^{g} [\lambda_n - \mu_j(t,x)] \prod_{\substack{\ell=1 \\ \ell \neq j}}^{g} [\mu_\ell(t,x) - \mu_j(t,x)]^{-1}$$

$$= - \sum_{j,\ell=1}^{g} U_{0,j} \partial^2_{z_j z_\ell} \log \theta(\underline{z})\Big|_{\underline{z}=\underline{\xi}_{P_0} - \underline{A}_{P_0}(P_\infty) + \underline{\alpha}_{P_\infty}(\underline{\hat{\mu}}(t,x))} \bullet \tag{2.32}$$

$$\bullet\, 2 \sum_{m=1}^{g} c_{\ell,m} \sum_{k=1}^{g} \mu_k(t,x)^{m-1} \prod_{\substack{n=1 \\ n \neq k}}^{g} [\mu_n(t,x) - \mu_k(t,x)]^{-1}, \quad (t,x) \in \mathbb{R}^2.$$

Invoking (2.15) and

$$\sum_{j=1}^{g} \prod_{n=1}^{g} (\lambda_n - \mu_j) \prod_{\substack{\ell=1 \\ \ell \neq j}}^{g} (\mu_\ell - \mu_j)^{-1} = \sum_{j=1}^{g} (\lambda_j - \mu_j), \ \lambda_j, \mu_j \in \mathbb{C}, \ 1 \le j \le g \tag{2.33}$$

then yields (2.31). ∎

Theorem 2.8. $\psi(P,x,x_0,t,t_0)$ and $\psi(P^*,x,x_0,t,t_0)$ satisfy the one-dimensional Schrödinger equation

$$- \psi_{xx}(P^{(*)},x,x_0,t,t_0) + V_g(t,x)\psi(P^{(*)},x,x_0,t,t_0) = z\psi(P^{(*)},x,x_0,t,t_0),$$
$$x, x_0, t, t_0 \in \mathbb{R}, \ P = \left(z, R_0(z)^{1/2}\right), \tag{2.34}$$

where the finite-gap potential V_g is given by the Its-Matveev formula (2.35)

$$V_g(t,x) = E_0 + \sum_{j=1}^{g} (E_{2j-1} + E_{2j} - 2\lambda_j)$$
$$- 2\partial_x^2 \ln \theta \left(\underline{\zeta}_{P_0} - \underline{A}_{P_0}(P_\infty) + \underline{\alpha}_{P_0}\left(\underline{\hat{\mu}}(t_0,x_0)\right) + ((x-x_0)/2\pi)\underline{U}_0 + (6(t-t_0)/\pi)\underline{U}_2\right) \tag{2.35}$$

$$= E_0 + \sum_{j=1}^{g} \left[E_{2j-1} + E_{2j} - 2\mu_j(t,x)\right], \ (t,x) \in \mathbb{R}^2. \tag{2.36}$$

The spectrum of the corresponding finite-gap Schrödinger operator in $L^2(\mathbb{R})$

$$H_g(t) := -\frac{d^2}{dx^2} + V_g(t,.) \text{ on } \mathfrak{D}(H_g(t)) = H^2(\mathbb{R}), \ t \in \mathbb{R} \tag{2.37}$$

is purely absolutely continuous, of multiplicity two and given by

$$\sigma\left(H_g(t)\right) = \bigcup_{j=1}^{g} \left[E_{2(j-1)}, E_{2j-1}\right] \cup \left[E_{2g}, \infty\right), \ t \in \mathbb{R}. \tag{2.38}$$

V_g is a real-valued, quasi-periodic function of $t \in \mathbb{R}$ and $x \in \mathbb{R}$. V_g is periodic in x with fundamental period $a > 0$ iff

$$a = \min\left\{\omega > 0 | 2\omega|c_{j,g}| \in \mathbb{N}, \ 1 \le j \le g\right\} = \min\left\{\omega > 0 | (\omega|U_{0,j}|/2\pi) \in \mathbb{N}, \ 1 \le j \le g\right\}. \tag{2.39}$$

Moreover, (2.35) describes all almost periodic potentials in x that yield a finite-gap spectrum of the type (2.38).

Proof. Abbreviating

$$\mathcal{C} := \left[E_0 + \sum_{j=1}^{g} (E_{2j-1} + E_{2j} - 2\lambda_j)\right]/2 \tag{2.40}$$

we get from (2.27)

$$\omega_0^{(2)} = \left\{\zeta_\alpha^{-2} + \mathcal{C} + 0(\zeta_\alpha^2)\right\} d\zeta_\alpha \tag{2.41}$$

near P_∞. Thus one infers from (2.23) that

$$\begin{aligned} \psi(P^{(*)}, x, x_0, t, t_0) = \exp\left\{ \underset{(-)}{+} i(x - x_0)\left[\zeta_\alpha^{-1} - \mathcal{C}\zeta_\alpha + 0(\zeta_\alpha^3)\right] \underset{(-)}{+} 4i(t - t_0)\left[\zeta_\alpha^{-3} + 0(\zeta_\alpha)\right]\right\} \bullet \\ \bullet \left\{1 + c_1(x,t)\zeta_\alpha + c_2(x,t)\zeta_\alpha^2 + 0(\zeta_\alpha^3)\right\}, \\ P \in K_g \backslash \{P_\infty\}, \ x, x_0, t, t_0 \in \mathbb{R} \end{aligned} \tag{2.42}$$

and hence

$$\psi_{xx} = \left\{-\zeta_\alpha^{-2} + 2\mathcal{C} + 2ic_{1,x}(x,t) + 0(\zeta_\alpha)\right\}\psi \tag{2.43}$$

near P_∞. Since $-\psi_{xx} + \left\{2\mathcal{C} + 2ic_{1,x}(x,t) - \zeta_\alpha^{-2}\right\}\psi = 0(\zeta_\alpha)\psi$ is also a B–A-function with the same essential singularity near P_∞ and the same pole-divisor as ψ, uniqueness of the B–A-function in Theorem 2.2 proves

$$-\psi_{xx} + \{2\mathcal{C} + 2ic_{1,x}(x,t) - z\}\psi = 0. \tag{2.44}$$

Next we compute c_1, c_2 directly. (We also compute c_2 since we shall need it in Theorem 2.9.) By (C.38), (C.43) we have near P_∞ (see (C.45))

$$\begin{aligned}\theta\left(\underline{\xi} - \underline{A}_{P_0}(P)\right) &= \theta\left(\underline{\xi} - \underline{A}_{P_0}(P_\infty)\right) - (2\pi i)^{-1}\zeta_\alpha \sum_{j=1}^{g} U_{0,j}\partial_{z_j}\theta(\underline{z})\Big|_{\underline{z}=\underline{\xi}-\underline{A}_{P_0}(P_\infty)} \\ &- (8\pi^2)^{-1}\zeta_\alpha^2 \sum_{j,\ell=1}^{g} U_{0,j}U_{0,\ell}\partial^2_{z_j z_\ell}\theta(\underline{z})\Big|_{\underline{z}=\underline{\xi}-\underline{A}_{P_0}(P_\infty)} + 0(\zeta_\alpha^3).\end{aligned} \tag{2.45}$$

Inserting (2.45) into (2.18), expanding the ratio of theta functions yields

$$\begin{aligned}&\frac{\theta(\dots)\theta(\dots)}{\theta(\dots)\theta(\dots)} \\ &= 1 - i\zeta_\alpha\partial_y \ln\theta\left(\underline{\zeta}_{P_0} - \underline{A}_{P_0}(P_\infty) + \underline{\alpha}_{P_0}\left(\underline{\hat{\mu}}(t_0,x_0)\right) + (y/2\pi)\underline{U}_0\right)\Big|_{y=0} \\ &+ i\zeta_\alpha\partial_x \ln\theta\left(\underline{\zeta}_{P_0} - \underline{A}_{P_0}(P_\infty) + \underline{\alpha}_{P_0}\left(\underline{\hat{\mu}}(t_0,x_0)\right) + ((x-x_0)/2\pi)\underline{U}_0 + (6(t-t_0)/\pi)\underline{U}_2\right) \\ &- (1/8\pi^2)\zeta_\alpha^2 \sum_{j,\ell=1}^{g} U_{0,j}U_{0,\ell}\left[\partial^2_{z_j z_\ell}\theta(\underline{z})\right]\theta(\underline{z})^{-1}\Big|_{\underline{z}=\underline{\zeta}_{P_0}-\underline{A}_{P_0}(P_\infty)+\underline{\alpha}_{P_0}(\underline{\hat{\mu}}(t_0,x_0))} \\ &+ \zeta_\alpha^2\left[\partial_x \ln\theta\left(\underline{\zeta}_{P_0} - \underline{A}_{P_0}(P_\infty) + \underline{\alpha}_{P_0}\left(\underline{\hat{\mu}}(t_0,x_0)\right) + ((x-x_0)/2\pi)\underline{U}_0 + (6(t-t_0)/\pi)\underline{U}_2\right)\right] \bullet \\ &\quad\bullet\left[\partial_y \ln\theta\left(\underline{\xi}_{P_0} - \underline{A}_{P_0}(P_\infty) + \underline{\alpha}_{P_0}\left(\underline{\hat{\mu}}(t_0,x_0)\right) + (y/2\pi)\underline{U}_0\right)\Big|_{y=0}\right] \\ &- \zeta_\alpha^2\left[\partial_y \ln\theta\left(\underline{\zeta}_{P_0} - \underline{A}_{P_0}(P_\infty) + \underline{\alpha}_{P_0}\left(\underline{\hat{\mu}}(t_0,x_0)\right) + (y/2\pi)\underline{U}_0\right)\Big|_{y=0}\right]^2 + 0(\zeta_\alpha^3)\end{aligned} \tag{2.46}$$

$$= 1 + c_1(x,t)\zeta_\alpha + c_2(x,t)\zeta_\alpha^2 + 0(\zeta_\alpha^3). \tag{2.47}$$

Thus

$$\begin{aligned}&c_{1,x}(x,t) \\ &= i\partial_x^2 \ln\theta\left(\underline{\zeta}_{P_0} - \underline{A}_{P_0}(P_\infty) + \underline{\alpha}_{P_0}\left(\underline{\hat{\mu}}(t_0,x_0)\right) + ((x-x_0)/2\pi)\underline{U}_0 + (6(t-t_0)/\pi)\underline{U}_2\right), \\ &\qquad (t,x)\in\mathbb{R}^2\end{aligned} \tag{2.48}$$

and hence

$$V_g = 2\mathcal{C} + 2ic_{1,x} \tag{2.49}$$

yields (2.34) and (2.35). The trace formula (2.36) then follows from (2.31). Since $\psi(P^{(*)}, x, x_0, t, t_0)$ are polynomially bounded for $z \in \bigcup_{j=1}^{g}\left[E_{2(j-1)}, E_{2j-1}\right] \cup [E_{2g}, \infty)$ with respect to x and of exponential type for $z \in \mathbb{C}\backslash \left\{ \bigcup_{j=1}^{g} \left[E_{2(j-1),2j-1}\right] \cup [E_{2g}, \infty) \right\}$, one infers (2.38) by general spectral theory for Schrödinger-type operators (see e.g. [Sim 3], Ch. C). Absolute continuity of $\sigma(H_g(t))$ follows e.g. from Theorem XIII.20 of [Re-Si] (see also [Ko-Kr]). Moreover, $\psi(P)$ and $\psi(P^*)$ are linearly independent for $P \in K_g \backslash \left\{ \{(E_n, 0)\}_{n=0}^{2g} \cup \{P_\infty\} \right\}$ proving multiplicity two of $\sigma(H_g(t))$ (see also [De-Si]). Quasi-periodicity of V_g follows directly from the theta function representation (2.35) (or from the quasi-periodicity of the μ_j, $1 \le j \le g$ in (2.36)). That any almost periodic potential that gives rise to the finite-gap spectrum (2.38) is included in (2.35) is discussed e.g. in [DMN], [Joh], [Lev 4].

■

The trace formula (2.36) and extensions of it are discussed in detail in [Bu-Fi], [Cra], [Dub 1], [DMN], [Fla], [Ko-Kr], [Lev 1–4], [Mc-Mo], [Mos], [Tru].

Theorem 2.9. $\psi(P, x, x_0, t, t_0)$ and $\psi(P^*, x, x_0, t, t_0)$ satisfy

$$\psi_t(P^{(*)}, x, x_0, t, t_0) - \left(B_{V_g}(t)\psi\right)(P^{(*)}, x, x_0, t, t_0) = 0, \;\; x, x_0, t, t_0 \in \mathbb{R}, \;\; P \in K_g \backslash \{P_\infty\} \tag{2.50}$$

in the distributional sense, where

$$B_{V_g}(t) := -4\frac{d^3}{dx^3} + 6V_g(t,.)\frac{d}{dx} + 3V_{g,x}(t,.) \text{ on } \mathfrak{D}(B_{V_g}(t)) = H^3(\mathbb{R}), \;\; t \in \mathbb{R}, \tag{2.51}$$

or equivalently (using (2.34)),

$$\psi_t(P^{(*)}, x, x_0, t, t_0) - 2\left[V_g(t,x) + 2z\right]\psi_x(P^{(*)}, x, x_0, t, t_0) + V_{g,x}(t,x)\psi(P^{(*)}, x, x_0, t, t_0) = 0,$$
$$x, x_0, t, t_0 \in \mathbb{R}, \;\; P = (z, R_0(z)^{1/2}) \in K_g \backslash \{P_\infty\}. \tag{2.52}$$

Proof. Following the first part of the proof of Theorem 2.8 step by step, using the uniqueness of the B–A-function one arrives at

$$\psi_t - B_{V_g}\psi = \{2ic_{1,xx} - 4c_{2,x} + 4c_1c_{1,x}\}\psi. \tag{2.53}$$

(2.46) and (2.47) imply

$$
\begin{aligned}
&c_{2,x}(x,t)\\
&= -(1/2)\partial_x^3 \ln\theta\left(\underline{\zeta}_{P_0} - \underline{A}_{P_0}(P_\infty) + \underline{\alpha}_{P_0}\left(\hat{\underline{\mu}}(t_0,x_0)\right) + ((x-x_0)/2\pi)\underline{U}_0 + (6(t-t_0)/\pi)\underline{U}_2\right)\\
&- \left[\partial_x \ln\theta\left(\underline{\zeta}_{P_0} - \underline{A}_{P_0}(P_\infty) + \underline{\alpha}_{P_0}\left(\hat{\underline{\mu}}(t_0,x_0)\right) + ((x-x_0)/2\pi)\underline{U}_0 + (6(t-t_0)/\pi)\underline{U}_2\right)\right] \bullet\\
&\bullet \left[\partial_x^2 \ln\theta\left(\underline{\zeta}_{P_0} - \underline{A}_{P_0}(P_\infty) + \underline{\alpha}_{P_0}\left(\hat{\underline{\mu}}(t_0,x_0)\right) + ((x-x_0)/2\pi)\underline{U}_0 + (6(t-t_0)/\pi)\underline{U}_2\right)\right]\\
&+ \left[\partial_x^2 \ln\theta\left(\underline{\zeta}_{P_0} - \underline{A}_{P_0}(P_\infty) + \underline{\alpha}_{P_0}\left(\hat{\underline{\mu}}(t_0,x_0)\right) + ((x-x_0)/2\pi)\underline{U}_0 + (6(t-t_0)/\pi)\underline{U}_2\right)\right] \bullet\\
&\bullet \left[\partial_y \ln\theta\left(\underline{\zeta}_{P_0} - \underline{A}_{P_0}(P_\infty) + \underline{\alpha}_{P_0}\left(\hat{\underline{\mu}}(t_0,x_0)\right) + (y/2\pi)\underline{U}_0\right)\Big|_{y=0}\right]
\end{aligned}
\tag{2.54}
$$

and hence

$$
2ic_{1,xx}(x,t) + 4c_1(x,t)c_{1,x}(x,t) - 4c_{2,x}(x,t) = 0, \ (t,x) \in \mathbb{R}^2. \tag{2.55}
$$

■

Corollary 2.10. The finite-gap potential $V_g(t,x)$ in (2.35), (2.36) satisfies the KdV-equation

$$
KdV(V_g) = V_{g,t} - 6V_g V_{g,x} + V_{g,xxx} = 0. \tag{2.56}
$$

Moreover, (2.35) describes all spatially almost periodic solutions of the KdV-equation that give rise to the finite-gap spectrum (2.38).

Proof. Applying the Lax equation

$$
\frac{d}{dt}H_g(t) - \left[B_{V_g}(t), H_g(t)\right] = KdV(V_g) \tag{2.57}
$$

to $\psi(P,x,x_0,t,t_0)$ in the distributional sense, observing (2.34) and (2.50), yields (2.56). That all spatially almost periodic, finite-gap solutions of (2.56) are actually of the form (2.35) is discussed e.g. in [DMN], [Joh], [Lev 4]. ■

Remark 2.11. For brevity we have avoided a detailed discussion of Floquet theory and its relation to the formalism presented above. (Such a treatment, however, can be found e.g. in [Ges 3].) Introducing Floquet solutions $\psi_\pm$ of H_g in the standard manner (see e.g. [Fir], [Ges 3] or (3.5)), one can readily verify that (up to time-dependent multiples) these Floquet solutions are the two-branches of ψ on $K_g\backslash\{P_\infty\}$ (i.e., they are proportional to $\psi(P)$ and $\psi(P^*)$ for $P \in \Pi_+$). In particular, one verifies that $-\int_{P_0}^{P} \omega_0^{(2)}$ corresponds to the usual Floquet (Bloch) momentum.

Remark 2.12. For notational simplicity we have assumed that $\rho_1, \dots, \rho_g$ are nonvoid gaps in the spectrum of H_g. The general case of finitely many open (i.e., nonvoid) gaps in $\sigma(H_g(t))$ can be treated in complete analogy to the case discussed in these notes (though the amount of notational inconvenience increases considerably).

Finally we remark that while we concentrate on the finite-genus case $g < \infty$ throughout this paper, the case $g = \infty$ in connection with the KdV-equation has been dealt with in detail in [Mc-Tr 1,2], [McK 3]. These results can be transferred to the mKdV-equation in analogy to our treatment of the case $g < \infty$ in Section 2 and will be published elsewhere.

3. FINITE-GAP mKdV-SOLUTIONS IN TERMS OF THETA FUNCTIONS

Based on the B–A-function for the Schrödinger operator discussed in Section 2 and on Hirota's τ-function approach in connection with commutation methods studied in Appendix D we now derive the real-valued, quasi-periodic, finite-gap solutions of the mKdV-equation.

In order to avoid the trivial case $g = 0$, we again choose $g \in \mathbb{N}$ throughout this section. Let $H_{2,g}(t)$ be a finite-gap Schrödinger operator in $L^2(\mathbb{R})$ of the type

$$H_{2,g}(t) = -\frac{d^2}{dx^2} + V_{2,g}(t,.) \text{ on } \mathfrak{D}(H_{2,g}(t)) = H^2(\mathbb{R}), \; t \in \mathbb{R}, \tag{3.1}$$

where $V_{2,g}(t,x)$ is one of the quasi-periodic, finite-gap solutions (2.35) of the KdV-equation (see Corollary 2.10)

$$KdV(V_{2,g}) = 0. \tag{3.2}$$

We shall assume that

$$H_{2,g}(0) \geq 0, \text{ i.e., } E_0 \geq 0 \tag{3.3}$$

(and hence $H_{2,g}(t) \geq 0$ for all $t \in \mathbb{R}$) where

$$E_0 = \inf \left[\sigma\left(H_{2,g}(t)\right)\right] \tag{3.4}$$

according to (2.38). Following the strategy in Appendix D we consider the Floquet-type solutions

$$\begin{aligned} \psi_{2,+}(z,x,x_0,t,t_0) :&= \psi_2(P,x,x_0,t,t_0), \\ \psi_{2,-}(z,x,x_0,t,t_0) :&= \psi_2(P^*,x,x_0,t,t_0), \\ & x, x_0, t, t_0 \in \mathbb{R},\ P = (z, R_0(z)^{1/2}) \in \Pi_+ \backslash \overline{\rho_{H_{2,g}}}, \end{aligned} \tag{3.5}$$

where

$$\overline{\rho_{H_{2,g}}} := [-\infty, E_0] \cup \bigcup_{j=1}^{g} [E_{2j-1}, E_{2j}] \tag{3.6}$$

and $\psi_2(P)$ is defined as in (2.18). On $\overline{\rho_{H_{2,g}}}\backslash\{\infty\}$ we define

$$\psi_{2,\pm}(\lambda,x,x_0,t,t_0):=\lim_{\epsilon\downarrow 0}\psi_{2,\pm}(\lambda+i\epsilon,x,x_0,t,t_0),\ \lambda\in\overline{\rho_{H_{2,g}}},\ x,x_0,t,t_0\in\mathbb{R}. \tag{3.7}$$

Thus $\psi_{2,\pm}$ are the two branches of the meromorphic B–A-function ψ_2 on $K_g\backslash\{P_\infty\}$. From (2.18) and (2.23) we infer that

$$0<\psi_{2,\pm}(\lambda,x,x_0,t,t_0)\in L^2\left((R,\pm\infty);\ dx\right),\ \lambda<E_0,\ x,x_0,t,t_0,R\in\mathbb{R} \tag{3.8}$$

and, taking $\lambda\uparrow E_0$, also

$$0<\psi_{2,+}(E_0,x,x_0,t,t_0)=\psi_{2,-}(E_0,x,x_0,t,t_0),\ x,x_0,t,t_0\in\mathbb{R} \tag{3.9}$$

(since $H_{2,g}(t)$ is a 2nd-order differential operator). The functions $\psi_{2,\pm}(\lambda,x,x_0,t,t_0)$, $\lambda\in\mathbb{R}\backslash\{\mu_j(t_0,x_0)\}_{j=1}^g$ are C^∞ with respect to x,x_0,t,t_0 and smooth in $\lambda\in\mathbb{R}\backslash\{\mu_j(t_0,x_0)\}_{j=1}^g$. With the help of $\psi_{2,\pm}$ we also introduce

$$\psi_{2,\sigma}(t,x):=2^{-1}(1-\sigma)\psi_{2,-}(0,x,x_0,t,t_0)+2^{-1}(1+\sigma)\psi_{2,+}(0,x,x_0,t,t_0), \tag{3.10}$$

$$\phi_{g,\sigma}(t,x):=\partial_x\ln\psi_{2,\sigma}(t,x)=\psi_{2,\sigma,x}(t,x)/\psi_{2,\sigma}(t,x);\ \sigma\in[-1,1],\ (t,x)\in\mathbb{R}^2. \tag{3.11}$$

We recall Miura's transformation [Miu]

$$V_{2,g}(t,x)=\phi_{g,\sigma}(t,x)^2+\phi_{g,\sigma,x}(t,x),\ \sigma\in[-1,1],\ (t,x)\in\mathbb{R}^2 \tag{3.12}$$

and

$$\phi_{g,\sigma}\in C^\infty(\mathbb{R}^2),\ \partial_x^n\phi_{g,\sigma}\in L^\infty(\mathbb{R}^2),\ n\in\mathbb{N}_0,\ \sigma\in[-1,1] \tag{3.13}$$

by Lemma D.1. Next we introduce in accordance with the notation in Appendix D (see (2.18) and (D.31))

$$\tau_{1,\underset{(-)}{+}}(t,x):=\theta\left(\underline{\zeta}_{P_0}-\underline{A}_{P_0}(0^{(*)})+\underline{\alpha}_{P_0}\left(\underline{\hat{\mu}}(t_0,x_0)\right)+((x-x_0)/2\pi)\,\underline{U}_0+(6(t-t_0)/\pi)\,\underline{U}_2\right), \tag{3.14}$$

$$\tau_2(t,x):=\theta\left(\underline{\zeta}_{P_0}-\underline{A}_{P_0}(P_\infty)+\underline{\alpha}_{P_0}\left(\underline{\hat{\mu}}(t_0,x_0)\right)+((x-x_0)/2\pi)\,\underline{U}_0+(6(t-t_0)/\pi)\,\underline{U}_2\right);$$
$$P_0=(E_0,0),\ (t,x)\in\mathbb{R}^2, \tag{3.15}$$

where $0^{(*)}\in K_g$ is defined according to (3.7), i.e.,

$$0=(0,\lim_{\epsilon\downarrow 0}R_0(i\epsilon)^{1/2}),\ 0^*=(0,-\lim_{\epsilon\downarrow 0}R_0(i\epsilon)^{1/2}). \tag{3.16}$$

Then

$$V_{2,g}(t,x) = E_0 + \sum_{j=1}^{g}(E_{2j-1} + E_{2j} - 2\lambda_j) - 2\partial_x^2 \ln \tau_2(t,x), \ (t,x) \in \mathbb{R}^2 \tag{3.17}$$

by (2.35). Next we state

Lemma 3.1. ([Ma-Mo], Theorems 8,9)
$\phi_{g,\sigma}$ in (3.11) is almost periodic (and hence quasi-periodic) in x iff $\sigma = \pm 1$.

Together with Theorem D.2 this yields our main result on real-valued, quasi-periodic, finite-gap mKdV-solutions.

Theorem 3.2. Let $V_{2,g}$ in (3.16) be a real-valued, quasi-periodic, finite-gap solution of the KdV-equation. Define $\phi_{g,\pm 1}$ as in (3.11). Then

$$\begin{gathered}\phi_{g,\underset{(-)}{+}1}(t,x) = \underset{(+)}{-}\, i \int_{P_0}^{0} \omega_0^{(2)} \\ + \partial_x \ln \left[\frac{\theta\left(\underline{\zeta}_{P_0} - \underline{A}_{P_0}(0^{(*)}) + \underline{\alpha}_{P_0}\left(\underline{\hat{\mu}}(t_0,x_0)\right) + ((x-x_0)/2\pi)\underline{U}_0 + (6(t-t_0)/\pi)\underline{U}_2\right)}{\theta\left(\underline{\zeta}_{P_0} - \underline{A}_{P_0}(P_\infty) + \underline{\alpha}_{P_0}\left(\underline{\hat{\mu}}(t_0,x_0)\right) + ((x-x_0)/2\pi)\underline{U}_0 + (6(t-t_0)/\pi)\underline{U}_2\right)} \right], \\ 0 = \left(0, \lim_{\epsilon \downarrow 0} R_0(i\epsilon)^{1/2}\right), \ P_0 = (E_0, 0), \ (t,x) \in \mathbb{R}^2\end{gathered} \tag{3.18}$$

is a real-valued, quasi-periodic (in t and x), finite-gap solution of the mKdV-equation,

$$mKdV(\phi_{g,\pm 1}) = \phi_{g,\pm 1,t} - 6\phi_{g,\pm 1}^2 \phi_{g,\pm 1,x} + \phi_{g,\pm 1,xxx} = 0. \tag{3.19}$$

The spectrum of the corresponding finite-gap Dirac operator in $\left[L^2(\mathbb{R})\right]^2$

$$\begin{gathered}Q_{g,\pm 1}(t) := \begin{pmatrix} 0 & -\dfrac{d}{dx} + \phi_{g,\pm 1}(t,.) \\ \dfrac{d}{dx} + \phi_{g,\pm 1}(t,.) & 0 \end{pmatrix} \\ \text{on } \mathfrak{D}\left(Q_{g,\pm 1}(t)\right) = \left[H^1(\mathbb{R})\right]^2, \ t \in \mathbb{R}\end{gathered} \tag{3.20}$$

is purely absolutely continuous and of multiplicity two

$$\begin{gathered}\sigma\left(Q_{g,\pm 1}(t)\right) = \bigcup_{j=-g}^{g} \sum\nolimits_j \cup (-\infty, -E_{2g}^{1/2}] \cup [E_{2g}^{1/2}, \infty), \ t \in \mathbb{R}, \\ \sum\nolimits_j := \left[E_{2(j-1)}^{1/2}, E_{2j-1}^{1/2}\right], \ \sum\nolimits_{-j} = -\sum\nolimits_j, \ 1 \le j \le g.\end{gathered} \tag{3.21}$$

Moreover, (3.18) describes all spatially almost periodic solutions of the mKdV-equation that give rise to the finite-gap spectrum (3.21). Finally, defining

$$V_{1,g,\pm 1}(t,x) := \phi_{g,\pm 1}(t,x)^2 - \phi_{g,\pm 1,x}(t,x),\ (t,x) \in \mathbb{R}^2 \tag{3.22}$$

one obtains

$$V_{1,g,\pm 1}(t,x) = E_0 + \sum_{j=1}^{g} (E_{2j-1} + E_{2j} - 2\lambda_j) - 2\partial_x^2 \ln \tau_{1,\pm 1}(t,x),\ (t,x) \in \mathbb{R}^2. \tag{3.23}$$

In particular, $V_{1,g,\pm 1}$ satisfies the KdV-equation

$$\mathrm{KdV}\ (\mathrm{V}_{1,g,\pm 1}) = 0 \tag{3.24}$$

and is isospectral to $V_{2,g}$, i.e.,

$$\sigma\left(H_{1,g,\pm 1}(t)\right) = \sigma\left(H_{2,g}(t)\right) = \bigcup_{j=1}^{g} \left[E_{2(j-1)}, E_{2j-1}\right] \cup [E_{2g}, \infty),\ t \in \mathbb{R} \tag{3.25}$$

since

$$H_{1,g,\pm 1}(t) := -\frac{d^2}{dx^2} + V_{1,g,\pm 1}(t,.) \text{ on } \mathfrak{D}\left(H_{1,g,\pm 1}(t)\right) = H^2(\mathbb{R}),\ t \in \mathbb{R} \tag{3.26}$$

in $L^2(\mathbb{R})$ is unitarily equivalent to $H_{2,g}(t)$. Finally, $V_{2,g}$, $V_{1,g,\pm 1}$, $\phi_{g,\pm 1}$ are periodic in x with fundamental period $a > 0$ iff

$$a = \min\left\{\omega > 0 | 2\omega |c_{j,g}| \in \mathbb{N},\ 1 \le j \le g\right\} = \min\left\{\omega > 0 | (\omega |U_{0,j}|/2\pi) \in \mathbb{N},\ 1 \le j \le g\right\}. \tag{3.27}$$

Proof. One only needs to combine Theorems 2.8, 2.9, D.1, Corollary 2.10 and the facts

$$Q_{g,\pm 1}(t)^2 = H_{1,g,\pm 1}(t) \oplus H_{2,g}(t),\ t \in \mathbb{R}, \tag{3.28}$$

$$\sigma_3 Q_{g,\pm 1}(t) \sigma_3 = -Q_{g,\pm 1}(t),\ t \in \mathbb{R}, \sigma_3 = \begin{pmatrix} 1 & 0 \\ 0 & -1 \end{pmatrix}, \tag{3.29}$$

$$\begin{gathered} \frac{d}{dt} Q_{g,\pm 1}(t) - [B_{g,\pm 1}(t),\ Q_{g,\pm 1}(t)] = \mathrm{mKdV}(\phi_{g,\pm 1}) \begin{pmatrix} 0 & 1 \\ 1 & 0 \end{pmatrix}, \\ B_{g,\pm 1}(t) := B_{V_{1,g,\pm 1}}(t) \oplus B_{V_{2,g}}(t),\ t \in \mathbb{R}, \end{gathered} \tag{3.30}$$

$$\begin{gathered} H_{1,g,\pm 1}(t) = A_{\pm 1}(t)^* A_{\pm 1}(t),\ H_{2,g}(t) = A_{\pm 1}(t) A_{\pm 1}(t)^*, \\ A_{\pm 1}(t) := \frac{d}{dx} + \phi_{g,\pm 1}(t,.) \text{ on } \mathfrak{D}(A_{\pm 1}(t)) = H^1(\mathbb{R}),\ t \in \mathbb{R} \end{gathered} \tag{3.31}$$

(see [Dei], [Ges 2,3], [G-S-S]). ■

Periodic (not necessarily finite-gap) solutions of the mKdV-equation, related to periodic KdV - solutions by means of Miura's transformation (3.12), (3.22) have been studied in [Ges 2,3], [G-S-S]. The special case $g = 1$ (the elliptic one) in Theorem 3.2 is also discussed in detail in [Ges 2,3], [G-S-S].

Similarly to (2.36) in Theorem 2.8 one can also express $\phi_{g,\pm 1}$ in terms of the Dirichlet eigenvalues of $H_{2,g}(t)$.

Theorem 3.3. Let $\{\mu_{2,n}(t,x)\}_{n\in\mathbb{N}}$ be the Dirichlet eigenvalues associated with $V_{2,g}$ such that

$$V_{2,g}(t,x) = E_0 + \sum_{j=1}^{g} [E_{2j-1} + E_{2j} - 2\mu_{2,j}(t,x)], \ (t,x) \in \mathbb{R}^2 \tag{3.32}$$

in accordance with (2.36). Then

$$\phi_{g,\pm 1}(t,x) = \left\{ \pm \lim_{\epsilon \downarrow 0} R_0(i\epsilon)^{1/2} + 2^{-1}\partial_x \left[\prod_{j=1}^{g} \mu_{2,j}(t,x) \right] \right\} \prod_{\ell=1}^{g} \mu_{2,\ell}(t,x)^{-1}, \ (t,x) \in \mathbb{R}^2. \tag{3.33}$$

Proof. Given Lemma 2.6, the proof of [Ges 3] given in the periodic case (based on results of [Bu-Fi]) extends to the quasi-periodic situation considered here. ■

Remark 3.4. Iff $E_0 = 0$, i.e., iff $H_{2,g}(t)$, $t \in \mathbb{R}$ is critical in the terminology of [Sim 1,2] (see also [Ge-Zh], [Mur]) then $H_{2,g}\psi = 0$ has a unique (up to multiple of constants) positive solution $\psi_{2,0} := \psi_{2,\pm 1} > 0$ and hence $\phi_{g,0} := \phi_{g,\pm 1}$ is the unique quasi-periodic solution of (3.12) given $V_{2,g}$. Moreover, $V_{1,g,+1} = V_{1,g,-1}$ in this case, $\lim_{\epsilon\downarrow 0} R_0(i\epsilon)^{1/2} = 0$ in (3.33) and we may take $\underline{A}_{P_0}(0^{(*)}) = 0$ in (3.14) (implying $\tau_{1,+1} = \tau_{1,-1}$) simultaneously with $\int_{P_0}^{0} \omega_0^{(2)} = 0$ in (3.18). Iff $E_0 > 0$, i.e., iff $H_{2,g}(t)$, $t \in \mathbb{R}$ is subcritical, then $V_{1,g,+1} \neq V_{1,g,-1}$ and $\phi_{g,+1} \neq \phi_{g,-1}$ are different solutions of the (m)KdV-equations respectively.

Remark 3.5. While (2.35), (2.36) and (3.18), (3.33) describe all real-valued almost periodic (a.p.) (and hence quasi-periodic) finite-gap KdV and mKdV-solutions, Theorem D.2 in the subcritical case $E_0 > 0$ also yields a one-parameter family of real-valued, finite-gap KdV and mKdV-solutions $V_{1,g,\sigma}$ and $\phi_{g,\sigma}$, $\sigma \in (-1,1)$ that are not almost periodic: Define in $L^2(\mathbb{R})$

$$\begin{aligned} H_{1,g,\sigma}(t) &:= A_\sigma(t)^* A_\sigma(t) = -\frac{d^2}{dx^2} + V_{1,g,\sigma}(t,.) \\ &\text{on } \mathfrak{D}(H_{1,g,\sigma}(t)) = H^2(\mathbb{R}), \ \sigma \in (-1,1), \ t \in \mathbb{R}, \end{aligned} \tag{3.34}$$

$$V_{1,g,\sigma}(t,x) = \phi_{g,\sigma}(t,x)^2 - \phi_{g,\sigma,x}(t,x), \ \sigma \in (-1,1), \ (t,x) \in \mathbb{R}^2. \tag{3.35}$$

We also note

$$H_{2,g}(t) = A_\sigma(t)A_\sigma(t)^*, \ A_\sigma(t) := \frac{d}{dx} + \phi_{g,\sigma}(t,.), \ \mathfrak{D}(A_\sigma(t)) = H^1(\mathbb{R}), \ \sigma \in [-1,1], \ t \in \mathbb{R}. \tag{3.36}$$

Then $\phi_{g,\sigma}$, $\sigma \in (-1,1)$ is not a.p. by Lemma 3.1 and $V_{1,g,\sigma}$, not being quasi-periodic for $\sigma \in (-1,1)$ by inspection, is not a.p. too. In fact, $H_{1,g,\sigma}(t)$ and $Q_{g,\sigma}(t)$ have a simple eigenvalue at zero for $\sigma \in (-1,1)$,

$$\sigma(H_{1,g,\sigma}(t) = \{0\} \cup \sigma(H_{2,g}(t)), \ \sigma(Q_{g,\sigma}(t)) = \{0\} \cup \sigma(Q_{g,\pm 1}(t)), \ \sigma \in (-1,1), \ t \in \mathbb{R} \tag{3.37}$$

and the essential spectra of $H_{1,g,\sigma}(t)$, $Q_{g,\sigma}(t)$ are purely absolutely continuous and of multiplicity two. (In general, $A^*A\Big|_{\mathrm{Ker}(A)^\perp}$ and $AA^*\Big|_{\mathrm{Ker}(A^*)^\perp}$ are unitarily equivalent for any densely defined, closed linear operator A in a complex, separable Hilbert space [Dei].) $V_{1,g,\sigma}$ (resp. $\phi_{g,\sigma}$), $\sigma \in (-1,1)$ describe the simplest one-soliton (m)KdV-solutions on finite-gap, quasi-periodic backgrounds $V_{0,2,g}$ (resp. $\phi_{0,g}$) associated with the spectrum (2.38) (resp. (3.21)). ($V_{0,2,g}$ and $V_{1,g,\sigma}$ (and similarly $\phi_{0,g}$ and $\phi_{g,\sigma}$) are related by double commutation.) The general N-soliton (m)KdV-solutions on finite-gap, quasi-periodic background solutions will appear elsewhere. This illustrates the complexity of the general, finite-gap inverse problem for Schrödinger and Dirac operators on $\mathbb{R}$. The situation gets worse if one drops the assumption of uniform spectral multiplicity two of the essential spectra involved since then two "half-crystals" appropriately patched together (i.e.,

$$V(x) = \begin{cases} V_1(x), \ x \geq 0 \\ V_2(x), \ x \leq 0, \quad V_1(0) = V_2(0), \end{cases} \tag{3.38}$$

$V_j(x)$ quasi-periodic, finite-gap potentials, $j = 1,2$) in general yield finite-gap spectra with intervals of multiplicity one and two [Da-Si], [Ges 1], [Rod 1].

We also mention that the results of Sections 2 and 3 extend to complex-valued, finite-gap solutions of the (m)KdV-equation by considering nonreal branch points $(E_n, 0)$, $0 \leq n \leq 2g$. This has been discussed in great detail for the KdV-case in [Bir 1,2].

Darboux transformations for a.p. one-dimensional (t-independent) Schrödinger operators related to $V_{2,g}(0,x)$, $V_{1,g,\pm 1}(0,x)$ appeared in [Trl]. The algebro-geometric content of the Bäcklund transformation relating $V_{2,g}(t,x)$, $V_{1,g,\pm 1}(t,x)$ is studied e.g. in [Eh-Kn], [Er-Fl], [Er-Mc], [McK 5,6], [Wil].

Obviously all results of Sections 2 and 3 extend to the entire (m)KdV-hierarchy in a straightforward manner.

APPENDIX A: ABELIAN DIFFERENTIALS ON HYPERELLIPTIC RIEMANN SURFACES

The material in this appendix is classical and can be found e.g. in [Be-So], [Fa-Kr], [Gri], [Gr-Ha], [Hu-Co], [Mum], [Rey], [Rod 2], [Sie], [Spr].

Let $R = ((\mathcal{M}, \tau), \mathcal{A} = (U_\alpha, z_\alpha)_{\alpha \in J})$ be an abstract Riemann surface, where $(\mathcal{M}, \tau)$ is a connected, second countable, compact Hausdorff space, $\mathcal{A}$ an atlas for $(\mathcal{M}, \tau)$, i.e., $\mathcal{M} = \bigcup_{\alpha \in J} U_\alpha$, $U_\alpha \in \tau$, $z_\alpha : U_\alpha \to \mathbb{C}$ homeomorphisms, $z_\alpha(U_\alpha)$ open in $\mathbb{C}$, $z_\alpha \circ z_\beta^{-1} :\ z_\beta(U_\alpha \cap U_\beta) \to z_\alpha(U_\alpha \cap U_\beta)$ biholomorphic whenever $U_\alpha \cap U_\beta \neq \emptyset, \alpha, \beta \in J$ (an index set). Moreover, let $F :\ R \to \mathbb{C}_\infty$ be holomorphic (i.e., F meromorphic on R) and nonconstant with multiplicity two and a set of branch points $\mathcal{B}(F)$ of F be given by

$$\mathcal{B}(F) = \{e_\infty, e_0, e_1, \dots, e_{2g}\},$$
$$e_n \neq e_m \text{ for } n \neq m,\ e_n \neq e_\infty,\ 0 \leq m, n \leq 2g,\ g \in \mathbb{N}_0. \quad \text{(A.1)}$$

(Here $\mathbb{C}_\infty = \mathbb{C} \cup \{\infty\}$ denotes the one-point compactification of $\mathbb{C}$ usually identified with the Riemann sphere.) We also assume

$$F :\ R_g \backslash \{e_\infty\} \to \mathbb{C},\ F(e_\infty) = \infty. \quad \text{(A.2)}$$

Since by hypothesis F has multiplicity two one gets

$$F_{\alpha,j} := (w_j \circ F \circ z_\alpha^{-1})(\zeta_\alpha) = \begin{cases} \zeta_\alpha, & p \in R \backslash \mathcal{B}(F) \\ \zeta_\alpha^2, & p \in \mathcal{B}(F) \end{cases}, \quad \zeta_\alpha \in z_\alpha(U_\alpha),\ \alpha \in J,\ j = 1, 2 \quad \text{(A.3)}$$

for appropriate charts (U_α, z_α) near $p \in R$ with $z_\alpha(p) = 0$ and w_1, w_2 the usual charts near $F(p) \in \mathbb{C}_\infty$ with $w_j(F(p)) = 0,\ j = 1, 2$,

$$w_1 : \begin{cases} \mathbb{C} \to \mathbb{C} \\ w \to w - F(p), \quad F(p) \neq \infty, \end{cases}$$
$$w_2 : \begin{cases} (\mathbb{C} \backslash \{0\}) \cup \{\infty\} \to \mathbb{C} \\ w \longrightarrow w^{-1}, \quad F(p) = \infty. \end{cases} \quad \text{(A.4)}$$

Since the covering map F is a local homeomorphism near any point $p_0 \in R \backslash \mathcal{B}(F)$, and twice covers a neighborhood of $F(p_0)$ if $p_0 \in \mathcal{B}(F)$, the mappings

$$z_\alpha(p) := [F(p) - F(p_0)]^{1/m_0(p_0)},\ p_0 \in R \backslash \{e_\infty\},\ p \in U_\alpha,$$
$$z_\alpha(p) := [F(p)]^{-1/m_0(p_0)},\ p_0 = e_\infty,\ p \in U_\alpha, \quad \text{(A.5)}$$
$$m_0(p_0) = \begin{cases} 1, & p_0 \in R \backslash \mathcal{B}(F) \\ 2, & p_0 \in \mathcal{B}(F) \end{cases}$$

are local homeomorphisms from U_α onto $z_\alpha(U_\alpha)$ (assuming the single-valued branches of (A.5) are chosen consistently in the usual way if $m_0(p_0) = 2$). We denote by $S_g = ((\mathcal{M}, \tau), \mathcal{A}_m(F))$ the covering Riemann surface with maximal atlas $\mathcal{A}_m(F)$ given by the coordinate maps (A.5) and U_α, $\alpha \in J$. S_g is hyperelliptic for $g \geq 2$. The elliptic Riemann surface S_1 is conformally equivalent to a torus $\mathbb{C}/L_1$ (L_1 a two-dimensional lattice, i.e., a discrete subgroup of $\mathbb{C}$ isomorphic to $\mathbb{Z}^2$), S_0 is conformally equivalent to $\mathbb{C}_\infty$.

Introducing

$$E_n := F(e_n),\ 0 \leq n \leq 2g \tag{A.6}$$

(recalling $\infty = F(e_\infty)$) and

$$\mathcal{R}_0(p)^{1/2} := \left\{ \prod_{n=0}^{2g} [E_n - F(p)] \right\}^{1/2},\ p \in S_g \tag{A.7}$$

one checks that $\mathcal{R}_0^{1/2} : S_g \to \mathbb{C}_\infty$ is meromorphic on S_g with the only pole at e_∞ of order $2g+1$. For notational simplicity we assume in the following that $E_n \in \mathbb{R}$ and that they are ordered as

$$E_0 < E_1 < \ldots < E_{2g}$$

since this special case is the one used in Sections 2 and 3.

Topologically, S_g is realized as follows. Consider $\mathbb{C} \cup \{\infty\}$ and introduce the cuts $\overline{\rho_n}$, $0 \leq n \leq g$,

$$\rho_0 := (-\infty, E_0),\ \rho_n := (E_{2n-1}, E_{2n}),\ 1 \leq n \leq g. \tag{A.8}$$

Take two copies $\Pi_\pm$ of the cut plane $\Pi_0 := \{\mathbb{C} \cup \{\infty\}\} \setminus \bigcup_{n=0}^{g} \rho_n$, and join the upper and lower rims of the cuts $\overline{\rho_n}$ of $\Pi_\pm$ crosswise. The resulting surface is then homeomorphic to a sphere with g handles.

Using the simplified notation

$$z := F(p),\ p \in S_g \text{ and } R_0(z)^{1/2} := \left[\prod_{n=0}^{2g} (E_n - z) \right]^{1/2} \tag{A.9}$$

one is led to yet another description of S_g in terms of the hyperelliptic curve K_g consisting of points

$$P := \left(z, R_0(z)^{1/2}\right),\ z \in \mathbb{C},\ P_\infty := (\infty, \infty) \tag{A.10}$$

with branch points

$$(E_n, 0),\ 0 \leq n \leq 2g,\ P_\infty = (\infty, \infty). \tag{A.11}$$

Here $z = F(P)$ is the projection Π of P onto $\mathbb{C} \cup \{\infty\}$

$$\Pi : \begin{cases} K_g \longrightarrow \mathbb{C} \cup \{\infty\} \\ P = (z, R_0(z)^{1/2}) \to z \end{cases} \tag{A.12}$$

The involution $*$ (the sheet exchange map) is defined by

$$* : \begin{cases} K_g \longrightarrow K_g \\ P = (z, R_0(z)^{1/2}) \to P^* = (z, -R_0(z)^{1/2}). \end{cases} \tag{A.13}$$

The upper sheet Π_+ is declared as follows: We define $\lim_{\epsilon \downarrow 0} R_0(\lambda + i\epsilon)^{1/2} = -\left|R_0(\lambda)\right|^{1/2}$ for $\lambda < E_0$ on Π_+ and then continue analytically with respect to λ. The local parameters at $P_0 = \left(z_0, R_0(z_0)^{1/2}\right) \in K_g$ according to (A.5) are given by

$$\zeta_\alpha := \begin{cases} z - z_0, & z_0 \in \mathbb{C} \backslash \left\{\{E_n\}_{n=0}^{2g} \cup \{\infty\}\right\} \\ (z - E_n)^{1/2}, & z_0 = E_n, \ 0 \le n \le 2g \\ z^{-1/2}, & z_0 = \infty. \end{cases} \tag{A.14}$$

A homology basis $\{a_j, b_j\}_{j=1}^{g}$ on $K_g, g \ge 1$ is then chosen as follows: The cycle a_j surrounds the cut $\overline{\rho_j} = [E_{2j-1}, \ E_{2j}], \ 1 \le j \le g$ clockwise on Π_+ while b_j starts at the lower rim of ρ_j on Π_+, intersects a_j, then encircles E_0 clockwise thereby changing to the lower sheet Π_- and returns on Π_- to its initial point on ρ_j. The cycles are chosen in such a way that their intersection matrix reads

$$a_j \circ b_\ell = \delta_{j,\ell}, \ 1 \le j, \ell \le g \tag{A.15}$$

(with a_j, b_j intersecting like a right handed coordinate system).

Next we briefly need to consider meromorphic differentials (1-forms) on K_g.

Theorem A.1. (Riemann's period relations)
Let ω, ν be closed C^1 1-forms on $K_g, g \ge 1$. Then
(i).

$$\iint_{K_g} \omega \wedge \nu = \sum_{j=1}^{g} \Big[\Big(\int_{a_j} \omega\Big)\Big(\int_{b_j} \nu\Big) - \Big(\int_{b_j} \omega\Big)\Big(\int_{a_j} \nu\Big) \Big]. \tag{A.16}$$

If in addition ω, ν are holomorphic 1-forms on K_g then

$$\sum_{j=1}^{g} \Big[\Big(\int_{a_j} \omega\Big)\Big(\int_{b_j} \nu\Big) - \Big(\int_{b_j} \omega\Big)\Big(\int_{a_j} \nu\Big) \Big] = 0. \tag{A.17}$$

(ii). If ω is a nonzero holomorphic 1-form on K_g then

$$\operatorname{Im}\Big[\sum_{j=1}^{g}\Big(\int_{a_j}\omega\Big)\Big(\int_{b_j}\omega\Big)\Big] > 0. \tag{A.18}$$

The proof of Theorem A.1 is standard and involves Stokes theorem and a canonical dissection of K_g along its cycles yielding the simply connected interior $\hat{K}_g$ of the fundamental polygon $\partial\hat{K}_g$ given by

$$\partial\hat{K}_g = a_1 b_1 a_1^{-1} b_1^{-1} a_2 b_2 a_2^{-1} b_2^{-1} \dots a_g^{-1} b_g^{-1}. \tag{A.19}$$

A basis for the space of holomorphic differentials (also called Abelian differentials of the first kind, DFK) on K_g is given by

$$\eta_j := R_0(z)^{-1/2} z^{j-1} dz,\ 1 \le j \le g. \tag{A.20}$$

In terms of the local coordinates (A.14) one computes

$$\eta_j = \begin{cases} \dfrac{(z_0+\zeta_\alpha)^{j-1} d\zeta_\alpha}{\Big[\prod\limits_{n=0}^{2g}(E_n - z_0)\Big]^{1/2}\Big[1 + 2^{-1}\zeta_\alpha \sum\limits_{n=0}^{2g}(z_0 - E_n)^{-1} + 0(\zeta_\alpha^2)\Big]}, \\ \qquad\qquad z_0 \in \mathbb{C}\backslash\Big\{\{E_m\}_{m=0}^{2g} \cup \{\infty\}\Big\} \\ \dfrac{2(E_m + \zeta_\alpha^2)^{j-1} d\zeta_\alpha}{\Big[-\prod\limits_{\substack{n=0\\ n\neq m}}^{2g}(E_n - E_m)\Big]^{1/2}\Big[1 - 2^{-1}\zeta_\alpha^2 \sum\limits_{\substack{n=0\\ n\neq m}}^{2g}(E_n - E_m)^{-1} + 0(\zeta_\alpha^4)\Big]}, \\ \qquad\qquad z_0 = E_m,\ 0 \le m \le 2g \\ \dfrac{2i(-1)^g \zeta_\alpha^{2(g-j)} d\zeta_\alpha}{\Big[1 - 2^{-1}\zeta_\alpha^2 \sum\limits_{n=0}^{2g} E_n + 0(\zeta_\alpha^4)\Big]},\ z_0 = \infty, \\ \qquad\qquad 1 \le j \le g. \end{cases} \tag{A.21}$$

We shall employ the usual normalization

$$\omega_j := \sum_{\ell=1}^{g} c_{j,\ell}\eta_\ell,\ \int_{a_j}\omega_\ell = \delta_{j,\ell},\ 1 \le j,\ell \le g. \tag{A.22}$$

Define the b-periods of ω_ℓ by

$$\tau_{j,\ell} := \int_{b_j}\omega_\ell,\ 1 \le j,\ell \le g. \tag{A.23}$$

Then Theorem A.1 implies

Theorem A.2. τ is symmetric, i.e.,

$$\tau_{j,\ell} = \tau_{\ell,j}, \ 1 \leq j, \ell \leq g \tag{A.24}$$

and purely imaginary with positive definite imaginary part

$$\tau = iT, \ T > 0. \tag{A.25}$$

(We note that (A.25) is a consequence of our simplifying assumption $E_n \in \mathbb{R}$, $0 \leq n \leq 2g$. In general, if $E_n \in \mathbb{C}$ then τ has a positive definite imaginary part.)

Abelian differentials of the second kind (DSK) $\omega^{(2)}$ are characterized by the property that all their residues vanish. They are normalized by the vanishing of all their a-periods

$$\int_{a_j} \omega^{(2)} = 0, \ 1 \leq j \leq g \tag{A.26}$$

which determines them uniquely. (We will always assume that the poles of DSK's on K_g lie in $\hat{K}_g$, i.e., do not lie on $\partial \hat{K}_g$. This can always be achieved by an appropriate choice of the cycles a_j, b_j.) We may add in this context that the sum of the residues of any meromorphic differential ν on K_g vanishes, the residue at a pole $P_0 \in K_g$ of ν being defined by

$$\operatorname*{Res}_{P_0}(\nu) = (2\pi i)^{-1} \int_{\gamma_{P_0}} \nu, \tag{A.27}$$

where γ_{P_0} is a counter clockwise oriented, smooth, simple, closed contour encircling P_0 but no other pole of ν.

Theorem A.3. Let $\omega_{P_1}^{(2)}$ be a DSK on K_g, $g \in \mathbb{N}$ whose only pole is $P_1 \in \hat{K}_g$ with principle part $\zeta_\alpha^{-n} d\zeta_\alpha$, $n \geq 2$ and $\omega^{(1)}$ a DFK on K_g of the type $\omega^{(1)} = \sum_{m=0}^{\infty} c_m \zeta_\alpha^m d\zeta_\alpha$ near P_1. Then

$$\sum_{j=1}^{g} \Big[\Big(\int_{a_j} \omega^{(1)} \Big) \Big(\int_{b_j} \omega_{P_1}^{(2)} \Big) - \Big(\int_{a_j} \omega_{P_1}^{(2)} \Big) \Big(\int_{b_j} \omega^{(1)} \Big) \Big] = 2\pi i (n-1)^{-1} c_{n-2}, \ n \geq 2. \tag{A.28}$$

In particular, taking $\omega^{(1)} = \omega_j = \Big(\sum_{m=0}^{\infty} c_{j,m} \zeta_\alpha^m \Big) d\zeta_\alpha = f_{\alpha,j}(\zeta_\alpha) d\zeta_\alpha$, (A.28) implies

$$\int_{b_j} \omega_{P_1}^{(2)} = 2\pi i (n-1)^{-1} c_{j,n-2} = \frac{2\pi i}{(n-1)!} \frac{d^{n-2} f_{\alpha,j}}{d\zeta_\alpha^{n-2}}(0), \ n \geq 2, \ 1 \leq j \leq g. \tag{A.29}$$

A basis for DSK's at $P_\infty = (\infty, \infty)$ on K_g, $g \in \mathbb{N}_0$ is provided by

$$\omega_{2k}^{(2)} := R_0(z)^{-1/2} z^{g+k} dz = \Big[\prod_{n=0}^{2g} (1 - E_n \zeta_\alpha^2))\Big]^{-1/2} 2i(-1)^g \zeta_\alpha^{-2-2k} d\zeta_\alpha, \ k \in \mathbb{N}_0. \quad (A.30)$$

APPENDIX B: DIVISORS, JACOBI VARIETY

For basic references to the material of this appendix see e.g. [Be-So], [Fa-Kr], [For], [Gri], [Gr-Ha], [Gun], [Hu-Co], [Mum], [Rey], [Rod 2], [Sie], [Spr].

Let $\mathcal{M}(K_g)$ and $\mathcal{M}^1(K_g)$ denote the set of meromorphic functions (0-forms) and meromorphic 1-forms on K_g, $g \in \mathbb{N}_0$.

Definition B.1 Let $f \in \mathcal{M}(K_g)$, $\omega = g_\alpha(\zeta_\alpha) d\zeta_\alpha \in \mathcal{M}^1(K_g)$, $g \in \mathbb{N}_0$ and (U_α, z_α) a chart near $P_0 \in K_g$.

(i). If $(f \circ z_\alpha^{-1})(\zeta_\alpha) = \sum_{n=m_0}^{\infty} a_n \zeta_\alpha^n$ for some $m_0 \in \mathbb{Z}$ (which turns out to be independent of the chosen chart), the order $\nu_f(P_0)$ of f at P_0 is defined by

$$\nu_f(P_0) := m_0. \quad (B.1)$$

One defines $\nu_f(P) = \infty$ for all $P \in K_g$ iff $f \equiv 0$ on K_g.

(ii). If $g_\alpha(\zeta_\alpha) = \sum_{n=m_0}^{\infty} b_n \zeta_\alpha^n$ for some $m_0 \in \mathbb{Z}$ (which again is independent of the chart chosen), the order $\nu_\omega(P_0)$ of ω at P_0 is defined by

$$\nu_\omega(P_0) := m_0. \quad (B.2)$$

Definition B.2. Let $g \in \mathbb{N}_0$.

(i). A divisor $\mathfrak{D}$ on K_g is a map $\mathfrak{D} : K_g \to \mathbb{Z}$, where $\mathfrak{D}(P) \neq 0$ for only finitely many $P \in K_g$. On the set of all divisors $\mathrm{Div}(K_g)$ on K_g one introduces the partial ordering

$$\mathfrak{D} \geq \mathcal{E} \text{ iff } \mathfrak{D}(P) \geq \mathcal{E}(P), \ P \in K_g. \quad (B.3)$$

(ii). The degree $\deg(\mathfrak{D})$ of $\mathfrak{D} \in \mathrm{Div}(K_g)$ is defined by

$$\deg(\mathfrak{D}) := \sum_{P \in K_g} \mathfrak{D}(P). \quad (B.4)$$

(iii). $\mathfrak{D} \in \mathrm{Div}(K_g)$ is called positive (or effective) iff

$$\mathfrak{D} \geq 0, \quad (B.5)$$

where $\mathcal{O}$ denotes the zero divisor

$$\mathcal{O}(P) = 0, \quad P \in K_g. \tag{B.6}$$

(iv). $\mathfrak{D}, \mathcal{E} \in \operatorname{Div}(K_g), \mathfrak{D}$ is called a multiple of $\mathcal{E}$ iff

$$\mathfrak{D} \geq \mathcal{E} \tag{B.7}$$

(i.e., iff $\mathfrak{D} - \mathcal{E} \geq \mathcal{O}$ is positive). $\mathfrak{D}$ and $\mathcal{E}$ are called relatively prime iff

$$\mathfrak{D}(P)\mathcal{E}(P) = 0, \quad P \in K_g. \tag{B.8}$$

(v). If $f \in \mathcal{M}(K_g)\backslash\{0\}$, $\omega \in \mathcal{M}^1(K_g)\backslash\{0\}$ then the divisor (f) of f is defined by

$$(f) : \begin{cases} K_g \to \mathbb{Z} \\ P \to \nu_f(P) \end{cases} \tag{B.9}$$

(thus f is holomorphic iff $(f) \geq 0$) and the divisor of ω is defined by

$$(\omega) : \begin{cases} K_g \to \mathbb{Z} \\ P \to \nu_\omega(f) \end{cases} \tag{B.10}$$

(thus ω is a DFK iff $(\omega) \geq 0$). (f) is called a principal divisor, (ω) a canonical divisor.

(vi). $\mathfrak{D}, \mathcal{E} \in \operatorname{Div}(K_g)$ are called equivalent $\mathfrak{D} \sim \mathcal{E}$ iff

$$\mathfrak{D} - \mathcal{E} = (f) \text{ for some } f \in \mathcal{M}(K_g)\backslash\{0\}. \tag{B.11}$$

The divisor class $[\mathfrak{D}]$ of $\mathfrak{D}$ is defined by

$$[\mathfrak{D}] := \{\mathcal{E} \in \operatorname{Div}(K_g) |\ \mathcal{E} \sim \mathfrak{D}\}. \tag{B.12}$$

Clearly $\operatorname{Div}(K_g)$ forms an Abelian group with respect to addition of divisors. The principal divisors form a subgroup $\operatorname{Div}_p(K_g)$ of $\operatorname{Div}(K_g)$. The quotient group $\operatorname{Div}(K_g)/\operatorname{Div}_p(K_g)$ consists of cosets of divisors, the divisor classes defined in (B.12). (The equivalence $\sim$ of divisors is an equivalence relation.) Also the set of divisors of degree zero $\operatorname{Div}_0(K_g)$ forms a subgroup of $\operatorname{Div}(K_g)$. Since $\operatorname{Div}_p(K_g) \subset \operatorname{Div}_0(K_g)$, one can introduce the quotient group $\operatorname{Pic}(K_g) := \operatorname{Div}_0(K_g)/\operatorname{Div}_p(K_g)$ called the Picard group of K_g.

Theorem B.3. Let $f \in \mathcal{M}(K_g)\backslash\{0\}$, $\omega \in \mathcal{M}^1(K_g)\backslash\{0\}$, $g \in \mathbb{N}_0$. Then

$$\deg((f)) = 0, \tag{B.13}$$

$$\deg((\omega)) = 2(g-1). \tag{B.14}$$

Definition B.4. Let $g \in \mathbb{N}_0$ and define

$$\mathcal{L}(\mathfrak{D}) := \{f \in \mathcal{M}(K_g) | \, (f) \geq \mathfrak{D}\}, \tag{B.15}$$

$$\mathcal{L}^1(\mathfrak{D}) := \{\omega \in \mathcal{M}^1(K_g) | \, (\omega) \geq \mathfrak{D}\}. \tag{B.16}$$

Both, $\mathcal{L}(\mathfrak{D})$ and $\mathcal{L}^1(\mathfrak{D})$ are linear spaces over $\mathbb{C}$. We denote their (complex) dimensions by

$$\ell(\mathfrak{D}) := \dim \mathcal{L}(\mathfrak{D}), \tag{B.17}$$

$$i(\mathfrak{D}) := \dim \mathcal{L}^1(\mathfrak{D}). \tag{B.18}$$

$i(\mathfrak{D})$ is also called the index of speciality of $\mathfrak{D}$.

Lemma B.5. $\mathfrak{D} \in \mathrm{Div}(K_g)$, $g \in \mathbb{N}_0$. Then $\deg(\mathfrak{D})$, $\ell(\mathfrak{D})$, and $i(\mathfrak{D})$ only depend on the divisor class $[\mathfrak{D}]$ of $\mathfrak{D}$ (and not on the particular representative $\mathfrak{D}$). Moreover, for $\omega \in \mathcal{M}^1(K_g) \backslash \{0\}$,

$$i(\mathfrak{D}) = \ell(\mathfrak{D} - (\omega)), \ \mathfrak{D} \in \mathrm{Div}(K_g). \tag{B.19}$$

Theorem B.6. (Riemann-Roch)
Let $g \in \mathbb{N}_0$, $\mathfrak{D} \in \mathrm{Div}(K_g)$. Then $\ell(-\mathfrak{D})$ and $i(\mathfrak{D})$ are finite and

$$\ell(-\mathfrak{D}) = \deg(\mathfrak{D}) + i(\mathfrak{D}) - g + 1. \tag{B.20}$$

In particular, Riemann's inequality

$$\ell(-\mathfrak{D}) \geq \deg(\mathfrak{D}) - g + 1 \tag{B.21}$$

holds.

Next we turn to the Jacobi variety and the Abel map.

Definition B.7. Let $g \in \mathbb{N}$ and define the period lattice L_g in $\mathbb{C}^g$ by

$$L_g := \{\underline{z} \in \mathbb{C}^g \, | \, \underline{z} = \underline{N} + (\tau \underline{M}), \ \underline{N}, \underline{M} \in \mathbb{Z}^g\}. \tag{B.22}$$

Then the Jacobi variety $J(K_g)$ of K_g is defined by

$$J(K_g) := \mathbb{C}^g / L_g \tag{B.23}$$

and the Abel map is defined by

$$\underline{A}_{P_0} : \begin{cases} K_g \to J(K_g) \\ P \to \underline{A}_{P_0}(P) = \left\{ A_{P_0,j}(P) := \int_{P_0}^{P} \omega_j \right\}_{j=1}^{g} \pmod{L_g}, \end{cases} \tag{B.24}$$

respectively by

$$\underline{\alpha}_{P_0} : \begin{cases} \mathrm{Div}(K_g) \to J(K_g) \\ \mathfrak{D} \longrightarrow \underline{\alpha}_{P_0}(\mathfrak{D}) := \sum_{P \in K_g} \mathfrak{D}(P) \underline{A}_{P_0}(P), \end{cases} \tag{B.25}$$

where $P_0 \in K_g$ is a fixed base point and the same path is chosen from P_0 to P for all $1 \leq j \leq g$.

Clearly $\underline{A}_{P_0}$ is well-defined since changing the path from P_0 to P amounts to adding a closed cycle whose contribution in the integral (B.24) consists in adding a vector in L_g. Moreover, $\underline{\alpha}_{P_0}$ is a group homomorphism and $J(K_g)$ is a complex torus of (complex) dimension g that depends on the choice of the homology basis $\{a_j, b_j\}_{j=1}^{g}$. However, different homology bases yield isomorphic Jacobians [Fa-Kr], p. 137, [Gun], Section 8(b).

Theorem B.8. (Abel's theorem)
Let $g \in \mathbb{N}$. Then $\mathfrak{D} \in \mathrm{Div}(K_g)$ is principal iff

$$\begin{aligned} &(i). \ \deg(\mathfrak{D}) = 0. \\ &(ii). \ \underline{\alpha}_{P_0}(\mathfrak{D}) = 0. \end{aligned} \tag{B.26}$$

Theorem B.9. (Jacobi inversion theorem)
Let $g \in \mathbb{N}$. Then $\underline{\alpha}_{P_0}$ is surjective.

We shall describe a constructive approach to Theorem B.9 in terms of Riemann theta functions in Appendix C.

APPENDIX C: RIEMANN THETA FUNCTIONS, JACOBI INVERSION PROBLEM

For basic references to the subject of this appendix we refer e.g. to [Dub 3], [Fa-Kr], [Fay], [Gr-Ha], [Kra], [Lew], [McK 1], [Mum], [Ra-Fa], [Rey], [Rod 2], [Sie].

Given the hyperelliptic curve K_g (we assume $g \in \mathbb{N}$ throughout this appendix), the homology basis $\{a_j, b_j\}_{j=1}^{g}$, and the matrix τ of the b-periods of the DFK's $\{\omega_j\}_{j=1}^{g}$ introduced in Appendix A, the Riemann theta function associated with K_g is defined as

$$\theta(\underline{z}) := \sum_{\underline{m} \in \mathbb{Z}^g} \exp\left[2\pi i(\underline{m}, \underline{z}) + \pi i(\underline{m}, \tau \underline{m})\right], \ \underline{z} \in \mathbb{C}^g, \tag{C.1}$$

where $(.,.)$ denotes the scalar product in $\mathbb{C}^g$. Because of (A.25), θ is well-defined and represents an entire function on $\mathbb{C}^g$.

Elementary properties of θ are e.g.,

$$\theta(z_1, \ldots, z_{j-1}, -z_j, z_{j+1}, \ldots, z_g) = \theta(\underline{z}), \ \underline{z} = (z_1, \ldots, z_g) \in \mathbb{C}^g, \tag{C.2}$$

$$\theta(\underline{z} + \underline{e} + \tau\underline{e}') = \exp[-2\pi i(\underline{e}', \underline{z}) - \pi i(\underline{e}', \tau\underline{e}')]\, \theta(\underline{z}), \ \underline{e}, \underline{e}' \in \mathbb{Z}^g. \tag{C.3}$$

Lemma C.1. Let $\underline{\xi} \in \mathbb{C}^g$ and define

$$F : \begin{cases} \hat{K}_g \to \mathbb{C} \\ P \to \theta\left(\underline{\hat{A}}_{P_0}(P) - \underline{\xi}\right), \end{cases} \tag{C.4}$$

where

$$\underline{\hat{A}}_{P_0} : \begin{cases} \hat{K}_g \to \mathbb{C}^g \\ P \to \underline{\hat{A}}_{P_0}(P) = \left\{\hat{A}_{P_0,j}(P) := \int_{P_0}^{P} \omega_j\right\}_{j=1}^{g}. \end{cases} \tag{C.5}$$

Suppose $F \not\equiv 0$ on $\hat{K}_g$. Then F has precisely g zeros on $\hat{K}_g$ counting multiplicity.

Lemma C.1 is proven in the usual way by integrating $d\log F$ along $\partial\hat{K}_g$.

Theorem C.2. (Riemann)
Let $\underline{\xi} \in \mathbb{C}^g$ and define F as in (C.4). Suppose $F \not\equiv 0$ on $\hat{K}_g$ and let $P_1, \ldots, P_g \in \hat{K}_g$ be the zeros of F (multiplicities included) given by Lemma C.1. Define

$$\mathfrak{D}_{P_1+\ldots+P_g} : \begin{cases} K_g \to \mathbb{Z} \\ P \to \mathfrak{D}_{P_1+\ldots+P_g}(P) = \begin{cases} 1, & P \in \{P_1, \ldots, P_g\} \\ 0, & P \in K_g \backslash \{P_1, \ldots P_g\} \end{cases} \end{cases} \tag{C.6}$$

and recall the Abel map $\underline{\alpha}_{P_0}$ in (B.25). Then there exists a vector $\underline{\zeta}_{P_0} \in \mathbb{C}^g$, the vector of Riemann constants, such that

$$\underline{\alpha}_{P_0}\left(\mathfrak{D}_{P_1+\ldots+P_g}\right) = \left[\underline{\xi} - \underline{\zeta}_{P_0}\right] (\mathrm{mod}\, L_g). \tag{C.7}$$

$\underline{\zeta}_{P_0} = (\zeta_{P_0,1}, \ldots, \zeta_{P_0,g})$ is given by

$$\zeta_{P_0,j} = \left[(1 + \tau_{j,j})/2 - \sum_{\substack{\ell=1 \\ \ell \neq j}}^{g} \int_{a_\ell} \omega_\ell(P) \int_{P_0}^{P} \omega_j\right], \ 1 \leq j \leq g. \tag{C.8}$$

For the proof of Theorem C.2 one integrates $\hat{A}_{P_0,j}(P)d\log F(P)$ along $\partial\hat{K}_g$. Clearly $\underline{\zeta}_{P_0}$ depends on the base point P_0 and on the choice of the homology basis $\{a_j, b_j\}_{j=1}^g$. In the special case, where $P_0 = P_\infty = (\infty, \infty)$ (C.8) simplifies to

$$\zeta_{P_\infty,j} = \left[j + \sum_{\ell=1}^{g} \tau_{j,\ell}\right] /2, \ 1 \leq j \leq g. \tag{C.9}$$

Remark C.3. Theorem C.2 yields a partial solution to Jacobi's inversion problem which can be stated as follows: Given $\underline{\xi} \in \mathbb{C}^g$, find a divisor $\mathfrak{D}_{P_1+\ldots+P_g} \in \mathrm{Div}(K_g)$ such that

$$\underline{\alpha}_{P_0}(\mathfrak{D}_{P_1+\ldots+P_g}) = \underline{\xi}(\mathrm{mod}\, L_g). \tag{C.10}$$

Indeed, if $\tilde{F}(P) := \theta(\underline{\zeta}_{P_0} - \underline{\hat{A}}_{P_0}(P) + \underline{\xi}) \not\equiv 0$ on $\hat{K}_g$, the zeros $P_1, \ldots, P_g \in \hat{K}_g$ of $\tilde{F}$ (guaranteed by Lemma C.1) satisfy Jacobi's inversion problem by (C.7). Thus it remains to specify conditions such that $\tilde{F} \not\equiv 0$ on $\hat{K}_g$.

Remark C.4. While $\theta(\underline{z})$ is well-defined (even entire) for $\underline{z} \in \mathbb{C}^g$, it is not well-defined on $J(K_g) = \mathbb{C}^g/L_g$ because of (C.3). Nevertheless θ is a "multiplicative function" on $J(K_g)$ since the multipliers in (C.3) are exponentials that cannot vanish. In particular, if $\underline{z}_1 = \underline{z}_2(\mathrm{mod}\, L_g)$ then $\theta(\underline{z}_1) = 0$ iff $\theta(\underline{z}_2) = 0$. Hence it is meaningful to state that θ vanishes at points of $J(K_g)$. Since the Abel map $\underline{A}_{P_0}$ maps K_g into $J(K_g)$, $\theta(\underline{A}_{P_0}(P) - \underline{\xi})$, $\underline{\xi} \in \mathbb{C}^g$ becomes a multiplicative function on K_g. Again it makes sense to say that $\theta(\underline{A}_{P_0}(.) - \underline{\xi})$ vanishes at points of K_g.

We use the obvious notions

$$\begin{aligned} X + Y &:= \{(\underline{x} + \underline{y}) \in J(K_g) \,|\, \underline{x} \in X,\ \underline{y} \in Y\}, \\ -X &:= \{-\underline{x} \in J(K_g) \,|\, \underline{x} \in X\}, \\ X + \underline{z} &:= \{(\underline{x} + \underline{z}) \in J(K_g) \,|\, \underline{x} \in X\} \end{aligned} \tag{C.11}$$

for $X, Y \subseteq J(K_g)$, $\underline{z} \in J(K_g)$ and identify $\sigma^n K_g$, the n-th symmetric power of K_g, with the set of positive divisors of degree $n, n \in \mathbb{N}$ on K_g.

Definition C.5.

(i). We define

$$\underline{W}_0 := \{\underline{0}\} \subset J(K_g),\ \underline{W}_n := \underline{\alpha}_{P_0}(\sigma^n K_g),\ n \in \mathbb{N}. \tag{C.12}$$

(ii). A positive divisor $\mathfrak{D} \in \mathrm{Div}(K_g)$ is called special iff $i(\mathfrak{D}) \geq 1$, otherwise $\mathfrak{D}$ is called nonspecial.

(iii). $\tilde{P} \in K_g$ is called a Weierstrass point of K_g iff $i(\mathfrak{D}_{g\tilde{P}}) \geq 1$, where

$$\mathfrak{D}_{g\tilde{P}} : \begin{cases} K_g \to \mathbb{N}_0 \\ P \to \mathfrak{D}_{g\tilde{P}}(P) := \begin{cases} g, & P = \tilde{P} \\ 0, & P \in K_g \backslash \{\tilde{P}\}. \end{cases} \end{cases} \tag{C.13}$$

Remark C.6. (i). Since $i(\mathfrak{D}_p) = 0$ for all $P \in K_1$, K_1 has no Weierstrass points. For $g \geq 2$ the Weierstrass points of K_g are given precisely by the branch points $(E_n, 0)$, $0 \leq n \leq 2g$, $P_\infty = (\infty, \infty)$ of K_g.

(ii). The special divisors of the type $\mathfrak{D}_{P_1+\ldots+P_N}$ (see (C.6)) with $\deg(\mathfrak{D}_{P_1+\ldots+P_N}) = N \geq g$ are precisely the critical points of the Abel map $\underline{\alpha}_{P_0} : \sigma^N K_g \to J(K_g)$, i.e., the set of points $\mathfrak{D}$ at which the rank of the differential $d\underline{\alpha}_{P_0}$ is less than g (cf. (C.20)).

(iii). While $\sigma^m K_g \not\subset \sigma^n K_g$ for $m < n$, one has $\underline{W}_m \subset \underline{W}_n$ for $m < n$. Thus $\underline{W}_n = J(K_g)$ for $n \geq g$ by Theorem B.9.

Theorem C.7. $\left[\underline{W}_{g-1} + \underline{\zeta}_{P_0}\right] \subset J(K_g)$ is the complete set of zeros of θ on $J(K_g)$, i.e.,

$$\theta(X) = 0 \text{ iff } X \in \underline{W}_{g-1} + \underline{\zeta}_{P_0} \tag{C.14}$$

(i.e., iff $X = \left[\underline{\alpha}_{P_0}(\mathfrak{D}) + \underline{\zeta}_{P_0}\right] (\operatorname{mod} L_g)$ for some $\mathfrak{D} \in \sigma^{g-1} K_g$). The set $\left[\underline{W}_{g-1} + \underline{\zeta}_{P_0}\right]$ has complex dimension $g - 1$.

Theorem C.8. (Riemann's vanishing theorem)
Let $\underline{\xi} \in \mathbb{C}^g$.
(i). If $\theta(\underline{\xi}) \neq 0$ then there exists a unique $\mathfrak{D} \in \sigma^g K_g$ such that

$$\underline{\xi} = \left[\underline{\alpha}_{P_0}(\mathfrak{D}) + \underline{\zeta}_{P_0}\right] (\operatorname{mod} L_g) \tag{C.15}$$

and

$$i(\mathfrak{D}) = 0. \tag{C.16}$$

(ii). If $\theta(\underline{\xi}) = 0$ and $g = 1$ then

$$\underline{\xi} = \underline{\zeta}_{P_0} (\operatorname{mod} L_1) = [(1+\tau)/2] (\operatorname{mod} L_1), \; L_1 = \mathbb{Z} + \tau\mathbb{Z}, \; (-i\tau) > 0. \tag{C.17}$$

(iii). If $\theta(\underline{\xi}) = 0$ and $g \geq 2$, let $s \in \mathbb{N}$, $1 \leq s \leq (g-1)$ be the smallest integer such that $\theta(\underline{W}_s - \underline{W}_s - \underline{\xi}) \neq 0$ (i.e., there exist $\mathfrak{E}, \mathfrak{F} \in \sigma^s K_g$, $\mathfrak{E} \neq \mathfrak{F}$ such that $\theta(\underline{\alpha}_{P_0}(\mathfrak{E}) - \underline{\alpha}_{P_0}(\mathfrak{F}) - \underline{\xi}) \neq 0$). Then there exists a $\mathfrak{D} \in \sigma^{g-1} K_g$ such that

$$\underline{\xi} = \left[\underline{\alpha}_{P_0}(\mathfrak{D}) + \underline{\zeta}_{P_0}\right] (\operatorname{mod} L_g) \tag{C.18}$$

and

$$i(\mathfrak{D}) = s. \tag{C.19}$$

All partial derivatives of θ with respect to $A_{P_0,j}$, $1 \le j \le g$ of order strictly less than s vanish at $\underline{\xi}$, whereas at least one partial derivative of θ of order s is nonzero at $\underline{\xi}$. Moreover, $1 \le s \le (g+1)/2$ and the integer s is the same for $\underline{\xi}$ and $-\underline{\xi}$.

Note that there is no explicit reference to the base point P_0 in the formulation of Theorem C.8 since the set $[\underline{W}_s - \underline{W}_s] \subset J(K_g)$ is independent of the base point while $\underline{W}_s$ alone is not.

This yields a complete solution of Jacobi's inversion problem (see Remark C.3, Theorem B.8).

Theorem C.9. (Jacobi inversion theorem)
$\underline{\alpha}_{P_0}$ is surjective. More precisely, given $\underline{\tilde{\xi}} = \left(\underline{\xi} + \underline{\zeta}_{P_0}\right) \in \mathbb{C}^g$, the divisors $\mathfrak{D}$ in (C.15), (C.18) (resp. $\mathfrak{D} = \mathfrak{D}_{P_0}$ if $g = 1$) solve the Jacobi inversion problem for $\underline{\xi} \in \mathbb{C}^g$.

Moreover, one has

Theorem C.10. Let $g \ge 2$ and $P_1, \dots, P_g \in K_g$. Then $\mathfrak{D}_{P_1 + \dots + P_g}$ is special and hence

$$\operatorname{rank}\left[d\underline{\alpha}_{P_0}(\mathfrak{D}_{P_1+\dots+P_g})\right] = g - i(\mathfrak{D}_{P_1+\dots+P_g}) = g - \nu \tag{C.20}$$

for some $\nu \ge 1$ iff there is at least one pair $(P, P^*) \subseteq \{P_1, \dots, P_g\}$ or (as a limiting case) a Weierstrass point occurs at least twice in $\{P_1, \dots, P_g\}$. (Here $*$ denotes the hyperelliptic involution (A.13).) Otherwise

$$\operatorname{rank}\left[d\underline{\alpha}_{P_0}(\mathfrak{D}_{P_1+\dots+P_g})\right] = g \tag{C.21}$$

and hence $\mathfrak{D}_{P_1+\dots+P_g}$ is nonspecial.

We summarize some of this analysis in

Remark C.11. Consider on $\hat{K}_g$ the function

$$G(P) := \theta\Big(\underline{\zeta}_{P_0} - \underline{\hat{A}}_{P_0}(P) + \sum_{j=1}^{g} \underline{\hat{A}}_{P_0}(P_j)\Big), \; P, P_j \in \hat{K}_g, \; 1 \le j \le g. \tag{C.22}$$

Then

$$G(P_\ell) = \theta\Big(\underline{\zeta}_{P_0} + \sum_{\substack{j=1 \\ j \ne \ell}}^{g} \underline{\hat{A}}_{P_0}(P_j)\Big) = \theta\left(\underline{\zeta}_{P_0} + \underline{\alpha}_{P_0}(\mathfrak{D}_{P_1+\dots+P_{\ell-1}+P_{\ell+1}+\dots+P_g})\right) = 0, \; 1 \le \ell \le g \tag{C.23}$$

by Theorem C.7. Moreover, by Lemma C.1 and Theorem C.8, $P_1, \ldots, P_g$ are the only zeros of G on $\hat{K}_g$ iff $\mathfrak{D}_{P_1+\ldots+P_g}$ is nonspecial, i.e., iff

$$i\left(\mathfrak{D}_{P_1+\ldots+P_g}\right) = 0. \tag{C.24}$$

Conversely, $G \equiv 0$ on $\hat{K}_g$ iff $\mathfrak{D}_{P_1+\ldots+P_g}$ is special, i.e., iff $i(\mathfrak{D}_{P_1+\ldots P_g}) \geq 1$.

We also mention the elementary change in the Abel map and in Riemann's vector if one changes the base point:

$$\begin{aligned} \underline{A}_{Q_0} &= \left[\underline{A}_{P_0} - \underline{A}_{P_0}(Q_0)\right] (\operatorname{mod} L_g), \\ \underline{\zeta}_{Q_0} &= \left[\underline{\zeta}_{P_0} + (g-1)\underline{A}_{P_0}(Q_0)\right] (\operatorname{mod} L_g), \;\; P_0, Q_0 \in K_g. \end{aligned} \tag{C.25}$$

Remark C.12. The L_g-quasi-periodic, holomorphic function θ on $\mathbb{C}^g$ can be used to construct L_g-periodic, meromorphic functions f on $\mathbb{C}^g$ as follows:

$$1.\;\; f(\underline{z}) := \frac{\prod_{j=1}^{N} \theta(\underline{z} + \underline{c}_j)}{\prod_{\ell=1}^{N} \theta(\underline{z} + \underline{d}_\ell)}, \quad \underline{z}, \underline{c}_j, \underline{d}_j \in \mathbb{C}^g, \;\; 1 \leq j \leq N, \tag{C.26}$$

where

$$\sum_{j=1}^{N} \underline{c}_j = \sum_{\ell=1}^{N} \underline{d}_\ell (\operatorname{mod} L_g).$$

$$2.\;\; f(\underline{z}) := \partial_{z_j} \log \left[\frac{\theta(\underline{z} + \underline{e})}{\theta(\underline{z} + \underline{f})}\right], 1 \leq j \leq g, \underline{z} = (z_1, \ldots, z_g) \in \mathbb{C}^g, \underline{e}, \underline{f} \in \mathbb{C}^g. \tag{C.27}$$

$$3.\;\; f(\underline{z}) := \partial^2_{z_j z_\ell} \log \theta(\underline{z}), \;\; \underline{z} \in \mathbb{C}^g, \;\; 1 \leq j, \ell \leq g. \tag{C.28}$$

Then indeed

$$f\left(\underline{z} + \underline{m} + (\tau \underline{n})\right) = f(\underline{z}), \;\; \underline{z} \in \mathbb{C}^g, \;\; \underline{m}, \underline{n} \in \mathbb{Z}^g \tag{C.29}$$

by (C.3).

Remark C.13. Let $\underline{\xi} \in J(K_g)$ be given, assume that $\theta\left(\underline{\zeta}_{P_0} - \underline{A}_{P_0}(.) + \underline{\xi}\right) \not\equiv 0$ on K_g and suppose that $\underline{A}_{P_0}^{-1}(\underline{\xi}) = (P_1, \ldots, P_g) \in \sigma^g K_g$ is the unique solution of Jacobi's inversion problem. Let $f \in \mathcal{M}(K_g) \backslash \{0\}$ and suppose $f(P_j) \neq \infty$, $1 \leq j \leq g$. Then $\underline{\xi}$ uniquely determines the values $f(P_1), \ldots, f(P_g)$. Moreover, any symmetric function of these values is a single-valued meromorphic function of $\underline{\xi} \in J(K_g)$, i.e., an Abelian function on $J(K_g)$. Any such meromorphic function on $J(K_g)$ can be expressed in terms of the Riemann theta function on K_g. E.g., for the elementary symmetric functions of the 2nd kind (Newton polynomials) one obtains from the residue theorem in analogy to the proof of Lemma C.1,

$$\begin{aligned} \sum\nolimits_{n,f}(\underline{\xi}) &:= \sum_{j=1}^{g} f(P_j)^n \\ &= \sum_{j=1}^{g} \int_{a_j} f(P)^n \omega_j(P) - \sum_{\substack{Q_r \in \hat{K}_g \\ f(Q_r)=\infty}} \operatorname*{Res}_{P=Q_r} \left\{ f(P)^n d \log \theta \left(\underline{\zeta}_{P_0} - \underline{A}_{P_0}(P) + \underline{\xi} \right) \right\}, \end{aligned} \tag{C.30}$$

where an appropriate homology basis $\{a_j, b_j\}_{j=1}^{g}$ with $\partial \hat{K}_g = a_1 b_1 a_1^{-1} b_1^{-1} \dots a_g^{-1} b_g^{-1}$ avoiding $\{P_1, \dots, P_g\}$ and the poles $\{Q_r\}$ of f has been chosen. (We also note that Lemma C.1 just corresponds to the case $n = 0$ in (C.30).)

Since we need $\sum_{1,\Pi}$ in Section 2 we prove

Theorem C.14. Consider the projection Π in (A.12) (with a pole of multiplicity two at P_∞) and the unique solution $(P_1, \dots, P_g) \in \sigma^g K_g$, $P_j := \left(z_j, R_0(z_j)^{1/2}\right)$, $1 \le j \le g$ of the Jacobi inversion problem

$$\underline{\alpha}_{P_\infty}\left(\mathfrak{D}_{P_1 + \dots + P_g}\right) = \underline{\xi} \in J(K_g) \tag{C.31}$$

given $\underline{\xi} \in J(K_g)$ such that $\theta\left(\underline{\zeta}_{P_\infty} - \underline{A}_{P_\infty}(.) + \underline{\xi}\right) \not\equiv 0$ on K_g. Then

$$\begin{aligned} \sum\nolimits_{1,\Pi}(\underline{\xi}) &= \sum_{j=1}^{g} \Pi(P_j)(\underline{\xi}) = \sum_{j=1}^{g} z_j(\underline{\xi}) \\ &= \sum_{j=1}^{g} \int_{a_j} \Pi(P) \omega_j(P) + (4\pi^2)^{-1} \sum_{j,\ell=1}^{g} U_{0,j} U_{0,\ell} \partial^2_{\xi_j \xi_\ell} \log \theta \left(\underline{\zeta}_{P_\infty} + \underline{\xi} \right) \end{aligned} \tag{C.32}$$

with

$$U_{0,j} = 2\pi i f_{\alpha,j}(0) = 4\pi(-1)^{g+1} c_{j,g} = \int_{b_j} \omega_0^{(2)}, \ 1 \le j \le g, \tag{C.33}$$

where

$$\omega_j = f_{\alpha,j}(\zeta_\alpha) d\zeta_\alpha, \ 1 \le j \le g, \ \zeta_\alpha = z_\alpha^{-1/2} \text{ near } P_\infty = (\infty, \infty) \tag{C.34}$$

and $\omega_0^{(2)}$ is a normalized DSK with a single pole of multiplicity two at P_∞ of the type

$$\omega_0^{(2)} = \left[\zeta_\alpha^{-2} + 0(1)\right] d\zeta_\alpha. \tag{C.35}$$

Proof. Because of (C.30) it suffices to compute $\operatorname*{Res}_{P=P_\infty}\left\{\Pi(P)d\log\theta\left(\underline{\zeta}_{P_\infty}-\underline{A}_{P_\infty}(P)+\underline{\xi}\right)\right\}$. Choosing $\zeta_\alpha = z_\alpha^{-1/2}$ as the local coordinate near P_∞ we get

$$d\log\theta\left(\underline{\zeta}_{P_\infty}-\underline{A}_{P_\infty}(P)+\underline{\xi}\right)=\sum_{j=1}^{g}\left\{\partial_{z_j}\log\theta(\underline{z})\Big|_{\underline{z}=\underline{\zeta}_{P_\infty}-\underline{A}_{P_\infty}(P)+\underline{\xi}}\right\}f_{\alpha,j}(\zeta_\alpha)d\zeta_\alpha \quad (C.36)$$

since

$$\frac{d}{d\zeta_\alpha}\left(A_{P_\infty,j}\circ z_\alpha^{-1}\right)(\zeta_\alpha)=f_{\alpha,j}(\zeta_\alpha),\ 1\le j\le g. \quad (C.37)$$

The expansion of $\underline{A}_{P_\infty}$ near P_∞ yields

$$\left(\underline{A}_{P_\infty}\circ z_\alpha^{-1}\right)(\zeta_\alpha)=(\underline{U}_0/2\pi i)\,\zeta_\alpha+(\underline{V}/2\pi i)\,\zeta_\alpha^2+0(\zeta_\alpha^3), \quad (C.38)$$

where

$$\underline{U}_0=(U_{0,1},\dots,U_{0,g}),U_{0,j}=2\pi i f_{\alpha,j}(0),\ 1\le j\le g, \quad (C.39)$$

$$\underline{V}=(V_1,\dots,V_g),V_j=2\pi i f'_{\alpha,j}(0)/2,\ 1\le j\le g. \quad (C.40)$$

From (A.21) we infer

$$\eta_j:=g_{\alpha,j}(\zeta_\alpha)d\zeta_\alpha=\left\{2(-1)^g i\delta_{j,g}+(-1)^g i\left[\Lambda\delta_{j,g}+2\delta_{j,g-1}\right]\zeta_\alpha^2+0(\zeta_\alpha^4)\right\}d\zeta_\alpha,\ 1\le j\le g \quad (C.41)$$

and thus

$$g'_{\alpha,j}(0)=0,\ 1\le j\le g \quad (C.42)$$

implying

$$\underline{V}=0 \quad (C.43)$$

by (A.22). Moreover, (A.22) implies

$$U_{0,j}=2\pi i f_{\alpha,j}(0)=4\pi(-1)^{g+1}c_{j,g},\ 1\le j\le g. \quad (C.44)$$

A comparison with (A.29) reveals that $U_{0,j}$ are indeed the b-periods of a normalized DSK $\omega_0^{(2)}$ with principal part $\zeta_\alpha^{-2}d\zeta_\alpha$ near P_∞. Inserting $\underline{V}=0$ in (C.38) and (C.36) yields

$$\begin{aligned}d\log\theta\left(\underline{\zeta}_{P_\infty}-\underline{A}_{P_\infty}(P)+\underline{\xi}\right)&=-(2\pi i)^{-1}\sum_{\ell=1}^{g}U_{0,\ell}\partial_{z_\ell}\log\theta(\underline{z})\Big|_{\underline{z}=\underline{\zeta}_{P_\infty}+\underline{\xi}}d\zeta_\alpha\\&-(4\pi^2)^{-1}\zeta_\alpha\sum_{j,\ell=1}^{g}U_{0,j}U_{0,\ell}\partial^2_{z_j z_\ell}\log\theta(\underline{z})\Big|_{\underline{z}=\underline{\zeta}_{P_\infty}+\underline{\xi}}d\zeta_\alpha+0(\zeta_\alpha^2)d\zeta_\alpha.\end{aligned} \quad (C.45)$$

Using $\Pi(P) = z_\alpha = \zeta_\alpha^{-2}$ near P_∞ one finally infers

$$\operatorname*{Res}_{P=P_\infty}\left\{\Pi(P) d\log\theta\left(\underline{\zeta}_{P_\infty} - \underline{A}_{P_\infty}(P) + \underline{\xi}\right)\right\} = -(4\pi^2)^{-1}\sum_{j,\ell=1}^{g} U_{0,j}U_{0,\ell}\partial^2_{z_j z_\ell}\log\theta(\underline{\zeta}_{P_\infty} + \underline{\xi}). \tag{C.46}$$

■

Theorem C.15. Let $P_0 = (E_0, 0)$, $(P_1, \ldots, P_g) \in K_g^g$, $P_j \neq P_\infty$, $P_j \neq P_\ell^*$, $1 \leq j, \ell \leq g$. Then

$$\begin{aligned} &2\pi i\sum_{j=1}^{g}\int_{P_0}^{P_j}\omega_0^{(2)} + \sum_{j=1}^{g} U_{0,j}\int_{a_j} d\log\theta\left(\underline{\zeta}_{P_0} - \underline{A}_{P_0}(.) + \underline{\alpha}_{P_0}\left(\mathfrak{D}_{P_1+\ldots+P_g}\right)\right) \\ &= -\sum_{j=1}^{g} U_{0,j}\partial_{z_j}\log\theta\left(\underline{\zeta}_{P_0} - \underline{A}_{P_0}(P_\infty) + \underline{\alpha}_{P_0}\left(\mathfrak{D}_{P_1+\ldots+P_g}\right)\right), \end{aligned} \tag{C.47}$$

where $\omega_0^{(2)}$ is the DSK (C.35) and $\underline{U}_0$ the vector (C.33) of its b-periods. Moreover, let $\underline{\xi} \in \mathbb{C}^g$ then the constants c_j

$$c_j := \int_{a_j} d\log\theta\left(\underline{\xi} - \underline{A}_{P_0}(.)\right), \; 1 \leq j \leq g \tag{C.48}$$

are independent of $\underline{\xi}$.

Proof. Applying Stokes theorem and Theorem A.1. (i) with $\omega = \omega_0^{(2)}$, $\nu = d\log\theta\left(\underline{\zeta}_{P_0} - \underline{A}_{P_0}(.) + \underline{\alpha}_{P_0}\left(\mathfrak{D}_{P_1+\ldots+P_g}\right)\right)$ yields with $\Omega(P) := \int_{P_0}^{P}\omega_0^{(2)}$ by the residue theorem

$$\begin{aligned} \int_{\partial\hat{K}_g}\Omega\nu &= -\sum_{j=1}^{g} U_{0,j}\int_{a_j} d\log\theta\left(\underline{\zeta}_{P_0} - \underline{A}_{P_0}(.) + \underline{\alpha}_{P_0}\left(\mathfrak{D}_{P_1+\ldots+P_g}\right)\right) \\ &= 2\pi i\sum_{P\in K_g}\operatorname*{Res}_{P}\left(\int_{P_0}^{P}\omega_0^{(2)}\right)\left(d\log\theta\left(\underline{\zeta}_{P_0} - \underline{A}_{P_0}(P) + \underline{\alpha}_{P_0}\left(\mathfrak{D}_{P_1+\ldots+P_g}\right)\right)\right) \end{aligned} \tag{C.49}$$

since $\int_{a_j}\omega_0^{(2)} = 0$, $1 \leq j \leq g$. By Theorem C.10, the zeros of $\theta\left(\underline{\zeta}_{P_0} - \underline{A}_{P_0}(.) + \underline{\alpha}_{P_0}\left(\mathfrak{D}_{P_1+\ldots+P_g}\right)\right)$ are all simple and given by $P_1, \ldots, P_g$. Thus

$$(\text{C.49}) = 2\pi i\sum_{j=1}^{g}\int_{P_0}^{P_j}\omega_0^{(2)} + 2\pi i\operatorname*{Res}_{P=P_\infty}\left(\int_{P_0}^{P}\omega_0^{(2)}\right)d\log\theta\left(\underline{\zeta}_{P_0} - \underline{A}_{P_0}(P) + \underline{\alpha}_{P_0}\left(\mathfrak{D}_{P_1+\ldots+P_g}\right)\right). \tag{C.50}$$

Using $\zeta_\alpha = z_\alpha^{-1/2}$,

$$\omega_0^{(2)} = \left[\zeta_\alpha^{-2} + 0(1)\right] d\zeta_\alpha, \quad \int_{P_0}^{P} \omega_0^{(2)} = -\zeta_\alpha^{-1} + 0(1) \tag{C.51}$$

near P_∞, one infers from (C.45) that

$$2\pi i\Big(\int_{P_0}^{P} \omega_0^{(2)}\Big) d\log\theta\Big(\underline{\zeta}_{P_0} - \underline{A}_{P_0}(P) + \underline{\alpha}_{P_0}\big(\mathfrak{D}_{P_1+\ldots+P_g}\big)\Big)$$

$$= \left[-\zeta_\alpha^{-1} + 0(1)\right]\Big[-\sum_{j=1}^{g} U_{0,j}\partial_{z_j}\log\theta\Big(\underline{\zeta}_{P_0} - \underline{A}_{P_0}(P_\infty) + \underline{\alpha}_{P_0}\left(\mathfrak{D}_{P_1+\ldots+P_g}\right)\Big) + 0(\zeta_\alpha)\Big] d\zeta_\alpha \tag{C.52}$$

for P near P_∞ implying (C.47). (C.48) follows from (C.3) since

$$\int_{a_j} d\log\theta\left(\underline{\xi} - \underline{A}_{P_0}(.)\right)$$

$$= \log\left\{\frac{\theta\left(\underline{\xi} - \underline{A}_{P_0}\left((E_{2j},0)\right)\right)\theta\left(\underline{\xi} + \underline{A}_{P_0}\left((E_{2j-1},0)\right)\right)}{\theta\left(\underline{\xi} - \underline{A}_{P_0}\left((E_{2j-1},0)\right)\right)\theta\left(\underline{\xi} + \underline{A}_{P_0}\left((E_{2j},0)\right)\right)}\right\}, \quad 1 \le j \le g \tag{C.53}$$

and

$$2\underline{A}_{P_0}\left((E_n,0)\right) \in L_g, \quad 0 \le n \le 2g. \tag{C.54}$$

■

APPENDIX D: τ-FUNCTIONS AND COMMUTATION METHODS

In this appendix we combine results developed by Hirota [Hir] (see also [Hi-It], [Mat], Sect. 2.3, [Nak], [Wil] and the references therein) and commutation methods [Dei] to recall the auto-Bäcklund transformation for the KdV-equation.

Suppose that

$$\phi \in C^\infty(\mathbb{R}^2) \text{ is real-valued}, \ \partial_x^n\phi \in L^\infty(\mathbb{R}^2), \ n \in \mathbb{N}_0 \tag{D.1}$$

and define in $L^2(\mathbb{R})$ the operators

$$A(t) := \frac{d}{dx} + \phi(t,.) \text{ on } \mathfrak{D}\left(A(t)\right) = H^1(\mathbb{R}), \ t \in \mathbb{R} \tag{D.2}$$

and

$$H_1(t) := A(t)^* A(t), \ H_2(t) := A(t)A(t)^*, \ t \in \mathbb{R}, \tag{D.3}$$

i.e.,

$$H_j(t) = -\frac{d^2}{dx^2} + V_j(t,.) \text{ on } \mathfrak{D}(H_j(t)) = H^2(\mathbb{R}),\ t \in \mathbb{R},\ j = 1,2 \tag{D.4}$$

with

$$V_j = \phi^2 + (-1)^j \phi_x,\ j = 1,2. \tag{D.5}$$

Assume that $0 < \psi_2 \in C^\infty(\mathbb{R}^2)$ is a distributional solution of

$$H_2(t)\psi_2(t) = 0,\ t \in \mathbb{R},\ \text{i.e., } A(t)^*\psi_2(t) = 0,\ t \in \mathbb{R} \tag{D.6}$$

implying $\phi = \partial_x \ln \psi_2$. Making the ansatz

$$\begin{aligned} \psi_2(t,x) :&= F\, e^{Dx+Et} \tau_1(t,x)/\tau_2(t,x), \\ V_2(t,x) :&= C - 2\partial_x^2 \ln \tau_2(t,x), \\ & (t,x) \in \mathbb{R}^2, C, D, E, F \in \mathbb{R},\ 0 < \tau_j \in C^\infty(\mathbb{R}^2), j = 1,2 \end{aligned} \tag{D.7}$$

yields

$$\phi = \partial_x \ln \psi_2 = D + \tau_1^{-1}\tau_{1,x} - \tau_2^{-1}\tau_{2,x}, \tag{D.8}$$

$$\begin{aligned} V_2 = \phi^2 + \phi_x &= D^2 + 2D\tau_1^{-1}\tau_{1,x} - 2D\tau_2^{-1}\tau_{2,x} - 2\tau_1^{-1}\tau_2^{-1}\tau_{1,x}\tau_{2,x} \\ &\quad + 2\tau_2^{-2}\tau_{2,x}^2 + \tau_1^{-1}\tau_{1,xx} - \tau_2^{-1}\tau_{2,xx} \\ &= C - 2\partial_x^2 \ln \tau_2 = C + 2\tau_2^{-2}\tau_{2,x}^2 - 2\tau_2^{-1}\tau_{2,xx}, \end{aligned} \tag{D.9}$$

$$\begin{aligned} V_1 = \phi^2 - \phi_x &= D^2 + 2D\tau_1^{-1}\tau_{1,x} - 2D\tau_2^{-1}\tau_{2,x} - 2\tau_1^{-1}\tau_2^{-1}\tau_{1,x}\tau_{2,x} \\ &\quad + 2\tau_1^{-2}\tau_{1,x}^2 + \tau_2^{-1}\tau_{2,xx} - \tau_1^{-1}\tau_{1,xx}, \end{aligned} \tag{D.10}$$

$$\begin{aligned} V_2 - V_1 &= 2\tau_2^{-2}\tau_{2,x}^2 - 2\tau_1^{-2}\tau_{1,x}^2 + 2\tau_1^{-1}\tau_{1,xx} - 2\tau_2^{-1}\tau_{2,xx} \\ &= 2\partial_x^2 \ln \tau_1 - 2\partial_x^2 \ln \tau_2. \end{aligned} \tag{D.11}$$

Thus

$$V_1(t,x) = C - 2\partial_x^2 \ln \tau_1(t,x),\ (t,x) \in \mathbb{R}^2 \tag{D.12}$$

and

$$C - D^2 = 2D\tau_1^{-1}\tau_{1,x} - 2D\tau_2^{-1}\tau_{2,x} - 2\tau_1^{-1}\tau_2^{-1}\tau_{1,x}\tau_{2,x} + \tau_1^{-1}\tau_{1,xx} - \tau_2^{-1}\tau_{2,xx},\ (t,x) \in \mathbb{R}^2. \tag{D.13}$$

Next we assume that ψ_2 satisfies in addition

$$\psi_{2,t}(t) = B_{V_2}(t)\psi_2(t),\ t \in \mathbb{R} \tag{D.14}$$

in the distributional sense, where

$$B_{V_2}(t) := -4\frac{d^3}{dx^3} + 6V_2(t,.)\frac{d}{dx} + 3V_{2,x}(t,.) \text{ on } \mathfrak{D}(B_{V_2}(t)) = H^3(\mathbb{R}),\ t \in \mathbb{R}. \quad \text{(D.15)}$$

Since $\psi_{2,xx} = V_2\psi_2$ by (D.6), (D.14) is equivalent to

$$\psi_{2,t}(t,x) = 2V_2(t,x)\psi_{2,x}(t,x) - V_{2,x}(t,x)\psi_2(t,x),\ (t,x) \in \mathbb{R}^2. \quad \text{(D.16)}$$

(Existence of ψ_2 satisfying $\psi_{2,xx} = V_2\psi_2$ and $\psi_{2,t} = 2V_2\psi_{2,x} - V_{2,x}\psi_2$ can easily be shown [Ge-Si], [GSS].) We note that $H_2(t)$, $B_{V_2}(t)$ constitute a Lax pair for the KdV-equation, i.e.,

$$\frac{d}{dt}H_2(t) - [B_{V_2}(t), H_2(t)] = KdV(V_2). \quad \text{(D.17)}$$

Applying (D.17) to ψ_2, taking into account (D.6) and (D.14) shows that V_2 satisfies the KdV-equation

$$KdV(V_2) = V_{2,t} - 6V_2V_{2,x} + V_{2,xxx} = 0. \quad \text{(D.18)}$$

By (D.7) this is equivalent to

$$\partial_x\Big[\tau_2^{-2}(\tau_2\tau_{2,tx} - \tau_{2,t}\tau_{2,x} + 3\tau_{2,xx}^2 - 6C\tau_2\tau_{2,xx} + 6C\tau_{2,x}^2 - 4\tau_{2,x}\tau_{2,xxx} + \tau_2\tau_{2,xxxx})\Big] = 0. \quad \text{(D.19)}$$

Taking into account (D.7), (D.13) and (D.16) one gets

$$\partial_x\Big[\tau_1^{-1}\tau_{1,t} - \tau_2^{-1}\tau_{2,t} + 2\tau_1^{-3}\tau_{1,x}^3 - 2\tau_2^{-3}\tau_{2,x}^3 - 2(D + \tau_1^{-1}\tau_{1,x} - \tau_2^{-1}\tau_{2,x})^3 - 3\tau_1^{-2}\tau_{1,x}\tau_{1,xx}$$
$$+ 3\tau_2^{-2}\tau_{2,x}\tau_{2,xx} + \tau_1^{-1}\tau_{1,xxx} - \tau_2^{-1}\tau_{2,xxx}\Big] = 0, \quad \text{(D.20)}$$

or equivalently,

$$\text{mKdV}(\phi) = \phi_t - 6\phi^2\phi_x + \phi_{xxx} = 0 \quad \text{(D.21)}$$

by (D.8). This in turn implies

$$\text{KdV}(V_1) = 0 \quad \text{(D.22)}$$

because of Miura's identity [Miu]

$$\text{KdV}(V_j) = \Big[2\phi + (-1)^j\partial_x\Big]\ \text{mKdV}(\phi),\ j = 1,2. \quad \text{(D.23)}$$

Next we recall

Lemma D.1. ([Har], Corollary XI. 6.5, [Ge-Si], [GSS])
Assume

$$V_2 \in C^\infty(\mathbb{R}^2) \text{ is real-valued },\ \partial_x^n V_2 \in L^\infty(\mathbb{R}^2),\ n \in \mathbb{N}_0 \quad \text{(D.24)}$$

and let $0 < \psi_2 \in C^\infty(\mathbb{R}^2)$ be a distributional solution of

$$\psi_{2,xx} = V_2 \psi_2. \tag{D.25}$$

Define

$$\phi(t,x) := \partial_x \ln \psi_2(t,x) = \psi_{2,x}(t,x)/\psi_2(t,x), \ (t,x) \in \mathbb{R}^2. \tag{D.26}$$

Then ϕ also satisfies (D.24), i.e., (D.1) is valid. Moreover,

$$V_2(t,x) = \phi(t,x)^2 + \phi_x(t,x), \ (t,x) \in \mathbb{R}^2 \tag{D.27}$$

and V_1 defined by

$$V_1(t,x) = \phi(t,x)^2 - \phi_x(t,x), \ (t,x) \in \mathbb{R}^2 \tag{D.28}$$

also satisfies (D.24).

Given Lemma D.1 we summarize the main result of this appendix.

Theorem D.2. Suppose V_2 satisfies (D.24) and

$$\mathrm{KdV}(V_2) = 0. \tag{D.29}$$

Let $0 < \psi_2 \in C^\infty(\mathbb{R}^2)$ be a distributional solution of (D.25) and of

$$\psi_{2,t} = 2V_2\psi_{2,x} - V_{2,x}\psi_2 \tag{D.30}$$

and make the ansatz

$$\psi_2(t,x) = F\, e^{Dx+Et}\tau_1(t,x)/\tau_2(t,x), \ D,E,F \in \mathbb{R}, \ (t,x) \in \mathbb{R}^2, \tag{D.31}$$

$$V_2(t,x) = C - 2\partial_x^2 \ln \tau_2(t,x), \ C \in \mathbb{R}, \ (t,x) \in \mathbb{R}^2. \tag{D.32}$$

Define

$$\phi(t,x) := \partial_x \ln \psi_2(t,x) = \psi_{2,x}(t,x)/\psi_2(t,x), \ (t,x) \in \mathbb{R}^2 \tag{D.33}$$

and

$$V_1(t,x) := \phi(t,x)^2 - \phi_x(t,x), \ (t,x) \in \mathbb{R}^2. \tag{D.34}$$

Then ϕ and V_1 satisfy (D.24). Moreover,

$$V_1(t,x) = C - 2\partial_x^2 \ln \tau_1(t,x), \ (t,x) \in \mathbb{R}^2 \tag{D.35}$$

and

$$\mathrm{KdV}(V_1) = 0, \ \mathrm{mKdV}(\phi) = 0. \tag{D.36}$$

(Actually, as discussed in the beginning, one does not need to assume (D.29) since it follows from (D.25) and (D.30).)

The above remarks illustrate the auto-Bäcklund transformation $V_2 \to V_1$ for the KdV-equation. Its space part is Miura's transformation $V_1 = \phi^2 - \phi_x$ and its time part is the mKdV-equation (D.21).

In general, the relation between the constants C and D in (D.7) is rather intricated as shown by the periodic case in Section 2 (see (2.18) and (2.35)). (In the pure soliton case, however, $\partial_x^n \tau_j \to 0$ as $|x| \to \infty$, $n \in \mathbb{N}$ implying $C = D^2$.)

The connection between Baker-Akhiezer functions and Hirota's τ-functions has been studied in detail in [Se-Wi] in connection with the KdV-equation and has more recently been exploited by [Du-Na] in the KP-context. (For an alternative formulation of Theorem D.2 (without use of τ-functions) see [Ges 2,3], [Ge-Si], [GSS] and the references therein.)

Finally, we emphasize that singular solutions $\phi, V_j, j = 1, 2$ of the (m)KdV-equation can be treated in exactly the same way. One only needs to drop the positivity and regularity conditions on ψ_2 (and hence on τ_1, τ_2) in (D.7). As a result one obtains meromorphic (m)KdV-solutions.

ACKNOWLEDGEMENTS

It is a pleasure to thank W. Bulla resp. W. Schweiger and K. Unterkofler for discussions on the material of Appendices A–C resp. D.

I am indebted to S. Albeverio, L. Streit and H. Holden for their kind invitations to the BiBoS-Research Center at the University of Bielefeld, FRG (May 17–June 30, 1990) and to the Division of Mathematical Sciences of the University of Trondheim, Norway (July 1–31, 1990) respectively. The stimulating atmosphere and the extraordinary hospitality provided at both institutions is gratefully acknowledged. This research was supported by BiBoS and the Norwegian Research Council for Science and the Humanities (NAVF).

REFERENCES

[Ahi] N.I. Ahiezer, (1961), Sov. Math. Dokl. **2B**, 1409–1412.

[Be-So] H. Behnke, F. Sommer, (1965), "Theorie der analytischen Funktionen einer komplexen Veränderlichen", 3rd ed., Springer, Berlin.

[Bir 1] B. Birnir, (1986), Commun. Pure Appl. Math. **39**, 1–49.

[Bir 2] B. Birnir, (1986), Commun. Pure Appl. Math. **39**, 283–305.

[Bu-Fi] M. Buys, A. Finkel, (1984), J. Diff. Eqs. **55**, 257–275.

[Che] I.V. Cherednik, (1978), Funct. Anal. Appl. **12**, 195–203.

[Co-Jo] C. DeConcini, R.A. Johnson, (1987), Ergod. Th. Dyn. Sys. **7**, 1–24.

[Cra] W. Craig, (1989), Commun. Math. Phys. **126**, 379–407.

[Da-Ta] E. Date, S. Tanaka, (1976), Progr. Theoret. Phys. Suppl. **59**, 107–125.

[Da-Si] E.B. Davies, B. Simon, (1978), Commun. Math. Phys. **63**, 277–301.

[Dei] P.A. Deift, (1978), Duke Math. J. **45**, 267–310.

[De-Si] P. Deift, B. Simon, (1983), Commun. Math. Phys. **90**, 389–411.

[Dub 1] B.A. Dubrovin, (1975), Funct. Anal. Appl. **9**, 215–223.

[Dub 2] B.A. Dubrovin, (1977), Funct. Anal. Appl. **11**, 265–277.

[Dub 3] B.A. Dubrovin, (1981), Russ. Math. Surv. **36:2**, 11–92.

[Dub 4] B.A. Dubrovin, (1983), Rev. Sci. Tech. **23**, 20–50.

[Du-Na] B.A. Dubrovin, S.M. Natanzon, (1989), Math. USSR Izv. **32**, 269–288.

[Du-No] B.A. Dubrovin, S.P. Novikov, (1975), Sov. Phys. JETP **40**, 1058–1063.

[DMN] B.A. Dubrovin, V.B. Matveev, S.P. Novikov, (1976), Russ. Math. Surv. **31:1**, 59–146.

[Eh-Kn] F. Ehlers, H. Knörrer, (1982), Comment. Math. Helvetici **57**, 1–10.

[Er-Fl] N.M. Ercolani, H. Flaschka, (1985), Phil. Trans. Roy. Soc. London **A315**, 405–422.

[Er-Mc] N. Ercolani, H.P. McKean, (1990), Invent. math. **99**, 483–544.

[Fa-Kr] H.M. Farkas, I. Kra, (1980), "Rieman Surfaces", Springer, New York.

[Fay] J.D. Fay, (1973), "Theta Functions on Riemann Surfaces", Lecture Notes in Math. **352**, Springer, Berlin.

[Fir] N.E. Firsova, (1989), Math. USSR Sbornik **63**, 257–265.

[Fla] H. Flaschka, (1975), Arch. Rat. Mech. Anal. **59**, 293–309.

[For] O. Forster, (1981), "Lectures on Riemann Surfaces", Springer, New York.

[Ges 1] F. Gesztesy, (1986), in "Schrödinger Operators, Aarhus 1985", E. Balslev (ed.), Lecture Notes in Math. **1218**, Springer, Berlin, p. 93–122.

[Ges 2] F. Gesztesy, (1989), in "Schrödinger Operators", H. Holden, A. Jensen (eds.), Lecture Notes in Physics **345**, Springer, Berlin, p. 93–117.

[Ges 3] F. Gesztesy, to appear in "Differential Equations with Applications in Biology, Physics and Engineering", J. Goldstein, F. Kappel, W. Schappacher (eds.), Marcel Dekker.

[Ge-Si] F. Gesztesy, B. Simon, (1990), J. Funct. Anal. **89**, 53–60.

[Ge-Zh] F. Gesztesy, Z. Zhao, J. Funct. Anal. (to appear).

[GSS] F. Gesztesy, W. Schweiger, B. Simon, Trans. Amer. Math. Soc. (to appear).

[Gi-Jo] R. Giachetti, R. Johnson, (1984), Nuovo Cim. **82B**, 125–168.

[Gr-Gu] B. Grebert, J.C. Guillot, (1990), *Le probleme spectral inverse pour les systemes AKNS periodiques sur la droite reelle*, preprint, Centre de Mathematiques, Ecole Polytechnique, Palaiseau, France.

[Gri] P.A. Griffiths, (1989), "Introduction to Algebraic Curves", Transl. Math. Monographs **76**, Amer. Math. Soc., Providence, RI.

[Gr-Ha] P. Griffiths, J. Harris, (1978), "Principles of Algebraic Geometry", Wiley, New York.

[Gun] R.C. Gunning, (1972), "Lectures on Riemann Surfaces, Jacobi Varieties", Princeton Univ. Press, Princeton.

[Har] P. Hartman, (1982), "Ordinary Differential Equations", 2nd ed., Birkhäuser, Boston.

[Hir] R. Hirota, (1980), in "Solitons", R.K. Bullough, P.J. Caudrey (eds.), Springer, Berlin, p. 157–176.

[Hi-It] R. Hirota, M. Ito, (1981), J. Phys. Soc. Japan **50**, 338–342.

[Hu-Co] A. Hurwitz, R. Courant, (1964), "Funktionentheorie", 4th ed., Springer, Berlin.

[It-Ma1] A.R. Its, V.B. Matveev, (1975), Funct. Anal. Appl. **9**, 65–66.

[It-Ma2] A.R. Its, V.B. Matveev, (1975), Theoret. Math. Phys. **23**, 343–355.

[Joh] R.A. Johnson, (1982), J. Diff. Eqs. **46**, 165–193.

[Ko-Kr] S. Kotani, M. Krishna, (1988), J. Funct. Anal. **78**, 390–405.

[Kra] A. Krazer, (1970), "Lehrbuch der Thetafunktionen", Chelsea, New York.

[Kri 1] I.M. Kričever, (1976), Sov. Math. Dokl. **17**, 394–397.

[Kri 2] I.M. Krichever, (1977), Funct. Anal. Appl. **11**, 12–26.

[Kri 3] I.M. Krichever, (1977), Russ. Math. Surv. **32:6**, 185–213.

[Kri 4] I.M. Krichever, (1983), Rev. Sci. Tech. **23**, 51–90.

[Lax] P.D. Lax, (1975), Commun. Pure Appl. Math. **28**, 141–188.

[Lev 1] B.M. Levitan, (1982), Math. USSR Izv. **18**, 249–273.

[Lev 2] B.M. Levitan, (1983), Math. USSR Izv. **20**, 55–87.

[Lev 3] B.M. Levitan, (1985), Math. USSR Sbornik **51**, 67–89.

[Lev 4] B.M. Levitan, (1987), "Inverse Sturm-Liouville Problems", VNU Science Press, Utrecht.

[Lew] J. Lewittes, (1964), "Acta Math. **111**, 37–61.

[Mar 1] V.A. Marchenko, (1986), "Sturm-Liouville Operators and Applications", Birkhäuser, Basel.

[Mar 2] V.A. Marchenko, (1988), "Nonlinear Equations and Operator Algebras", Reidel, Dordrecht.

[Ma-Mo] L. Markus, R.A. Moore, (1956), Acta Math. **96**, 99–123.

[Mat] Y. Matsuno, (1984), "Bilinear Transformation Method", Academic Press, New York.

[McK 1] H.P. McKean, (1979), in "Global Analysis", Lecture Notes in Math. **755**, M. Grmela, J.E. Marsden (eds.), Springer, Berlin, p. 83–200.

[McK 2] H.P. McKean, (1979), in "Partial Differential Equations and Geometry", C.I. Byrnes (ed.), Marcel Dekker, New York, p. 237–254.

[McK 3] H.P. McKean, (1981), in "Geometry and Analysis", Springer, Berlin, p. 81–94.

[McK 4] H.P. McKean, (1985), Commun. Pure Appl. Math. **38**, 669–678.

[McK 5] H.P. McKean, (1986), Rev. Mat. Iberoamericana **2**, 235–261.

[McK 6] H.P. McKean, (1987), J. Stat. Phys. **46**, 1115–1143.

[Mc-Mo] H.P. McKean, P. van Moerbeke, (1975), Invent. math. **30**, 217–274.

[Mc-Tr1] H.P. McKean, E. Trubowitz, (1976), Commun. Pure Appl. Math. **29**, 143–226.

[Mc-Tr2] H.P. McKean, E. Trubowitz, (1978), Bull. Amer. Math. Soc. **84**, 1042–1085.

[Miu] R.M. Miura, (1968), J. Math. Phys. **9**, 1202–1204.

[Mos] J. Moser, (1981), in Academia Nazionale Dei Lincei, Scuola Normale Superiore, Lezioni Fermiane, Pisa.

[Mum] D. Mumford, (1983), "Tata Lectures on Theta I", Birkhäuser, Boston.

[Mur] M. Murata, (1986), Duke Math. J. **53**, 869–943.

[Nak] A. Nakamura, (1979), J. Phys. Soc. Japan **47**, 1701–1705.

[Nov] S.P. Novikov, (1974), Funct. Anal. Appl. **8**, 236–246.

[NMPZ] S. Novikov, S.V. Manakov, L.P. Pitaevskii, V.E. Zakharov, (1984), "Theory of Solitons", Consultants Bureau, New York.

[Ra-Fa] H.E. Rauch, H.M. Farkas, (1974), "Theta Functions with Applications to Riemann Surfaces", Williams and Wilkins, Baltimore.

[Re-Si] M. Reed, B. Simon, (1978), "Methods of Modern Mathematical Physics IV, Analysis of Operators", Academic Press, New York.

[Rey] E. Reyssat, (1989), "Quelques Aspects des Surfaces de Riemann", Birkhäuser, Boston.

[Rod 1] Yu.L. Rodin, (1988), Lett. Math. Phys. **15**, 1–6.

[Rod 2] Yu.L. Rodin, (1988), "The Riemann Boundary Problem on Riemann Surfaces", Reidel, Dordrecht.

[Se-Wi] G. Segal, G. Wilson, (1985), Publ. Math. IHES **61**, 5–65.

[Sie] C.L. Siegel, (1988), "Topics in Complex Functions II", Wiley, New York.

[Sim 1] B. Simon, (1980), J. Funct. Anal. **35**, 215–229.

[Sim 2] B. Simon, (1981), J. Funct. Anal. **40**, 66–83.

[Sim 3] B. Simon, (1982), Bull. Amer. Math. Soc. **7**, 447–526.

[Spr] G. Springer, "Introduction to Riemann Surfaces", 2nd ed., Chelsea, New York, 1981.

[Trl] L. Trlifaj, (1989), in "Resonances", E. Brändas, N. Elander (eds.), Lecture Notes in Physics **325**, Springer, Berlin, p. 57–75.

[Tru] E. Trubowitz, (1977), Commun. Pure Appl. Math. **30**, 321–337.

[Wil] G. Wilson, (1985), Phil. Trans. Roy. Soc. London **A315**, 393–404.

A NEW REPRESENTATION OF SOLITON SOLUTIONS OF THE KADOMTSEV–PETVIASHVILI EQUATION

F. GESZTESY, H. HOLDEN

ABSTRACT. We express the N-soliton solutions of the Kadomtsev–Petviashvili equation as $V_N = -2\sum_{n=1}^{N}(p_n+q_n)\psi_{-1,n}\psi_{+1,n}$ where $\psi_{\pm 1,n}$ are particularly normalized eigenfunctions of the operator $-\partial_x^2 \pm \partial_y + V_N$ associated with the eigenvalue $-\frac{1}{2}(p_n^2+q_n^2)$, generalizing a well-known result for the Korteweg–de Vries equation.

1. Introduction.

For the Korteweg–de Vries (KdV) equation

$$KdV(V) := V_t - 6VV_x + V_{xxx} = 0, \quad (t,x) \in \mathbb{R}^2 \tag{1}$$

(here subscript means derivative with respect to the corresponding variable) one has the well-known representation formula for the N-soliton solutions (Gardner *et al.* (1970)),

$$V_N(t,x) = -4\sum_{j=1}^{N} \kappa_j \psi_j(t,x)^2, \tag{2}$$

where $\psi_j(t,x)$ and κ_j satisfy

$$(-\partial_x^2 + V_N)\psi_j = -\kappa_j^2\psi_j, \quad \|\psi_j(t)\|_2 = 1, \quad j = 1,\ldots,N. \tag{3}$$

The aim of this note is to generalize this formula to the two-dimensional analogue of the KdV equation, the Kadomtsev–Petviashvili (KP) equation (Kadomtsev and Petviashvili (1970)).

Recall that the KP equation is given by

$$KP(V) := V_t - 6VV_x + V_{xxx} + 3\int_{\infty}^{x} dx' V_{yy} = 0, \quad (t,x,y) \in \mathbb{R}^3. \tag{4}$$

The KdV equation is obtained from (4) by considering y-independent solutions V. The KP equation exists in two different types, denoted by I and II, depending on the sign of

1980 *Mathematics Subject Classification* (1985 *Revision*). 35C05, 35Q20.

Key words and phrases. Solitons, Kadomtsev–Petviashvili equation.

F.G. is grateful for support from The Norwegian Research Council for Science and the Humanities (NAVF).

the integral term in (4). We will here only discuss type II given by (3), but the results described will also be valid for type I.

The Lax-pair for the KP equation reads (Dryuma (1974), Konopelchenko (1982))

$$\text{(5)}\qquad \begin{aligned} L_\epsilon &= -\partial_x^2 + \epsilon\partial_y + V, \\ B_{\epsilon,\lambda} &= -4\partial_x^3 + 6(V+\lambda)\partial_x + V_x + 3\epsilon\int_\infty^x dx' V_{yy}, \quad \epsilon = \pm 1, \end{aligned}$$

so that

$$\text{(6)}\qquad L_{\epsilon,t} - [B_{\epsilon,\lambda}, L_\epsilon] = KP(V+\lambda).$$

For $\epsilon = 0$ this reduces to the KdV equation. The N-soliton solutions are most easily constructed using the 'dressing method' due to Zakharov and Shabat (1974), where one constructs the full Lax pair L_ϵ, $B_{\epsilon,\lambda}$ from the 'free operators' L_ϵ^0 and B_λ^0 given by (5) with $V = 0$ using Volterra operators $K_\pm$ so that

$$\text{(7)}\qquad \begin{aligned} L_\epsilon &= (1+K_\pm)L_\epsilon^0(1+K_\pm)^{-1}, \\ \partial_t - B_{\epsilon,\lambda} &= (1+K_\pm)(\partial_t - B_\lambda^0)(1+K_\pm)^{-1}. \end{aligned}$$

The operators $K_\pm$ satisfy a Gelfand–Levitan–Marchenko (GLM) equation (14) where one uses a particular finite rank operator F as input, see (18).

By manipulating the GLM equation and involving both operators $L_{\pm 1}$ we find that

$$\text{(8)}\qquad V_N(t,x,y) = -2\sum_{j=1}^N (p_j+q_j)\psi_{-1,j}(t,x,y)\psi_{+1,j}(t,x,y),$$

where

$$\text{(9)}\qquad L_\epsilon\psi_{\epsilon,j} = -\frac{1}{2}(p_j^2+q_j^2)\psi_{\epsilon,j}, \quad j = 1,\ldots,N, \quad \epsilon = \pm 1$$

and the $\psi_{\epsilon,j}$, $j = 1,\ldots,N$ satisfy the system (21) which determines the normalization as x tends to $+\infty$ (see Theorem 1).

2. The representation formula.

We will construct the N-solitons using the 'dressing method' (Zakharov and Shabat (1974), Manakov *et al.* (1977), Pöppe and Sattinger (1988), Gesztesy *et al.* (1991)). Consider the operator F satisfying

$$\text{(10)}\qquad [F, L_\epsilon^0] = [F, \partial_t - B_\lambda^0] = 0,$$

where

$$\text{(11)}\qquad L_\epsilon^0 = -\partial_x^2 + \epsilon\partial_y, \quad B_\lambda^0 = -4\partial_x^3 + 6\lambda\partial_x.$$

Define Volterra operators $K_\pm$ with

$$(K_\pm f)(t,x,y) = \pm \int_x^{\pm\infty} dz K_\pm(t,x,z,y) f(t,z,y) \tag{12}$$

and assume that

$$1 + F = (1 + K_+)^{-1}(1 + K_-) \tag{13}$$

Then the integral kernels satisfy the GLM equation

$$K_+(t,x,z,y) + F(t,x,z,y) + \int_x^\infty dz' K_+(t,x,z',y) F(t,z',z,y) = 0, \quad z > x. \tag{14}$$

By explicit computations one verifies that

$$\begin{aligned} L_\epsilon &= (1 + K_+) L_\epsilon^0 (1 + K_+)^{-1}, \\ \partial_t - B_{\epsilon,\lambda} &= (1 + K_+)(\partial_t - B_\lambda^0)(1 + K_+)^{-1}, \end{aligned} \tag{15}$$

with

$$V(t,x,y) = -2\frac{d}{dx} K(t,x,x,y) = -2(\frac{\partial}{\partial x} + \frac{\partial}{\partial z}) K(t,x,x,y). \tag{16}$$

The relations (6) and (10) then imply

$$KP(V + \lambda) = 0. \tag{17}$$

To obtain N-soliton solutions of (3) we consider solutions of (10) given by

$$\begin{aligned} F_{\epsilon,p,q}^N(t,x,z,y) &= \sum_{n=1}^N c_n(t,y) e^{-(p_n x + q_n z)}, \\ p &= (p_1, \ldots, p_N), \quad q = (q_1, \ldots, q_N) \in (0,\infty)^N \end{aligned} \tag{18}$$

with

$$\begin{aligned} c_n(t,y) &= c_n \exp[\frac{\epsilon}{2}(p_n^2 - q_n^2) y + [2(p_n^3 + q_n^3) - 3\lambda(p_n + q_n)] t], \\ &\lambda \in \mathbb{R}, \quad c_n \in \mathbb{R} - \{0\}, \quad n = 1, \ldots, N. \end{aligned} \tag{19}$$

We suppress the λ dependence in the following. Then one verifies by explicit computation (see Gardner *et al.* (1974)) that

$$K_+(t,x,z,y) = \sum_{j=1}^N c_j(t,y) \psi_{\epsilon,j}(t,x,y) e^{-q_j z} \tag{20}$$

satisfies (14) if $\psi_{\epsilon,p,q} = (\psi_{\epsilon,1}, \ldots, \psi_{\epsilon,N})$ fulfills

$$(1 + \Lambda^N_{\epsilon,p,q})\psi_{\epsilon,p,q} = d_{\epsilon,p,q}, \quad \epsilon \in \{0, \pm 1\}, \tag{21}$$

where

$$\Lambda^N_{\epsilon,p,q}(t,x,y) = \left[c_n c_m \frac{e^{-(p_n+q_m)x}}{p_n + q_m}\right]^N_{n,m=1}, \quad N \in \mathbb{N}, \epsilon \in \{0, \pm 1\}, p_n, q_m > 0 \tag{22}$$

and

$$d_{\epsilon,p,q}(t,x,y) = (-c_1(t,y)e^{-p_1 x}, \ldots, -c_N(t,y)e^{-p_N x}). \tag{23}$$

Equations (16) and (20) imply that

$$V^N_{\epsilon,p,q}(t,x,y) = -2\frac{\partial}{\partial x}\sum_{j=1}^N c_j(t,y)\psi_{\epsilon,j}(t,x,y)e^{-q_j x}. \tag{24}$$

satisfies

$$KP(V^N_{\epsilon,p,q} + \lambda) = 0. \tag{25}$$

A further matrix manipulation (see Kay and Moses (1956), Gardner *et al.* (1974)) then gives the familiar expression for the N-soliton solution of the KP equation, viz.

$$V^N_{\epsilon,p,q}(t,x,y) = -2\partial_x^2 \ln\det(1 + \Lambda^N_{\epsilon,p,q}(t,x,y)). \tag{26}$$

We can now state and prove the alternative representation formula for $V^N_{\epsilon,p,q}$ which generalizes (2) to the KP equation.

Theorem 1. *Let $V^N_{\epsilon,p,q}$ denote the N-soliton solution of the KP equation given by*

$$V^N_{\epsilon,p,q}(t,x,y) = -2\partial_x^2 \ln\det(1 + \Lambda^N_{\epsilon,p,q}(t,x,y)) \tag{27}$$

which satisfies

$$KP(V^N_{\epsilon,p,q} + \lambda) = 0. \tag{28}$$

Then

$$V^N_{\epsilon,p,q}(t,x,y) = -2\sum_{j=1}^N (p_j + q_j)\psi_{\epsilon,j}(t,x,y)\psi_{-\epsilon,j}(t,x,y), \tag{29}$$

where

$$L_\epsilon \psi_{\epsilon,j} = -\frac{1}{2}(p_j^2 + q_j^2)\psi_{\epsilon,j}, \quad j = 1, \ldots, N, \quad \epsilon \in \{0, \pm 1\} \tag{30}$$

and $\psi_{\epsilon,j}$ satisfies the system of equations (21) (which determines their normalizations as $x \to +\infty$).

Proof. Consider the N-soliton solutions $V^N_{\epsilon,p,q}$ and $V^N_{-\epsilon,q,p}$ constructed using $F^N_{\epsilon,p,q}$ and $F^N_{-\epsilon,q,p}$ respectively. Since

$$\Lambda^N_{\epsilon,p,q} = (\Lambda^N_{-\epsilon,q,p})^* \tag{31}$$

where $*$ denotes the adjoint operation (observe that $c_n(t, y, -\epsilon, q_j, p_j) = c_n(t, y, \epsilon, p_j, q_j)$, $n = 1, \ldots, N$), we see that actually

$$V^N_{\epsilon,p,q} = V^N_{-\epsilon,q,p}. \tag{32}$$

To simplify the notation we write

$$\psi_{\epsilon,p,q} = (\psi_1, \ldots, \psi_N), \quad \psi_{-\epsilon,q,p} = (\psi_1^*, \ldots, \psi_N^*). \tag{33}$$

Then (24) and (32) imply

$$\begin{aligned} V^N_{\epsilon,p,q}(t,x,y) &= -\sum_{j=1}^N \frac{d}{dx}(c_j(t,y)\psi_j(t,x,y)e^{-q_j x} + c_j(t,y)\psi_j^*(t,x,y)e^{-p_j x}) \\ &= -\sum_{j=1}^N c_j(t,y)[(\psi_j(t,x,y)' - q_j\psi_j(t,x,y))e^{-q_j x} \\ &\quad + (\psi_j^*(t,x,y)' - p_j\psi_j^*(t,x,y))e^{-p_j x}] \\ &= A - B + \tilde{A} - \tilde{B}, \end{aligned} \tag{34}$$

where $'$ denotes derivative with respect to x, and

$$\begin{aligned} A &:= \sum_{j=1}^N c_j q_j \psi_j e^{-q_j x}, \quad B := \sum_{j=1}^N c_j \psi_j e^{-q_j x}, \\ \tilde{A} &:= \sum_{j=1}^N c_j p_j \psi_j^* e^{-p_j x}, \quad \tilde{B} := \sum_{j=1}^N c_j \psi_j^* e^{-p_j x}, \end{aligned} \tag{35}$$

(here we temporarily have suppressed the (t, x, y)-dependence in the notation). Relation (21) implies

$$\begin{aligned} A &= -\sum_{j=1}^N \left[q_j\psi_j\psi_j^* + q_j c_j \psi_j \sum_{m=1}^N c_m \frac{e^{-(q_j+p_m)x}}{q_j + p_m}\psi_m^* \right], \\ \tilde{A} &= -\sum_{j=1}^N \left[p_j\psi_j\psi_j^* + p_j c_j \psi_j^* \sum_{m=1}^N c_m \frac{e^{-(p_j+q_m)x}}{p_j + q_m}\psi_m^* \right], \\ B &= -\sum_{j=1}^N \left[\psi_j^*\psi_j' + c_j \psi_j' \sum_{m=1}^N c_m \frac{e^{-(q_j+p_m)x}}{q_j + p_m}\psi_m^* \right], \\ \tilde{B} &= -\sum_{j=1}^N \left[\psi_j\psi_j^{*\prime} + c_j \psi_j^{*\prime} \sum_{m=1}^N c_m \frac{e^{-(p_j+q_m)x}}{p_j + q_m}\psi_m \right]. \end{aligned} \tag{36}$$

The derivative of (21) with respect to x reads

(37)
$$\psi'_n + c_n \sum_{m=1}^{N} c_m \frac{e^{-(p_n+q_m)x}}{p_n + q_m}\psi'_m - c_n \sum_{m=1}^{N} c_m e^{-(p_n+q_m)x}\psi_m = p_n c_n e^{-p_n x}, \qquad n = 1, \ldots, N,$$

which multiplied by ψ'_n and inserted in the definition (35) of $\tilde{A}$ yields

(38)
$$\begin{aligned}\tilde{A} &= \sum_{n=1}^{N}\left(\psi'_n\psi^*_n + \sum_{m=1}^{N}\frac{c_n c_m}{p_n+q_m}e^{-(p_n+q_m)x}\psi'_m\psi^*_n - \sum_{m=1}^{N} c_n c_m e^{-(p_n+q_m)x}\psi_m\psi^*_n\right)\\ &= \sum_{n=1}^{N}\psi'_n\psi^*_n + \sum_{n,m=1}^{N}\frac{c_n c_m}{p_m+q_n}e^{-(p_m+q_n)x}\psi'_n\psi^*_m - \sum_{n,m=1}^{N} c_n c_m e^{-(p_n+q_m)x}\psi_m\psi^*_n\\ &= -B - \sum_{n,m=1}^{N} c_n c_m e^{-(p_n+q_m)x}\psi_m\psi^*_n.\end{aligned}$$

Similarly we find

(39)
$$A = -\tilde{B} - \sum_{n,m=1}^{N} c_n c_m e^{-(p_m+q_n)x}\psi^*_m\psi_n.$$

Hence

(40)
$$\begin{aligned}A + \tilde{A} &= -\sum_{n=1}^{N}(p_n+q_n)\psi_n\psi^*_n - \sum_{n,m=1}^{N} c_n c_m e^{-(p_n+q_m)x}\psi^*_n\psi_m\\ &= -\sum_{n=1}^{N}(p_n+q_n)\psi_n\psi^*_n + \frac{1}{2}(A+\tilde{B}+\tilde{A}+B)\end{aligned}$$

or

(41)
$$\frac{1}{2}(A+\tilde{A}-B-\tilde{B}) = -\sum_{n=1}^{N}(p_n+q_n)\psi_n\psi^*_n$$

which proves (29) using (36).

In order to prove (30) we define

(42)
$$M_n = -\partial_x^2 + \epsilon\partial_y + V + \frac{1}{2}(p_n^2 + q_n^2)$$

which we apply to each component of (21), viz.

(43)

$$
\begin{aligned}
0 &= M_n\psi_n + M_n(c_n e^{-p_n x}) + \sum_{m=1}^{N} M_n\Big(c_n c_m \frac{e^{-(p_n+q_m)x}}{p_n+q_m}\psi_m\Big) \\
&= M_n\psi_n + V c_n e^{-p_n x} + \sum_{m=1}^{N} [M_m + \frac{1}{2}(p_n^2 + q_n^2 - p_m^2 - q_m^2)] c_n c_m \frac{e^{-(p_n+q_m)x}}{p_n+q_m}\psi_m \\
&= M_n\psi_n + \sum_{m=1}^{N} c_n c_m \frac{e^{-(p_n+q_m)x}}{p_n+q_m}(M_m\psi_m) + V c_n e^{-p_n x} \\
&\quad + 2c_n e^{-p_n x}\sum_{m=1}^{N} c_m e^{-q_m x}[\psi_m' - q_m\psi_m] \\
&= M_n\psi_n + \sum_{m=1}^{N} c_n c_m \frac{e^{-(p_n+q_m)x}}{p_n+q_m}(M_m\psi_m), \quad n = 1,\ldots,N
\end{aligned}
$$

using (24). Since $\det(1+\Lambda^N_{\epsilon,p,q}) \neq 0$ we conclude that

(44) $$M_n\psi_n = 0, \quad n = 1,\ldots,N$$

which proves (41) with the + sign. The other case is similar. □

If we ignore the y-dependence we recover the KdV result:

Corollary 2. *If $\epsilon = 0$ and $\kappa_j := p_j = q_j$ for $j = 1,\ldots,N$ then*

(45) $$KdV(V^N_{0,\kappa} + \lambda) = 0$$

and

(46) $$V^N_{0,\kappa}(t,x) = -4\sum_{j=1}^{N} \kappa_j \psi_{0,j}(t,x)^2,$$

where $\psi_{0,j}$ satisfies

(47) $$(-\partial_x^2 + V_{0,N})\psi_{j,0} = -\kappa_j^2\psi_{j,0}, \quad j = 1,\ldots,N$$

and the system (30) with $\epsilon = 0$. (One can show in addition that $\|\psi_{j,0}\|_2 = 1$, $j = 1,\ldots,N$, see Gardner et al. (1974).)

References

Druyma, V.S. (1974), *Analytical solution of the two-dimensional Korteweg–de Vries (KdV) equation*, Sov. Phys. JETP Lett. **19**, 387–388.

Gardner, C.S., Greene, J.M., Kruskal, M.D., and Miura, R.M. (1974), *Korteweg–de Vries equation and generalization. VI. Methods for exact solution*, Comm. Pure Appl. Math. **27**, 97–133.

Gesztesy, F., Holden, H., Saab, E., and Simon, B. (1991), *Explicit construction of solutions of the modified Kadomtsev–Petviashvili equation*, J. Func. Anal. (in print).

Kadomtsev, B.B., Petviashvili, V.I. (1970), *On the stability of solitary waves in weakly dispersing media*, Sov. Phys. Dokl. **15**, 539–541.

Kay, I., Moses, H.E. (1956), *Reflectionless transmission through dielectrics and scattering potentials*, J. Appl. Phys. **27**, 1503–1508.

Konopelchenko, B.G. (1982), *On the gauge-invariant describtion of the evolution integrable by Gelfand–-Dikij spectral problems*, Phys. Lett. **92A**, 323–327.

Manakov, S.V., Zakharov, V.E., Bordag, L.A., Its, A.R., and Matveev, V.B. (1977), *Two-dimensional solitons of the Kadomtsev–Petviashvili equation and their interaction*, Phys. Lett. **63A**, 205–206.

Pöppe, C., Sattinger, D.H. (1988), *Fredholm determinants and the τ function for the Kadomtsev–-Petviashvili hierarchy*, Publ. RIMS, Kyoto Univ. **24**, 505–538.

Zakharov, V.E., Shabat, A.B. (1974), *A scheme for integrating the nonlinear equations of mathematical physics by the method of the inverse scattering problem I*, Func. Anal. Appl. **8**, 226–235.

Department of Mathematics
University of Missouri
Columbia, MO 65211
USA

E-mail: mathfg@umcvmb.bitnet

Division of Mathematical Sciences
The Norwegian Institute of Technology
The University of Trondheim
N-7034 Trondheim
Norway

E-mail: holden@imf.unit.no

On Scalar Conservation Laws in One Dimension

HELGE HOLDEN[1], LARS HOLDEN[2,*]

1 Division of Mathematical Sciences, The Norwegian Institute of Technology, The University of Trondheim, N–7034 Trondheim, Norway

2 Norwegian Computing Center, P.O.Box 114, Blindern, N–0314 Oslo 3, Norway

Abstract

We present a new numerical method for the general Cauchy problem for first order scalar conservation laws in one space dimension, using an idea by Dafermos. Explicit error estimates are provided for the numerical method. The error in the method is far smaller than in any other first order method. The method is furthermore used to give an alternative proof of existence, uniqueness and stability of the solution of the differential equation.

* Supported in part by the Royal Norwegian Council for Technical and Industrial Research (NTNF)

1 INTRODUCTION

In this paper we suggest a numerical method for the Cauchy problem

$$u_t + f(u)_x = 0 \tag{1}$$

$$u(x,0) = u_0(x) \tag{2}$$

where f is absolutely continuous with f' bounded, and u_0 is assumed to be bounded and of locally bounded variation in $\boldsymbol{R}$. Using the convergence of the method it is easy to prove existence, uniqueness and stability of the solution of the Cauchy problem.

It is well-known that there exists a unique weak solution of (1) and (2), see e.g. Lax (1954), Oleinik (1957) and Oleinik (1959), Vol'pert (1967) and Kruzkov (1970). The usual approach to the equation is by finite differences.

The method we use here was introduced in the mathematics literature by Dafermos (1972) as a new method to study the initial value problem, although it had been employed already in 1963 by Barker, see Barker (1963). See also Kale (1982). Using this method Dafermos proved some properties of the solution. He stated that it may be used as a numerical method for f convex or f concave. See also Hedstrom (1972), Swartz and Wendroff (1986). But unaware of Corollary 2.3 in this paper he stated that there is no guarantee that there is a finite number of steps in the algorithm. In fact Corollary 2.3 in this paper is necessary for the numerical method to be well-defined. Lucier (1985) proved that the method may be used as a numerical scheme for a special f and initial value and that the method has optimal convergence. LeVeque (1982) used the method and for each timestep projected the solution back on a grid. Most other numerical methods for (1) and (2) are not optimal, see Lucier (1986).

Dafermos' observation is that with f piecewise linear and the initial value piecewise constant, then the solution is piecewise constant. There are no rarefaction waves, only shocks. The solution u(.,t) is piecewise constant for all t and the solution is found by solving Riemann problems and tracking the shocks.

The numerical method is to approximate f by a piecewise linear function and the inital value by a piecewise constant function. Solving the perturbed problem, we get an approximation to the solution of the original problem. Since the perturbed system is solved exactly, the numerical method is completely stable and introduces no diffusion.

This may be used as a numerical scheme if there is a finite number of constant states. In this paper we prove that there is only a finite number of constant states even in infinite time. Therefore it is possible with a finite algorithm to approximate the solution in infinite time. Usually a numerical method is only finite in finite time. Thus we may describe the method as superfast.

This method may be used to prove existence and uniqueness of the Riemann problem in several space dimensions, see Høegh-Krohn and Risebro (1988). A new type of reservoir simulators, based on the front tracking concept, uses this numerical method, see Bratvedt *et al.* (1989), (1991).

The results of this paper with some numerical examples were announced in Holden, Holden and Høegh-Krohn (1988), see also L. Holden (1990).

It is possible to prove existence, uniqueness and stability of this method by using the convergence of the numerical method. Since the numerical method solves the perturbed problem exactly, the stability results are sharp.

A weak solution of (1) and (2) is by definition a bounded and measurable function u(x,t) which satisfies

$$\int\int [u(x,t)\phi_t(x,t) + f(u(x,t))\phi_x(x,t)]\,dx\,dt + \int u_0(x)\,\phi(x,0)\,dx = 0 \tag{3}$$

for all $\phi \in C_0^1$.

(3) does not define a unique weak solution. To get the physically right solution a so-called entropy condition is necessary in addition to (3). With this entropy condition the solution is unique.

In this paper the entropy condition given in Ballou (1970) is used. To state this condition some definitions are necessary. Assume that u is discontinuous at (x_0, t). Define u_+ and u_- by $u_- = \lim_{x \to x_0 -} u(x,t)$ and $u_+ = \lim_{x \to x_0 +} u(x,t)$. Let $x(t)$ be any curve of discontinuity of the weak solution, and let v be any number lying between u_+ and u_-. We say that u *satisfies the entropy condition along* $x(t)$ *iff*

$$\frac{f(v) - f(u_-)}{v - u_-} \geq \frac{f(u_+) - f(u_-)}{u_+ - u_-} \tag{4}$$

except possibly at a finite number of t. This entropy condition is equivalent with the entropy condition in Oleinik (1957).

It is easy to show (see Smoller (1982)) that if u satisfies (3) and is continuous on each side of a discontinuity the speed $s = x'(t)$ of the discontinuity satisfies the Rankine-

Hugoniot condition

$$s[u] = [f] \tag{5}$$

where $[u]$ and $[f]$ are defined by

$$[u] = u_+ - u_- \text{ and } [f] = f(u_+) - f(u_-).$$

In this paper we prove that there exists a unique weak solution of (1) and (2) satisfying the entropy condition if u_0 is bounded and measurable and f absolutely continuous and f' is bounded.

Since u locally is in $L_1(\boldsymbol{R})$, u_+ and u_- are not necessarily well-defined. Therefore u only satisfies the entropy condition in the weak sense. It is proved that it u is continuous except at a finite number of discontinuities, then u satisfies the entropy condition in the sense of (4).

2 THE NUMERICAL METHOD

In this and the following section we assume f to be continuous and piecewise linear. In the final section we will consider general flux functions.

The simplest problem of the form (1) and (2) is the Riemann problem where

$$u_0(x) = \begin{cases} u_-, & \text{for} \quad x < 0 \\ u_+, & \text{for} \quad x > 0. \end{cases}$$

Some notation is needed before we are able to describe the well-known exact solution. Assume that $u_- < u_+$. First we define f_c to be the convex envelope of f relative to the interval (u_-, u_+). See Figure 2.1 for an example. Since f is continuous and piecewise linear, f_c is also continuous and piecewise linear. We define u_i, $i = 1, 2, ..., N$ by $u_1 = u_-$, $u_N = u_+$ and $u_i < u_{i+1}$ such that f_c is linear in the interval (u_1, u_{i+1}). Then the exact solution of the Riemann problem is

$$u(x,t) = u_i \quad \text{for} \quad t\, f_c'(u_{i-1}+) = t\, f_c'(u_i-) \leq x < t\, f_c'(u_i+) = t\, f_c'(u_{i+1}-), \;\; i = 1, ..., N$$

where $f_c'(u_0+) = f_c'(u_1-) = -\infty$ and $f_c'(u_N+) = f_c'(u_{N+1}) = \infty$. See Figure 2.2. If $u_- > u_+$, the solution is found by replacing x with $-x$ and f with $-f$.

Observe that we do not encounter rarefaction waves with our assumptions on f. As we have seen it is easy to solve the Riemann problem exactly and represent it numerically with our restriction on f. This is the cornerstone of the numerical method.

Given arbitrary u and f, we may approximate u with a piecewise constant function and f with a continuous and piecewise linear function. Then the solution of the Riemann problem on each discontinuity in u is used. It is then possible to solve (1) and (2) until two discontinuity lines collide. $u(.,t)$ is still piecewise constant and it is possible to start over again. See Figure 2.3 for a typical solution $u(x,t)$.

To prove that this is a well-defined procedure it is necessary to prove that if u_0 has a finite number of jumps, then $u(x,t)$ is constant on a finite number of domains for $t < T$. In fact we are going to prove that $u(x,t)$ is constant on a finite number of domains for all t.

First we need a lemma.

Lemma 2.1 *Assume u_0 to be a step function with a finite number of jumps and assume that f is continuous and piecewise linear with a finite number of breakpoints, a breakpoint being a point where the first derivate does not exist. Then for fixed t, $u(x,t)$ is a step function and*

$$u(x,t) \in \{u_0(x);\ x \in \boldsymbol{R}\} \cup \{u;\ f' \text{ is discontinuous at } u\}.$$

Proof. The only place new values can arise are in the Riemann problems. With our assumption on f, the only values that may arise in the Riemann problems are the values where f' is discontinuous.

□

We now need some definitions. Using Lemma 2.1 we let $w_1 < w_2 < \ldots < w_M$ be the values u can take.

A curve of discontinuity for u is called a *shock front.* A point where two or more shock fronts collide is called a *shock collision.* A shock front where u has the value w_i on one side and w_j on the other side is said to contain $\mid i - j \mid$ *shock lines.* The total number of shock lines in $u(.,t)$ is the measure of the total variation of $u(.,t)$. See Figure 2.4 for examples of shock fronts, shock collisions and shock lines.

Proposition 2.2 *Let N be the number of intervals where f is linear, let $L(t)$ be the number of shock lines in $u(.,t)$ and $F(t)$ be the number of shock fronts in $u(.,t)$. Then the function*

$$G(t) = L(t)N + F(t)$$

is strictly decreasing for every shock collision for $t > 0$.

Proof. Assume that $u_1, u_2, ..., u_M$ are the values of $u(x,t)$ which meet in a shock collision. Then $M-1$ shock fronts meet in the collision. See Figure 2.5. We distinguish two cases. In the first case at least one of $u_2, ..., u_{M-1}$ is not between u_1 and u_M, and the second case all $u_2, ..., u_{M-1}$ are between u_1 and u_M.

In the first case at least one of the values $u_2, ..., u_{M-1}$ is not between u_1 and u_M. After the shock collision we obtain at most N shock fronts and the value to the left of the left shock front is u_1 and the value to the right of the right shock front is u_M. If there is more than one shock front after the collision, then the values between the shock fronts form a monotone sequence. Thus the values which were not between u_1 and u_M have disappeared. Then the number of shock lines has decreased and the number of shock fronts has at most increased with $N-2$. Thus $G(t)$ has decreased.

In the second case we may ignore the values which make the sequence $\{u_i\}$ non-monotone. For a monotone sequence $\{u_i\}$ it is easy to prove that after the shock collision all the $M-1$ shock fronts are united into one shock front. Then the number of shock fronts decreases and the number of shock lines does not increase, thus $G(t)$ decreases.

□

Corollary 2.3 *If u_0 has a finite number of jumps and f is linear on a finite number of intervals, then $u(x,t)$ is constant on a finite number of domains.*

Since the maximum speed of a shock in a Riemann problem problem is M, the numerical solution at (x_0, t) is independent of $u_0(x)$ for $x < x_0 - Mt$ and for $x > x_0 + Mt$ where $| f' | < M$.

3 STABILITY

In this section we will prove two theorems. First u_0 is replaced by another step function and secondly, f is replaced by another continuous and piecewise linear function. In both cases it is proved that if the substituted function is near the original function, then the error is small in the L_1 norm.

Theorem 3.1 *Let $u(x,t)$ and $v(x,t)$ be solutions of (1) and (2) with the initial condition $u_0(x)$ and $v_0(x)$ respectively, u_0 og v_0 being step functions with a finite number of* *jumps. Assume f to be continuous and piecewise linear on a finite number of intervals. Then*

$$\int | u(x,t) - v(x,t) | \, dx \leq \int | u_0(x) - v_0(x) | \, dx.$$

Proof. Assume that u_0 and v_0 are constant on the intervals $I_i = (a_1, a_{i+1})$ for $i = 1, ..., M$ where $a_1 = -\infty$ and $a_M = \infty$. For arbitrary i we assume that $u |_{I_i} = w_s$ and $v |_{I_i} = w_t$. Define the increasing sequence s_i, $i = 1, ..., M+1$, by $s_1 = 1$ and $s_{i+1} - s_i = | s - t |$. Let $H = 3s_{M+1}$. Define furthermore the sequence $u_{0,n}$ (see Figure 3.1) by

$$u_{0,n}(x) = \begin{cases} u_0, & \text{for} \quad x \in I_k, \quad k > i \\ v_0, & \text{for} \quad x \in I_k, \quad k < i \\ w_{s+sgn(t-s)[(n-3s_i+2)/3]}, & \text{for } x \in (a_i, (2a_i + a_{i+1})/3) \\ w_{s+sgn(t-s)[(n-3s_i+2)/3]}, & \text{for } x \in ((2a_i, +a_{i+1})/3, (a_i + 2a_{i+1})/3) \\ w_{s+sgn(t-s)[(n-3s_i)/3]}, & \text{for } x \in ((a_i + 2a_{i+1})/3, a_{i+1}) \end{cases}$$

when $3s_i \leq n < 3s_{i+1}$ for $i = 1, 2, ..., M$ and $u_{0,H}(x) = v_0(x)$. Notice that the sequence is defined for $n = 3, 4, ..., H$ and $u_{0,3} = u_0$, $u_{0,3} \equiv u_0$ and

$$u_{0,3s_i}(x) = \begin{cases} u_0, & \text{for} \quad x \in I_k, \ k \geq i \\ v_0, & \text{for} \quad x \in I_k, \ k \leq i+1 \end{cases}$$

for $i = 1, 2, ..., M$. $u_{0,i}$ and $u_{0,i+1}$ are identical except on a third of a I_k interval and the distance between the values is the smallest possible, w_j and w_{j+1} are identical except on a third of a I_k interval and the distance between the values is the smallest possible, w_j and w_{j+1}. Observe that

$$| u_0 - v_0 | = \sum_{i=3}^{H-1} | u_{0,i} - u_{0,i+1} | .$$

Let $u_i(x,t)$ be defined by (1) with initial value $u_{0,i}(x)$.

Assuming

$$\int | u_i(x,t) - u_{i+1}(x,t) | \, dx \leq \int | u_{0,i}(x) - u_{0,i+1}(x) | \, dx \tag{6}$$

we get

$$\int | u(x,t) - v(x,t) | \, dx \leq \sum_{i=3}^{H-1} \int | u_i(x,t) - u_{i+1}(x,t) | \, dx$$
$$\leq \sum_{i=3}^{H-1} \int | u_{0,i}(x) - u_{0,i+1}(x) | \, dx = \int | u_0(x) - v_0(x) | \, dx.$$

Hence it suffices to prove (6).

$u_i(x,t)$ is piecewise constant in the (x,t)-plane. The shock fronts, i.e. the lines where $u_i(x,t)$ is discontinuous, are straight lines between the shock collisions. Since the shock fronts are not parallel with the x-axis, $\int \mid u_i(x,t) - u_{i+1}(x,t) \mid dx$ is continuous and piecewise linear. To prove (3.1) it suffices to prove that for t the right derivative

$$\frac{d}{dt_+} \int \mid u_i(x,t) - u_{i+1}(x,t) \mid dx \Bigg|_{t = \tilde{t}} \leq 0 \tag{7}$$

We may assume $\tilde{t} = 0$.

In order to prove (7) we have to distinguish several cases. First we separate (7) into two main cases depending on whether $u_{0,i}$ and $u_{0,i+1}$ are equal on the middle third of an interval or are equal on the left or right third of an interval. If the two functions are different on the left third this is equivalent to the two functions being different on the right third if we substitute x with $-x$ and f with $-f$. If the two functions are different on the left third, we partition this main case in four subcases depending on the common value of the two functions on the interval to the left of the interval where the two functions differ. The three subcases are: (i) The value to the left is larger than the values of $u_{0,i}$ and $u_{0,i+1}$ on the left third, (ii) the value equals one of $u_{0,i}$ and $u_{0,i+1}$ and (iii) the value is less than $u_{0,i}$ and $u_{0,i+1}$.
In each of the cases it is easy to prove that

$$\frac{d}{dt_+} \int \mid u_i(x,t) - u_{i+1}(x,t) \mid dx \Bigg|_{t = \tilde{t}} = 0$$

□

In this theorem we substitute the initial condition with another step function and prove that the error in L_1 norm does not increase. In the next theorem we substitute the function f with another continuous and piecewise linear function.

First we need a definition.

Definition 3.2 Assume that the step function $u(.,t)$ takes the values $u(x,t) = u_i$ for $x \in (a_i, a_{i+1})$, $i = 1,2,\ldots,M$, for a fixed t. Then we define $u_c(.,t)$ by

$$u_c(x,t) = \begin{cases} u_i, & \text{for } x \in (a_i, a_{i+1} - \varepsilon) \\ u_i + \frac{(x - a_{i+1} + \varepsilon)}{\varepsilon}(u_{i+1} - u_i), & \text{for } x \in (a_{i+1} - \varepsilon, a_{i+1}) \end{cases}$$

where $\varepsilon = \frac{1}{3} \min_i \{a_{i+1} - a_i\}$. $u_c(x,t)$ is a continuous piecewise linear function. See Figure 3.2.

Theorem 3.3 *Let $u(x,t)$ and $v(x,t)$ be solutions of $u_t + f(u)_x = 0$ and $v_t + g(v)_x = 0$ respectively, with initial condition $u(x,0) = v(x,0) = u_0(x)$ where f and g are* *continuous and piecewise linear functions with a finite number of intervals where f and g are linear and u_0 is a step function with a finite number of jumps. Then*

$$\begin{aligned}\frac{d}{dt}\int \mid u(x,t)-v(x,t)\mid dx &\leq T.V._x(f(u_c(x,t))-g(u_c(x,t)))\\ &\leq T.V._x(f(u_{0,c}(x))-g(u_{0,c}(x))).\end{aligned}$$

Remark. $T.V._x(f(u(x)))$ is the total variation of f when x run through the real numbers, i.e.

$$T.V._x(f(u(x))) = \sup_{\{x_i\}}\sum_{i=1}^{N} \mid f(u(x_{i+1}))-f(u(x_i))\mid$$

where $\{x_i\}$ is a finite set of real numbers. Here it is essential that u is continuous.

Before we prove the theorem, we prove two lemmas.

Lemma 3.4 *Let $u(x,t)$ and $v(x,t)$ be solutions of the two Riemann problems $u_t + f(u)_x = 0$ and $v_t + g(v)_x = 0$ respectively, with initial condition*

$$v(x,0) = u(x,0) = \begin{cases} w_n, & \text{for} \quad x<0 \\ w_m, & \text{for} \quad x>0 \end{cases}$$

where f and g are continuous and piecewise linear functions and either $w_n < w_m$ and f and g are convex or $w_n > w_m$ and f and g are concave. Then

$$\frac{d}{dt}\int \mid u(x,t)-v(x,t)\mid dx = \int_{\min(w_n,w_m)}^{\max(w_n,w_m)} \mid f'(u)-g'(u)\mid du.$$

Proof. By substituting if necessary f by $-f$, g by $-g$ and x by $-x$ we can assume that f and g are convex.

f and g are linear in the intervals (w_i, w_{i+1}) and we define s_i and t_i by $s_i = f'\mid_{(w_i,w_{i+1})}$ and $t_i = g'\mid_{(w_i,w_{i+1})}$ for $i = 1,2,...,N-1$. Then

$$A = \int_{w_1}^{w_N} f'(u) - g'(u)\mid dx = \sum_{i=1}^{N-1}(w_{i+1}-w_i)\mid s_i - t_i\mid .$$

The solution of the Riemann problem is constant in sectors centered at $(x,t) = (0,0)$. See Figure 3.3. These sectors are separated by straight lines, shock fronts. Then also $u(x,t) - v(x,t)$ is constant in sectors with center $(x,t) = (0,0)$. The speeds of the shock fronts are s_i and t_i, $i = 2,...,N-1$. Then $\int \mid u(x,t) - v(x,t)\mid dx$ is linear in t, and it is then easily shown that

$$\frac{d}{dt}\int \mid u(x,t) - v(x,t)\mid dx = A.$$

□

Before we are able to state the next lemma we need another definition. We define $f_{c(a,b)}(x)$ to be the convex envelope to f in the interval (a, b).

Lemma 3.5 *Assume f and g to be continuous and piecewise linear functions on (a, b). Then*

$$\int_a^d \mid f'_{c(a,d)}(x) - g'_{c(a,d)}(x) \mid dx \leq \int_a^d \mid f'(x) - g'(x) \mid dx.$$

A corresponding lemma with the same inequality holds for the concave envelope.

Proof. We may assume that f and g have common breakpoints. It suffices to prove the lemma when (a, d) contains two intervals where f and g are linear on each interval, hence we assume that f and g are linear on (a, b) and (b, d). Let $F(x) = f(x) - g(x)$. Then

$$\int_a^d \mid f'(x) - g'(x) \mid dx = \mid F(b) - F(a) \mid + \mid F(b) - F(d) \mid .$$

Define

$$G(x) = f_{c(a,d)}(x) - g_{c(a,d)}(x).$$

Notice that $G(a) = F(a)$ and $G(d) = F(d)$.
Assume first that both f and g are concave. Then f_c and g_c are linear on (a, d). In this case it is easy to prove the lemma since

$$\begin{aligned}\int_a^d \mid f'_{c(a,d)}(x) - g'_{c(a,d)}(x) \mid dx &= \mid G(d) - G(a) \mid = \mid F(d) - F(a) \mid \\ &\leq \mid F(b) - F(a) \mid + \mid F(d) - F(b) \mid = \int_a^d \mid f'(x) - g'(x) \mid dx.\end{aligned}$$

It then remains to prove the lemma when one function is convex and the other is concave. Assume f is concave and g is convex. By definition we find

$$G(x) = f_c(x) - g_c(x) \leq f(x) - g_c(x) = f(x) - g(x) = F(x).$$

G is concave since f_c and $-g$ are concave. Then

$$G(b) \geq \frac{k_1}{k_1 + k_2} F(a) + \frac{k_2}{k_1 + k_2} F(d)$$

where

$$k_1 = b - a \quad \text{and} \quad k_2 = d - b.$$

Thus

$$\begin{aligned} G(b) &= \gamma\left[\tfrac{k_1}{k_1+k_2}F(a) + \tfrac{k_2}{k_1+k_2}F(d)\right] + (1-\gamma)F(b) \\ &= \alpha F(x) + \beta F(d) + (1-\alpha-\beta)F(b) \end{aligned}$$

for $\alpha,\ \beta,\ \gamma \in [0,1]$. Hence we are able to prove the lemma with

$$\begin{aligned} \int_a^b \mid f_c'(x) - g_c'(x) \mid dx &= \mid G(b) - G(a) \mid + \mid G(b) - G(d) \mid \\ &= \mid \beta(F(d) - f(b)) + (1-\alpha)F(b) - F(a)) \mid \\ &+ \mid (1-\beta)(F(b) + F(d)) + \alpha(F(a) - F(b)) \mid \end{aligned}$$

$$\leq \mid F(b) - F(a) \mid + \mid F(d) - F(b) \mid = \int_a^d \mid f'(x) - g'(x) \mid dx.$$

□

We can now prove the theorem.

Proof of Theorem 3.3. We may without loss of generality assume $t = 0$. For sufficiently small ϵ, the differential equation is a series of independent Riemann problems when $t \in (0,\epsilon)$.

Assume u_0 to be constant on the intervals $I_i,\ i = 1,2,...,M$ and $u_0(x) = u_i$ for $x \in I_i\ i = 1,2,...,M$.

Define f_{c_i} as follows: If $u_i < u_{i+1}$, then f_{c_i} is the convex envelope of $f\mid_{(u_i,u_{i+1})}$ and if $u_i > u_{i+1}$, then f_{c_i} is the concave envelope of $f\mid_{(u_{i+1},u_i)}$. Then the three preceeding lemmas may be used to conclude that

$$\begin{aligned} \frac{d}{dt}\int \mid u(x,t) - f(x,t) \mid dx &= \sum_{i=1}^{M-1}\int_{\min(u_i,u_{i+1})}^{\max(u_i,u_{i+1})} \mid f_{c_i}'(u) - g_{c_i}'(u) \mid du \\ \leq \sum_{i=1}^{M-1}\int_{\min(u_i,u_{i+1})}^{\max(u_i,u_{i+1})} \mid f'(u) - g'(u) \mid du &= \sum_{i=1}^{M-1} T.V._{\cdot u \in (\min(u_i,u_{i+1}),\max(u_i,u_{i+1}))}(f(u) - g(u)) \\ = \sum_{i=1}^{M-1} T.V._{\cdot x \in (a_i,a_{i+1})}(f(u_c(x,t)) - g(u_c(x,t))) &= T.V._{\cdot x}(f(u_c(x,t)) - g(u_c(x,t))). \end{aligned}$$

It remains to prove

$$T.V._{\cdot x}(f(u_c(x,t)) - g(u_c(x,t))) \leq T.V._{\cdot x}(f(u_{0,c}(x)) - g(u_{0,c}(x))).$$

This inequality is proved by using that the only new values for u arise in the Riemann problems, i.e. in following situation

$$u_0(x,0) = \begin{cases} u_1, & x < 0 \\ u_n, & x > 0. \end{cases}$$

Hence the solution of the Riemann problem is of the form

$$u(x,t) = \begin{cases} u_1, & x < s_1 t \\ u_i, & s_i t < x < s_{i+1} t \quad \text{for} \quad i = 1,2,...,n-1. \end{cases}$$

where $u_1 < u_2 < ... < u_n$. Then u_c and $u_{0,c}$ run through the same values, which proves the inequality.

□

This theorem may be used together with Theorem 3.1 if both u_0 and f are changed. In the next section both theorems are used to prove existence and uniqueness for (1) and (2) and error estimates for the numerical method.

4 EXISTENCE, UNIQUENESS AND ERROR ESTIMATES

In this section the results from the previous sections are used to prove existence and uniqueness for (1) and (2). Finally we state the stability result.

Theorem 4.1 *Let u_0 be a bounded and measurable and f absolute continuous with f' bounded. Then there exists a unique weak entropy solution of*

$$u_t + f(u)_x = 0$$

with initial condition

$$u(x,0) = u_0(x)$$

which satisfies the following: For all piecewise constant functions v_0, all continuous and piecewise linear functions g,

$$\int_{a+M\,t}^{b-M\,t} | u(x,t) - v(x,t) | \, dx \tag{8}$$

$$\leq \int_a^b | u_0(x) - v_0(x) | \, dx + t\, T.V._{x\in[a,b]}(f(v_{0,c}(x)) - g(v_{0,c}(x)))$$

where $a < b$ and $t > 0$ are arbitrary and where $| f' | < M$, $| g' | < M$ and $v(x,t)$ is the piecewise constant solution of

$$v_t + g(v)_x = 0$$

with initial condition

$$v(x,0) = v_0(x),$$

which satisfies the entropy condition.
$u(.,t)$ *is locally in* $L_1(\boldsymbol{R})$ *for all* t *and locally in* $L_1(\boldsymbol{R} \times \boldsymbol{R}^+)$.

Furthermore there exists a sequence $\{(v_{0,i}, g_i)\}$ *such that* $v_i(x,t)$ *converges to* $u(x,t)$ *locally in* L_1 *norm and both terms at the right side of (4.1) vanish when* $i \to \infty$ *and* g *converges uniformly to* f.

In Smoller (1982) it is proved that it is not possible to get uniqueness without using the entropy condition. But the entropy condition in the form in the first section is not well-defined in L_1.

We get around this problem by approximating u by functions where the entropy condition is well-defined and then use the stability condition (8).

In the next theorem we will prove that if u is piecewise smooth, then u satisfies the entropy condition.

First we need the following simple lemma.

Lemma 4.2 *Assume that the measurable function* f *is approximated by a sequence of measurable uniformly bounded functions* $\{g_n\}$ *satisfying*

$$| g_n(x) - f(x) | < \frac{1}{na_n} \quad \text{for} \quad x \in [a,b] \setminus A_n$$

where the Lebesgue measure of $A_n, m(A_n)$, *satisfies* $m(A_n) < \frac{1}{na_n}$, $\{a_n\}$ *being an increasing sequence of real numbers.*
Then for $m > n$, *the sequence* $\{g_n\}$ *satisfies the following Cauchy criterion*

$$\int_a^b | g_n(x) - g_m(x) | \, dx \leq \frac{2}{na_n}(b - a) + \frac{4M}{na_n}$$

where M *is such that* $| g_n | < M$.

Proof. The lemma is proved by dividing the interval (a,b) into three subintervals:

$$\begin{aligned} & \int_a^b | g_n(x) - g_m(x) | \, dx \\ \leq \; & \int_{(a,b)\setminus(A_n \cup A_m)} | g_n(x) - g_m(x) | \, dx \end{aligned}$$

$$+ \int_{A_n} | g_n(x) - g_m(x) | \, dx + \int_{A_m} | g_n(x) - g_m(x) | \, dx$$
$$\leq (b-a) \left[\tfrac{1}{n a_n} + \tfrac{1}{m a_m} \right] + \frac{2M}{n a_n} + \frac{2M}{m a_m} \leq (b-a)\frac{2}{n a_n} + \frac{4M}{n a_n}.$$

□

We can now prove Theorem 4.1.

Proof of Theorem 4.1 The maximum speed of information is less than M. Thus in the domain

$$\Omega = \{(x,t);\, x \in [a + Mt, b - Mt], \ \ t < T\}$$

$u(x,t)$ is independent of u_0 outside [a,b]. Since we study $u(x,t)$ in Ω, we may assume $u_0(x) = 0$ for $x < a$ and for $x > b$. Then $u_0 \in L_1(\boldsymbol{R})$.

First we will prove existence of a special $u(x,t)$, and then prove that it satisfies the weak formulation (3). Finally we will prove that $u(x,t)$ satisfies (8) for all entropy solutions $v(x,t)$.

In order to prove existence of $u(x,t)$ it is sufficient to define a Cauchy sequence $u_i(x,t)$ in $L_1(\boldsymbol{R} \times \boldsymbol{R}^+)$, and define $u(x,t)$ as the limiting function. This is possible since L_1 is complete.

Given u_0 and f, we define the sequence of functions $u_{0,n}$ as follows: u_n equals the solution of $u_{n_t} + f(u_n)_x = 0$ with initial condition $u_n(x,0) = u_{0,n}(x)$ where f_n and $u_{0,n}$ remain to be defined. In Royden (1968) it is proved that there exists a sequence $\{u_{0,n}\}$ which satisfies

$$| u_{0,n}(x) - u_0(x) | \leq \frac{1}{n} \quad \text{for all} \quad x \in \boldsymbol{R} \setminus A_n \quad \text{where} \quad m(A_n) < \frac{1}{n}.$$

We may assume that $u_{0,n}$ is bounded and $u_{0,n}(x) = 0$ for $x < a$ and $x > b$, viz. $| u_0(x) | < M$ and $| u_{0,n}(x) | < M$ for all n. Let N_n be the number of intervals where $u_{0,n}(x)$ is constant. Then the continuous function $u_{0,n,c}(x)$ is defined from $u_{0,n}(x)$ using Definition 3.1.

The sequence $\{C_n\}$ is defined by $C_n = \max\{N_1, N_2, ..., N_n\}$. At this moment $\{u_{0,n}\}$ and some constants and functions depending on this sequence are defined. Now we can define a sequence of continuous and piecewise linear functions which approximate f. In the definition we use the fact that f' is a bounded and measurable function. Using Theorem 3.22 in Royden (1968) there is a sequence of step functions $\{g_n\}$ such that

$$| g_n(u) - f'(u) | < \frac{1}{n\, M C_n} \quad \text{for} \quad u \in \boldsymbol{R} \setminus A_n \quad \text{where} \quad m(A_n) < \frac{1}{n\, M C_n}.$$

This is possible if we assume $f'(u) = 0$ for $|u| > M$. We also assume $M > b - a$. Then f_n is defined by

$$f_n(u) = f(-M) + \int_{-M}^{u} g_n(t)\,dt.$$

Since $u_{0,i}(x)$ is a step function and f_i is continuous and piecewise linear, the theorems in Sec. 3 may be used to prove that u_i is a Cauchy sequence. Let $j > i$. Then

$$\int |u_i(x,t) - u_j(x,t)|\,dx \le \int |u_{0,i}(x) - u_{0,j}(x)|\,dx + t\,T.V._x(f_i(u_{0,i,c}(x)) - f_j(u_{0,i,c}(x))).$$

Both terms on the right vanish when i increases. Lemma 4.2 gives

$$\int |u_{0,i}(x) - u_{0,j}(x)|\,dx \le 2\frac{b-a}{i} + \frac{4M}{i} \le \frac{6M}{i}.$$

For the second term we get

$$T.V._x(f_i(u_{0,i,c}(x)) - f_j(u_{0,i,c}(x))) \le N_i\,T.V._u(f_i(u) - f_j(u)) = N_i \int_{-M}^{M} |f_i'(u) - f_j'(u)|$$
$$= N_i \int_{-M}^{M} |g_i(u) - g_j(u)|\,du \le N_i\left[\frac{2M}{iMC_i} + \frac{4M}{iMC_i}\right] \le \frac{6}{i}$$

where Lemma 4.2 is used in the second inequality. Then

$$\int_0^T \int_R |u_i(x) - u_j(x)|\,dx$$
$$\le T\int_R |u_{0,j}(x) - u_{0,i}(x)|\,dx + \frac{1}{2}T^2\,T.V._x(f_i(u_{0,i,c}(x)) - f_j(u_{0,i,c}(x)))$$
$$\le T\frac{6M}{i} + \frac{1}{2}T^2\frac{6M}{i} \to 0 \quad \text{when} \quad i \to \infty.$$

$u(x,t)$ is then defined as the limit of the Cauchy sequence $u_i(x,t)$. Next we will prove that $u(x,t)$ is a weak solution of (1) and (2), i.e.

$$\int_0^T \int [u(x,t)\varphi_t(x,t) + f(u(x,t))\varphi_x(x,t)]\,dx\,dt + \int u_0(x)\varphi(x,0)\,dt = 0.$$

Since $u_i(x,t)$, $i = 1,2,\ldots$ are weak solutions they satisfy

$$\int_0^T \int [u_i(x,t)\varphi_t(x,t) + f_i(u(x,t))\varphi_x(x,t)]\,dx\,dt + \int u_{0,i}(x)\varphi(x,0)\,dt = 0$$

for all i. Then

$$\left|\int_0^T \int [u(x,t)\varphi_t(x,t) + f(u(x,t))\varphi_x(x,t)]\,dx\,dt + \int u_0(x)\varphi(x,0)\,dt\right|$$
$$= \left|\int_0^T \int [(u(x,t) - u_i(x,t))\varphi_t(x,t) + (f(u(x,t)) - f_i(u_i(x,t)))\varphi_x(x,t)]\,dx\,dt\right.$$
$$\left. + \int (u_0(x) - u_{0,i}(x))\varphi(x,0)dt\right|$$

$$\leq \int_0^T \int [| (u(x,t) - u_i(x,t))\varphi_t(x,t) | + | (f(u(x,t)) - f(u_i(x,t)))\varphi_x(x,t) |] \, dx \, dt$$
$$+ \int_0^T \int | (f(u_i(x,t)) - f_i(u_i(x,t)))\varphi_x(x,t) | \, dx \, dt$$
$$+ \int | (u_0(x) - u_{0,i}(x))\varphi(x,0) | \, dt.$$

Next we will be bound each term. Assume that $M > \max\{\varphi, \varphi_x, \varphi_t\}$. Then

$$\int_0^T \int | (u(x,t) - u_i(x,t))\varphi_t(x,t) | \, dx \, dt \leq M \int_0^T \int | (u(x,t) - u_i(x,t)) | \, dx \, dt \to 0$$

using the definition of $u(x,t)$. Similarly

$$\int | (u_0(x) - u_{0,i}(x)) | \varphi(x,0) \, dx \leq M \int | (u_0(x) - u_{0,i}(x)) | \, dx \to 0$$

using the definition of $u_{0,i}(x,t)$.

Since f is continuous in $[-M, M]$, f is uniformly continuous. Hence for sufficiently large M

$$\int_0^T \int | (f(u(x,t)) - f(u_i(x,t)))\varphi_x(x,t) | \, dx \, dt$$
$$\leq M \int_0^T \int | (u(x,t) - u_i(x,t))\varphi_x(x,t) | \, dx \, dt$$
$$\leq M^2 \int_0^T \int | (u(x,t) - u_i(x,t)) | \, dx \, dt \to 0$$

from the definitionen of $u(x,t)$. Since $f(u) = f(-M) + \int_{-M}^{u} f'(t) \, dt$ and $f_i(u) = f(-M) + \int_{-M}^{u} g_i'(t) \, dt$,

$$| f(u) - f_i(u) | \leq \int_{-M}^{u} | f'(t) - g_i(t) | \, dt \leq \frac{C}{i}$$

for a constant C, f_i converges uniformly towards f. Then

$$\int_0^T \int | (f(u(x,t)) - f(u(x,t)))\varphi_x(x,t) | \, dx \, dt$$
$$\leq \frac{C}{i} \int_0^T \int | \varphi_x(x,t) | \, dx \, dt \to 0$$

when $i \to \infty$.

It remains to prove that $u(x,t)$ is the unique function satisfying (8).

Let $v(x,t)$ be as in the theorem and u_i as in the begining of this proof. Then Theorems 3.1 and 3.2 give

$$\int | u(x,t) - v(x,t) | \, dx$$

$$
\begin{aligned}
&\leq \int | u(x,t) - u_i(x,t) | \, dx + \int | u_i(x,t) - v(x,t) | \, dx \\
&\leq \int | u(x,t) - u_i(x,t) | \, dx + \int |_{0,i} (x) - v_0(x) | \, dx + t\, T.V._x(f_i(v_{0,c}(x)) - g(v_{0,c}(x))) \\
&\leq \int | u(x,t) - u_i(x,t) | \, dx + \int |_{0,i} (x) - u_0(x) | \, dx + \int | u_0(x) - v_0(x) | \, dx \\
&+ \; t\, T.V._x(f_i(v_{0,c}(x)) - g(v_{0,c}(x))) + t\, T.V._x(f(v_{0,c}(x)) - g(v_{0,c}(x))).
\end{aligned}
$$

In order to prove (8), it is necessary to prove that the first, second and fourth term vanish when i increases.

Using the definitions of $u(x,t)$ and $u_{0,i}(x)$ it is easy to prove that the first and the second term vanish. Let $N(v_0)$ be the number of jumps in v_0 on the interval [a,b]. Given $\epsilon > 0$, there exists a partition $\{v_j\}$ such that

$$
\begin{aligned}
& T.V._x(f_i(v_{0,c}(x)) - f(v_{0,c}(x))) \leq N(v_0) T.V._v(f_i(v) - f(v)) \\
\leq \; & N(v_0) \sum_{j=1}^{N(v_0)} | f_i(v_j) - (f_i(v_{j+1}) - f(v_{j+1})) | + \epsilon \\
= \; & N(v_0) \sum_{i=1}^{N(v_0)} | \int_{v_i}^{v_{i+1}} f_i'(u) - f'(u) \, du | + \epsilon \leq N(v_0) \int_a^b | f_i'(u) - f'(u) | \, du + \epsilon.
\end{aligned}
$$

Since $\epsilon > 0$ is arbitrary, we have

$$
T.V._x(f_i(v_{0,c}(x)) - f(v_{0,c}(x))) \leq N(v_0) \int_a^b | f_i'(u) - f'(u) | \, du. \tag{9}
$$

Using the definition of f_i, we may get this expression arbitrary small. Thus we have proved that $u(x,t)$ satisfies (8). It remains to prove that $u(x,t)$ is unique. Assume $v(x,t)$ is another function which satisfies (1) and (8). Using (8) on v, then (9), the definitions of f_i' and finally Lemma 4.2, we obtain

$$
\begin{aligned}
& \int | u(x,t) - v(x,t) | \, dx \\
\leq \; & \int | u(x,t) - u_i(x,t) | \, dx + \int | u_i(x,t) - v(x,t) | \, dx \\
\leq \; & \int | u(x,t) - u_i(x,t) | \, dx + \int | u_{0,i}(x) - u_0(x) | \, dx \\
+ \; & t\, T.V._x(f(u_{0,i,c}(x)) - f(u_{0,i,c}(x))) \\
\leq \; & \int | u(x,t) - u_i(x,t) | \, dx + \int | u_{0,i}(x) - u_0(x) | \, dx \\
+ \; & t\, N(u_{0,i}) \int_a^b | f_i'(u) - f'(u) | \, du \\
\leq \; & \int | u(x,t) - u_i(x,t) | \, dx + \int | u_{0,i}(x) - u_0(x) | \, dx \\
+ \; & t\, N(u_{0,i})((b-a)\frac{2}{i\,MC_i} + \frac{4M}{i\,MC_i}) \\
\leq \; & \int | u(x,t) - u_i(x,t) | \, dx + \int | u_{0,i}(x) - u_0(x) | \, dx + \frac{6t}{i}.
\end{aligned}
$$

Using the definitions of $u_{0,i}$ and $u(x,t)$ it is easy to prove that these two terms vanish when i increases. Thus $u(x,t) = v(x,t)$ almost everywhere.

□

This estimate may also be used as an estimate for the error in the numerical method described in section 2. The following theorem is a weaker error estimate, however it is easier to use.

Corollary 4.3 *Let $N(v_0)$ be the number of jumps in the step function $v_0(x)$, and assume that $u_0(x)$ and $v_0(x)$ are bounded below and above by c and d respectively. Then*

$$\int_{a+Mt}^{b-Mt} | u(x,t) - v(x,t) | \, dx \tag{10}$$

$$\leq \int_a^b | u_0(x) - v_0(x) | \, dx + t\, N(v_0)\, T.V._{u\in[c,d]}(f(u) - g(u)),$$

where we have used the same definitions as in Theorem 4.1.

Proof. Writing

$$u_{0,i}(x) = u_j \quad \text{for} \quad x \in (x_j, x_{j+1}) \quad \text{for} \quad j = 1, 2, ..., N(u_{0,i})$$

where $x_1 = a$ and $x_{N(u_{0,i})} = b$, we find

$$T.V._{x\in[a,b]}(f(u_{0,i,c}(x)) - g(u_{0,i,c}(x))) \leq \sum_{j=1}^{N(u_{0,i})} T.V._{u\in[u_j,u_{j+1}]}(f(u) - g(u))$$

$$\leq N(u_{0,i})\, T.V._{u\in[c,d]}(f(u) - g(i)).$$

□

Theorem 4.4 *If $u(x,t)$ is the solution of (1.1) and (1.2) constructed in Theorem 4.1 and if $u(x,t)$ is continuous except along a finite number of piecewise continuous differentiable curves, then $u(x,t)$ satisfies the entropy condition except at a finite number of points.*

Remark. See Schaeffer (1973) for a similar result.

Proof. Assume P to be a point on a curve of discontinuity where the entropy condition is not satisfied. We will prove that if $u(x,t)$ is continuous on both sides of this curve, then this curve will be separated by two or more curves of discontinuity leaving P in the direction of increasing t. ¿From the smoothness assumptions on $u(x,t)$ this happens only a finite number of times. So there is only a finite number of such points P.

This is proved by assuming there is only one curve of discontinuity leaving P and prove that this gives a contradiction.

Assume that $P = (0,0)$. We define $u_+ = \lim_{x\to 0+} u(x,0)$ and $u_- = \lim_{x\to 0-} u(x,0)$. Let $u_+ = u_- + A$. Without loss of generality assume $A > 0$. Since the entropy condition is not satisfied on P, there exists a z between u_- and u_+ such that

$$\frac{f(z) - f(u_-)}{x - u_-} = \frac{f(u_+) - f(u_-)}{u_+ - u_-} + B$$

for $B > 0$. For this z

$$f(x) < f(u_-) + \frac{x - u_-}{u_+ - u_-}(f(u_+) - f(u_-)) - C$$

for $C > 0$. See Figure 4.1.

u is approximated by two functions v and w. In the definition of v we use that u is continuous and in the defintion of w we use the numerical method described in Sec. 2.

Given $\epsilon > 0$, there exists $\delta > 0$ such that

$$|u(x,t) - u_-| < \epsilon \quad \text{for} \quad |x| < 2\delta \quad \text{and} \quad |t| < \delta$$

to the left of the curve of discontinuity and

$$|u(x,t) - u_+| < \epsilon \quad \text{for} \quad |x| < 2\delta \quad \text{and} \quad |t| < \delta$$

to the right of the curve of discontinuity. See Figure 4.2a. Define v_0 by

$$v_0(x) = \begin{cases} u_-, & \text{for} \quad x < 0 \\ u_+, & \text{for} \quad x > 0. \end{cases}$$

Let the direction of the curves discontinuity in the point P be s. See Figure 4.2b. Then we define

$$v(x,t) = \begin{cases} u_-, & \text{for} \quad x < st \\ u_+, & \text{for} \quad x > st. \end{cases}$$

For sufficiently small ϵ

$$\int_{-\delta}^{\delta} |u(x,t) - v(x,t)|\, dx \le 2\delta\epsilon + 0(t^2).$$

Let $w(x,t)$ be the solution of $w_t + g(w)_x = 0$ with initial condition $w(x,0) = v_0(x)$ where g is continuous and piecewise linear. Then Theorem 4.1. gives

$$\int_{-\delta}^{\delta} | u(x,t) - w(x,t) | \, dx \leq \int_{-\delta}^{\delta} | u_0(x) - v_0(x) | \, dx + t\, T.V._{x\in[-\delta,\delta]}(f(v_{0,c}(x) - g(v_{0,c}(x)).$$

Using Theorem 4.1, we may choose g such that

$$\int_{-\delta}^{\delta} | u(x,t) - w(x,t) | \, dx \leq 2\delta\epsilon + t\epsilon.$$

$v(x,t)$ is also the solution of the differential equation $v_t + h(v)_x = 0$ with initial condition $v(x,0) = v_0(x)$ for a $h(v)$ linear between the values u_- and u_+ with derivative s and $h(u_-) = f(u_-)$. Lemma 3.3 then gives

$$\int_{-\delta}^{\delta} | u(x,t) - w(x,t) | \, dx = t \int_{u}^{u_+} | g'(u) - h'(u) | \, du$$
$$\geq t \,| g(x) - h(z) | \geq t(| f(z) - h(z) | - | g(z) - f(z) |) \geq t(C - | g(z) - f(z) |)$$

where g_c is the convex envelope to g. We may choose g arbitrarily near f_c uniformly so

$$\int_{-\delta}^{\delta} | v(x,t) - w(x,t) | \, dx \geq t\,C. \tag{11}$$

However

$$\begin{aligned}\int_{-\delta}^{\delta} | v(x,t) - w(x,t) | \, dx &\leq \int_{-\delta}^{\delta} | v(x,t) - u(x,t) | \, dx + \int_{-\delta}^{\delta} | u(x,t) - w(x,t) | \, dx \\ &\leq 4\delta\epsilon + \epsilon t + 0(t^2).\end{aligned}$$

This contradicts (11) and proves the theorem.

□

ACKNOWLEDGEMENTS

This paper is based on joint work with the late Raphael Høegh-Krohn. The results were presented in Holden, Holden, and Høegh-Krohn (1988), but the full proofs appear here for the first time.

5 REFERENCES

Ballou, D. (1970), *Solutions to nonlinear hyperbolic Cauchy problems without convexity conditions,* Amer. Math. Soc. Trans., **152**, 441-460.

Barker, L.M. (1963), *SWAP – A computer program for shock wave analysis*, Technical report no SC 4796(RR), Sandia National Laboratories, Albuquerque, New Mexico.

Bratvedt, F., Bratvedt, K., Buchholz, C.F., Gimse, T., Holden, H., Holden, L., and Risebro, N.H. (1991), *Front tracking for petroleum reservoirs*, These proceedings.

Bratvedt, F., Bratvedt, K., Buchholz, C.F., Holden, H., Holden, L. and Risebro, N.H. (1989), *A new front-tracking method for reservoir simulation,* presented at 64th Annual Technical Conference and Exhibition, SPE, San Antonio, October 1989. To appear in SPE Reservoir Engineering.

Dafermos, C.M. (1972), *Polygonal approximation of solutions of the initial value problem for a conservation law,* J. Math. Anal. Appl., **38**, 33-41.

Hedstrom, G.W. (1972), *Some numerical experiments with Dafermos's method for nonlinear hyperbolic problem,* in: Numerische Lösung nichtlinearer partieller Differential- und Integrodifferentialgleichungen (Eds. R. Ansorge, W. Törnig), Lecture Notes in Mathematics, Volume 267, Springer-Verlag, Berlin.

Høegh-Krohn, R. and Risebro, N.H. (1988), *The Riemann problem for hyperbolic conservation laws in several space dimensions,* Preprint, University of Oslo, Norway.

Holden, L. (1990), *Some problems related to conservation laws and front tracking,* Dr. Philos. Dissertation, University of Oslo.

Holden, H., Holden, L. and Høegh-Krohn, R. (1988), *A numerical method for first order nonlinear scalar hyperbolic conservation laws in one dimension,* Comput. Math. Applic., **15**, 595-601.

Kale, D. M. (1982), *Dynamics of moving interfaces in porous media: III. Extension of Buckley–Leverett theory,* Society of Petroleum Engineers, preprint no 11249, 2-11.

Kruzkov, S.N. (1970), *First order quasilinear equations in several independent variables,* Math. USSR Sbornik, **10**, 217-243.

Lax, P.D. (1954), *Weak solutions of hyperbolic equations and their numerical computation,* C omm. Pure Appl. Math., **7**, 159-193.

LeVeque, R.J. (1982), *A large step shock-capturing techniques for scalar conservaton laws,* SIAM J. Numer. Anal., **19**, 1051-1073.

Lucier, L.J. (1985), *Error bounds for the methods of Glimm, Godunov and LeVeque,* SIAM J. Numer. Anal., **22**, 1074-1081.

Lucier, L.J. (1986), *A moving mesh numerical method for hyperbolic conservation laws,* Math. Comp., **46**, 59-69.

Oleinik, O.A. (1957), *Discountinuous solutions of non-linear differential equations,* Usp. Mat. Nauk (N.S.), **12**, 3-73, English transl. Amer. Math. Soc. Transl. Ser. 2, **26** (1963) 95-172.

Oleinik, O. A. (1959), *Uniqueness and a stability of the generalized solution of the Cauchy problem for a quasilinear equation,* Usp. Mat. Nauk., **14**, 165-170, English transl. Amer. Math. Soc. Transl. Ser. 2, **33** (1964) 285-290.

Royden, H.L. (1968), *Real Analysis,* MacMillan Publ.

Schaeffer, D.G. (1973), *A regularity theorem for conservation laws,* Adv. Math., **11**, 368-386.

Smoller, J. (1982), *Shock Waves and Reaction-Diffusion Equations,* Springer-Verlag, New-York-Heidelberg-Berlin.

Swartz, B.K., Wendroff, B. (1986), *AZTEC: A front tracking code based on Godunov's method,* Appl. Num. Math., **2**, 385-397.

Vol'pert, A.I. (1967), *The spaces BV and quasilinear equations,* Math. USSR Sbornik, **2**, 225-267.

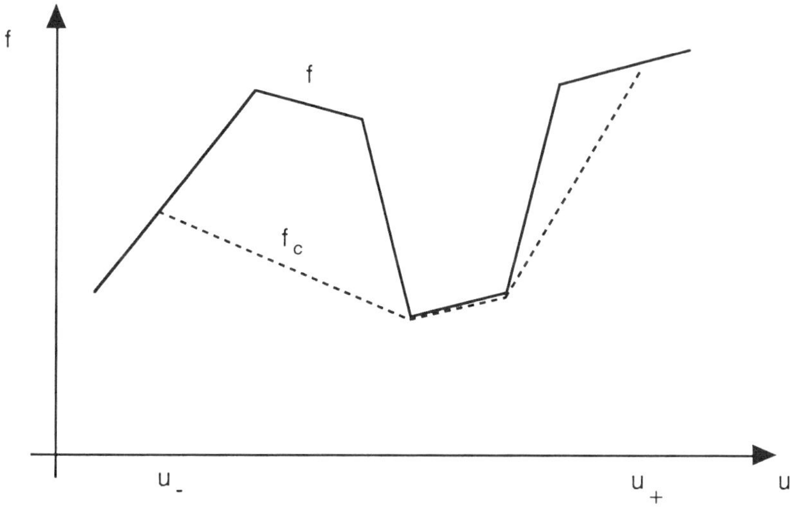

Figure 2.1

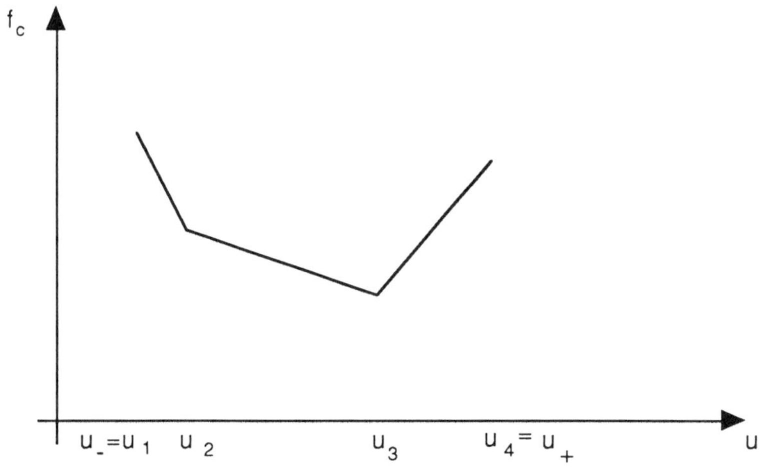

Figure 2.2a

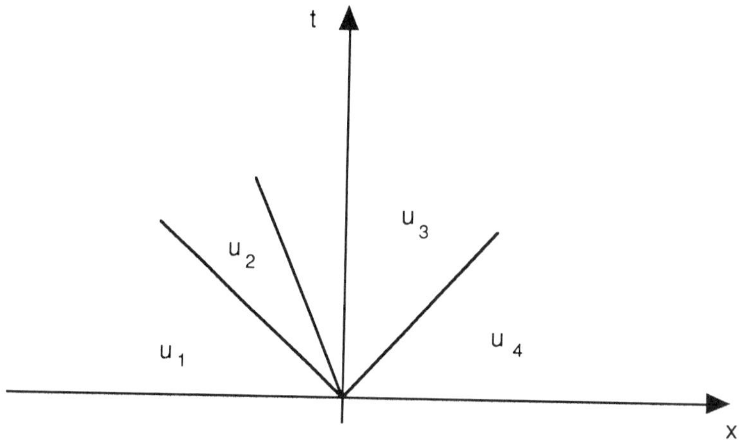

Figure 2.2b

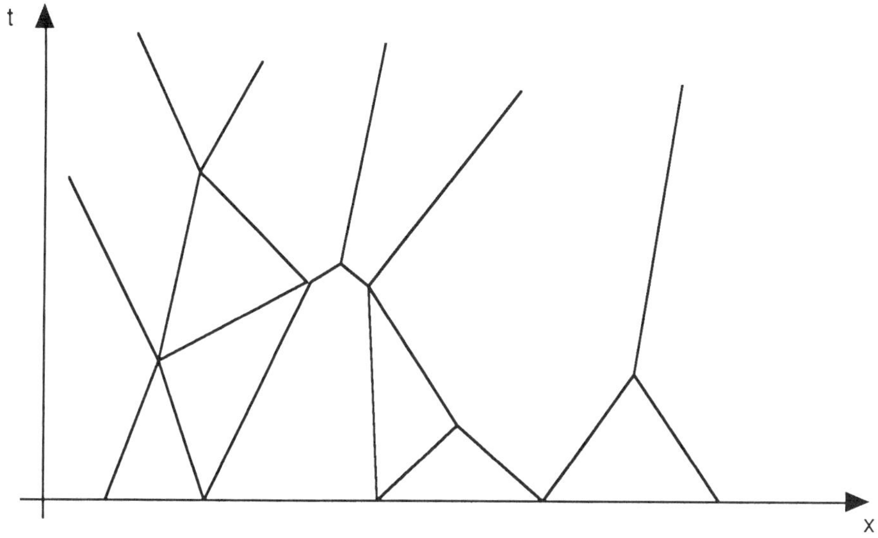

Figure 2.3 Regions where u(x,t) is constant

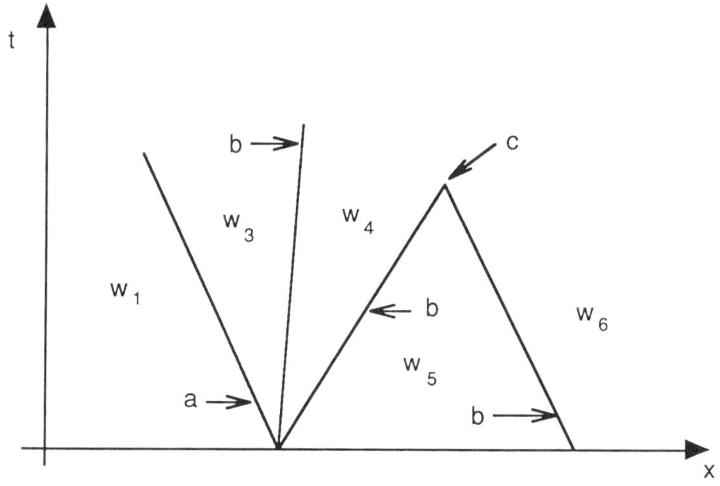

Figure 2.4

a: shock front with two shock lines
b: shock front with one shock line
c: shock collision

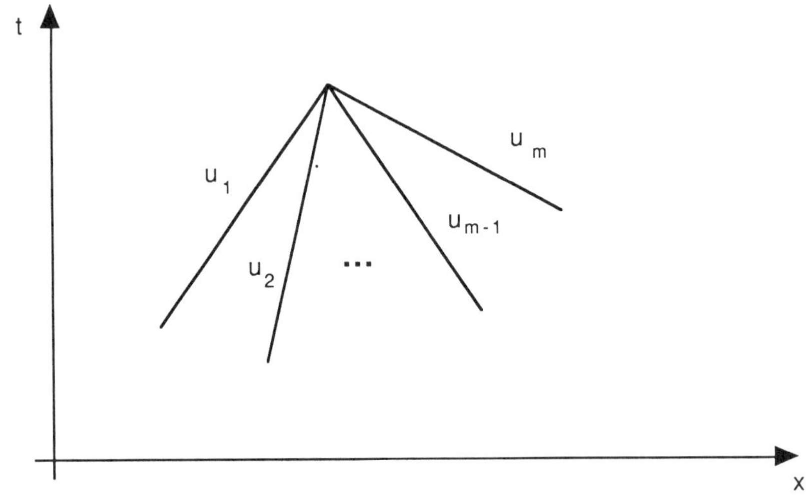

Figure 2.5

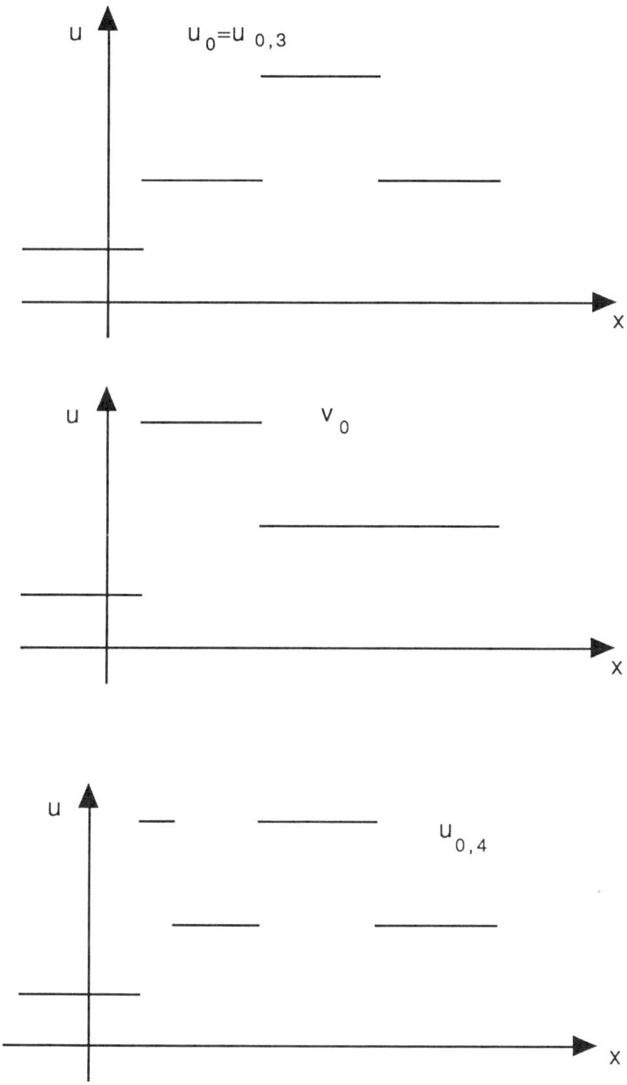

Figure 3.1

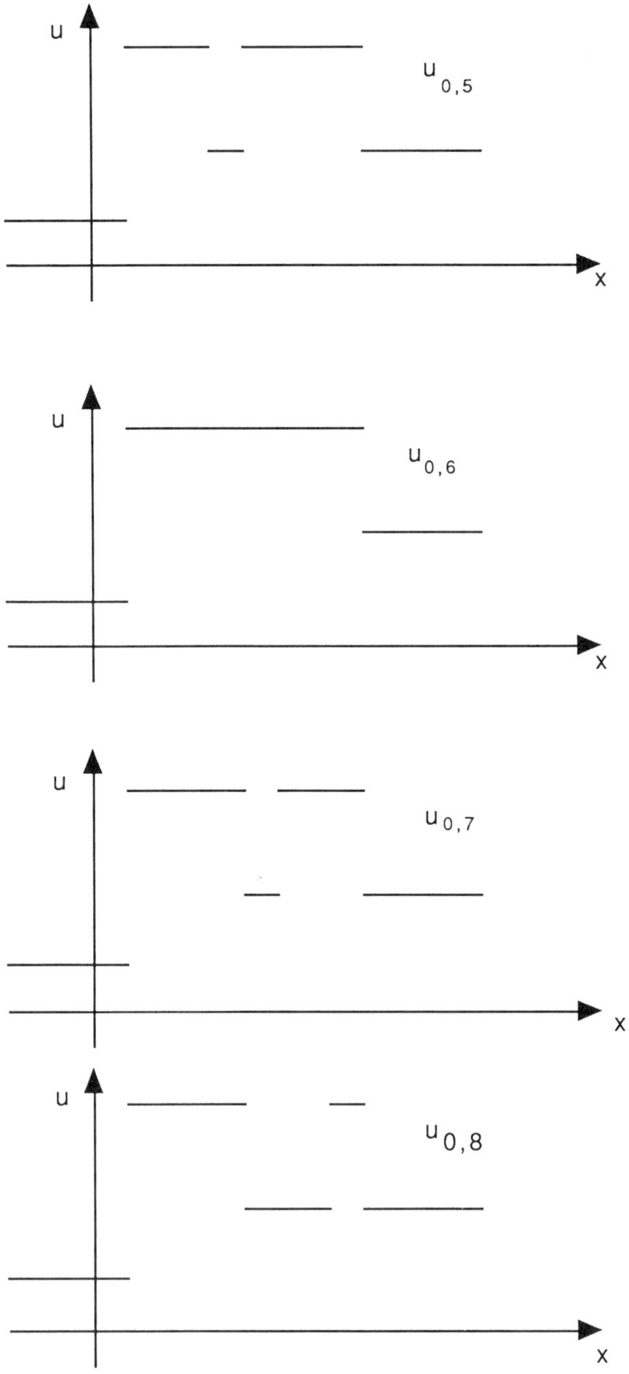

Figure 3.1 (cont.)

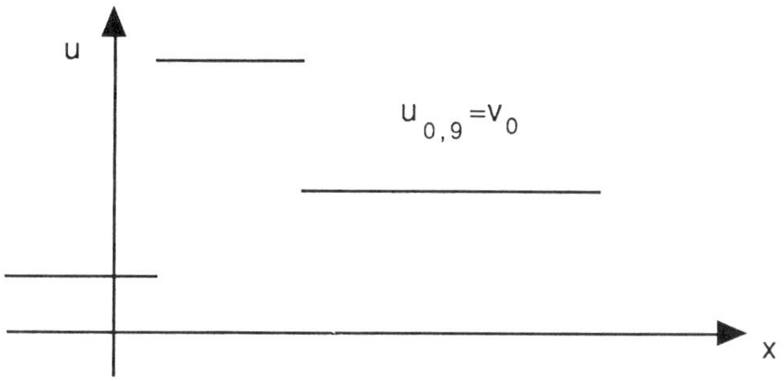

Figure 3.1 (cont.)

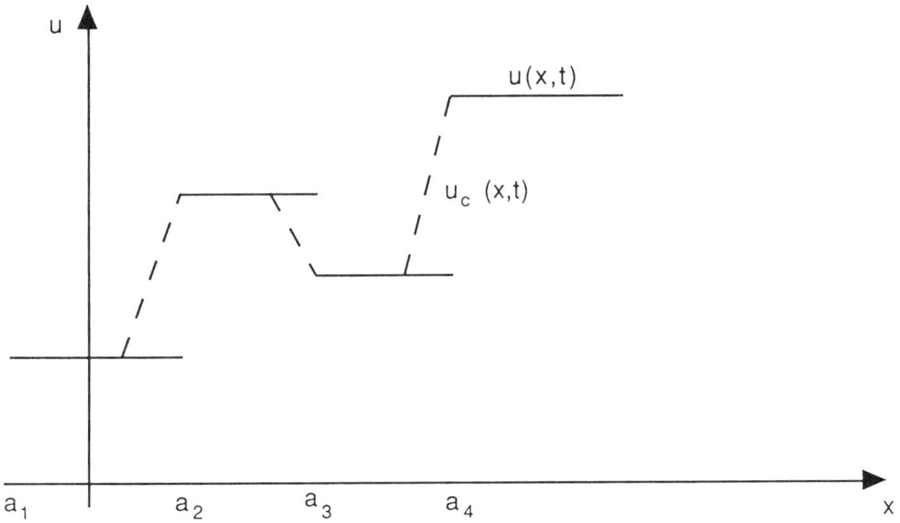

Figure 3.2

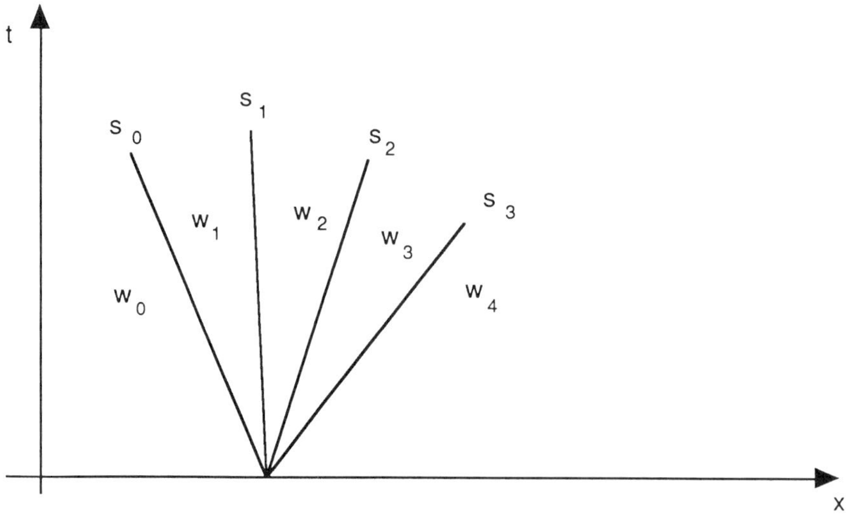

Figure 3.3

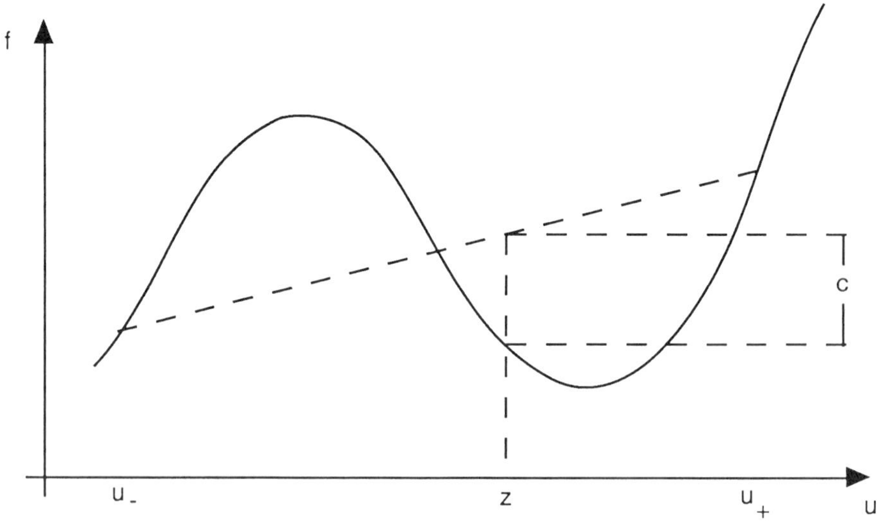

Figure 4.1

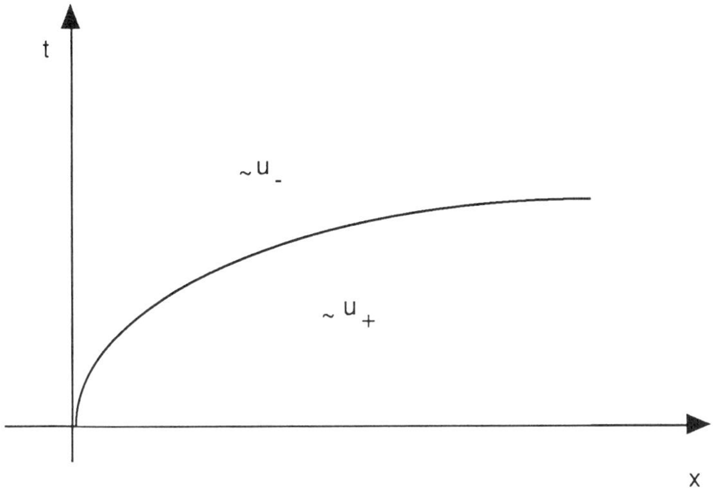

Figure 4.2a

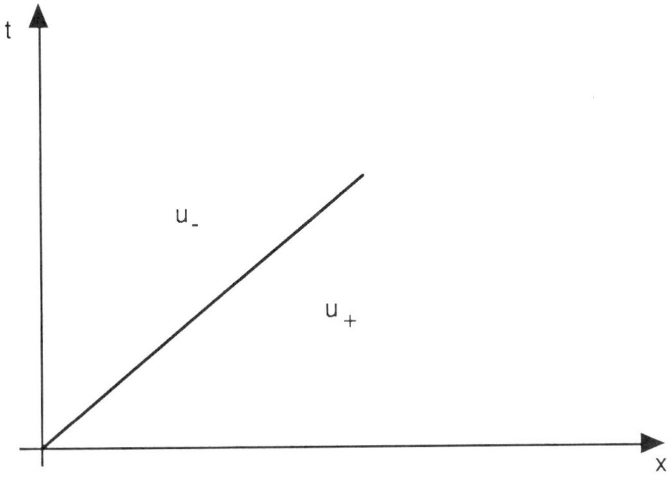

Figure 4.2b